Functional Food and Health

ACS SYMPOSIUM SERIES **993**

Functional Food and Health

Takayuki Shibamoto, Editor
University of California at Davis

Kazuki Kanazawa, Editor
Kobe University

Fereidoon Shahidi, Editor
Memorial University of Newfoundland

Chi-Tang Ho, Editor
Rutgers, The State University of New Jersey

Sponsored by the
ACS Division of Agricultural and Food Chemistry, Inc.

American Chemical Society, Washington, DC

ISBN: 978–0–8412–6982–8

The paper used in this publication meets the minimum requirements of American National Standard for Information Sciences—Permanence of Paper for Printed Library Materials, ANSI Z39.48–1984.

Distributed by Oxford University Press

Foreword

The ACS Symposium Series was first published in 1974 to provide a mechanism for publishing symposia quickly in book form. The purpose of the series is to publish timely, comprehensive books developed from ACS sponsored symposia based on current scientific research. Occasionally, books are developed from symposia sponsored by other organizations when the topic is of keen interest to the chemistry audience.

Before agreeing to publish a book, the proposed table of contents is reviewed for appropriate and comprehensive coverage and for interest to the audience. Some papers may be excluded to better focus the book; others may be added to provide comprehensiveness. When appropriate, overview or introductory chapters are added. Drafts of chapters are peer-reviewed prior to final acceptance or rejection, and manuscripts are prepared in camera-ready format.

As a rule, only original research papers and original review papers are included in the volumes. Verbatim reproductions of previously published papers are not accepted.

ACS Books Department

Contents

Overview and Perspectives

Bioavailability and Metabolism

Cereal

Antioxidants

Carcinogenesis and Anticarcinogenesis

Other Health Effects

Indexes

Preface

It has been more than two decades since the term functional food was first introduced in Japan. Numerous definitions for functional foods have also been advanced in various countries. Generally, *Functional Foods* are foods containing some demonstrated health benefits beyond their basic nutritional value that arises from their constituent proteins, lipids, and carbohydrates as well as from vitamins. Therefore, there are many names for this type of food in addition to *functional foods*, such as designer foods, pharmafoods, nutraceuticals, medical foods, and biofoods. However, *Nutraceuticals* are often referred to as materials from foods that also have demonstrated physiological benefits and are used in the medicinal form of pills, capsules, potions, and liquid. The words *Food Factors* are also commonly used to describe functional foods. After many reports on the potential health benefits and biological as well as chemical activities of foods have been published, consumers have become interested in the relationship between diet and health. Consequently, consumers have been learning about the importance of their choice of foods in maintaining a healthy body and for preventing diseases or reducing disease risk.

Just about 10 years ago, the first symposium on *Functional Foods* was held at the American Chemical Society (ACS) Division of Agricultural and Food Chemistry, Inc. during its 213th National Meeting in San Francisco, California, April 13–17, 1997. The objective of this symposium was to provide a forum for interaction among food chemists, nutritionists, medical doctors, students, policy makers, and interested personnel from the food industry. The symposium was a great success with presentation of some 60 scientific papers associated with functional foods. Since then, there have been many symposia and congresses on functional foods. The symposium entitled *Functional Foods and Health* was held at the ACS Division of Agricultural and Food Chemistry, Inc. 232th National Meeting that was held in San Francisco, California, September 10–14, 2006 to update the research conducted on functional foods since the 213th ACS national meeting. The contributors were experts in different areas of functional foods and were selected from many countries, including Australia, Austria, Canada, China, Germany, Japan, Korea, Mexico, New Zealand, Spain, Sweden, Taiwan, and the United States.

This book is based on the presentations of the 2006 ACS meeting and provides valuable information about the overall reviews and perspectives on functional foods, the bioavailability and metabolism of plant and cereal constituents, natural antioxidants found in fruits and vegetables, carcinogenesis and anticarcinogenesis of food constituents, and other miscellaneous food components with health effects. This book would serve as a useful research tool for a diverse range of backgrounds for scientists, including, biologist, biochemists, chemists, dietitians, food scientists, nutritionists, pharmacologists, and toxicologists as well as medical doctors, from academic institutions, governmental agencies, and private industries.

We are grateful to the contributors of this book for their efforts in preparation of their manuscripts and to the participants of the Symposium for their comments and encouragements. We are also grateful for the continuous support of the officers of the ACS Division of Agricultural and Food Chemistry, Inc. for their assistance in providing the necessary logistics for this symposium.

Acknowledgments

Finally, we are most grateful to the following companies for their donations in support of the symposium: Hannon R&D Center, Fuji Oil Company Ltd., Nippon Suisan Kaisha Ltd., Sunstar Inc., L.S. Factory Company Ltd., Oryza Oil and Fat Chemicals Company, Ltd., Sugar Lady Company, Ltd. and Mr. Tetsuro Nishiyama.

Takayuki Shibamoto
Department of Environmental Toxicology
University of California at Davis
One Shields Avenue
Davis, CA 95616

Kazuki Kanazawa
Graduate School of Science and Technology
Department of Biofunctional Chemistry
Kobe University
1–1 Rokkodai
Nada, Kobe, Hyogo 657–8501
Japan

Fereidoon Shahidi
Department of Biochemistry
Memorial University of Newfoundland
St. John's, NFL A1B 3X9
Canada

Chi-Tang Ho
Department of Food Science
Rutgers, The State University of New Jersey
65 Dudley Road
New Brunswick, NJ 08901–8520

Functional Food and Health

Chapter 1

Functional Food and Health: An Overview

Takayuki Shibamoto[1], Kazuki Kanazawa[2], Fereidoon Shahidi[3], and Chi-Tang Ho[4]

[1]Department of Environmental Toxicity, University of California, Davis, CA 95616
[2]Laboratory of Food and Nutritional Chemistry, Faculty of Agriculture, Kobe University, Rokkodai, Nada-ku, Kobe 657–8501, Japan
[3]Department of Biochemistry, Memorial University of Newfoundland, St. John's, Newfoundland A1B 3X9, Canada
[4]Department of Food Science, Rutgers, The State University of New Jersey, New Brunswick, NJ 08903

Functional food ingredients and nutraceutical products are important in health promotion and disease risk reduction. Such foods are expected to render physiological benefits beyond their traditional nutritional value. Although presence of phytochemicals and bioactives in foods is important, their absorption, bioavailability and metabolism are also critical factors that influence in their beneficial effects. Processing, however, may affect such bioactives and hence process-induced changes in functional food ingredients must be carefully examined. Furthermore, the mechanisms of action of bioactivies in food are numerous, but include antioxidant activity, among others. Important aspects of functional foods and their health benefits are presented.

It has been over 20 years since the concept of "Functional Foods" was first introduced as a third factor in the analysis of foods after nutrients, such as proteins, lipids, and carbohydrate, and flavors, such as sugars, acids, and volatile compounds. Not only food chemists but also many scientists from diverse areas, including biology, biochemistry, botany, plant/animal sciences, and nutrition as well as medical science, have become involved in research associated with functional foods today. Moreover, consumers are now deeply interested in the food factors which provide beneficial effects to humans in terms of health promotion and disease risk reduction. They also demand more detailed information about food factors in order to obtain appropriate functional food products.

This Proceeding book comprises six major subject areas: 1. Overview and Perspectives (4 chapters), 2. Bioavailability and Metabolism (5 chapters), 3. Cereals (3 chapters), 4. Antioxidants (12 chapters), 5. Carcinogenesis and Anti-carcinogenesis (6 chapters), and 6. Other Health Effects (10 chapters).

Overview and Perspectives

The concept of functional foods is relatively new to the United States compared to the countries of Asia. Some foodstuffs, such as herbs and spices, have been used as folk medicine. However, since the modern concept of functional foods was advanced nearly 20 years ago, there have been numerous studies and symposiums focused on functional foods or food factors in all parts of the world (*1-3*). In order to satisfy the consumers' interest in the relationship between foods and health benefits as well as the growing interest in the up-to-date research on functional foods among researchers in various fields, many excellent overview articles on and articles offering different perspectives on functional foods have been published.

The Overview and Perspectives section covers the bioavailability of dietary phytochemicals, such as flavonoids. A single cell gel electrophoresis assay to determine DNA migration associated with DNA damage is introduced. An overview of the biological and pharmacological effects of flavonoids is outlined. The current development of the guidelines of good agricultural practices related to herbal medicines is summarized.

Bioavailability and Metabolism

A tremendous number of food components with medicinal activities beside nutrients, such as proteins, lipids, and carbohydrates, have been found in various natural plants. These so-called phytochemicals include terpenoids, xanthophylls, phenolic compounds, and some sulfur-containing compounds (*4*). Also, many studies on the metabolic pathways of functional phytochemicals, such as flavonoids, have been reported (*5*).

The Bioavailability and Metabolism section contains research results from the bioavailability and metabolic fates of various naturally occurring chemicals, including anthocyanins, phenolic acids, and flavonoids—soybean isoflavones, including daidzein, and quercetin. The general metabolic pathway of anthocyanins in rats was found to occur at the *O*-methylation of the B ring and glucuronidation of both anthocyanidin and anthocyanin. The effect of phenolic acid, gallic acid, on the expression of the phase II P-form of phenol sulfotransferase in human hepatoma HepG2 cells is described. The role of the acylated moiety in *in vivo* uptake and in the antioxidant activity of anthocyanins is reported. Quercetin was found to prevent the early stage of carcinogenesis in a study using male F344 rats.

Cereal

Cereals or grains are the most important source of nutrients, such as protein and carbohydrates, for people throughout the entire world, and particularly in the developing countries (*6*). It is known that cereals contain various health beneficial compounds such as phytoestrogens and phenolics (*7*).

Cereal and grains are one of the most important food sources. In particular, wheat is consumed by people in almost all countries. Its world production reached over 600 tons in 2005. It is rich in carbohydrate and proteins. Recently, the presence of health beneficial phenolic compounds in cereal has been recognized. In the Cereal section, wheat grain is evaluated for its antioxidant, anti-proliferation, cyclooxygenase-2-inhibition, and phytochemical properties. A comprehensive summary of phenolic compounds, such as caffeic acid and ferulic acid, found in grains and their antioxidant activities is presented.

Antioxidants

Research related to antioxidants represents nearly 70% of the studies focused on functional foods today. Most of these studies involve natural products, such as fruits, vegetables, whole grains, herbs, and leaves as well as some seafood. The classic concept of antioxidants present in foods is that of vitamins (A, C, and E) and minerals. Later, potent antioxidant activity was discovered in various naturally occurring chemicals in foodstuffs. Examples are carotenoids (source of vitamin A), polyphenols, flavonoids, anthocyanins, and lycopene (*2,3*). In addition, most recently, some aroma and flavor chemicals, such as certain terpenes (*8*) and volatile heterocyclic compounds (*9*), have been found to possess appreciable antioxidant activity. Consumption of antioxidants has been recommended to protect oxidative damage, which promotes various diseases including cancer, cardiovascular disease, cognitive impairment, Alzheimer's disease, immune dysfunction, cataract, and various inflammatory diseases (*10*).

The Antioxidant section is one of the core parts of this proceedings. The results of antioxidant studies on many naturally occurring chemicals in various plant foodstuffs are reported. The antioxidants present were turmerone, curlone, and α-terpineol in curcumin-free turmeric oil; caffeic acid in parsley; ferulic acid, methyl ferulate, isoferulic acid, sinapic acid, and methyl sinapate in radish sprout; volatile chemicals in herbs, spices, and teas; proanthocyanidins in grape, pine, and witch hazel; flavonoids in young green barley leaves and agrofood by-products; and polyphenols in citrus. Antioxidant activities of various terpenoids, coumarins, and flavonoids from Mexican and Chilean plants, such as Meliaceae and Zygophyllaceae families, were also reported. The role of various polyphenol antioxidants, such as tannins, catechin, chlorogenic acid, and gallic acid, in the inhibition of aflatoxin biosynthesis is presented here for the first time. In addition to these chemicals, some polymers, such as phosphopeptides from hen egg yolk, were discovered to suppress oxidative stress-induced gut inflammatory disorders.

Carcinogenesis and Anti-carcinogenesis

One of people's major concerns is how to prevent cancer incidences because cancer has been ranked among the top three causes of death in developing countries. Therefore, consumption of certain anti-cancer chemicals present in foods and beverages has begun to receive much attention as a promising strategy to prevent cancer (*11*).

The section on Carcinogenesis and Anti-carcinogenesis reports studies performed using biological systems. Chemopreventive effects of natural anthocyanins were investigated using animal and human cells. A study on cancer preventive mechanisms was conducted using epigallocatechin 3-gallate. Human leukemia cancer cells were used to examine the growth inhibitory effects of black tea polyphenol, theaflavin. A significant relationship between 8-hydroxydeoxyguanosine formation and oxidized dietary oils in feed was found using mice. The suppressant effects of flavonoids on AhR transformation induced by dioxin were observed in a system consisting of rat hepatic cytosol fraction.

Other Health Effects

Comprehensive epidemiological studies indicate that cancer risk in people consuming diets high in fruits and vegetables is much lower than in those consuming few of these foods (*12*). However, the effects of phytochemicals present in foods, including fruits and vegetables, not only preventing cancer but also preventing various other diseases. Therefore, there have been many reports on food factors associated with diseases beside cancer.

In addition to the topics presented in the above five sections, miscellaneous studies on food factors have also been reported. Many health beneficial activities of a variety of food components were demonstrated: the antimicrobial activity of the pungent components of *wasabi* isothiocyanates against periodontal pathogens, immunomodulatory activities of β-glucan found in mushroom, antidepressant effects of *Ginkgo biloba*, Serum cholesterol-lowering effects of a broccoli and cabbage juice mixture, inhibition of angiotensin converting enzyme (ACE) by Tahitian Noni juice, and inhibitory activities of vialinins A and B against TNF-α production.

The topics introduced in this book are all original studies, were prepared and presented by experts in the area of functional foods. Therefore, this proceeding book provides up-to-date research results and a cutting-edge review of the status of functional food research in various countries.

References

1. Ohigashi, H.; Osawa, T.; Terao, J.; Watanabe, S.; Yoshikawa, T. (Eds.) *Food Factors for Cancer Prevention.* Springer-Verlag Tokyo: Tokyo, Japan, 1997.
2. Shibamoto, T.; Terao, J.; Osawa, T. (Eds). *Functional Foods for Disease Prevention I: Fruits, Vegetables, and Teas. and II: Medicinal Plants and Other Foods.* ACS Symposium Series 701, ACS: Washington, DC, 1998.
3. Shahidi, F.; Ho, C.T. (Eds.) *Antioxidant Measurement and Applications.* ACS Symposium Series 956, ACS: Washington, DC, 2007.
4. Sakakibara, H.; Honda, Y.; Nakagawa, S.; Ashida, H.; Kanazawa, K. *J. Agric. Food Chem.* **2003**, *51*, 571-581.
5. van der Woude, H.; Boersma, M. G.; Vervoort, J.; Rietjens, I. M. *Chem. Res. Toxicol.* **2004**, *17,* 1520-1530.
6. Pedersen, B.; Knudsen, K.E.; Eggum, B.O. *World Rev. Nutr. Diet* **1989**, *60*, 1-91.

7. Slavin, J.L.; Jacobs, D.; Marquart, L.; Wiemer, K. *J. Am. Diet. Assoc.* **2001**, *101*, 780-785.
8. Lee, K.-G.; Shibamoto, T. *J. Agric. Food Chem.* **2002,** *50,* 4947–4952.
9. Yanagimoto, K.; Lee, K.-G.; Ochi, H.; Shibamoto, T. *J. Agric. Food Chem.* **2002,** *50,* 5480-5484.
10. Hasler, C. M. *Functional Foods: Their Role in Disease Prevention and Health Promotion.* http://www.nutriwatch.org/04Foods/ff.html (accessed on August 30, 2006).
11. Surh, Y. J. *Nat. Rev. Cancer* **2003**, *3*, 768-780.
12. Block, G.; Patterson, B.; Subar, A. Nutr. Cancer 1992, 18, 1-29.

Overview and Perspectives

Chapter 2

Finding Bioavailable Phytochemicals Which Express a Beneficial Effect on Health

Kazuki Kanazawa

Laboratory of Food and Nutritional Chemistry, Faculty of Agriculture, Kobe University, Rokkodai, Nada-ku, Kobe 657–8501, Japan

Various phytochemicals in the daily diet have been found to possess biofunctions. Most of them, however, are eliminated in the intestinal absorption process, by excretion to the digestive lumen after conjugation with glucuronide and/or sulfate. A better understanding to prevent disease with the diet would involve finding factors which can escape from conjugation systems. One of the candidates is the prenyl compounds, because the prenyl group makes conjugation enzymes difficult to approach. Another group is chemicals incorporated into the body in unchanged forms, such as the xanthophylls. These dietary phytochemicals that escape conjugation probably exhibit their beneficial activity as antioxidants and protein function modulators.

Our daily food is composed of ingredients beneficial to our health. They are nutrients that are used in energy-producing systems and non-nutrients (Table I). Nutrients such as sugars, lipids, and proteins, have been recognized to play an important role in preserving our health. Among the non-nutrients, vitamins, minerals and dietary fibers have been somewhat understood regarding their beneficial roles. Other non-nutrients have recently been found to possess interesting effects on the health. They are phytochemicals such as terpenoids including 600 species of carotenoid and xanthophylls, sulfur-containing compounds such as isothiocyanate and sulforaphane, and a large group of phenolics. Around 2.7 million species of phenolics are identified in daily food and are classified into phenylpropanoids, flavonoids, anthraquinones, and lignans (*1*). Their functions are antioxidative potency and protein function-modulating activity. Antioxidative potency is an action eliminating reactive oxygen species (ROS) and results in prevention of degenerative diseases. Protein function-modulating activity is a modulation of metabolic activity through the interaction of phytochemicals with enzymes, membranous transporters, cellular or intracellular receptors, or signaling proteins. This activity modifies our body functions dynamically accompanying the prevention of degenerative diseases including cancers. Most of the dietary phytochemicals, however, are removed by conjugation in the intestinal absorption processes (*2-4*).

Table I. Food Ingredients and the Roles

Ingredients	Roles
Nutrients	
Sugars	Energy source
Lipids	Membrane construction, Eicosanoid production, Energy source
Proteins	Nitrogen source, Functional peptide production
Non-nutrients	
Vitamins	Support of metabolism
Minerals	Support of metabolism, Signals for metabolism, Skeleton construction
Dietary fibers (β-Glucan, Oligosaccharides, Bacterial membrane)	Modulation of immunoresponse, Prebiotics for intestinal Bifidobacteria, Modulation of cholesterol metabolism, Modulation of insulin secretion, Prevention of overeating, Facilitation of a bowel movement, Evacuation of intestinal xenobiotics
Terpenoids (Terpenoids, Carotenoids and Xanthophylls)	Antioxidative, Anticarcinogenic, Antiobesity, Sweetener
Sulfur-containing compounds	Antioxidative, Anticarcinogenic
Phenolics (Phenylpropanoids, Flavonoids, Anthraquinones, Lignans)	Antioxidative, Modulation of protein functions

In order to find bioavailable phytochemicals that play preventive roles in our body, chemicals which can escape the conjugation systems should be selected. Here, a strategy for finding bioavailable factors will be discussed.

Bioavailable Antioxidants

Degenerative diseases such as cancers, atherosclerosis and diabetes mellitus can be induced by oxidative damage of DNA by ROS (*5-7*). To prevent this event from intracellularly generating ROS, the antioxidants should remain in the cells and nuclei. The strategy for finding bioavailable antioxidants in the dietary phytochemicals is simple. An antioxidant possessing a greater ability to release reducing agents should be selected, because dietary phytochemicals are diluted when absorbed into the body and further incorporated into the cells from the bloodstream. The candidates are catechol polyphenols such as the flavonol quercetin (Figure 1).

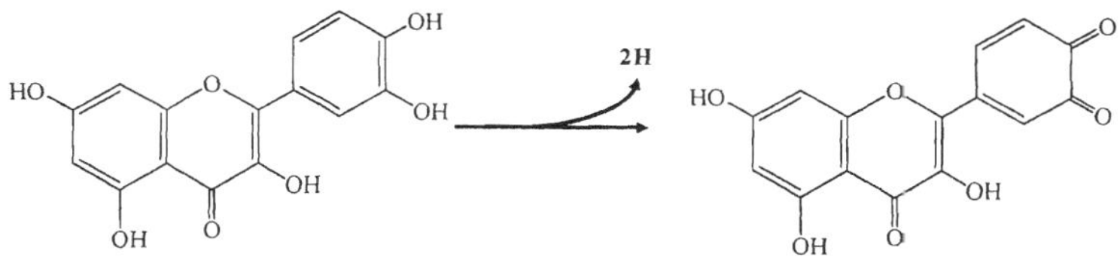

Figure 1. Antioxidative ability of a catechol flavonoid.

Every bond in quercetin can construct a conjugated double bond and thereby easily release the reducing agents transforming stable quinone. Quercetin significantly suppressed the formation of oxidation products of cellular DNA, e.g., 8-hydroxy-2'-guanosine (8-OHdG), even though quercetin was incorporated into the nuclei at a few nmol of the concentration when 10 μM of quercetin was incubated with 2.5×10^7 HepG2 cells for 30 min (*8*). Other candidates, anthocyanin delphinidin glycoside (*9*) and epigallocatechin gallate (*10*), can also be candidates for bioavailable antioxidants.

Protein Function-Modulating Activity

Dietary phytochemicals can also express dynamic modulation of protein functions. This action is attributed to the interactions of phytochemicals with the corresponding proteins, and thus the stereochemical structures of the phytochemicals select their own target proteins depending on whether they can fit into the pocket of the proteins. In the stereo structure, phytochemicals are classified into three types, the phenylpropanoid type, the coplanar type of flavones, flavonols and anthraquinones, and the non-coplanar type of flavanones including catechins (Figure 2). The coplanar group constructs a plate-like shape three-dimensionally, a non-coplanar constructs a ball-like shape, and phenylpropanoids create a pendulum-like structure. The members of the coplanar group, flavones, flavonols and anthraquinones, strongly suppress the transformation of the arylhydrocarbon receptor (AhR) induced by 2,3,7,8-tetrachlorodibenzo-*p*-dioxin (TCDD) (*11, 12*) and the enzyme activity of cytochrome P450 (CYP) 1As (*13*) at very low IC_{50} values that are near the physiological levels after ingestion. The pocket size of AhR to induce transformation is 14x12x5 Å and favors hydrophobicity, and TCDD is 13.5x6x3.5 Å and is a hydrophobic molecule (*14*). The size of flavones, flavonols and anthraquinones is slightly bigger 14.5x8x4 Å. These antagonistically interfere with the attack of TCDD on AhR, and the activity is stronger in lower hydroxyl apigenin, galangin and emodin than in myricetin and morin having 5 or more hydroxyl groups (*15*). Coplanar phytochemicals can also interact with sodium and calcium channel proteins, which is probably accompanied by prevention of hypertension (*16, 17*). The inhibitory action of coplanars has also been reported on the expressions of cyclooxygenase-2 (COX-2) and inducible NO synthase (iNOS) that involve cancer promotion (*18*).

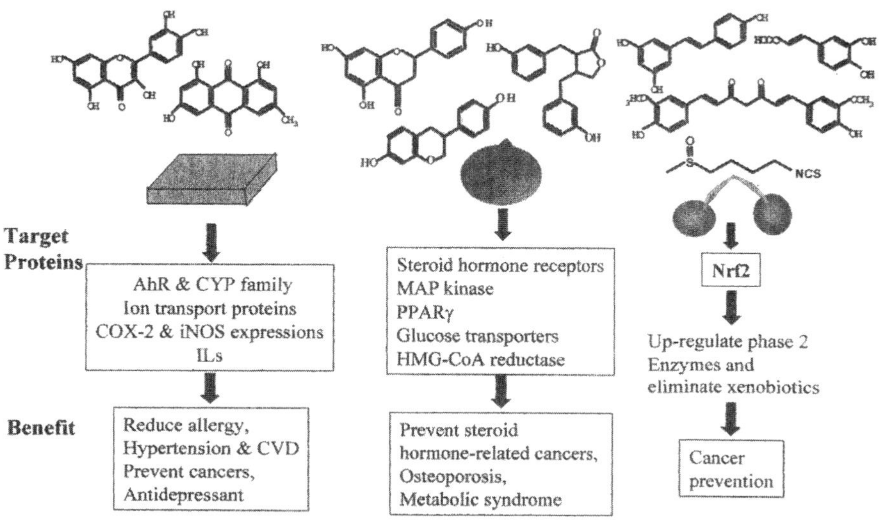

Figure 2. Target proteins for dietary phytochemicals.

On the contrary, non-coplanar chemicals like flavanones and catechins possess weak or no activity for interacting with AhR, CYP, ion transport proteins, and COX-2 and iNOS. They, however, have other target proteins.

Catechins modulate mitogen-activated protein (MAP) kinase activity and induce tumor cells to undergo G_2/M arrest of the cell cycle through suppression of the phosphorylation of extracellular signal-regulated kinase (ERK)1/2 (*19*) and also suppress the differentiation of adipocytes by down-regulation of peroxisome proliferator-activated receptor (PPAR) (*20*). Catechins are classified into catechin gallates and non-gallates. Catechin gallates inhibit the translocation of glucose transporter 4 and suppress the incorporation of blood glucose into adipocytes (*21*). Non-gallates act on glucose transporter 1 and stimulate glucose incorporation. Both events result in modulation of blood glucose concentrations and prevent diabetes mellitus. Flavanone hesperidin has been suggested to act on HMG-CoA reductase and thereby inhibits bone loss in ovariectomized mice (*22*).

Isoflavones daidzein and genistein have been well recognized to prevent steroid hormone-related cancers and osteoporosis. This effect is understood to be due to an agonistic action on the estrogen receptor (ER). Isoflavones are coplanars similar to flavones and anthraquinones. However, isoflavones required a high concentration to modulate AhR and CYPs, and flavones and anthraquinones exhibit almost no effect on ER. On the other hand, another study reveals that ingested daidzein is metabolized in the intestinal micro flora to equol and is absorbed into body, and resulting in the action on ER and preventing diseases (*23*). As shown in Figure 3, equol is non-coplanar. ER may be an exceptional protein favoring isoflavones alone.

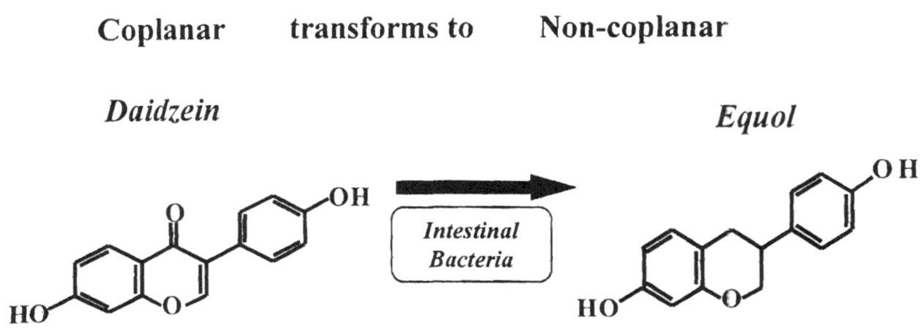

Figure 3. Metabolic transformation of isoflavone daidzein in the intestines.

Phenylpropanoids can construct a pendulum-like structure sterochemically, and isothiocyanates also form it because the S atom has a large ionic diameter (Figure 2). They act on the Nrf2-Keap1 system and stimulate the expression of phase 2 enzymes such as glutathione *S*-transferase and quinone reductase (*24, 25*). We consume various kinds of food xenobiotics, and they are detoxified and excreted with oxidation by phase 1 enzymes, CYPs, following by conjugations with phase 2 enzymes. In our body, however, the enzyme activity of phase 2 is not at levels similar to that of phase 1. The phase 2 activity is generally lower than phase 1, and the oxidation products of phase 1 accumulate. Most of the oxidation products are strong carcinogens (*26*). One better means of preventing food-derived cancers is to maintain high levels of the phase 2 enzyme activity. The pendulum-like phytochemicals in food have been considered to induce the phase 2 enzyme activity.

Selection of Bioavailabe Phytochemicals

Most of the nutrients are metabolized in the liver after incorporation into the body. On the contrary, phytochemicals are metabolized in intestinal cells during the absorption process (*2, 3*). The incorporated phytochemicals immediately undergo conjugation to functional groups such as hydroxyls through UDP-glucuronosyltransferase and/or phenol sulfotransferase in the intestinal cells (*4*). The conjugates are excreted into the intestinal lumen while a small part is transported to the bloodstream and circulates, and thereby the endogenous levels of the conjugates are 1.5 μM or less (*27, 28*). Under healthy conditions, the conjugates in the blood cannot be incorporated into the body cells. Consequently, the conjugates do not interact with cellular proteins and are excreted through the urine within 25 hr after ingestion (*29*). Thus, most of the dietary phytochemicals lose their bioactivity in the absorption process. In order to find functional ingredients, the strategy is a selection of refractory phytochemicals which escape metabolic conversions in the absorption process.

Prenyl Compounds

One of candidates for the refractory phytochemicals is the prenyl compounds (Figure 4).

For example, artepillin C included in Brazilian propolis possesses two prenyls attached hydroxyl group that usually undergoes conjugation. The prenyls may have difficulty in approaching the conjugation enzymes and thus result in a higher bioavailability. Indeed, artepillin C was easily incorporated into intestinal cells and released at serosal sites without undergoing conjugation when determined with a Caco-2 monolayer (*30*). Artepillin C activated an antioxidant response element and induced phase 2 enzyme activity in the liver when dosed orally into a colon carcinogen azoxymethane-challenged mouse and significantly suppressed the formation of aberrant crypt foci by around 40% (*31*, *32*). Other prenyl candidates are licorice glabridin and pueraria miroestrol (*Figure 4*). Both are known to possess a strong estrogen-like activity (*33, 34*).

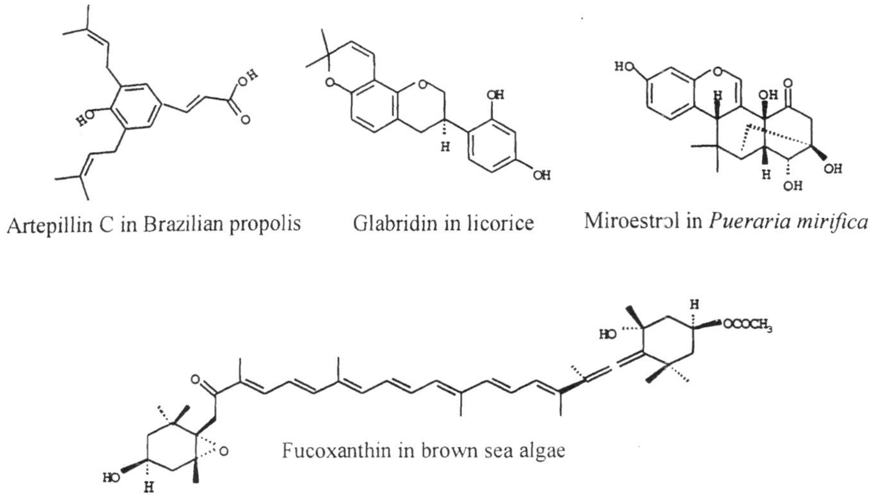

Figure 4. Phytochemicals assumed to be highly bioavailable.

Xanthophylls

Other candidates for refractory phytochemicals which escape metabolic conversion are the xanthophylls, fucoxanthin, astaxanthin, and canthaxanthin. They undergo only minor conversions in the absorption process. Especially, fucoxanthin in edible brown sea algae is incorporated into the body after acetyl side chain is hydrolyzed to alcohol (Figure 4), remaining unchanged in skeleton form (*35*). Fucoxanthin exhibits conspicuous activity such as anti-carcinogenicity and antiobesity. Anti-carcinogenic mechanisms have been found in which the hydrolyzed fucoxanthin stimulates an expression of $p21^{WAF/Cip1}$ in tumor cells and suppresses phosphorylation of retinoblastoma protein. This event inhibits a release of transcription factor E2F for cell proliferation and then induces cell arrest at the G_0/G_1 phase (*36, 37*). The antiobesity action has been reported to be due to stimulating consumption of fat through upregulation of UCP1 expression in white adipose tissue by the hydrolyzed fucoxanthin (*38*).

Protocol for Finding Bioavailable Functional Phytochemicals

Figure 5 is a proposal scheme for finding bioavailable functional phytochemicals. First, major phytochemicals in sample food should be identified and second classified by their stereostructure into three types, phenylpropanoid, coplanar and non-coplanar. On the basis of the classification, third, the target proteins will be assumed, and fourth the proof is provided by animal experiments. Next, the mechanisms should be clarified with cell-line experiments, and finally, human trials can be done. In this scheme, the third and fourth steps usually require various types of experiments and are expensive. The classification of phytochemical type and assumption regarding target proteins can simplify the third and fourth experiments and save time for animal experiments.

1. Identify the ingredients in sample foods

↓

2. Classify the ingredients by stereo structure into
Phenylpropanoid-type
Coplanar-type
Non-Coplanar-type

↓

3. Assume the species of target proteins

↓

4. Prove the activity with animal tests

↓

5. Clarify the mechanism with cell line experiments

↓

6. Human trials

Figure 5. A proposal scheme for finding bioavailable functional phytochemicals.

Acknowledgments

This research was supported by Research and Development Program for New Bio-industry Initiatives (2006-2010) of Bio-oriented Technology Research Advancement Institution (BRAIN), Japan.

References

1. Sakakibara, H.; Honda, Y.; Nakagawa, S.; Ashida, H.; Kanazawa, K. *J. Agric. Food Chem.* **2003**, *51*, 571-581.
2. Murota, K.; Terao, J. *Arch. Biochem. Biophys.* **2003**, *417*, 12-17.

3. Németh, K.; Plumb, G .W.; Berrin, J.-G.; Juge, N.; Jacob, R.; Naim, H. Y.; Williamson, G.; Swallow, D. M.; Kroon, P. A. *Eur. J. Nutr.* **2003**, *42*, 29-42.
4. Day, A. J.; Bao, Y.; Morgan, M. R. A.; Williamson, G. *Free Radical Biol. Med.* **2000**, *29*, 1234-1243.
5. Guyton, K. Z.; Kensler, T. W. *Br. Med. Bull.* **1993**, *49*, 523-544.
6. Emerit, I. *Free Radical. Biol. Med.* **1994**, *16*, 99-109.
7. Marnett, L. J. *Carcinogenesis* **2000**, *21*, 361-370.
8. Kanazawa, K.; Uehara, M.; Yanagitani, H.; Hashimoto, T. *Arch. Biochem. Biophys.* **2006**, *455*, 197-203.
9. Ichiyanagi, T.; Rahman, M. M.; Kashiwada, Y.; Ikeshiro, Y.; Shida, Y.; Hatano, Y.; Matsumoto, H.; Hirayama, M.; Tsuda, T.; Konishi, T. *Free Radical Biol. Med.* **2004**, *36*, 930-937.
10. Hagerman, A. E.; Dean, R. T.; Davies, M. J. *Arch. Biochem. Biophys.* **2003**, *414*, 115-120.
11. Ashida, H.; Yamashita, T.; Fukuda, I.; Kanazawa, K. *FEBS Lett.* **2000**, *476*, 213-217.
12. Fukuda, I.; Kaneko, A.; Yabushita, Y.; Sakane, I.; Kakudam, T.; Kanazawa, K.; Danno, G.; Ashida, H. In: *Food Factors in Health Promotion and Disease Prevention.* Shahidi, F.; Ho, C.-T.; Watanabe, S.; Osawa, T., Eds.; ACS Symposium Series 851; American Chemical Society: Washington, D.C., 2003, pp.119-127.
13. Kanazawa, K.; Yamashita, T.; Ashida, H.; Danno, G. *Biosci. Biotechnol. Biochem.* **1998**, *62*, 970-977.
14. Safe, S. H. *Annu. Rev. Pharmacol. Toxicol.* **1986**, *26*, 371-399.
15. Kanazawa, K.; Ashida, H.; Danno, G. In: *Functional Foods for Disease Prevention II: Medicinal Plants and other Food,* Shibamoto, T.; Terao, J. Osawa, T., Eds.; ACS Symposium Series 702; American Chemical Society, Washington, D.C., 1998, pp. 67-82.
16. Montero, M.; Lobaton, C.D.; Hernandez-Sanmiguel, E.; Santodomingo, J.; Vay, L.; Moreno, A.; Alvarez, J. *Biochem. J.* **2004**, *384*, 19-24.
17. Aoi, W.; Niisato, N.; Miyazaki, H.; Muranaka, Y. *Biochem. Biophys. Res. Commun.* **2004**, *315*, 892-896.
18. Chun, K. S.; Surh, Y. J. *Biochem. Pharmacol.* **2004**, *68*, 1089-1100.
19. Maeda-Yamamoto, M.; Suzuki, N.; Sawai, Y.; Miyase, T.; Sano, M.; Hashimoto-Ohta, A.; Isemura, M. *J. Agric. Food Chem.* **2004**, *51*, 1858-1863.
20. Furuyashiki, T.; Nagayasu, H.; Aoki, Y.; Bessho, H.; Hashimoto, T.; Kanazawa, K.; Ashida, H. *Biosci. Biotechnol. Biochem.* **2004**, *68*, 2353-2359.
21. Furuyashiki, T.; Terashima, S,; Nagayasu, H.; Kaneko, A.; Sakane, I.; Kakuda, T.; Kanazawa, K.; Danno, G.; Ashida, H. In: *Food Factors in*

Health Promotion and Disease Prevention, Shahidi, F.; Ho, C.-T.; Watanabe, S.; Osawa, T., Eds. ACS Symposium Series 851; American Chemical Society, Washington, D.C., 2003, pp. 224-234.
22. Chiba, H.; Uehara, M.; Wu, J.; Wang, X.; Masuyama, R.; Suzuki, K.; Kanazawa, K.; Ishimi, Y. *J. Nutr.* **2003**, *133*, 1892-1897.
23. Fujioka, M.; Uehara, M.; Wu, J.; Adlercreutz, H.; Suzuki, K.; Kanazawa, K.; Takeda, K.; Yamada, K.; Ishimi, Y. *J. Nutr.* **2004**, *134*, 2623-2627.
24. Morimitsu, Y.; Nakagawa, Y.; Hayashi, K.; Fujii, H.; Kumagai, T.; Nakamura, Y.; Osawa, T.; Horio, F.; Itoh, K.; Iida, K.; Yamamoto, M.; Uchida, K. *J. Biol. Chem.* **2002**, *277*, 3456-3463.
25. Nakamura, Y.; Miyamoto, M.; Murakami, A.; Ohigshi, H.; Osawa, T.; Uchida, K. *Biochem. Biophys. Res. Commun.* **2003**, *302*, 593-600.
26. Wakabayashi, K.; Nagao, M.; Esumi, H.; Sugiura, T. *Cancer Res.* **1992**, *52*, 2092s-2098s.
27. Manach, C.; Morand, C.; Texier, O.; Favier. M.-L.; Agullo, G.; Demigné, C.; Régérat, F.; Rémésy, C. *J. Nutr.* **1995**, *125*, 1911-1922.
28. Conquer, J. A.; Maiani, G.; Azzini, E.; Raguzzini, A.; Holub, B. J. *J. Nutr.* **1998**, *128*, 593-597.
29. Hollman, P. C. H.; van der Gaag, M.; Mengelers, M. J. B.; van Trup, J. M. P.; de Vries, J. H. M.; Katan, M. B. *Free Radical Biol. Med.* **1996**, *21*, 703-707.
30. Shimuzu, K.; Ashida, H.; Matsuura, Y.; Kanazawa, K. *Arch. Biochem. Biophys.* **2004**, *424*, 181-188.
31. Shimizu, K.; Das, S. K.; Baba, M.; Matsuura, Y.; Kanazawa, K. *Cancer Lett.* **2006**, *240*, 135-142.
32. Shimizu, K.; Das, S.K.; Hashimoto, T.; Sowa, Y.; Yoshida, T.; Sakai, T.; Matsuura, Y.; Kanazawa, K. *Mol. Carcinog.* **2005**, *44*, 293-299.
33. Somjen, D.; Katzburg, S.; Vaya, J.; Kaye, A.M.; Hendel, D.; Posner, G.H.; Tamir, S. *J. Steroid Biochem. Mol. Biol.* **2004**, *91*, 241-246.
34. Matsumura, A.; Ghosh, A.; Pope, G. S.; Darbre, P. D. *J. Steroid Biochem. Mol. Biol.* **2005**, *94*, 431-443.
35. Sugawara, T.; Baskaran, V.; Tsuzuki, W.; Nagao, A. *J. Nutr.* **2002**, *132*, 946-951.
36. Das, S. K.; Hashimoto, T.; Shimizu, K.; Yoshida,T.; Sakai, T.; Sowa, Y.; Komoto, A.; Kanazawa, K. *Biochim. Biophys. Acta* **2005**, *1726*, 328-335.
37. Das, S.K.; Hashimoto, T.; Baba, M.; Nishino, H.; Komoto, A.; Kanazawa, K. *J. Clin. Biochem. Nutr.* **2006**, *38*, 119-125.
38. Maeda, H.; Hosokawa, M.; Sashima, T.; Funayama, K.; Miyashita, K. *Biochem. Biophys. Res. Commun.* **2005**, *332*, 392-397.

Chapter 3

An Overview of Single-Cell Gel Electrophoresis-Based Dietary Human Intervention Trials for the Detection of DNA Protective Food Components

Veronika A. Ehrlich[1], Christine Hoelzl[1], Franziska Ferk[1], Julia Bichler[1], Armen Nersesyan[1], Michael Kundi[2], and Siegfried Knasmüller[1]

[1]**Medical University of Vienna, Department of Medicine I, Division Institute of Cancer Research, Borschkegasse 8a, 1090 Vienna, Austria**
[2]**Medical University of Vienna, Center for Public Health, Kinderspitalgasse 15, 1090 Vienna, Austria**

Single cell gel electrophoresis (SCGE- or comet-) assay is based on the determination of DNA-migration in an electric field and can be used in human intervention studies to investigate the impact of dietary factors on DNA stability. The use of this technique has yielded new and unexpected results concerning the antioxidant and chemoprotective properties of food components.

Integritiy and intact structure is a prerequisite for the optimal function of the DNA. Impaired DNA stability due to damage caused by reactive oxygen species (ROS), chemicals and lack of specific micronutrients plays a key role in a number of human diseases, such as cancer and neurodegenerative diseases as well as ageing (*1-3*). Therefore efforts have been made to identify dietary components, which protect against DNA damage and their consequences. However, one of the shortcomings of currently used *in vitro* approaches is that they are mainly conducted under unphysiological conditions and do not provide information if chemoprotective chemicals are absorbed and metabolized in the body (*4,5*). Therefore, the results obtained in such trials cannot be extrapolated to the human situation. In order to draw conclusions if dietary components are effective in man, a number of techniques have been developed. Conventional methods such as chromosomal aberration analyses of metaphase cells and monitoring of micronuclei in peripheral blood cells are laborious and time consuming. Therefore, the single cell gel electrophoresis (SCGE- or comet-) assay, which is based on the measurement of DNA migration in an electric field as indicator for DNA damage is increasingly used in human studies. One of the advantages of this technique, in comparison to other genotoxicity assays is, that no cell divisions are required and that the experiments can be conducted with few cells. The development of automated computer aided analysis programs has facilitated the evaluation of comet formation (*6,7*) (commonly used endpoints are tail lengths, tail moments and percentage DNA in the tail). An overview on the methodological principle and guidelines for SCGE tests can be found in the articles of Tice (*8*) and Speit and Hartmann (*9*). The present paper will give a short overview on different comet assay protocols currently used in human studies and provide information about the current state of knowledge on the effects of dietary factors on DNA stability that were detected with these approaches.

Endpoints and Experimental Design of Human Studies

The "classical" endpoint used in human SCGE studies is based on the protocol developed by Tice *et al.* (*8*). This procedure enables to measure endogenous formation of single and double strand breaks (SSBs and DSBs) and apurinic sites. Subsequently, Collins *et al.* (*10*) developed a modified protocol of the alkaline comet assay which enables the detection of oxidised DNA-bases by use of bacterial repair endonucleases. These enzymes, namely formamidopyrimidine glycosylase (FPG) and endonuclease III (ENDO III), recognize and remove oxidized purines and pyrimides by introducing additional breaks in the DNA and thereby increasing the tail intensity of the comets. The authors showed that the FPG-induced increase of tail intensity corresponds well with the formation of 8-hydroxydeoxyguanosine (8-OHdG) which is frequently used in human studies to monitor prevention of oxidative DNA damage (*10*).

In order to draw conclusions on alterations of DNA sensitivity to reactive oxygen species (ROS) it is possible to treat isolated peripheral lymphocytes with ROS inducers (e.g. hydrogen peroxide or α-radiation) before and after the intervention with putative protective food components (*11-15*). Furthermore, modifications of the SCGE - assay were developed, which provide information on the impact of dietary and other factors on DNA-repair processes (*11,16*).

In the last few years, we developed additional protocols for SCGE experiments with humans, which provide information if specific dietary factors affect DNA damage caused in lymphocytes by representatives of specific classes of genotoxic dietary carcinogens (e.g. acrylamide, furan, ochratoxin A, heterocyclic aromatic amines and nitrosamines). Due to lack of metabolizing (phase I) enzymes it is not possible to carry out studies with polycyclic aromatic hydrocarbons and aflatoxin B_1, but it is feasible to conduct the experiments with DNA-reactive metabolites of these compounds (e.g. aflatoxin B_1-epoxide, benzo(a)pyrene diolepoxide - BPDE). Figure 1 gives an overview on endpoints which can be monitored at present in human SCGE-studies.

Most human SCGE trials were carried out as intervention trials (cross over or parallel design) which have the advantage that inter-individual variations can be minimized. Therefore, the number of participants required to obtain meaningful results is much lower as in comparative studies. Moller and Loft (*17*) defined standards for SCGE-trials, and postulated that the use of placebos and wash-out periods would increase the quality of such studies. In addition, we

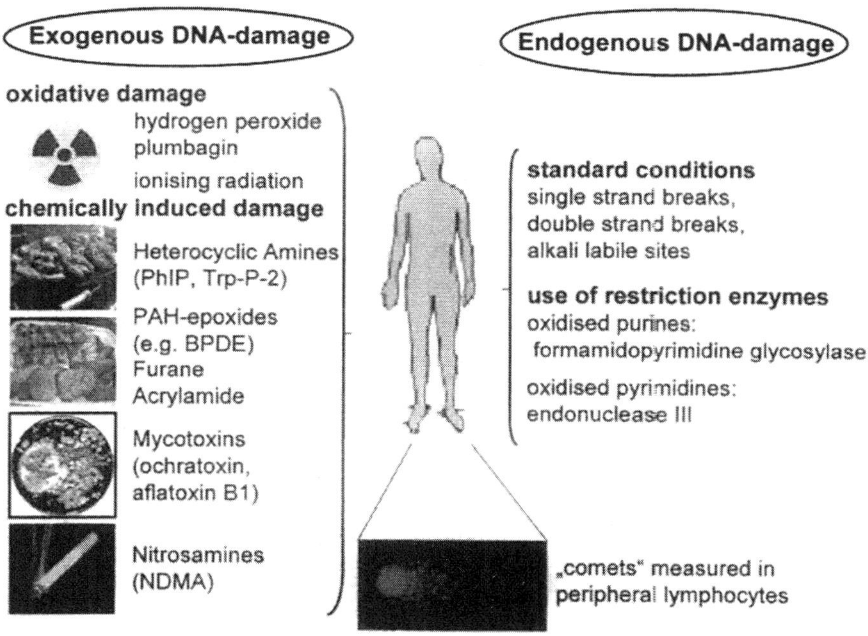

Figure 1. Endpoints used in SCGE-assays with humans.

include in our experiments run-in phases, in which the participants consume controlled diets (i.e. low amounts of foods that may interfere with the outcome of the study) and avoid excessive physical exercise. Other possible confounding factors which should be taken into consideration are age, sex, body mass index and smoking habits (*18*). In order to confirm the results of SCGE-measurements and to provide information about the mechanisms, responsible for the observed effects, additional parameters can be monitored: e.g. glutathione plasma levels, ROS-protective enzymes (SOD, GPx), total antioxidant capacity (TAC) and urinary isoprostane excretion in antioxidant studies. In experiments with adduct forming carcinogens, the activity of phase I and phase II enzymes can be determined by enzyme measurements or by use of microarrays (*19*).

The Current Database

In 1997, Pool-Zobel and coworkers published the first results of a dietary intervention study in which the SCGE technique was used (*20*). In the following years, a number of studies were published, the results of these trials (until 2004) are summarized in the reviews of Moller and Loft (*13, 17*). Also in the last two years further studies were conducted: for example with specific vegetables, red berry juice, spices and carotenoids (see Table I). In all these investigations, DNA migration was analyzed in peripheral lymphocytes or leukocytes under standard conditions. Additionally enzyme (FPG, ENDO III) and ROS treatment were used to investigate endogenous and exogenous DNA-damage by ROS. In some of the trials, the study population included patients with increased endogenous DNA damage (e.g. with Diabetes type II). Typical examples for recent studies (i.e. after 2004) for dietary factors which caused prevention of DNA damage are listed in Table 1 (upper part); studies, in which no effects were detected are listed in the lower.

Some of the trials yielded unexpected results: for example no protective effects were obtained in a large study conducted in Denmark (*33*), in which the participants consumed a mixed fruit and vegetable diet (600g/person/day). In contrast, significant protective effects were observed in a small trial with coffee (600ml/person/day), indicating that consumption of this beverage may contribute to a higher extent to prevention of oxidative DNA damage (*30, 34*). A very pronounced reduction of oxidised bases was also observed in a recent study with sumach (a traditional spice). Further experiments confirmed gallic acid as the active principle (Ferk et al., unpublished data). Strong protective effects were observed also with Brussels sprouts, and no correlation with their vitamin C contents was found (*31*), while in dietary interventions with blackcurrant juice or an anthocyanin-rich drink no reduction of oxidative damage was found (*21*); the lack of the effect may be due to the fact that anthocyans are not well absorbed in the GI tract.

Table I. Examples for DietaryIntervention Studies (2004 and newer) Based on the Comet Assay

Dietary factor[1]	*Design of the study*[2]	*Result*[3]	*Ref*
Vitamin C (plain or slow releasing) 2x250mg/p/d	blinded, placebo controlled parallel intervention study n = 48 ♂, 4 w	↓ endogenous damage, ENDO III- and FPG-sites (stronger in slow releasing)	(*28*)
quercetin and quercetin-rich fruit juice 1L/p/d (equals 18mg quercetin/p/d)	paired design: subjects act as their own control n = 8 ♂, 4 w	↓ H_2O_2-induced damage	(*12*)
carotenoids: lutein lycopene ß-carotene 12mg/p/d each or 4mg each in a mix/p/d	blinded, placebo controlled parallel intervention study n = 37 postmenopausal ♀ (divided in 5 groups) 56 d	↓ endogenous damage ↔ H_2O_2-induced damage	(*14*)
tomato-rich diet (8mg lycopene, 0.5mg ß-caro-tene, 11mg Vitamin C/p/d)	simple intervention design n = 12 ♀, 1 wk run in, 3 w intervention	↓ oxidative damage (induced by $FeSO_4$)	(*22*)
coffee 600ml/p/d (150ml metal filtered + 440ml paper filtered)	simple study design n = 8 ♂+♀, 5 d run in, 5 d intervention	↓ H_2O_2-, FPG-, ENDOIII- and Trp-P-2-induced damage ↑ PhIP- induced damage	(*30*)
coffee 1L/p/d metal filtered	simple study design n = 7 ♂+♀, 14 d run in, 6 d intervention	↓ BPDE-induced damage ↔ endogenous damage	(*32*)
Brussels sprouts 300g/p/d	simple study design n = 8 ♂+♀, 5 d run in, 6 d intervention	↓ H_2O_2-, FPG-, ENDOIII- and PhIP-induced damage ↔ Trp-P-2-induced damage	(*31*)
high phenol extra virgin oil 50g/p/d (equals 30mg phenols/p/d)	randomised cross over design (high phenol oil vs low phenol oil) n = 10 ♀, post-menopausal, 8 w	high phenol oil: ↓ oxidative DNA-damage	(*23*)
cruciferous and legume sprouts 113g/p/d	parallel design n = 20 ♂+♀, 2 w	↓ SSBs and FPG-induced damage only in response to H_2O_2-challenge	(*24*)

Table I. *Continued.*

Dietary factor[1]	*Design of the study*[2]	*Result*[3]	*Ref*
red wine and dealcoholized red wine 1 glass/p/d = 200ml/p/d	randomised placebo controlled intervention study n = 20 ♂+ 20♀, 6 w DNA damage analysed in leukocytes	↓ endogenous damage ↔ in H_2O_2-induced damage dealcoholized wine: ↔ in all endpoints	(*15*)
bread supplemented with prebiotics +/- antioxidants	design: n.i. n = 38 ♂ (smokers and non-smokers) duration: n.i.	↓ DNA damage in smokers ↔ in non-smokers	(*25*)
red mixed berry juice (rich in polyphenols TEAC 19:1) 700ml/p/d	placebo controlled parallel design n = 27 ♂, 3 w run in, 4 w intervention or placebo, 3 w wash out	↓ oxidative DNA-damage during the intervention ↔ after the intervention	(*26*)
tomato-drink (rich in careteniods) 250ml/p/d	blinded, placebo controlled cross over intervention study n = 13 ♂+13 ♀, 26 d per phase (placebo, wash out, intervention)	↔ endogenous damage	(*27*)
backcurrant juice or anthocyanin rich drink 475-1000ml/p/d (equals 397g or 365g antho-cyanins/p/d)	placebo controlled, parallel design n = 20 ♂ per group (blackcurrant juice, anthocyanin rich drink, placebo drink), 3 w	↑ FGP-sites in blackcurrant group ↔ all other endpoints	(*21*)
carotenoids: lutein, ß-carotene and lycopene; each 15mg/d/p	placebo controlled cross over design n = 28 ♂, 27 d intervention for each carotenoid (3 inde-pendent studies), 12 m break in between	lutein, lycopene: ↔endogenous or oxidative damage ß-carotene: ↑ SSBs in control cells	(*29*)

[1] p – person, d – day, L-liter, TEAC - trolox equivalent antioxidative capacity
[2] unless indicated otherwise, lymphocytes were used as indicator cells, n.i. - not indicated, d - days, w - weeks, m - months
[3] ↓ decrease, ↑ increase, ↔ no alteration, SSBs single strand breaks, ENDOIII – endonuclease III, FPG – formamidopyrimidine glycosylase, PhIP - 2-amino-1-methyl-6-phenylimidiazo[4;5-*b*]pyridine, Trp-P-2 - 3-amino-1-methyl-5*H*-pyrido[4;3-*b*]indole

Only few results have been reported from studies with adduct forming chemicals. Steinkellner *et al.* (*32*) found protective effects towards BPDE-induced DNA damage in an intervention trial with coffee. These results could be explained in a follow up study by induction of glutathione-S-transferase, which detoxifies the diol-epoxide. On the contrary, no effect on heterocyclic amine (PhIP, Trp-P-2) induced DNA-damage was found after coffee consuption (*30*), while a pronounced (almost complete) inhibition of PhIP was seen after consumption of Brussels sprouts. This effect is presumably due to inhibition of sulfotransferases, which play a role in the activation of this abundant carcinogenic heterocyclic amine (*31*).

As mentioned above, peripheral lymphocytes were used as indicator cells in most of the studies. Recently, Shaugnessy *et al.* (*35*) presented results of an intervention study (available only in abstract form) in which they used human colon biopsies in SCGE-experiments. The participants consumed fried meats under defined conditions which lead to a significant increase of DNA-migration. After consumption of the meat along with yogurt, cruciferous vegetables and chlorophyllin tablets, the meat-induced DNA damage was substantially decreased.

Future Prospects

The SCGE technique is a useful and valuable tool which enables the investigation of DNA-protective effects of dietary factors in humans. The data produced with this method in the last years show, that only specific vegetables are effective in humans and led to the identification of highly protective compounds, such as gallic acid. The combination of this approach with high throughput technologies, such as genomics and proteomics will lead to new insights concerning the efficacy ot putative chemoprotective compounds in the human diet.

References

1. Ames, B. N.; Shigenaga, M. K.; Hagen, T. M. *Proc. Natl. Acad. Sci.* **1993**, *90*, 7915-7922.
2. Culmsee, C.; Landshamer, S. *Curr. Alzheimer Res.* **2006**, *3*, 269-283.
3. Hussain, S. P.; Hofseth, L. J.; Harris, C. C. *Nat. Rev. Cancer* **2003**, *3*, 276-285.
4. Hoelzl, C.; Bichler, J.; Ferk, F.; Simic, T.; Nersensyan, A.; Elbling, L.; Ehrilich, V. J *Physiol. Pharmacol.* **2005**, *56*, 49-64.
5. Aruoma, O. I. *Mutat. Res.* **2003**, *523-524*, 9-20.
6. Helma, C.; Uhl, M. *Mutat Res.* **2000**, *466*, 9-15.

7. Frieauff, W.; Hartmann, A.; Suter, W. *Mutagenesis* **2001**, *16*, 133-137.
8. Tice, R.R.; Agurell, E.; Anderson, D.; Burlinson, B.; Hartmann, A.; Kobayashi, H.; Miyamae, Y.; Rojas, E.; Ryu, J. C.; Sasaki, Y. F. *Environ. Mol. Mutagen* **2000**, *35*, 206-221.
9. Speit, G.; Hartmann, A. *Methods Mol. Biol.* **1999**, *113*, 203-212.
10. Collins, A. R.; Dusinsika, M.; Gedik, C. M. Stetina, R. *Environ. Health Perspect.* **1996**, *104*, 465-469.
11. Chang, J. L.; Chen, G.; Lampe, J. W.; Ulrich, C. M. *Environ. Mol. Mutagen* **2006**, *47*, 503-508.
12. Wilms, L. C.; Hollman, P. C.; Boots, A. W.; Kleinjans, J. C. *Mutat. Res.* 2005, *582,* 155-162.
13. Moller, P.; Loft, S. *Mutat. Res.* **2004**, *551*, 79-89.
14. Zhao, X.; Aldini, G.; Johnson, E. J.; Rasmussen, H.; Kraemer, K.; Woolf, H.; Musaeus, N.; Krinsky, N. I.; Russell, R. M.; Yeum, K. J. *Am. J. Clin. Nutr.* **2006**, *83*, 163-169.
15. Arendt, B. M.; Ellinger, S.; Kekic, K.; Geus, L.; Fimmers, R.; Spengler, U.; Muller, W. U.; Goerlich, R. *Nutr. J.* **2005**, *4*, 33-38.
16. Collins, A. R.; Harrington, V.; Drew, J.; Melvin, R. Carcinogenesis **2003**, *24*, 511-555.
17. Moller, P.; Loft, S. *Am. J. Clin. Nutr.* **2002**, *76*, 303-310.
18. Anderson, D. *Mutat. Res.* **2001**, *480-481*, 337-347.
19. Raschke, M.; Rowland, I. R.; Magee, P. J.; Pool-Aobel, B. L. *Carcinogenesis* **2006**, *27*, 2322-2330.
20. Pool-Zobel, B. L.; Bub, A.; Muller, H.; Wollowski, I.; Rechkemmer, G. *Carcinogenesis* **1997**, *18*, 1847-1850.
21. Moller, P.; Loft, S.; Aolfthan, G.; Freese, R. *Mutat. Res.* **2004**, *551*, 119-126.
22. Riso, P.; Visioli, F.; Erba, D.; Testolin, G.; Porrini, M. *Eur. J. Clin. Nutr.* **2004**, *58*, 1350-1358.
23. Salvini, S.; Sera, F.; Caruso, D.; Giovannelli, L.; Visioli, F.; Saieva, C.; Masala, G.; Ceroti, M.; Giovacchini, V.; Pitozzi, V.; Galli, C.; Romani, A.; Mulinacci, N.; Bortolomeazzi, R.l; Dolara, P.; Palli8, D. *Br. J. Nutr.* **2006**, *95*, 742-751.
24. Gill, C. I.; Haldar, S.; Porter, S.; Matthews, S.; Sullivan, S.; Coulter, J.; McGlynn, H.; Rowland, I. *Cancer Epidemiol. Biomarkers Prev.* **2004**, *13*, 1199-1205.
25. Glei, M.; Habermann, N.; Osswald, K.; Seidel, C.; Person, C.; Jahreis, G.; Pool-Zobel, B. L. *Biomarkers* **2005**, *10*, 203-217.
26. Weisel, T.; Baum, M.; Eisenbrand, G.; Dietrich, H.; Willo, F.; Stockis, J. P.; Kulling, S.; Rufer, C.; Johannes, C.; Janzowski, C. *Biotechnol. J.* **2006**, *1*, 388-397.
27. Riso, P.; Visioli, F.; Grande, S.; Guarnieri, S.; Gardana, C.; Simonetti, P.; Porrini, M. *J. Agric. Food Chem.* **2006**, *54*, 2563-2566.

28. Moller, P.; Viscovich, M.; Lykkesfeldt, J.; Loft, S.; Jensen, A.; Poulsen, H. E. *Eur. J. Nutr.* **2004**, *43*, 267-274.
29. Astley, S. B.; Hughes, D. A.; Wright, A. J.; Elliott, R. M.; Southon, S. *J. Nutr.* **2004**, *91*, 53-61.
30. Bichler, J. In *Institute of Cancer Research.* University of Vienna: Vienna. 2005.
31. Hölzl, C., In *Institute of Cancer Research.* University of Vienna: Vienna. 2004, p. 72.
32. Steinkellner, H.; Hoelzl, C.; Uhl, M.; Cavin, C.; Haidinger, G.; Gsur, A.; Schmid, R.; Kundi, M.; Bichler, J.; Knasmuller, S. *Mutat. Res.* **2005**, *591*, 264-275.
33. Moller, P.; Vogel, U.; Pedersen, A.; Dragsted, L. O.; Sandstrom, S.; Loft, S. Cancer Epidemiol. Biomarkers Prev. **2003**, *12*, 1016-1022.
34. Bichler, J.; Cavin, C.; Simic, T.; Chakraborty, A.; Ferk, F.; Hoelzl, C.; Schulte-Hermann, R.; Kundi, M.; Haidinger, G.; Angelis, K.; Knasmueller, S. *Food Chem. Toxicol.* submitted.
35. Shaughnessy, D. T.; Gangarosa, L.; Schliebe, B.; DeMarini, D. M.; Xu, Z. L.; Umbach, D. M.; Sandler, R. S.; Taylor, J. A. San Francisco, CA: Environmental Mutagen Society. 2005.

Chapter 4

Identification and Characterization of Flavonoid Target Proteins

Herwig O. Gutzeit and Markus Böhl

Institut für Zoologie, TU Dresden, D–01062 Dresden, Germany

Flavonoids are known to interact with a host of cellular proteins *in vitro* and to affect their biological activity, but the relevance of the interaction *in vivo* has rarely been ascertained. To realistically assess the biological and pharmacological effects of flavonoids, a systematic approach to identify relevant cellular target proteins is required. We showed that the changing spectroscopic properties of target protein and ligand can be exploited to identify and characterize target proteins. In a proof-of-principle experiment actin was identified as a new and relevant flavonoid target protein.

The biological effects of flavonoids have been intensively studied due to their suspected beneficial effects on human health. The interaction of flavonoids with numerous cellular enzymes has been reported *(1)* and the picture emerged that each flavonoid is likely to affect a spectrum of cellular target proteins. However, the relevance of these studies for human health has often remained elusive. The complex interactions of quercetin with target proteins apparently result in profound physiological changes in the affected cells. This is reflected in an altered pattern of expressed proteins in human colon cancer cells in response to quercetin treatment as was shown recently in a proteomics approach *(2)*.

A causal analysis of flavonoid – induced effects on the cellular and tissue level has been hampered by lack of knowledge concerning the spectrum of relevant flavonoid-specific target proteins in a given cell or tissue. How can the flavonoid target proteins in a cell be identified systematically? The straightforward method is to immobilize the flavonoid of interest and to isolate and identify binding proteins by affinity chromatography *(3)*. The limitation of this procedure is that the biochemical coupling of the flavonoid to, for example,

Sepharose beads alters the structure of the flavonoid and hence altered binding properties to the target proteins cannot be excluded. An alternative method that requires no modification of the flavonoid is based on the changing spectroscopic properties of both target protein and flavonoid ligand upon their specific interaction. The basic principle and a proof-of-principle experiment will be outlined in this contribution.

Materials and Methods

Size Exclusion Chromatography

Chromatography and fractionation was performed using a Biorad Biologic Workstation (München, Germany) equipped with a Pharmacia (Uppsala, Sweden) Superdex 200 gel filtration column. At an isocratic flow of 0.4 mL/min (200 mM KCl, 1.5 mM $MgCl_2$, 200 mM Tris-HCl, pH 7.5) the absorption at 280nm was continuously measured and fractions of 0.5 mL collected. All standard proteins, buffer constituents, flavonoids and solvents were supplied by Sigma (Steinheim, Germany).

Fluorescence Spectroscopy

For each fraction obtained by chromatography the fluorescence intensity at Ex280nm/Em345nm was determined in 96 well testplates in the absence and in the presence of the tested flavonoids using a Tecan Infinite 200 microplate reader (Crailsheim, Germany). A more detailed description for this procedure is explained by Böhl *et al.* (*7*).

Results and Discussion

The changing spectroscopic properties as a result of protein/ligand interaction can best be illustrated using 3 model proteins which were separated by column chromatography and the fractions analysed by fluorescence spectroscopy (Figure 1).

Proteins typically show autofluorescence (e.g. BSA and lysozyme, Figure 1A) due to the presence of aromatic amino acids. In exceptional cases (e.g. ferritin, Figure 1A), the fluorescence may be suppressed by the presence of bound iron (Figure 1A). For the experimental analysis of protein autofluorescence the tryptophan (trp) residues are most relevant. When a flavonoid like genistein or quercetin binds to a trp-containing protein the autofluorescence is quenched (BSA, Figure 1B) and the absence of quenching is

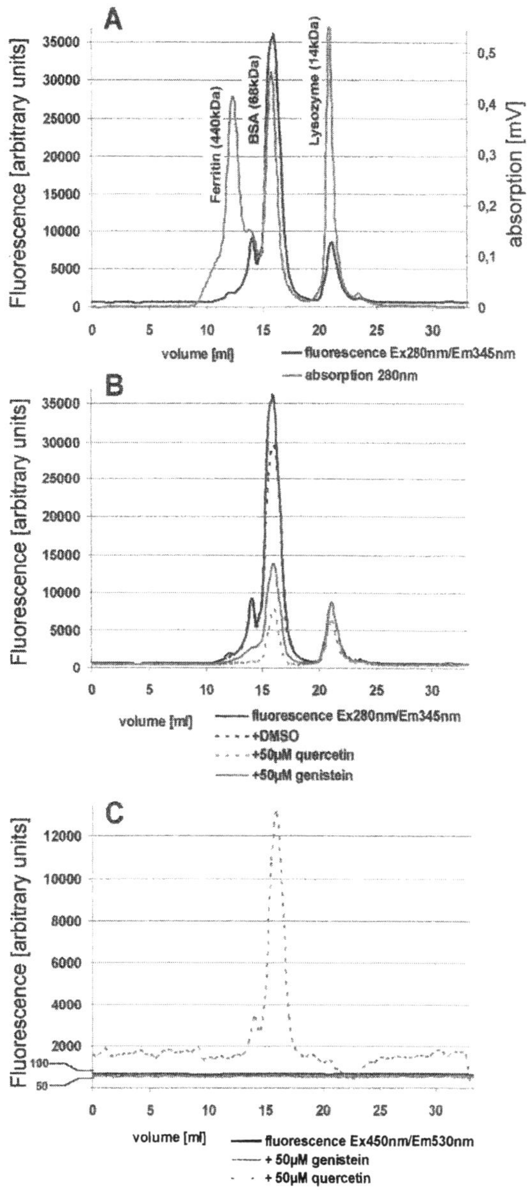

Figure 1. Fractionation of 3 proteins: ferritin (440kDa), BSA (68kDa) und lysozyme (14kDa). A: The absorption at 280 nm was continuously measured and the intrinsic protein fluorescence determined in the collected fractions (Ex280 nm/Em345 nm) B: Quenching of intrinsic protein fluorescence by quercetin and genistein (Ex280/Em345) C: Elicitation of flavonoid fluorescence by binding to BSA (Ex450/Em530 nm).

indicative for the absence of molecular interactions (lysozyme, Figure 1B). The degree of quenching depends on the spectral properties of the ligand, the strength of protein/ligand interaction, the number of trp residues in the protein and their location and distance from the ligand binding site. The solvent of the flavonoid stock solutions (DMSO) contributes little to the quenching effect (Figure 1B).

There is another interesting spectroscopic change when the flavonoid ligand interacts with a target protein: the binding of the flavonol quercetin and structurally related flavonoids may lead to the stabilisation of quercetin monoanion tautomers (*4*) and, as a result, to an elicitation of fluorescence in the visible spectrum (e.g. quercetin binding to BSA, Figure 1C). This property of quercetin allows to visualise target proteins in living cells *(5)* and, furthermore, to monitor quercetin metabolism in vital cells *(6)*. Other flavonoids like the isoflavone genistein do not show this property (Figure 1C).

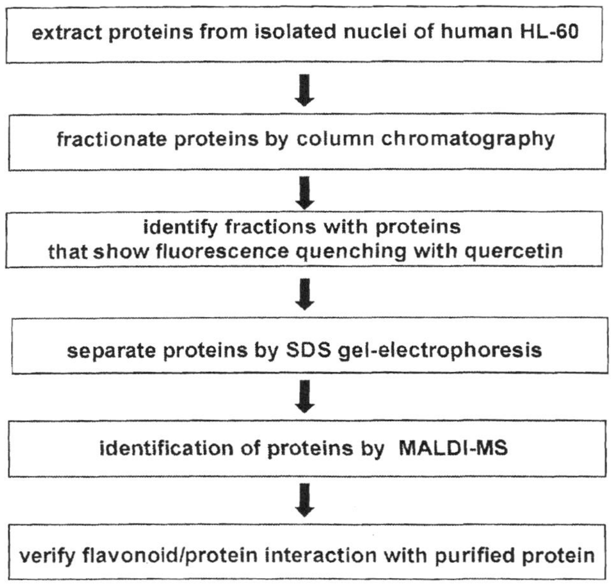

Figure 2. Outline of the experimental strategy to identify quercetin target proteins by means of the changing spectroscopic properties when proteins interact specifically with a flavonoid.

The quenching of autofluorescence by flavonoid ligands is a sensitive and a generally applicable method to analyze target/ligand interactions. Can target proteins be isolated and identified out of a complex protein mixture by virtue of fluorescence quenching? In a pilot experiment that is summarized in Figure 2, we showed that this is indeed possible *(7)*. As starting material we chose nuclear proteins of human leukemia cells (HL-60 cell line) since in previous experiments with HL-60 cells it had become clear that a high concentration of quercetin target proteins is present in cell nuclei. The extracted nuclear proteins were fractionated by column chromatography and each fraction analyzed by fluorescence spectroscopy. Fractions of nuclear proteins which showed strong fluorescence quenching after the addition of quercetin were subjected to SDS gel electrophoresis and, finally, single stained protein bands were cut out of the gel and the protein(s) identified by mass spectrometry (in collaboration with B. Hoflack, BIOTEC Dresden). Amongst the identified proteins were different DNA binding proteins like high mobility group proteins and also actin. Because this protein can easily be obtained in pure form, the molecular interaction with quercetin was analyzed further.

Actin is well-known for its cytoskeletal functions but the protein also plays an essential role in transcription (*8,9*). In a collaborative effort (*10*) we verified by fluorescence spectroscopy that quercetin binds specifically to actin. With respect to the actin activity we could show that actin polymerisation is inhibited by quercetin (25 μM) *in vitro* and this also applies to the nuclear function since transcription is inhibited using an *in vitro* transcription test system. If this inhibitory activity also leads to functional deficiencies in living cells remains to be shown.

How reliable is the method and what are the limitations? The identification of the flavonoid target proteins hinges, in part, on the sophistication of the applied fractionation procedure. Ideally, each fraction should contain only one predominant protein. As discussed above, not all proteins are amenable to an analysis by fluorescence spectroscopy and, for that reason, affinity chromatography of target proteins by immobilized flavonoids is a useful technique that may provide valuable additional information. The latter approach, however, suffers from the unavoidable modification of the flavonoid structure and there are many examples that small structural changes may substantially alter the biological activity in a given test system.

The flavonoid/protein interaction is specific and a discrete binding site could be defined in all cases that were analysed in sufficient detail. For example, the binding of some estrogenic flavonoids like genistein to the mammalian estradiol receptors (α and/or β) is well understood *(11)*. When known target proteins of flavonoids are compared, it is apparent that flavonoids do not necessarily target proteins with similar functions and ligand structures, but also interact with proteins which are functionally unrelated and possess different ligand binding sites (Table I).

For example, quercetin inhibits the activity of quite different proteins, like actin, phospholipase A2 and myosin II ATPase. In actin and myosin, quercetin appears to bind at the ATP binding site while in phospholipase A2 the binding site is in the phospholipid substrate pocket (*13*). The synthetic naringenin derivative 6-(1,1-dimethylallyl)-naringenin is not only highly estrogenic and

Table I. Examples for Different Biological Activities of Structurally Related Flavonoids at 25 μM Concentrations

	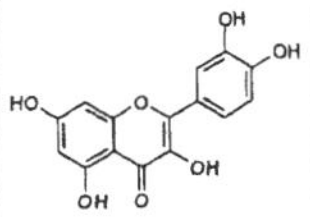 quercetin	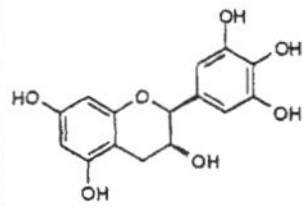 epigallocatechin	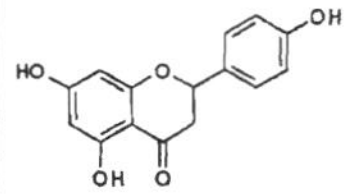 naringenin	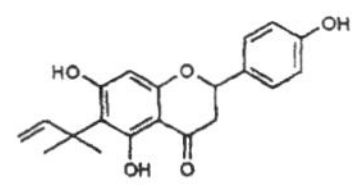 6-(1,1-dimethylallyl)-naringenin
phospholipase A2	inhibition	no effect	inhibition	inhibition
myosin II ATPase	inhibition	inhibition	no effect	inhibition
actin polymerisation	inhibition	stimulation	n.d.	n.d.
transcription (nuclear function of actin)	inhibition	stimulation	n.d.	n.d.

A single flavonoid may affect very different proteins and cellular functions while structurally similar flavonoids may not possess this activity or even produce opposite effects. n. d.: not determined.

binds to the steroid ligand pocket of the estradiol receptor (*14*) but also inhibits phospholipase A2 and myosin II ATPase. Other structurally related flavonoids do not show these effects or even have opposite effects and show stimulatory instead of inhibitory activity (compare quercetin with epigallocatechin concerning actin-dependent functions; Table I).

Such molecular promiscuity is pharmacologically considered undesirable. However, the numerous reports on beneficial effects of flavonoids should remind us that some flavonoids are apparently able to influence the cellular physiology in a positive way under certain pathological circumstances (e.g. anti-inflammatory, (anti)estrogenic, anti-oxidative etc.). While the exact number of relevant target proteins remains unknown it is clear that the number of affected proteins is not small. However, only a few of these target protein may be sufficient to produce the beneficial effects on human health.

Pharmacological research traditionally focuses on the specific inhibition of a single target protein, yet there is no *a priori* reason why a multi-hit approach as exemplified by flavonoids is inherently inferior to an attempted (but mostly not achieved) single-hit-approach. The systematic identification of flavonoid target proteins will give valuable information concerning the molecular mechanisms that give rise to the observed biological effects and it may also change our way of thinking about the pharmacological use of flavonoids.

References

1. Middleton E. Jr; Kandaswami, C.; Theoharides, T. C. *Pharmacol. Rev.* **2000**, *52*, 673-751.
2. Wenzel, U.; Herzog, A.; Kuntz, S.; Daniel, H. *Proteomics* **2004**, *4*, 2160-2174.
3. Ermakova, S.; Choi, B.Y.; Choi, H.S.; Kang, B.S.; Bode, A.M.; Dong, Z. *J. Biol. Chem.* **2005**, *280*, 16882-16890.
4. Dangles, O.; Dufour, C.; Bret, S. *J. Chem. Soc., Perkin Trans. 2*, **1999**, 737–744.
5. Gutzeit, H. O.; Henker, Y.; Kind, B.; Franz, A. *Biochem. Biophys. Res. Commun.* **2004**, *318*, 490-495.
6. Ludwig-Müller, J.; Tokalov, S.; Franz, A.; Gutzeit, H. O. *Biol. Chem.* **2005**, 386, 279-283.
7. Böhl, M.; Czupalla, C.; Tokalov, S. V.; Hoflack, B.; Gutzeit, H. O. *Anal. Biochem.* **2005**, *346*, 295-299.
8. Hofmann, W.A.; Stojiljkovic, L.; Fuchsova, B.; Vargas, G.M.; Mavrommatis, E.; Philimonenko, V.; Kysela, K.; Goodrich, J.A.; Lessard, J.L.; Hope, T.J.; Hozak, P.; de Lanerolle, P. *Nat. Cell Biol.* **2004**, *6*, 1094-1101.
9. Philimonenko, V.V.; Zhao, J.; Iben, S.; Dingova, H.; Kysela, K.; Kahle, M.; Zentgraf, H.; Hofmann, W.A.; de Lanerolle, P.; Hozak, P.; Grummt, I. *Nat. Cell Biol.* **2004**, *6*, 1165-1172.
10. Böhl, M.; Tietze, S.; Sokoll, A.; Madathil, S.; Pfennig, F.; Apostolakis, J.; Fahmy, K.; Gutzeit, H.O. *Biophys. J.* **2007**, *93*, 2767-2780.
11. Manas, E.S.; Xu, Z.B.; Unwalla, R.J.; Somers, W.S. *Structure* **2004**, *12*, 2197-2207.
12. Mewshaw, R.E.; Edsall, R.J., Jr.; Yang, C.; Manas, E.S.; Xu, Z.B.; Henderson, R.A.; Keith, J.C., Jr.; Harris, H.A. *J. Med. Chem.* **2005**, *48*, 3953-3979.
13. Böhl, M.; Ketscher, L; Gutzeit, H.O. *Analyt. Biochem.* **2006**, in press.
14. Zierau, O.; Hamann, J.; Tischer, S.; Schwab, P.; Metz, P.; Vollmer, G.; Gutzeit, H.O.; Scholz, S. *Biochem. Biophys. Res. Commun.* **2005**, *326*, 909-916.

Chapter 5

Good Agricultural Practice for the Quality Assurance of Traditional Chinese Herbs Used in Dietary Supplements

Frank S. C. Lee[1], Xiaoru Wang[1], and Lei Li[2]

[1]Key Lab of Analytical Technology Development and the Standardization of Chinese Medicines, QingDao, China
[2]First Institute of Oceanography, State Oceanic Administration, QingDao, China

The quality control and standardization of traditional Chinese herbs (TCHs) commonly used in functional or therapeutic food products is a challenging task. The guidelines of Good Agricultural Practice (GAP) and Good Manufacturing Practice (GMP) are designed to standardize respectively the different operations involved. Although GMP guidelines have been well established in the manufacturing processes, the operational specifics of GAP guidelines in the agricultural production of herbal plants are still in the developmental stage. This contribution reviews the current status of GAP development in China.

Traditional Chinese herbs (TCHs) are gaining increasing popularity worldwide in the development of dietary supplements and pharmaceutical products. The quality control and standardization of TCHs is a challenging task because of (1) the large variations in the sources and properties of raw herbs, and (2) the wide diversity in process types and manufacturing conditions leading to the products. The Good Agricultural Practice (GAP) and Good Manufacturing

Practice (GMP) guidelines are designed to address, respectively, the quality assurance issues involved in the above two areas. Although GMP guidelines have been well established in manufacturing processes, the scope and operational specifics of GAP guidelines in the agricultural production of herbal plants are still in the developmental stages. In recent years, the development of GAP-based farming in many countries including China is accelerating; and the adoption of GAP as an international standard for the marketing and trading of food or herbal products is gaining momentum in international communities.

GAP and the "5P" Quality Assurance System

The "5Ps" is the most widely accepted quality assurance system for the development and production of consumer products intended for therapeutic applications. These guidelines cover the entire lifecycle of a product from initial raw material supply, through the manufacturing processes, to the final stage of consumer consumption. Table I outlines the objectives and major activities of these guidelines. Of the 5Ps, Good Laboratory Practice (GLP) and Good Clinical Laboratory Practice (GcLP) safeguard the quality of laboratory and clinical testing. They are the gatekeeper at the front end of the product lifecycle during its research and developmental stages. At the other end is Good Supply Practice (GSP), which deals with product surveillance activities at the final stage. In between are Good Agricultural Practice (GAP) and Good Manufacturing Practice (GMP), which constitute the heart of the 5P system focusing on quality and safety issues during the manufacturing processes.

For the manufacturing of synthetic chemicals with well defined properties, GMP is sufficient to safeguard the quality involved in the entire production chain from raw materials to the final products. For the agricultural production of natural product such as herbal plants, however, a different set of problems exists. Aside from internal factors such as the intrinsic properties of the herbs, the quality of the TCHs are affected also by external factors including the genetic variations of the plant species, environmental conditions and climate fluctuation. These external variables are hard to control because they vary from grower to grower, from crop to crop, and with the geographic locations of the production site. Thus, stability of the produce is more difficult to monitor and control, and benchmarking standards are more difficult to establish compared to synthetic drugs. It is against this background that GAP has been evolved. In recent years, the adoption of GAP as a quality standard for agricultural or food products is gaining increasing attention both in China and internationally (1-6).

Although the overall 5Ps were designed originally for drug manufacturing, the general principle of these quality guidelines should be applicable to dietary supplement or health food products as well. By implementing quality standards from the farm to the factory, TCHs would be better prepared to meet the needs of

Table I. "5 Ps" Quality Assurance System for the Production of Medicinal Herbal Products

QC/QA Guidelines	*GAP-Good Agriculture Practice*	*GLP—Good Laboratory Practice*	*GCP-Good Clinical Practice*	*GMP-Good Manufacturing Practice*	*GSP-Good sales(Supply) Practice*
Product development Stage	preparation of raw herbs or crude drugs	Laboratory biochemical assay and animal tests	clinical testing for product registration	In-plant Product formulation and manufacturing	Surveillance of Finished products
Major activities	Species Authentication, plant cultivation, and the processing of crude herbs	Bioassay, Chemical analysis and animal tests	clinical trials of pre-marketing products	Quality control of raw materials, in-plant process and final products	Implementation of inspection, monitoring and reporting activities for commercial products
Objectives	Ensure the sustainable production of contamination-free herbs with controllable bioactivity and yield	Examination, Confirmation and quantification of active components in herbs.	Acquire clinical data to quantify dose/response relationship	Ensure the quality, efficacy and safety of manufacturing products	Enforcement of regulatory activities for commercial OTC or prescription products

Priority Chinese medicinal herbs selected for 5Ps by Chinese Ministry of Science and Technology(2000) include: Salvia Miltiorrhiza, Radix Astragoli, Radix Ophiopogonies, Bullbus Fritillariae Cirrhosea, Flos Chrysanthemi, Radix Glycyrrhizae, Radix Ginseng, Fructus Lycii, Herba Ephedrae, Radix Coicis, Radix Rehmanniae, Rhizoma Ligustici Chuanxiong, Radix Aconiti Praeparata, Rhizoma Gastrodiae, Cornu Cervi Pantotrichum, Radix Achyranthis Bidentatae, Rhizoma Pinelliae and Tuber Dioscoreae.

increasingly discerning domestic consumers and international demand for botanicals. **Table II** outlines the major GAP-related regulatory activities took place in different countries in recent years. In the eighties and early nineties, GLP, GcLP and GMP guidelines have all been officially promulgated by SFDA in China (State Food and Drug Administration of China) (*6*). Meanwhile, the concept of GAP has evolved in Europe, Japan and North America during this period. In 1998, the U.S. Food and Drug Administration (USFDA) published regulations to ensure the microbial safety of fresh produce by defining the GAP and GMP guidelines that producers and handlers should follow (*7*). In year 2000, EAEM (European Agency for the Evaluation of Medicinal Products) announced official GAP guidelines for botanical drugs and herbal products (*8*). Between year 2000 to 2002, scientists and governmental officials in China carried out extensive studies on the feasibility of implementing GAP in China (*9-11*). In June 2002, GAP for Chinese Crude Drugs (Interim) was passed into effect by SFDA (*6*). WHO (World Health Organization) also published guidelines on good agricultural and collection practices (GACP) for medicinal plants in 2003 (*12*).

GAP Certification System in China

The GAP guidelines of China addresses quality and safety requirements for TCHs in areas including: (i) ecological and environment conditions of the production site; (ii) germ-plasma and propagation material; (iii) management for cultivation of medicinal plants; (iv) packaging, transportation and storage; and (v) managerial and technical aspects of quality management. The official GAP program for Chinese medicinal herbs started its trial period in China from June 1, 2002 (9). The GAP certificate, which is usually awarded to a private enterprise, is valid for 5 years. The follow-up monitoring activities are carried out by the expert teams organized by SFDA. In addition, provincial, city and regional FDA offices of China provide the needed assistance to handle the application and monitoring activities at the provincial or local levels. In the original regulation set out by SFDA in 1998, all manufacturers must comply with the Good Manufacturing Practice by April, 2004; while farms producing raw ingredients have until 2007 to meet the guidelines specified in Good Agricultural Practices.

Work on GAP encompasses a wide range of topics in different disciplines. The type of work and major activities are described in ref *9*, and also outlined in **Table III.** More than simply the implementation of a quality assurance system on existing practices, the current GAP program in China also calls or the application and development of updated technology for the modernization of traditional herb farming practices. On the other hand, the developed techniques have to be user friendly enough that they can be practiced by farmers on a routine basis. All the methods developed thus have to be standardized and documented in Standard Operational Procedures (SOPs); and personnel trainings are necessary to facilitate technology transfer.

Table II. Regulatory Milestones Pertaining to GAP Development for Herbal Products

Regulatory Agency initiated (Year Published)	*Regulation/Guidelines Promulgated*
Ministry of Health of China (1982)	GMP Guidelines
Ministry of Health of China (1988)	GLP guidelines–Regulations on Experimental Animals
SFDA (State Food and Drug Administration) of China (1992)	Regulations on registration and approval of New CM drugs (GCP, GLP)
US FDA (1998)	guidelines for the microbial safety of fresh produce defining the Good Agricultural Practices (GAP) and Good Manufacturing Practices (GMP) guidelines that producers and handlers should follow.
SFDA of China (2000)	Article (2000) 157 published emphasizing the importance of well managed TCM (Traditional Chinese Medicines) herb Farms, and the requirements for Fingerprinting of CM Injection fluid products.
Secretary of States of China (2000)	published regulations prohibiting the collection and sales of wild licorice, ephedra sinica stapt
European Agency for the Evaluation of Medicinal Products (2000)	GAP guidelines for botanical drugs and herbal products
SFDA (2002)	Issuance of GAP regulation for Medicinal Plants and Animals
SFDA (2003)	GAP certification system starts operation–certificate valid for 5 years; status followed and monitored by SFDA
WHO (2003)	published guidelines on good agricultural and collection practices (GACP) for medicinal plants in 2003 (World Health Organization, 2003)

Table III. Scope of GAP Work

Area of work	Major Activities	Objectives
Environmental monitoring and impact assessment	Collection and analysis of environmental quality parameters for air, water and soil; the development and application of environmental friendly practices for pest prevention; environmental impact assessment of the production site.	Ensure the meeting of environmental quality standards, the absence of potential sources of contaminants including natural or man-made toxins/pollutants, heavy metals, and pesticides/herbicides residues, and the sustainable development of the production site and the surrounding area.
Selection, identification and authentication of plant species	Seed selection and preservation; Plant species identification through DNA fingerprinting and chemical compositional fingerprinting	Establish the authenticity and correct genetic identity of the plant species
Plantation, cultivation and harvesting technology	Standardized production based on modern science in combination with traditional wisdom; pest prevention; specified fertilizers; best harvest time determination.	Establish standardized farming practice to grow quality and safe produce
Processing, storage and transportation	harmonize traditional method with modern science for the field and factory processing of raw plant, and the storage and transport of crude products after primary treatment	Establish modernized and standardized field and factory processing technology for the preparation of contamination–free crude herbal products
Management, training and documentation	Establish SOPs (Standard Operation Procedures) for the technical operations including site selection, environmental monitoring, cultivation practices, quality control and primary processing; Establish management systems including product registration, personnel training; SOP documentation and updating, and facility maintenance	Establish standardized technical manuals, managerial systems quality control procedures and qualified personnel for GAP operation.

In the application of GAP certificate, the species should have completed at least one growth cycle. Documented information should include site selection and selection criteria, historical data, scale of production, and environmental conditions of the surrounding area. In cultivation practice, information should include species authentication/identification, speciation of wild or cultivated varieties, Details of the GAP program defined by SFDA can be found in ref (*14*) seeding and growth conditions, harvest practice, fertilization, pest prevention, and field and farm management practices. Also to be included are management and operation practices involved in quality control and assessment methods, personnel training and maps showing detailed cultivation area and experimental farms (scale, production yield and scope). Besides regular GAP studies, work also emphasize ways to: (1).maintain ecological balance and the sustainable development of environment and (2) develop and optimize the transition from wild to cultivated farming.

Since the inauguration of the GAP program in 2004, a series of TCH farms have already been awarded the GAP certificates issued by SFDA of China. Based on published information from SFDA (*6*), the location of these farms and the herbs which had received GAP certifications from 2004 to 2006 are summarized in **Table IV.** These certificates were awarded to the sponsoring party of the program which is usually a business enterprise. The technical work of the program is carried out by a working team of experts and professionals, generally from a research institution or university. For our purpose here, only the locations of the GAP farms are listed in the table while the names of the private companies are omitted. Also listed in Table IV are the locations of "genuine" herbs as specified in Chinese Pharmacopoeia (2005). The concept of "genuine herbs" (*13*) is rooted deeply in traditional Chinese medicines, meaning that only species grown in specific geographic locations are the authentic species with the best quality. The GAP farms are in general, but not always, located in sites with the reputed "genuine herbs".

Future Development and Challenges

With a full GMP/GAP certification scheme in place, and with both industrial and agricultural sectors understanding what is required of them, progress is being achieved at a rapid rate. Quality has been improved because raw materials via the GAP system are being controlled; and supply of raw materials also becomes more stable with less price fluctuation. High quality and contaminant-free raw material produced under GAP principles is a pre-requisite for the making of quality and safe Chinese medical products based on modern GMP production.

Table IV. GAP Herbs Certified in China**

Chinese name+ of Herb (pinYing)	Latin Name+ of herb	Functional part of herb	Location of genuine herb defined in Chinese Pharmacopoeia*	Location of GAP Farm in China*	Year SFDA GAP certificate Recv'd
Baizhi	Radix angelicae dahuricae	root	Hangzhou, Zhe Jiang province ; Sui Ning, Si Chuan province	Si Chuan province	2006
Banlangen	Radix isatidis	root	An Guo, He Bei province, Nan Tong, Jiang Su province	Bai Yun Shan , Fu Yang	2006
Chuanxiong	Rhizoma chuanxiong	root and stem	Guan Xian, Si Chuan, Yun Nan , He Bei	Si Chuan	2006
Danggui	Radix angelicae sinensis	root	Gan Su, Shan Xi	Gan Su	In process
Danshen	Radix salviae miltiorrhizae	root	Shan Xi, SiChuna	Tian Shi Li Co., Shan Xi	2004
Guanghuoxiang	Herba pogostemonis	entire grass	Guang Dong, Hai Nan	Guang Dong	2006
Huangqi	Radix astragali	root	Shan Xi, Inner Mongolia	Inner Mongolia	2006
Maidong	Radix ophiopogonis	root	Si Chuan, Zhe Jiang	Ya An San Jiu Co in SiChuna	2004
Renshen	Radix ginseng	root	Ji Lin, Hei Long Jiang	Ji Ling	2004
Saqi	Radix notoginseng	root	Yun Nan , Guan Xi	Yun Nan	2005
Taizishen	Radix pseudostellariae	root	Jiang Su	Gui Zhou	2006
Tianma	Rhizoma	root and	Yun Nan, Si Chuan	Shan Xi	2006

Continued on next page.

Table IV. Continued

Chinese name[+] of Herb (pinYing)	Latin Name[+] of herb	Functional part of herb	Location of genuine herb defined in Chinese Pharmacopoeia*	Location of GAP Farm in China*	Year SFDA GAP certificate Recv'd
Xiyangshen	Radix panacis quinquefolii	root	Unspecified or unknown	Ji Lin	2004
Yuxingcao	Herba houttuyniae	whole grass	Unspecified or unknown	Ya An San Jiu Co.	2004
Shanzhuyu	Fructus corni	fruit	He Nan	He Nan	2006
Jingjie	Fineleaf Schizonepeta Herb	whole grass	Jiang Su, Zhe Jiang, Jiang Xi, Hu Be, He Bei	He Bei	2006
Kudiding	Bungeanae Corydalis Herb	whole grass	Gan Su, Shan Xi, Shan Xi, Shan Dong	He Bei	2006
Yinxingye	Folium Ginkgo	leaf	Jiang Su	Jiang Su	2006
Heshouwu	Radix Polygoni Multiflori	root	He Nan, Hu Bei, Guang Xi, Guang Dong, Gui zhou, Si chuan , Jiang Su	Gui Zhou	2005
Jiegeng	Radix Platycodi		Shan Dong, Jiang Su, An Hui, Zhe Jiang, Si Chuan	Shan Dong	2005
Dangshen	Radix Codonopsis	root	Shan Xi	Shan Xi	2005
Yiyiren	Semen Coicis	fruit	Unspecified or unknown	Zhe Jiang (Zhe Jiang Tai Shun)	2005
Jiaogulan	Herba Gynostemmatis Pentaphylli	whole grass	Unspecified or unknown	Shan Xi	2005
Zhizi	Fructus Gardeniae	fruit	Jiang Xi	Jiang Xi	2004
Qinghao	Herba Artemisiae Annuae	whole grass	Unspecified or unknown	Chong Qing, Si Chuna	2004

Huanglian	Rhizoma Coptidis	root	Si Chuan, Hu Bei	Chong Qing, Si Chuna	2004
Chuanxinlian	Herba Andrographis	whole grass	Guang Dong, Fu Jian	Guang Dong	2004
Dengzhanxixin	Herba Erigerontis	whole grass	Unspecified or unknown	Yun Nan	2004
Yuxingcao	Herba Houttuyniae	whole grass	Unspecified or unknown	Si Chuan Ya An	in process
Xihonghua	Stigma Croci	Flower	Originated in Spain and Holand, now cultivated in Shanghai, Zhe Jiang,He Nan, Beijing and Xin Jiang	Shanghai	in process

[+]Pinging and Latin names are listed in Chinese Pharmacopoeia, 2005 edition
*For locations of "genuine herbs" and "GAP Farms", only the name of the province in China is listed
[**] Based on published data by SFDA from year 2004-2006 (*6*)

The critical challenge of GAP establishment is the difficulty involved in the quality control and standardization of herbal plants. The two main problems are the lack of scientific-based conventions to define and standardize quality, and the lack of comprehensive toxicological data. To date, although GMP guidelines have been well established in manufacturing processes, the GAP for efficacy assurance and safety of Chinese herbal plants used for functional foods and dietary supplements is still in the development stage requiring continuing research.

References

1. Lee, F.S.C., et al., Development of Good Agriculture Practice and Quality Index for Danshen (Radix Salviae Miliorrhizae) from Zhong Jiang, Sichuan", Program code UIM/6 funded by Hong Kong Innovation and Technology Fund (ITF) and New World Bioscience, Ltd., 2000-2003. For program details, refer to website www.itf.gov.hk.
2. Xiaoru Wang, et al, GAP programs on Licorice (Radix Glycyrrhiza,), Taizishen (Pseudostellaria heterophylla (Miq.) Pax.) and Alisma (Alisma gramineum Lej.). For program details, refer to SFDA website www.sfda.gov.cn.
3. Lee, F.S.C.; Wang, X. *Key Analytical Technologies for the Quality Control of Chinese Medicinal Herbs Produced according to GAP Guidelines*, Xiamen University Press, Xiamen, China, 2002.
4. Wang, X.; Lee, F.S.C.; Li, L. Eds., *Key Technologies for the Quality Control of Chinese Medicinal Herbs*, Chemical Industry Publishing House, Beijing, China, 2006.
5. *For general discussion of technical issues involved in Quality Control and GAP studies, see, for instance, relevant articles in J. Research and Practice of Chinese Medicines*, Special issue, **2004**, *Vol. 18.*
6. For regulatory and research activities, refer to official webside of SFDA: www.sfda.gov.cn.
7. For details, refer to U.S. FDA issued final GMP regulations, 2003
8. EMEA/HMPWG/18/99/ Draft on the document "Good Agriculture Practice (GAP) from the European Herbs Growers and Producers Association, 5 August 1998
9. Ren, D.Q.; Zhou, R.H. *An Operational Manual for the Implementation of GAP*, China Agriculture Publishing, 2003.
10. Zhou, R.H. *J. Med. Plants*, **1998**, *13*, 254.
11. Chang, N.P.; Shiao, X.; Lin, R.C. *Research and Information on Traditional Chinese Medicine*, **2001**, *3*, 15.

12. WHO Publication, Guidelines on Good Agricultural and Collection Practices for Medicinal Plants, http://www.who.int/medicines/library/trm/medicinalplants/agricultural.shtml
13. Hu, S.L. *Study of Chinese Genuine Herbs*, Chinese Medicine Publishing, Inc., 1997.

Bioavailability and Metabolism

Chapter 6

Bioavailability and Metabolic Fate of Anthocyanins

Takashi Ichiyanagi

Faculty of Pharmaceutical Sciences, Niigata University of Pharmacy and Applied Life Sciences, 265–1 Higashijima, Niigata 956–8603, Japan

In the present study, bioavailability and metabolic fate of naturally occurring anthocyanins with different structures were precisely examined for the understanding of their *in vivo* functionalities. Attached sugar types were the major modulating factor for the absorption of mono-glycosylated anthocyanins carrying the same aglycone. Bioavailability of bilberry anthocyanins (anthocyanin mixture) was 0.93%. It was further revealed that acylated anthocyanin such as nasunin showed better uptake than non-acylated analogues carrying the same aglycone. Interestingly, the pattern of anthocyanins distributed in tissues was completely different from that of administered anthocyanin mixture and *O*-methyl analogues were the major anthocyanins in liver and kidney, indicating these anthocyanins will play a critical role in tissues. From the precise study on the metabolism of six types of purified anthocyanins, it was clarified that the general metabolic path of anthocyanin in rats was *O*-methylation of B ring and glucuronidation of both anthocyanidin and anthocyanin (extended glucuronidation).

Anthocyanins are reddish pigment widely distributed in colored fruits (*1-4*) and vegetables (*5-9*). Glycoside or acyl glycoside of anthocyanidin is naturally occurring forms of anthocyanin. Acylated anthocyanins are further classified into two groups depending on the types of attached acyl moiety. They are aliphatic and aromatic types. Recently, numerous studies on functionality of anthocyanins such as improvement of vision (*10,11*), α-glucosidase inhibition (*12,13*) and antioxidant activity (*14*) have been reported. Anticancer activity of anthocyanins has also been reported both *in vivo* and *in vitro* (*15*). However, for

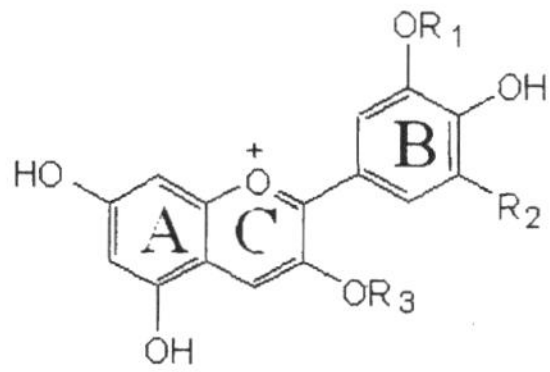

Anthocyanin	R_1	R_2	R_3	Peak
Delphinidin 3-*O*-β-D-galactopyranoside	H	OH	galactopyranose	1
Delphinidin 3-*O*-β-D-glucopyranoside	H	OH	glucopyranose	2
Delphinidin 3-*O*-α-L-arabinopyranoside	H	OH	arabinopyranose	4
Cyanidin 3-*O*-β-D-galactopyranoside	H	H	galactopyranose	3
Cyanidin 3-*O*-β-D-glucopyranoside	H	H	glucopyranose	5
Cyanidin 3-*O*-α-L-arabinopyranoside	H	H	arabinopyranose	7
Petunidin 3-*O*-β-D-galactopyranoside	H	CH_3	galactopyranose	6
Petunidin 3-*O*-β-D-glucopyranoside	H	CH_3	glucopyranose	8
Petunidin 3-*O*-α-L-arabinopyranoside	H	CH_3	arabinopyranose	10
Peonidin 3-*O*-β-D-galactopyranoside	CH_3	H	galactopyranose	9
Peonidin 3-*O*-β-D-glucopyranoside	CH_3	H	glucopyranose	11
Peonidin 3-*O*-α-L-arabinopyranoside	CH_3	H	arabinopyranose	13
Malvidin 3-*O*-β-D-galactopyranoside	CH_3	CH_3	galactopyranose	12
Malvidin 3-*O*-β-D-glucopyranoside	CH_3	CH_3	glucopyranose	14
Malvidin 3-*O*-α-L-arabinopyranoside	CH_3	CH_3	arabinopyranose	15

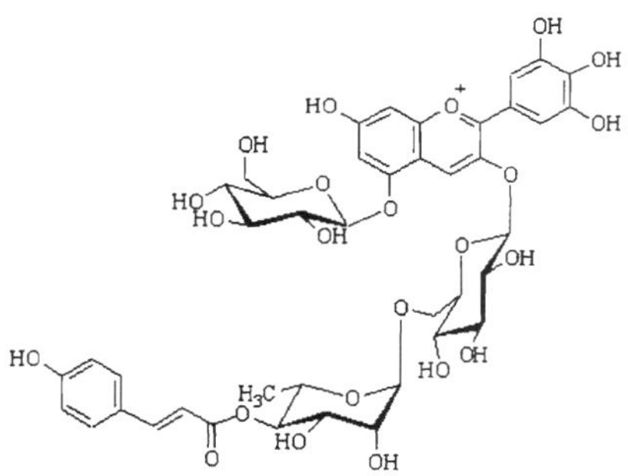

trans-nasunin

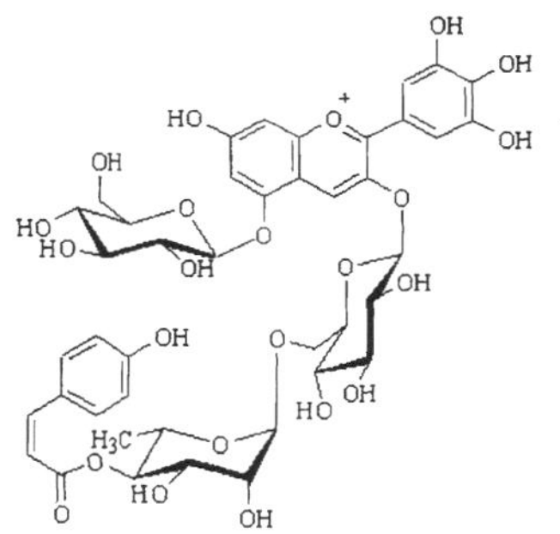

cis-nasunin

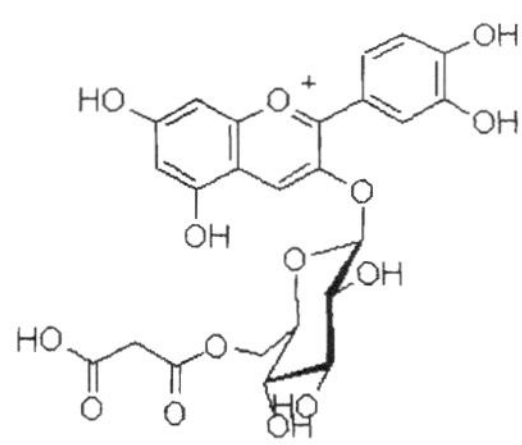

cyanidin 3-*O*-β-D-glucopyranoside mono-malonylate

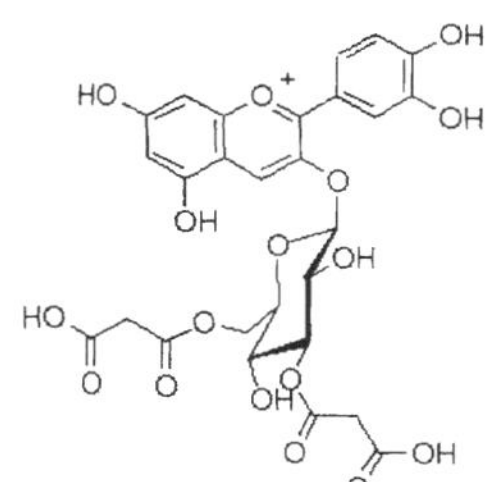

cyanidin 3-*O*-β-D-glucopyranoside di-malonylate

Figure 1. Structure of anthocyanins studied in the present report.

further discussion of *in vivo* functionality of anthocyanins, it is critical to know the physiological uptake and distribution of each anthocyanin in plasma and tissues together with their metabolic fate. Recently, we examined quantitative and systematic comparison of absorption and bioavailability of anthocyanins using bilberry (a wild type blueberry) extract because it contains fifteen types of anthocyanins comprised of five types of anthocyanidin (anthocyanin aglycone) and three types of attached sugar (Figure 1) (*16*). Further, metabolism of anthocyanins has been clarified using purified anthocyanins (six types of non-acylated and four types of acylated anthocyanins (Figure 1) by high sensitive semi-micro high performance liquid chromatography (HPLC), tandem-time of flight mass spectrometry (TOF MS) and nuclear magnetic resonance (NMR) techniques (*17-21*). In the present report, absorption and metabolism of anthocyanins are summarized together with their tissue distribution and discussed the relationship between the structural diversity and the biological behavior of anthocyanins *in vivo.*

Results and Discussion

Absorption and Bioavailability of Anthocyanins in Rats

Figure 2A showed typical HPLC chromatogram of anthocyanins in bilberry extract. Fifteen anthocyanins were detectable in bilberry extract together with minor two anthocyanidins (aglycone). Figure 2B showed HPLC chromatogram of rat blood plasma after 15 min of oral administration of bilberry extract (400 mg/kg body weight; 153.2 mg/kg as anthocyanins). Fourteen anthocyanins except peonidin 3-*O*-α-L-arabinopyranoside were detectable in rat blood plasma after oral administration of bilberry extract. The difference of anthocyanin profile on HPLC chromatogram in plasma from that of original bilberry extract indicates that absorption, excretion, and probably the metabolic fate are different among the anthocyanins. The pharmacokinetic parameters of each anthocyanin were summarized in Table I together with orally administered dose (mmol/kg) of each anthocyanin in bilberry extract (400 mg/kg body weight). The maximum plasma level of anthocyanins in the plasma was as follows; galactopyranoside > glucopyranoside > arabinopyranosides in any anthocyanins with the same aglycone except malvidin (Table I). When the bioavailability of anthocyanins carrying the same aglycone was compared, galactopyranoside tended to show a higher value than glucopyranoside. On the other hand, when the bioavailability of anthocyanins with the same sugar moiety was compared, peonidin tended to show the highest value. The bioavailability ofanthocyanin as mixture was 0.93%. These results indicated that both aglycone and attached sugar moiety are the modulator for gastrointestinal absorption of anthocyanins.

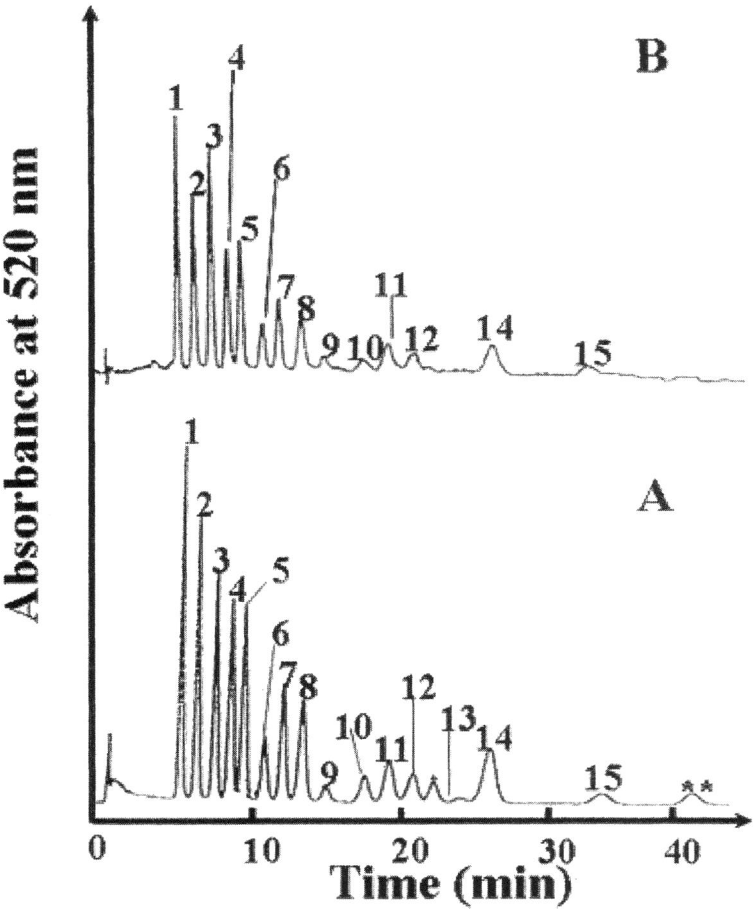

Figure 2. HPLC chromatogram of bilberry extract (Bilberon 25).
A: Bilbery extract, B: Rat blood plasma after 15 min of oral administration.

Table I. Pharmacokinetic Parameter of Bilberry anthocyanins

Peak	C_{max} (μM)	C_{max}/Dose (μM/mmol/kg)	t_{max} (min)	Bioavailability (%)	Dose (mmol/kg bw)
1	0.197 ± 0.024	4.664 ± 0.575	15	0.82989 ± 0.0973175	0.0422
2	0.154 ± 0.021	3.830 ± 0.510	15	0.77236 ± 0.0872718	0.0401
4	0.101 ± 0.001	2.677 ± 0.253	15	0.7107 ± 0.0750706	0.0379
3	0.188 ± 0.025	4.994 ± 0.625	15	0.98328 ± 0.1115948	0.0376
5	0.139 ± 0.022	3.828 ± 0.592	15	0.79207 ± 0.0922739	0.0363
7	0.075 ± 0.008	2.522 ± 0.282	15	0.74824 ± 0.077381	0.0296
6	0.055 ± 0.013	4.338 ± 1.031	30	1.21084 ± 0.1368601	0.0126
8	0.074 ± 0.013	3.055 ± 0.518	15	0.75921 ± 0.0996246	0.0241
10	0.015 ± 0.001	1.735 ± 0.104	15	1.8179 ± 0.3493325	0.0085
9	0.025 ± 0.008	7.288 ± 2.495	30	1.41852 ± 0.2357478	0.0034
11	0.050 ± 0.007	2.903 ± 0.396	15	0.83988 ± 0.1045179	0.0171
13	-	-	-	-	0.0022
12	0.039 ± 0.006	2.827 ± 0.454	30	0.90736 ± 0.0759843	0.0136
14	0.075 ± 0.009	3.069 ± 0.385	15	0.60804 ± 0.0581417	0.0245
15	0.010 ± 0.002	2.260 ± 0.358	15	1.01317 ± 0.1200774	0.0045
Anthocyanin as mixture	1.207 ± 0.131	3.571 ± 0.613	15	0.92823 ± 0.1045374	0.3343

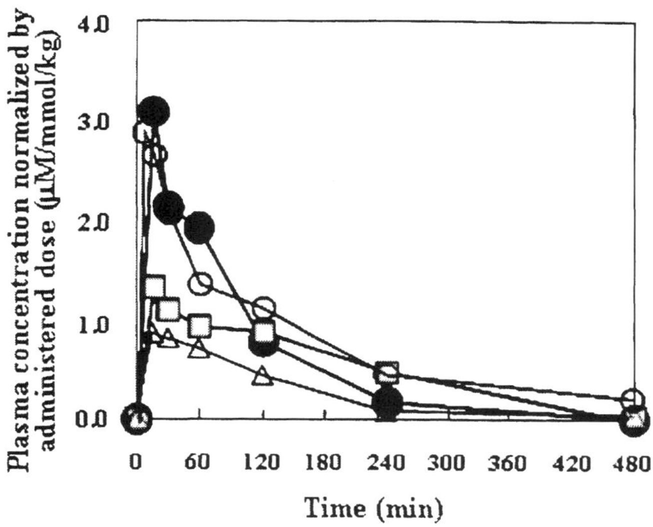

Figure 3. Plasma Concentration profile of acylated anthocyanins. ● = trans-nasunin, □ = Dp3G, ○ = mono-malonylated Cy3G, △ = Cy3G. Values are means of four rats.

Only limited study has been reported on absorption of acylated anthocyanins (*22*). In that study, it was reported that maximum plasma level of acylated anthocyanins from purple sweet potato was quite low level (i.e. 15 *n*M<). However, all acylated anthocyanins studied in the present study were detected in rat blood plasma at maximum concentration of 0.3 μM after 5 min of oral administration.

Figure 3 shows plasma concentration profile of acylated anthocyanins after normalization of orally administered dose. From the present result, it was clarified that absorption of acylated anthocyanins tended to higher than corresponding non-acylated analogues. Metabolism of acylated anthocyanins was also different depending on the type of attached acyl moiety. *O*-methyl metabolites that were produced in the case of non-acylated analogues were detectable in the case of aliphatic types. However, no metabolites were detected for anthocyanins carrying aromatic acyl moiety (*20*). Taking all the results obtained above together with the previous report into consideration, it was clarified that not only aglycone structure but also the type of attached acyl moiety are the modulator for absorption and metabolism of acylated anthocyanins.

Metabolism of Anthocyanins in Rats

Metabolism of anthocyanins was classified into three groups. The first one is *O*-methylation of B ring. Only glucopyranoside of delphinidin, cyanidin and petunidin underwent *O*-methylation. Thus, it was concluded that this is specific metabolism for anthocyanins carrying partial catechol structures on anthocyanidin B ring. Both 3′ and 4′ *O*-methyl metabolites were produced in the case of cyanidin 3-*O*-*β*-D-glucopyranoside (Cy3G). However, only 4–*O*-methyl metabolites were observed in the case of delphinidin 3-*O*-*β*-D-glucopyranoside (Dp3G) and petunidin 3-*O*-*β*-D-glucopyranoside (Pt3G). It was indicated that B ring structure is a modulator for the *O*-methylation site of anthocyanins *in vivo*. Further, *O*-methyl metabolites were also produced after intravenous administration of anthocyanins, thus it was estimated that this metabolism occurred mainly in liver and kidney by enzymatic reaction of catechol *O*-methyl transferase.

The second metabolism is glucuronidation of anthocyanin. This is what we call "extended glucuronidation". This metabolism is specific for anthocyanins which are absorbed as their intact glycoside forms. This metabolism was common for 6 non-acylated anthocyanins studied in the present work, although the number and the amount of these metabolites are largely different depending on the B ring structure of anthocyanins. From the fact that these metabolites could not be detected after intravenous administration of anthocyanins, this metabolism also occurred in liver.

The third metabolism is glucuronidation of anthocyanidin. This is preferable metabolism for other flavonoids such as quercetin 3-*O*-*β*-D-glucopyranoside. The amount of this metabolite was the highest for pelargonidin 3-*O*-*β*-D-glucopyranoside (Pg3G). Anthocyanins which undergo this metabolism are Pg3G, Cy3G and peonidin 3-*O*-*β*-D-glucopyranoside (Pn3G). Thus, it was concluded that no substitution on 5′ position of aglycone is necessary for this metabolism. Glucuronides of anthocyanidin could not be detected after intravenous administration, thus it was estimated that this metabolism mainly occurred in small intestine during absorption process.

Figure 4 summarized the anthocyanin metabolism in rats. Metabolism of anthocyanins was largely different depending on the B ring structure. Pg3G especially showed specific biological behaviors.

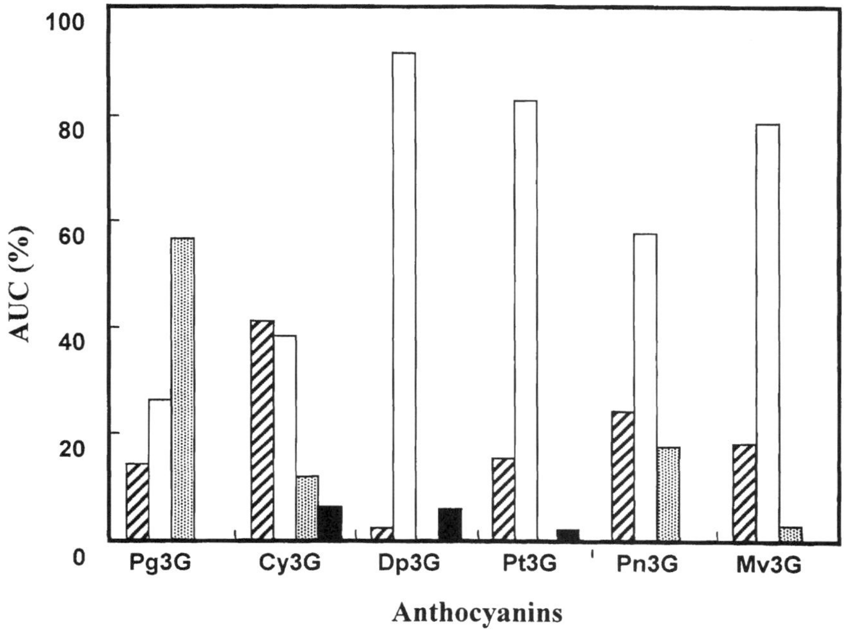

Figure 4. Summary of anthocyanin Metabolism in Rats. ▨ *= extended glucuronide,* □ *= intact,* ▦ *= anthocyanin-glucuronide,* ■ *=* O*-methylated.*

Tissue Distribution of Anthocyanins

Because the structure of anthocyanin metabolites has already been assigned by NMR and tandem TOF MS, 5 types of anthocyanin mixture and strawberry extract were used for further studies. Figure 5 shows HPLC chromatogram of rat liver and kidney after oral administration of anthocyanins. As shown in Figure 5 A, methylated anthocyanins such as Pn3G, Pt3G and malvidin 3-*O*-*β*-D-glucopyranoside (Mv3G) were major anthocyanins distributed in both tissues. High amount of Pg3G was also detected (Figure 5 B). However, the extent of Cy3G is lower than that in original extract. Moreover, Dp3G could not be detected especially in liver. On the other hand, there are newly detectable peaks in tissues (peak 1-3). These metabolites were assigned as 4′-*O*-methyl anthocyanins from HPLC retention time and *m/z* of tandem TOF MS. Dp3G especially was metabolized to 4′-*O*-methyl analogues and this metabolite was major anthocyanin distributed in liver. Further, several hydrophilic metabolites

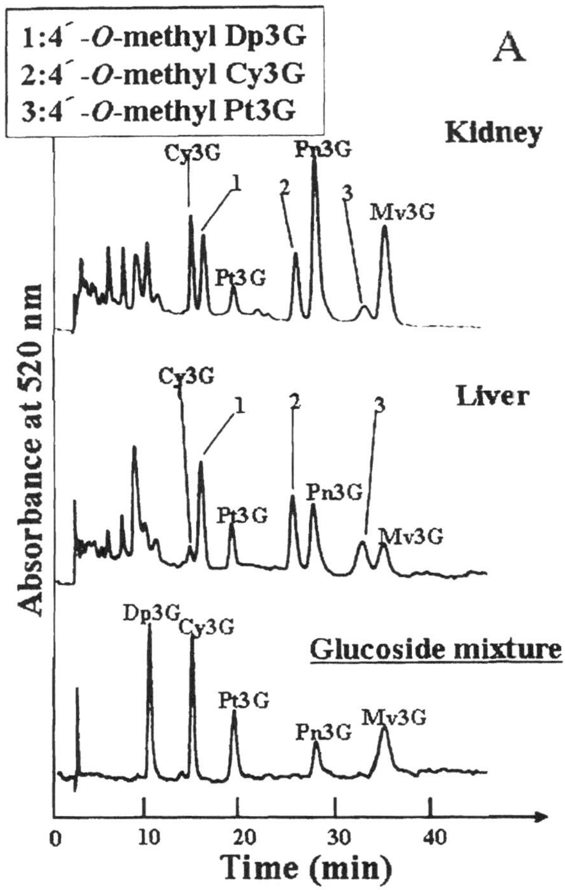

Figure 5. HPLC chromatogram of rat tissue after oral administration of anthocyanin mixture.

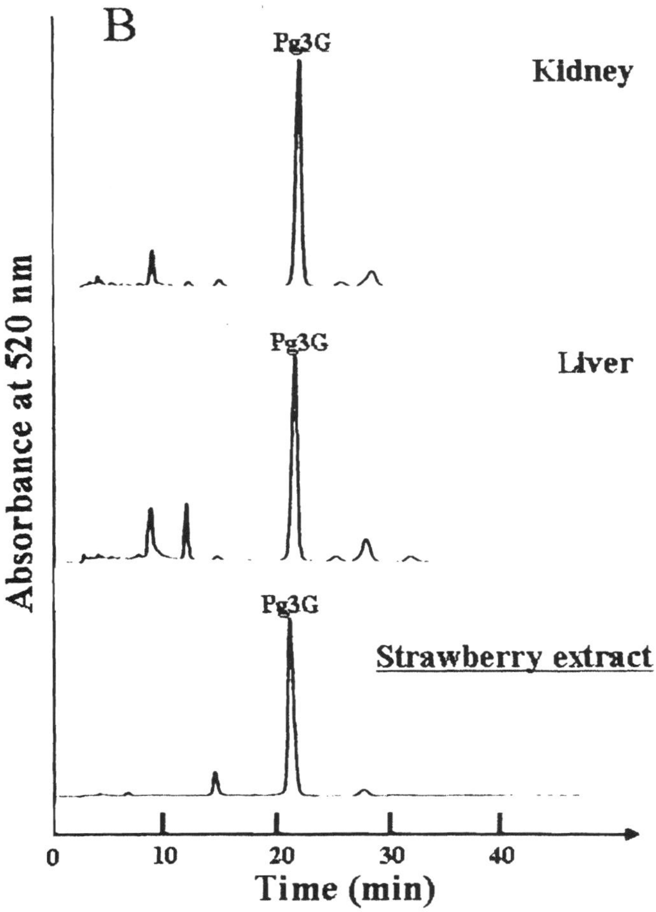

Figure 5. Continued.

were also detected in liver and kidney. These peaks were assigned as extended glucuronide of anthocyanins from their *m/z* values of tandem TOF MS as we have already shown in the previous reports (*16, 18*). The present results indicate that anthocyanins will function differently *in vivo* depending on their chemical structure and the type of tissue targeting.

Degradation of Anthocyanins *in vivo*

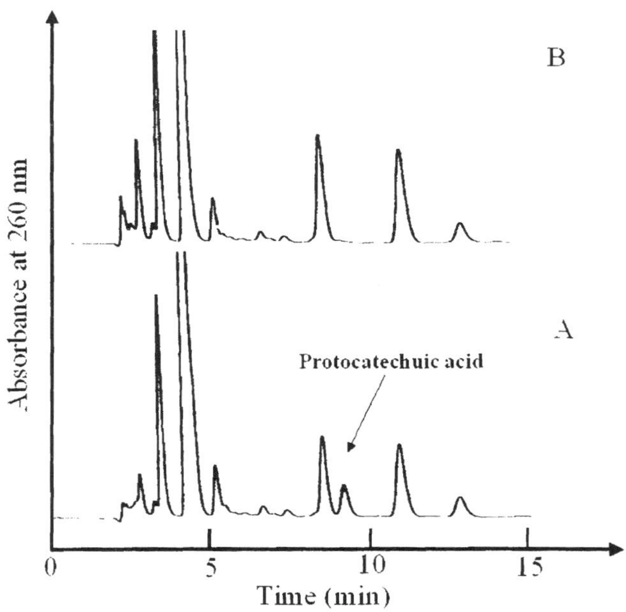

Figure 6. HPLC chromatogram of rat blood plasma detected at 260 nm.
A: Protocatechuic acid spiked rat blood plasma
B: 60 min after oral administration of Cy3G.

It is normally accepted that anthocyanins are quite unstable in physiological conditions. It has been previously reported that at least 8 times higher concentration of protocatechuic acid was detected in rat blood plasma after oral administration of Cy3G and it was concluded that protocatechuic acid was the major metabolite of Cy3G *in vivo* (*23*).

Figure 6 A shows HPLC chromatogram of protocatechuic acid spiked rat blood plasma detected at 260 nm and protocatechuic acid was clearly separated from plasma originated peaks. Thus, Cy3G is orally administered to rats. Figure 6 B shows HPLC chromatogram of rat blood plasma after 60 min of oral administration of Cy3G. However, no protocatechuic acid was detected in any rats at any time point. Thus, it was concluded that porotcatechuic acid was not a major metabolite of Cy3G in blood plasma, although its production in small intestine could not be excluded. Also, it was confirmed that the contents of rat small intestine were almost colorless after 4 hours of oral administration of Cy3G. This indicates that production of other ring fission metabolites must be taken into account.

Conclusions

These are a summary obtained from the present study.

1. The type of sugar moiety is a major modulator for absorption of anthocyanins.
2. Higher levels of acylated anthocyanins were detected in blood plasma than non-acylated anthocyanins.
3. The anthocyanin metabolism was classified into three groups and B ring structure was a major determinant for the metabolism.
4. *O*-methyl anthocyanins and *O*-methyl metabolites are major anthocyanins in tissues, which indicates these anthocyanins will play critical roles in tissues.
5. Taking all these facts obtained above together into consideration, it was deduced that anthocyanin will function differently depending on types of anthocyanins we ingest from our daily diet.

Acknowledgement

The author greatly thanks to Mr. Eiichi Sekiguchi, Agatsuma-machi, Japan for providing strawberry fruits. The author also thanks to Tokiwa Phytochemical. Co. Ltd. for providing Bilberon 25 (bilberry extract). Part of this study was supported by a Grant-in-Aid (No. 17780109) for Scientific Research from the Ministry of Education, Science and Culture, in Japan.

References

1. Ichiyanagi, T.; Hatano, Y.; Matsugo, Y.; Konishi, T. *Chem. Pharm.l Bull.* **2004**, *52*, 628-630.

2. Garcia-Viguera, C.; Zafrilla, P.; Tomas-Barveran, A. *Phytochem. Anal.* **1998**, *9*, 274-277.
3. Slimestad, R.; Solheim, H. *J. Agric. Food Chem.* **2002**, *50*, 3228-3231.
4. Ichiyanagi, T.; Kashiwada, Y.; Shida, Y.; Ikeshiro, Y.; Kaneyuki, T.; Konishi, T. *J. Agric. Food Chem.* **2005**, *53*, 9472-9477.
5. Otsuki, T.; Matsufuji, H.; Takeda, M.; Toyoda, M.; Goda, Y.; *Phytochem.* **2002**, *60*, 79-87.
6. Ichikawa, H.; Ichiyanagi, T.; Xu, B.; Yoshii, Y.; Nakajima, M.; Konishi, T. *J. Med. Food* **2001**, *4*, 211-218.
7. Terahara, N.; Shimizu, T.; Kato, Y.; Nakamura, M.; Maitani, T.; Yamaguchi, M.; Goda, Y. *Ipomoea batatas. Biosci. Biotechnol. Biochem.* **1999**, *63*, 1420-1424.
8. Yoshida, K.; Kondo, T.; Kameda, K.; Goto, T. *Agric. Biol. Chem.* **1990**, *54*, 1745-1751.
9. Stinzing, F. C.; Stintzing, A. S.; Carle, R.; Frei, B.; Wrolstad, R .E. *J. Agric. Food Chem.* **2002**, *50*, 6172-6181.
10. Mercier, A.; Perdriel. G.; Rozier, J.; Cheraleraud, J. *Bull. Soc. Ophthalmol. Fr.* **1965**, *65*, 1049-1053.
11. Matsumoto, H.; Nakamura, Y.; Tachibanaki, S.; Kawamura, S.; Hirayama, M. *J. Agric. Food Chem.* **2003**, *51*, 3560-3563.
12. Matsui, T.; Ueda, T.; Oki, T.; Sugita, K.; Terahara, N.; Matsumoto, K. *J. Agric. Food Chem.* **2001**, *49*, 1948-1951.
13. Matsui, T.; Ueda, T.; Oki, T.; Sugita, K.; Terahara, N.; Matsumoto, K. *J. Agric. Food Chem.* **2001**, *49*, 1952-1956.
14. Rahman, M. M.; Ichiyanagi, T.; Hatano, Y.; Komiyama, T.; Konishi, T. *Free Radic. Res.* **2006**, *40*, 993-1002.
15. Ding, M.; Feng, R.; Wang, S. Y.; Bowman, L.; Lu, Y.; Qian, Y.; Castranova, V.; Jiang, B. H.; Shi, X. *J. Biol. Chem.* **2006**, 17359-17368.
16. Ichiyangi, T.; Shida, Y.; Rahman, M. M.; Hatano, Y.; Konishi, T. *J. Agric. Food Chem.* **2006**, *54*, 6578-6587.
17. Ichiyanagi, T.; Rahman, M. M.; Shida, Y.; Hatano, Y.; Matsumoto, H.; Hirayama, M.; Konishi, T. *J. Agric. Food Chem.* **2005**, *53*, 145-150.
18. Ichiyanagi, T.; Shida, Y.; Rahman, M. M.; Hatano, Y.; Konishi, T. *J. Agric. Food Chem.* **2005**, *53*, 7312-7319.
19. Ichiyanagi, T.; Rahman, M. M.; Kashiwada, K.; Ikeshiro, Y.; Shida, Y.; Hatano, Y.; Matsumoto, H.; Hirayama, M.; Konishi, T. *Biofactors* **2004**, *21*, 411-413.
20. Ichiyanagi, T.; Rahman, M. M.; Kashiwada, Y.; Ikeshiro, Y.; Shida, Y.; Hatano, Y.; Matsumoto, H.; Hirayama, M.; Tsuda, T.; Konishi, T. *Free Radical Biol. Med.* **2004**, *36*, 930-937.

21. Ichiyanagi, T.; Terahara, N.; Rahman, M. M.; Konishi, T. *J. Agric. Food Chem.* **2006**, *54*, 5306-5312.
22. Suda, I.; Oki, T.; Masuda, M.; Nishiba, Y.; Furuta, S.; Matsugano, K.; Sugita, K.; Terahara, N. *J. Agric. Food Chem.* **2002**, *50*, 1672-1676.
23. Tsuda, T.; Horio, F.; Osawa, T. *FEBS Lett.* **1999**, *449*, 179-182.

Chapter 7

Modulation of Human Phenol Sulfotransferases Expression by Dietary Phenolic Acids

Gow-Chin Yen[1] and Chi-Tai Yeh[2]

[1]Department of Food Science and Biotechnology, National Chung Hsing University, 250 Kuokuang Road, Taichung 40227, Taiwan
[2]National Institute of Cancer Research, National Health Research Institutes, Miaoli County 350, Taiwan

Epidemiological studies have shown that consumption of fruits and vegetables is associated with the prevention of chronic diseases such as cancer and cardiovascular disease. Polyphenols in plants are a versatile group of phytochemicals with many potentially beneficial activities in terms of disease prevention. We recently showed that dietary plant polyphenols, namely, the phenolic acids, modulate expression of an important enzyme in both cellular antioxidant defenses and detoxification of xenobiotics, i.e. phenol sulfotransferases (PST). These enzymes are traditionally known as phase II drug-metabolizing or detoxifying enzymes that facilitate the removal of drugs and other xenobiotics compounds. We showed in vitro that p-hydroxybenzoic acid, gallic acid, gentisic acid, ferulic acid, and p-coumaric acid can increase the activities of both PST-P and PST-M. These phenolic acids also exhibit antioxidant activity as evaluated by ORAC and TEAC assays. Furthermore, in two- and three-compound combinations with other phenolic acids, gallic acid and gentistic acid exhibited synergistic effects in the promotion of

PST activities. Moreover, animal models were used to investigate the modulatory effect of phenolic acids on hepatic phase II PST and antioxidant status in vivo, our results suggest that phenolic acids might alter sulfate conjugation and antioxidant capacity in living systems. Evidently, PST is important in phase II detoxifying systems; regulation of intracellular PST gene expression may provide an efficient approach to understanding the chemopreventive mechanisms of dietary compounds.

Reactive oxygen species (ROS) play a very important role in normal cellular function. They are generated as a part of cellular metabolism and are important mediators in signal transduction pathways and cellular immunity (*1*). Due to their highly reactive nature, uncontrolled levels of ROS can be detrimental as they can damage vital structures within the cell (*2*). Oxidative stress occurs when the level of ROS remains high without appropriate mechanisms to counter their reactivity. This condition may result in ROS-mediated lipid peroxidation, DNA oxidation and protein carbonylation, which can functionally impair lipids, DNA and proteins (*3*). Thus, the proper balance of ROS generation and clearance is necessary to restrain the potentially damaging effects of these molecules. A possible mechanism for the protective effects of fruit and vegetables with respect to disease is that bioactive compounds in these foods reduce oxidative stress. Epidemiological studies have consistently shown an inverse association between consumption of vegetables and fruits and the risk of human cancers at many sites (*4*). Fruits and vegetables contain several thousands structurally diverse phytochemicals, of which large fractions are polyphenols (*5*). Interest in food phenolics, found mainly in fruits and vegetables, has recently been increasing, owing to their role as antioxidants, with implications for the prevention of pathologies such as cancer (*6*), cardiovascular diseases (*7*), and inflammatory disorders (*8*). There has been a growing interest in the naturally occurring anticarcinogenic substances found in plant foods. Plant phenols, for example, flavonoids and phenolic acids, are currently considered as one of the most promising groups of potential dietary anticarcinogens (*9*). Plant phenolic acids are of current interest, due to their important biological and pharmacological properties, especially for their anti-inflammatory (*10*), oxygen free radical scavenging (*11*), and antimutagenic and anticarcinogenic activities (*12*). The beneficial effects of phenolic acids can be traced to their antioxidant properties and, therefore, they may play a role in both cytoprotection and chemoprevention.

Chemoprevention is one of the most promising areas in cancer research. Potential chemopreventive agents may function by a variety of mechanisms directed at all major stages of carcinogenesis. One proposed mechanism for cellular protection, against the chemical and neoplastic effects of carcinogens, involves the induction of phase II detoxification enzymes (*13*). Many carcinogens are not in their full carcinogenic forms when first encountered. They are usually metabolized to proximate carcinogens by phase I enzymes, for example, cytochromes P450, that catalyze an oxidative reaction. The oxidized metabolites of potentially carcinogenic xenobiotics are then detoxified, by phase II metabolizing enzymes, into forms that are relatively inert and more easily excreted. There is considerable evidence that induction of phase II detoxification enzymes can modulate the threshold for chemical carcinogenesis, increasing cellular resistance to carcinogen exposure (*14*). Phenol sulfotransferases (PST) are the main phase II sulfoconjugation enzymes for catecholamines, thyroid hormones, and drugs, thereby facilitating biliary or urinary excretion and detoxification (*15, 16*). Sulfoconjugation plays not only an important role in xenobiotic metabolism but also a critical role in steroid biosynthesis, as well as modulating the biological activity and facilitating the inactivation and elimination of potent endogenous chemicals, including steroids, catecholamines, and thyroid hormones. The harmful substances may accumulate in the body when PST activity is inhibited (*17, 18*). Therefore, measuring the induction of PST activity may provide an efficient approach to understanding the chemopreventive mechanisms of dietary compounds.

This chapter will focus on the molecular mechanism underlying PST expression and the significance of targeted induction of PST by phenolic acids as a strategy to achieve chemoprevention and chemoprotection.

Chemopreventive Effect of Phenolic Acids

The existing literature provides a wealth of information that correlates a diet high in fruits and vegetables with the maintenance of health and disease prevention. Current thought links the high antioxidant content of fruits and vegetables with the inhibition of diseases caused by oxidative damage such as coronary heart disease, stroke, and cancers (*19*). Certain foods are even classified as functional foods, owing to their established healthful protective effects (*20*). Typically, when discussing phenolics in plants foods, flavonoids are the predominant class described, because they account for approximately two-thirds of dietary phenols (*21*). However, phenolic acids account for almost all of the remaining one-third, and there is an increasing awareness and interest in the antioxidant behavior and potential health benefits associated with these simple phenolic acids.

Phenolic acids have received considerable attention as potentially protective factors against caner and heart diseases in part because of their potent antioxidant properties and their ubiquity in a wide range of commonly consumed foods of plant origins (*22*). Phenolic acids belong to two major groups, hydroxybenzoic acids, and hydroxycinnamic acids (Figure 1).

Hydroxybenzoic acids

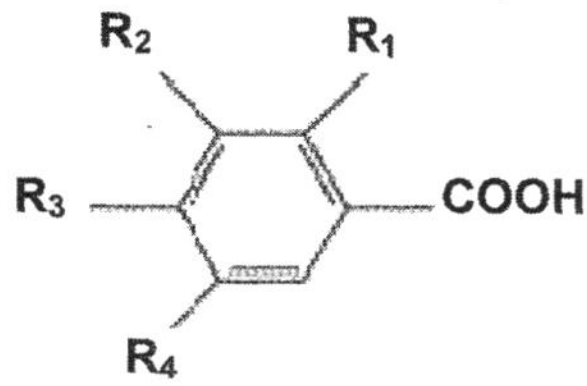

$R_1=R_2=R_4=H$, $R_3=OH$	*p*-Hydroxybenzoic acid
$R_1=R_4=H$, $R_2=R_3=OH$	Protocatechoic acid
$R_1=R_4=H$, $R_2=OCH_3$, $R_3=OH$	Vanillic acid
$R_1=H$, $R_2=R_3=R_4=OH$	Gallic acid
$R_1=H$, $R_2=R_4=OCH_3$, $R_3=OH$	Syringic acid
$R_1=R_4=OH$, $R_2=R_3=H$	Gentisic acid

Hydroxycinnamic acids

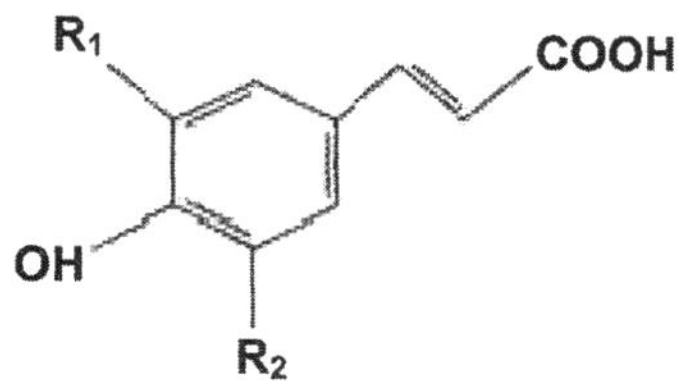

$R_1=R_2=H$	*p*-Coumaric acid
$R_1=OH$, $R_2=H$	Caffeic acid
$R_2=OCH_3$, $R_2=H$	Ferulic acid
$R_3=OCH_3$, $R_2=OCH_3$	Sinapic acid

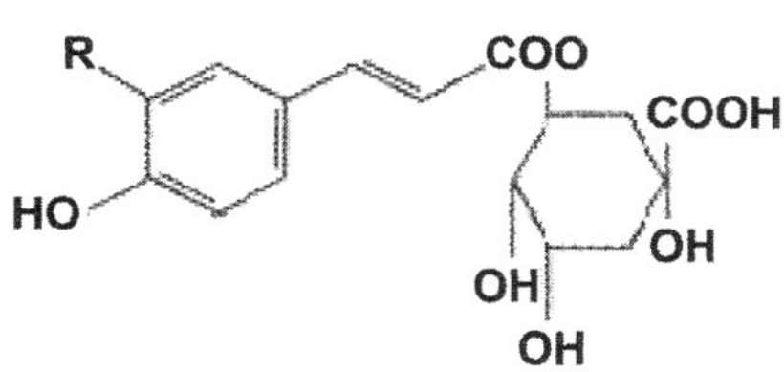

R=OH	Chlorogenic acid

Figure 1. Structural formulas of phenolic acids.

Hydroxybenzoic acid derivates include *p*-hydroxybenzoic, protocatechuic, vanillic, syringic, and gallic acids. They are commonly present in the bound form and are typically a component of a complex structure like lignins and hydrolysable tannins. They can also be found in the form of sugar derivatives and organic acids in plant foods. Hydroxycinnamic acid derivates include *p*-coumaric, caffeic, ferulic, and sinapic acids. They are mainly present in the bound form, linked to cell-wall structure components, such as cellulose, lignin, and proteins through ester bonds (*23*). Ferulic acid and *p*-coumaric acid have been reported to act as scavengers of thiol free radicals (*24*). *p*-Coumaric acid also possesses potent antioxidant activity, in enhancing the resistance of low-density lipoprotein to oxidation (*25*). As well, *p*-coumaric acid (50 mg/kg of body weight) significantly decreases the basal level of oxidative damage in rat colonic mucosa (*26*), whereas gentisic acid has been reported to have an inhibitory action in the myeloperoxidase system and is able to impair tyrosyl radical catalyzed low-density lipoprotein peroxidation (*27*). In addition, gallic acid is a potent natural antioxidant, exhibiting antimutagenic and anticarcinogenic activity, and is expected to reduce the risk of disease and to render health benefits through daily intake (*28*).

Because of their ubiquitous presence in plant-based foods, humans consume phenolic acids on a daily basis. The estimated range of consumption is 25 mg to 1 g on a daily basis, depending on the diet (fruit, vegetables, grains, teas, coffees, spices) consumed (*29*). Although the exact nature of the antioxidant or protective effects of phenolic acids *in vivo* is not well established, there exist some preliminary investigations. A great deal more has been done on the *in vitro* and chemical antioxidant behavior of phenolics. Only the metabolic fates of caffeic, ferulic, chlorogenic, and sinapic acid have been explored. Chlorogenic acid is most likely metabolized by the colonic microflora (*30*). Zhao *et al.* (*31*) reported that only 11-25% of ferulic acid ingested is excreted in urine as free ferulic acid or as its glucuronide conjugate. Certain common and recurring excreted metabolites reported are hippuric acid and ferulic acid and isoferulic acid, as well as glucuronides and sulfate conjugates of the phenolic acids (*32*). Beyond the protective antioxidant behavior, other biological activities of phenolic acids have been reported. Caffeic acid, one of the most prominent naturally occurring cinnamic acids, is known to selectively block the biosynthesis of leukotrienes, components involved in immunoregulation diseases, asthma, and allergic reactions (*33*). Other studies have reported that caffeic acid and some of its esters might possess antitumor activity against colon carcinogenesis (*34*). Recent investigations by Maggi-Capeyron *et al.* (*35*) have linked a series of phenolic acids with the inhibition of AP-1 transcriptional activity. Caffeic acid derivatives (e.g. dicaffeoylquinic and dicaffeolytartaric acids) have been shown to act as potent and selective inhibitors of human immunodeficiency virus type 1 (HIV-1) integrase (*36*). This enzyme catalyzes the integration of viral DNA into the host chromatin. Therefore, these hydroxycinnamate derivatives are currently investigated for their potential antiviral activity.

The total phenolics content and total antioxidant capacity (expressed as an $ORAC_{ROO^{\cdot}}$ value) of various vegetables were recently evaluated. Results (*37*) clearly demonstrated that all vegetables tested had antioxidant activities against peroxyl radical. The rank order based on ORAC mean values was broccoli > celery > asparagus, cauliflower > snap bean > eggplant > garlic > cabbage > potato, tomato > carrot, shiitake, lettuce > cucumber > soybean > sprouts > leek > onion > Chinese chive > spinach. Interestingly, broccoli high in antioxidant capacity was among the vegetables with the highest contents of total polyphenolics (*37*). This is intriguing, because broccolis have been shown to protect against oxidative stress-related pathologic conditions *in vivo*. For example, Heber and Bowerman (*38*) demonstrated that long-term feeding of broccolis to rats retarded and even reversed the onset of age-related neurologic dysfunctions, such as decline in neuronal signal transduction, and cognitive, behavioral, and motor deficits. Keum *et a.l* (*39*) showed that supplementation with broccoli in the diet reduces the multiplicity and incidence of esophageal tumors in *N*-nitrosomethylbenzylamine-treated rats. Thus, increased consumption of vegetables containing high levels of polyphenolics has been recommended to reduce cellular oxidative damage in the human body.

Human Sulfotransferases and Their Role in Chemoprevention

The blocking of oncogenic insult-induced genotoxic damage constitutes the first degree prevention of cancer. This can be achieved either by reducing the formation of reactive carcinogenic species or stimulating their detoxification. Induction of phase II detoxifying and antioxidant enzymes has been suggested as an effective and sufficient blocking strategy to protect cell tissues from the toxic and neoplastic effects of many carcinogens (*40*). It has also become apparent that the enhancement of detoxifying enzymes can promote cytoprotective mechanisms against carcinogens and reactive oxygen species (ROS) such as superoxide anion, hydrogen peroxide and hydroxyl radical. Thus inducers of detoxifying enzymes are potential candidates for preventing tumor development (*41*).

The sulfation system is a major component of the body's chemical defense armory and is directly involved in the metabolism, detoxification, and elimination of numerous xenobiotics (*42*). This system catalyzed by sulfotransferases is an important phase II conjugation reaction in the metabolism of xenobiotics, but also in modulating the biological activity and facilitating the inactivation and elimination of potent endogenous chemicals including thyroid hormones, steroids and monoamine neurotransmitters. Sulfotransferases are present in many tissues including the liver, kidney, brain, adrenals, gut and platelets. These enzymes catalyze the transfer of a sulfuryl group from the endogenous sulfate donor 3′-phosphoadenosine 5′-phosphosulfate (PAPS) to a variety of hydroxyl or amino groups of acceptor molecules as shown in Figure 2. In humans, there are two major subfamilies of cytosolic sulfotransferases involved in the metabolism of xenobiotics: phenol sulfotransferase (PST) and hydroxysteroid sulfotransferase (HST) (*43*). HST is responsible for the sulfation of steroids. Two functionally distinct forms of PST have been identified in human tissue. PST-P is relatively specific for the sulfation of small phenols and structurally related neutral compounds, whereas PST-M is primarily responsible for the sulfation of monoamines, such as dopamine. These cytosolic sulfotransferases are particularly active in platelets and are generally present in the intestinal wall, adrenal gland, and brain (*44*). However, the PST in human platelets, liver, and gut show great variation in their activities. PST-P levels in platelets are particularly interesting because they are highly correlated with the corresponding PST-P levels in the human liver, cerebral cortex, and small intestinal mucosa (*45*). Thus, it is feasible to use the PST activity in blood platelets to reflect any drug metabolizing activity in other tissues of interest. There is evidence that harmful substrates might accumulate in the body when the PST activity is inhibited (*46*).

Sulfotransferases are involved in a number of biological processes including molecular recognition, detoxification, hormone regulation, drug processing, and modulation of receptor binding. Generally, sulfation is a detoxification process that converts xenobiotics into less or non-toxic and more water soluble metabolites, thereby facilitating their excretion in the urine or bile. Since sulfotransferases are biologically important in the metabolism of endogenous as well as xenobiotic compounds, sulfotransferase inhibition may have important implications for human health. Many reports have shown that harmful substance might accumulate in the body when sulfotransferase activity is inhibited. Moreover, elevation of hepatic sulfotransferase activity in mice with resistance to cystic fibrosis (*47*) and in mice deficient in cerebroside sulfotrnasfease (*48*) exhibit a serious impairment of L-selectin metabolism, leading to liver and kidney oxidative damage and inflammation. Typically, this enzyme belongs to the phase II families of detoxification genes. The transcriptional activation of the phase II detoxifying and antioxidant enzymes has been traced to either the *cis*-acting transcriptional enhancer celled ARE or, alternatively, to the electrophile response element. The latter is defined by a specific consensus sequence of nucleotide and responds to substances with antioxidant properties (*49*). Most recent findings suggest and support the key role of the ARE/EpRE in the regulation of expression of some phase II gene such as NQO, GST, and UGT by phenolic antioxidants and other naturally occurring cancer chemopreventive agents (*50*).

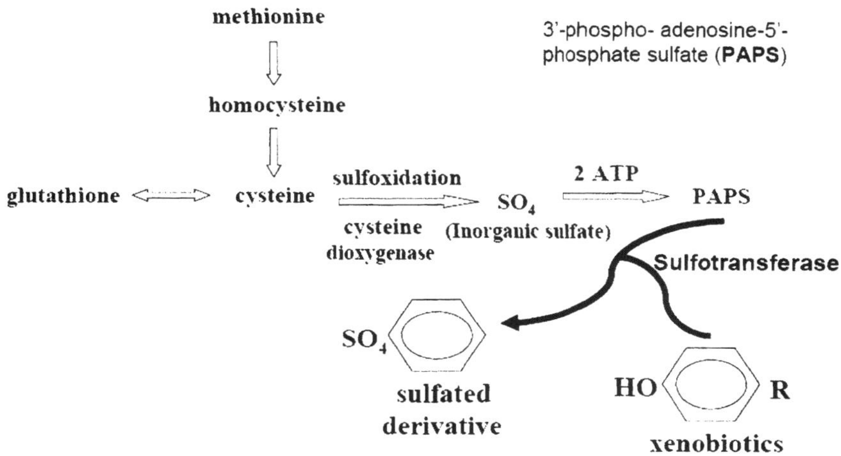

Figure 2. The general sulfotransferase-catalyzed with PAPS as the cosubstrate.

Plant phenols, for example, flavonoids and phenolic acids, are currently considered to be one of the most promising groups of potential dietary anticarcinogens. Furthermore, the beneficial effects of phenolic acids can be traced to their antioxidant properties and, therefore, they may play a role in both cytoprotection and chemoprevention (*51*). The mechanism of action of phenolic acids in cellular sulfation process is still not clearly understood. In a series of experiments, we showed that *p*-hydroxybenzoic acid, gentisic acid, ferulic acid, gallic acid, and *p*-coumaric acid can increase the activities of both P-form and M-form phenol sulfotransferases (*52*). These phenolic acids were also found to possess antioxidant activity in the ORAC and TEAC assays. Moreover, the activity of PST-P could be promoted by combinations of all of these phenolic acids (*53*). The overall effects of phenolic acids, on the activities of PST-P, are highly correlated to their ORAC values, suggesting that phenolic acids may alter sulfate conjugation. Furthermore, stimulation of PST-P expression by phenolic acids has been shown to occur primarily as a consequence of transcriptional regulation of the *PST-P* gene (*54*). Both broccoli extracts and pure phenolic acids transactivated sulfotransferase through antioxidant response elements in the promoter in $HepG_2$ cells, with gallic acid being the most potent phenolic acids (*37*). Structurally similar phenolic acids were not as potent; *p*-

hydroxybenzoic acid, with only two hydeoxyl groups less than gallic acid, was inactive, which emphasizes the apparent specificity of sulfotransferase induction.

Phenolic acids in foods are conjugated to various sugar molecules, which likely influence their intestinal absorption, transport, and entry into cells (*55*). To clarify the biological action of phenolic acids, *in vivo* sulfoconjuation experiment was carried out (*56*). Our experimental results also show that ferulic acid, gentisic acid, *p*-coumaric acid and gallic acid can increase sulfotransferase activities and mRNA expression in rat liver. This induction was seen at the gene expression levels. Interestingly, our results clearly demonstrated for the first time that orally administrated phenolic acids do significantly increase the $ORAC_{ROO^{\cdot}}$ values in rat plasma. In general, phenolic acids that induced sulfotransferase activity were found to have higher $ORAC_{ROO^{\cdot}}$ values. In addition, we also established that phenolic acids make a contribution in the induction of sulfotransferase activity and antioxidant *in vivo*. Two phenolic acids, gallic acid and *p*-coumaric acid, were orally administrated to rats at the same dosage (100 mg/kg of body weight), and these were found to be absorbed and distributed to the blood in the intact form (*56*). Comparison of the time course of changes in plasma concentrations of the two components showed that gallic acid and *p*-coumaric acid were directly absorbed and distributed in the blood, and the plasma concentrations increased up to 2.0 h post-administration and then gradually decreased, suggesting the phenolic acids, especially gallic acid and *p*-coumaric acid, were directly absorbed and distributed in the blood. These results indicate that phenolic acids are absorbed and their plasma metabolites may act as antioxidants in blood circulation (*56*).

According to the results of studies conducted in our laboratory, it was found that there was a significant correlation between the platelet content and the activities of both forms of PST in different patients with hypertension, migraine, or hypothyroidism ($r = 0.68$, $p < 0.01$; $r = 0.63$, $p < 0.01$). These data indicated that the activity of platelet PST varied among the patients and had a lower level of PST activity. Furthermore, we found that varied disorders may affect the activity of platelet PST. Therefore, the activities of both forms of PST in the platelet of these patients were significantly increased by gallic acid, ferulic acid, gentisic acid, and coumaric acid. These findings have clinical significance as the cytosolic phenolsulfotransferases inactivated phenols and amines by catalyzing their conjugations with sulfate donated from PAPS. The PST-M catalyzed the sulfate conjugation of micromolar concentrations of dopamine and other phenolic monoamines. Patients with migraine appear to have significantly lower levels of platelet PST-M activity, leading to raised central nervous system levels of catecholamines, which are believed to be a major factor in headaches (*57*). Friberg *et al.* (*58*) demonstrated that reduced metabolism of bioactive compounds such as dopamine or other monoamines following inhibition of PST-

M might result in an increase in the bioavailability of dopamine and the consequent altered pharmacological action such as migraine. Alternatively, changes in PST activity might be linked to a pharmacological response, a lower sulfating process that has been related to depressive patients (*59*). However, the capacity of these phenolic acids to modulate conjugation reactions may be a factor in the interindividual variation found in xenobiotic metabolism and could make it necessary for volunteers in drug metabolism studies to adhere to a common dietary regime.

Modulation of Phenol Sulfotransferases Gene Expression by Phenolic Acids

Transcriptional Induction of the PST Gene by Phenolic Acids

Phenol sulfotransfeases (PST) are Phase II enzymes that are transcriptionally regulated by a large variety of stimuli. These include thyroid hormone (*60*), oxidative stress (*61*), the cytokines tumor necrosis factor-alpha (TNF-alpha), transforming growth factor-beta (TGF-beta) and insulin (*62*), and phenolic compounds such as gallic acid and ferulic acid (*63*). Flavonoids and other dietary phenolics have been shown to strongly inhibit neoplastic transformation in mammary organ cultures and epithelial cells, inhibit benzo[a]pyrene DNA adduct formation and induce the Phase II metabolizing enzymes GST, NQO and UGT (*64*). Phase II enzymes induction may explain the chemopreventive effect of polyphenolic compounds in inhibiting heterocyclic amine-induced colonic aberrant crypt foci formation in rats (*65*). The synthetic flavonoid 4′-bromoflavone was the most potent *in vivo* inducer of NQO and reduced glutathione (GSH) synthesis enzymes and prevented mammary carcinogenesis in rats induced by polycyclic aromatic hydrocarbons (*66*). Dietary flavonoid/phenolic mediated induction of UGT may be important for the glucuronidation and detoxification of colon and other carcinogens, as well as for the metabolism of therapeutic drugs (*67*). Studies of the mechanisms, by which polyphenolic compounds are involved in chemoprevention, constitute an increasingly active area of research. The effects of polyphenolic compounds on Phase I enzymes, such as cytochromes P450, or on Phase II enzymes, such as GST and NQO-1, appear to involve multiple mechanisms (*68*).

Gallic acid, and its catechin derivatives, have demonstrated excellent chemopreventive effects in many target organs challenged with various carcinogens. A number of studies indicate that gallic acid is a potent inducer of Phase II drug metabolism enzymes; this molecular mechanism is thought to involve transcriptional upregulation of Phase II genes (*69*). Gallic acid stimulated a time-dependent PST-P mRNA and PST-P protein induction without significant toxicity in our experimental conditions (*63*). This is the first reported analysis of human PST-P expression in response to *in vitro* exposure to gallic acid. We further confirmed that the gallic acid-induced increases in PST-P mRNA and protein were accompanied by corresponding increases in PST-P enzyme activity. Thus, our data provide clear evidence that gallic acid is a potent inducer of PST-P in human hepatoma $HepG_2$ cells. Furthermore, stimulation PST-P expression by most inducers has been shown to occur primarily as a consequence of transcriptional regulation of the PST-P gene. In fact, *cis*-acting DNA sequences involved in induction by various agents have been identified in the PST-P gene from several species (*70*). We have demonstrated that activation of the transcription of PST-P promoter/luciferase reporter activity is induced in relation to increasing concentrations of gallic acid (*54*). In addition to the finding that transcription of PST-P is induced by gallic acid, we also found that both actinomycin D (a transcriptional inhibitor) and cycloheximide (a translation inhibitor) eliminated gallic acid-mediated PST-P mRNA expression. Based on our findings, we suggest that PST-P gene induction by gallic acid is primarily regulated at the transcription level (*54*).

NF-E2-related Factor 2 (Nrf2) and PST Expression

The transcriptional activation of the Phase II enzymes has been traced to either the cis-acting transcriptional enhancer called ARE (antioxidant response element) or, alternatively, to the electrophile response element. It has been shown that the transcription factor Nrf2 positively regulates the ARE-mediated expression of Phase II detoxification enzyme genes and stress-induced genes (*71*). The activity of Nrf2 is normally suppressed in the cytosol by specific binding to the chanperone Keap 1. Nrf2 and Keap 1 have been reported to be the primary sensors in the cellular response to oxidative stress (*72*). The cytoplasmic protein Keap 1 interacts with Nrf2 and inhibits its nuclear translocation. In addition, the disulfide bond in Keap 1 is thought to be a sensor for oxidative conditions. Signals associated with oxidative stress disrupt the binding between Nrf2 and Keap 1, resulting in the translocation of Nrf2 into the nucleus. In the nucleus, Nrf2 forms a heterodimer with small-Maf protein and activates the transcription of oxidative stress-related proteins via an ARE (*73*). However, upon stimulation by electrophilic agents or compounds that possess the ability to modify thiol groups, Keap 1 repression of Nrf2 activity is lost, allowing Nrf2 protein to translocate into the nucleus and potentate the ARE sequence (*74*). This mechanism of gene activation leads to the synthesis of highly specialized proteins that efficiently protect mammalian cells from various forms of stress and, consequently, reduce the propensity of tissues and organisms to develop disease or malignancy (*75*). Hence, activation of Nrf2, which controls constitutive and inducible expression of Phase II detoxifying genes, may be one of the protective mechanisms against xenobiotics (*76*).

The most significant finding in the study is the demonstration of the involvement of the Nrf2 pathway in gallic acid-mediated PST-P gene induction (*54*). The electrophoretic mobility shift assay revealed that the nuclear ARE binding activity was significantly increased by gallic acid treatment in HepG2 cells. Additionally, gallic acid also increased the Nrf2 nuclear translocation, suggesting that increased expression of the Nrf2 protein may play a key role in gallic acid-induced PST-P gene activation. Various polyphenol or flavone compounds, natural and synthetic, produce effects similar to gallic acid in the increase of Phase II enzymes activity. Such agents have been classified as monofunctional (Phase II) or bifunctional inducers with the capacity to increase both Phase I and Phase II enzymes (*77*).

Recent studies have demonstrated that gallic acid has mixed aryl hydrocarbon receptor (AhR) agonist/antagonist activities, allowing it to bind to the AhR and induce cytochrome P450 1A1 transcription as well as inhibit 3-methylcholanthrene induction of cytochrome P450 1A1 expression (*78*). These studies suggest that gallic acid can be a natural dietary ligand of the AhR. The ability to induce both cytochrome P450 1A1 and PST-P enzymes suggests the likelihood that gallic acid can operate as a bifucntional inducer. It has long been recognized that nearly all bifunctional inducers can activate xenobiotics responsive element (XRE) and ARE pathways concomitantly. The most common explanation is an indirect link. Briefly, bifunctional compounds first activate the AhR–XRE pathway, inducing Phase I enzymes, including cytochrome P450 1A1, which in turn metabolize these compounds. The resulting electrophilic intermediary metabolites further activate the Nrf2–ARE pathway.

Modulation of Mitogen-activated Kinases (MAPKs) by Phenolic Acids

The mitogen-activated protein kinases (MAPKs), characterized as praline-directed serine/threonine kinases, are important cellular signaling components which convert various extracellular signals into intracellular responses through serial phosphorylation cascade. At the present time, three distinct but parallel MAP kinase cascades (ERK, JNK and p38) have been identified in mammalian cells (*79*). Once activated, these three MAPKs (ERK, JNK, and p38) can phosphorylate many transcription factors and ultimately lead to changes in gene expression (*80*). Given the fact that MAPKs are activated by a wide range of factors, these signaling cascades may serve as a common mechanism and integrate with other signaling pathways to control cellular responses to various extracellular stimuli, including xenobiotics and pharmacological agents. Recently, studies on Phase II detoxifying enzyme induction by oxidative stress stimuli have shown that pathways involving MAPKs are responsible for the transduction of signals initiating gene activation (*81*). MAPKs signaling cascades are stimulated by many extracellular stimuli, such as growth factors, cytokines and various environmental stresses, and serve as a common signal transduction pathway shared by signals involved in proliferation, differentiation, functional activation and stress response. Activation of MAPKs may therefore mediate many or even opposite cellular processes and the specific outcome of these events may depend on the specific stimuli and cellular context. The p38 MAPK has been reported to participate in the activation of the ARE-mediated gene by inducing xenobiotics. Previous studies have shown that overexpression or activation of MAPKs differentially affects Nrf2 activity and Phase II detoxifying enzymes (*82*). In addition, it has been shown that induction of GST by green tea polyphenol extract (GTP) treatments is mediated by c-Jun N-terminal kinase pathways (*83*). To identify other upstream regulatory mechanisms involved in gallic acid-induced signaling events, MAPK pathways were also examined and p38 MAPK pathway was found to involved in the induction of PST-P expression by gallic acid in $HepG_2$ cells (*54*). Compared with the untreated $HepG_2$ cells, gallic acid-treated cells had higher levels of p-p38 while the levels of p-ERK did not change.

Cytoprotective Effects of Sulfotransferase on t-BHP-induced Oxidative Injury

t-BHP is a potent oxidizer capable of reacting with a wide range of biological molecules. Among the t-BHP-mediated deleterious alterations are oxidation of thiol-containing biomolecules, oxidation of proteins, lipid peroxidation as well as base modification and DNA single-strand breaks (*84*). These adverse effects are largely responsible for the cytoxicity mediated by t-BHP in target tissue/cells (*85*). However, whether upregulation of PST-P also affords cytoprotection against t-BHP-mediated toxicity in $HepG_2$ cells has not been investigated. The results presented in our study clearly indicate that incubation of $HepG_2$ cells with gallic acid results in high resistance to t-BHP-induced cell death (*54*). The involvement of PST-P in the cytoprotective action of gallic acid was examined using the PST-P and p38 MAPKs inhibitor of DCNP and SB203580, respectively. DCNP and SB203580 abrogated the protective effect of gallic acid on t-BHP-induced cell death. Plant polyphenols are well known antioxidants and have been shown to protect culture cells from oxidative stress. Furthermore, phenolic antioxidants exhibit anti-inflammatory, anti-atherosclerotic and anticarcinogenic activities (*86*). The anticarcinogenic activity is due to induction of the Phase II detoxifying gene through an ARE (*87*). Gallic acid, and its catechin derivatives, have been demonstrated to possess excellent chemopreventive effects in many target organs challenged with various carcinogens. Gallic acid, a naturally occurring plant phenolic acid present in fruits and vegetables, was the best inducer in PST-P (*52*). The above results indicated that the gallic acid-mediated cytoprotection against t-BHP toxicity was significantly reduced by the induction of PST-P. These observations strongly suggest that intracellular PST-P is an important factor in gallic acid-mediated cytoprotection against t-BHP toxicity in $HepG_2$ cells. Phenolic acids are widely distributed in plants and are present in considerable amounts in the human diet. The intake of hydroxybenzoic and hydroxycinnamic acids has been estimated to be ~11 and 211 mg/day, respectively (*88*). Moreover, a recent study demonstrated that plasma concentrations of gallic acid in humans given a single dose of gallic acid (two acidum gallicum tablets) could reach 52.3 μM (*89*). The effect of increased gallic acid levels in plasma on PST-P remains to be proven. Our study demonstrated that the ability of this phenolic acid to modulate PST-P at concentrations that may well be achievable in human plasma. Therefore, the significant induction of PST-P by the phenolic acids appear to be important and suggests that phenolic acids could be used as chemopreventive agents in sulfate conjugation.

Conclusions

Dietary polyphenols have long been suspected to scavenge ROS and thereby avert their deleterious effects on proteins, lipids, and nucleic acids in cells. This has been put forward as one of the major mechanisms for the disease-preventing effects of fruits and vegetables. Emerging findings suggest a variety of potential mechanisms of action of polyphenols in cytoprotection against oxidative stress, which may be independent of conventional antioxidant-reducing activities. Such mechanisms might entail the interaction of polyphenols with cell signaling and influence gene expression, with the consequent modulation of specific enzymatic activities that drive that intracellular response against oxidative stress. The available experimental evidence suggests that it is worth testing phenolic acids as cancer therapeutic agents. All published observations indicate that phenolic acids lead to a strong modification of cell signaling pathways including induction of Nrf2 transcription factors. As summarized in Figure 3, our study demonstrate that dietary polyphenols increase expression of Phase II enzymes, which, as a group, may effectively inhibit various forms of carcinogenesis and chemical-induced cellular damage. The observations indicate that PST-P gene expression in $HepG_2$ cells is enhanced via p38 MAPK and Nrf2. Thus, the inducibility of PST-P expression by gallic acid in $HepG_2$ cells appears to be determined by multiple signaling molecules, indicating that increased PST-P activity is an important element in gallic acid-mediated cytoprotection against oxidative stress. PST is a key enzyme in drug metabolism, bile acid detoxification and the regulation of intratissue active hormone levels; therefore, increased expression of PST will enhance the efficiency of detoxification. These findings provide some understanding of the antioxidant properties of phenolic acids.

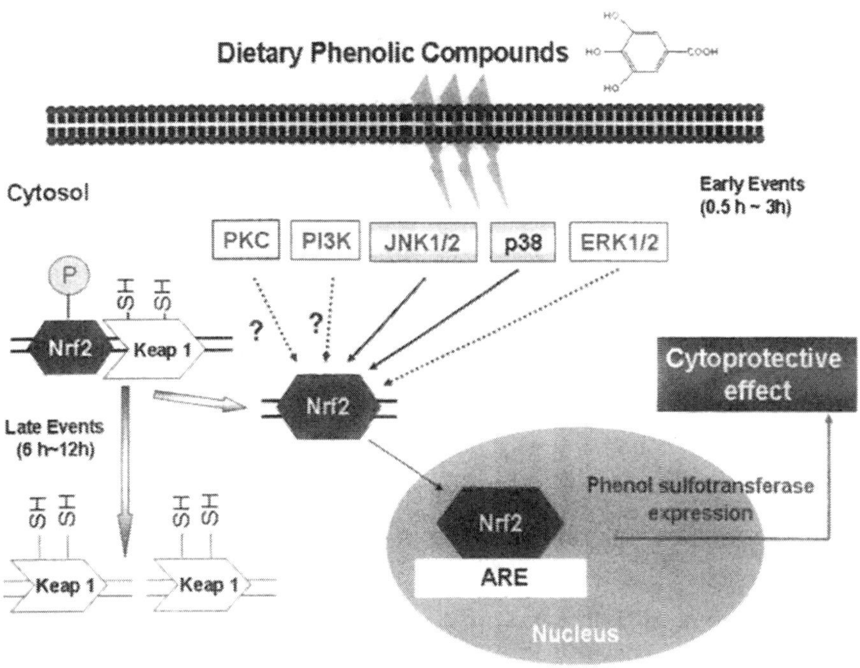

Figure 3. Schematic illustration of ARE-mediated regulation of sulfotranasferase. See text for details.

Acknowledgements

This research work was supported by the National Science Council (NSC94-2321-B005-005), Taiwan, Republic of China.

References

1. Galli, F.; Piroddi, M.; Annetti, C.; Aisa, C.; Floridi, E.; Floridi, A. *Contrib. Nephrol.* **2005**, *149*, 240-260.
2. Storz, P. *Front. Biosci.* **2005**, *10*, 1881-1896.
3. Higuchi, Y. *Biochem. Pharmacol.* **2003**, *66*, 1527-35.
4. He, F.J.; Nowson, C.A.; MacGregor, G. A. *Lancet* **2006**, *367*, 320-326.
5. Liu, R.H. *J Nutr.* **2004**, *134*, 3479S-3485S.
6. Nishino, H.; Murakoshi, M.; Mou, X.Y.; Wada, S.; Masuda, M.; Ohsaka, Y.; Satomi, Y.; Jinno, K. *Oncology* **2005**, *69*, 38-40.
7. Jiang, F.; Dusting, G.J. *Curr. Vasc. Pharmacol.* **2003**, *1*, 135-156.
8. Surh, Y.J.; Chun, K.S.; Cha, H.H.; Han, S.S.; Keum, Y.S.; Park, K.K.; Lee, S.S. *Mutat. Res.* **2001**, *480-481*, 243-268.
9. Yang, C.S.; Landau, J.M.; Huang, M.T.; Newmark, H.L. *Annu. Rev. Nutr.* **2001**, *21*, 381-406.
10. Rahman, I.; Biswas, S.K.; Kirkham, P.A. *Biochem. Pharmacol.* **2006**, *72*, 1439-1452.
11. Tapiero, H.; Tew, K.D.; Ba, G.N.; Mathe, G. *Biomed. Pharmacother.* **2002**, 56, 200-207.
12. Krizkova, L.; Nagy, M.; Polonyi, J.; Dobias, J.; Belicova, A.; Grancai, D.; Krajcovic, J. *Mutat. Res.* **2000**, *469*, 107-114.
13. Talalay, P.; Fahey, J.W.; Holtzclaw, W.D.; Prestera, T.; Zhang, Y. *Toxicol. Lett.* **1995**, *82-83*, 173-179.
14. Cuendet, M.; Oteham, C.P.; Moon, R.C.; Pezzuto, J.M. *J. Nat. Prod.* **2006**, 69, 460-463.
15. Brix, L. A.; Nicoll, R.; Zhu, X.; McManus, M.E. *Chem. Biol. Interact.* **1998**, *109*, 123-127.
16. Coughtrie, M.C.H.: Sharp, S.: Maxwell, K.; Innes, N.P. *Chem. Biol. Interact.* **1998**, *109*, 3-27.
17. Weinshiboum, R.M. *Fed. Proc.* **1986**, *45*, 2223-2228.
18. Falany, C.N. *Trends Pharmacol. Sci.* **1991**, *12*, 255-259.
19. Urquiaga, I.; Leighton, F. *Biol. Res.* **2000**, *33*, 55-64.
20. Agarwal, S.; Hordvik, S.; Morar, S. *Toxicology* **2006**, *221*, 44-49.
21. Galati, G.; O'Brien, P.J. *Free Radical Biol. Med.* **2004**, *37*, 287-303.
22. Croft, K.D. *Annu. NY Acad. Sci.* **1998**, *854*, 435-442.
23. Robbins, R.J. *J. Agric. Food Chem.* **2003**, *51*, 2866-2887.
24. Nardini, M.; D'Aquino, M.; Tomassi, G.; Gentili, V.; Di Felice, M.; Scaccini, C. *Free Radical Biol Med.* **1995**, *19*, 541-552.
25. Laranjinha, J.; Vierira, O.; Almeida, L.; Madeira, V.I. *Biochem Pharmacol.* **1996**, *51*, 395-402.
26. Guglielmi, F.; Luceri, C.; Giovannelli, L.; Dolara, P.; Lodovici, M. *Br. J. Nutr.* **2003**, *89*, 581-587.

27. Hermann, M.; Kapiotis, S.; Hofbauer, R.; Seelos, C.; Held, I.; Gmeiner, B. *Free Radical Biol. Med.* **1999**, *26*, 1253-1260.
28. Galati, G.; O'Brien, P.J. *Free Radical Biol. Med.* **2004**, *37*, 287-303.
29. Clifford, M.N. *J. Sci. Food Agric.* **2000**, *80*, 1033-1043.
30. Olthof, M.R.; Hollman, P.C.; Katan, M.B. *J Nutr.* **2001**, *131*, 66-71.
31. Zhao, Z.; Egashira, Y.; Sanada, H. *J Nutr.* **2003**, *133*, 1355-1361.
32. Rondini, L.; Peyrat-Maillard, M.N.; Marsset-Baglieri, A.; Fromentin, G.; Durand, P.; Tome, D.; Prost, M.; Berset, C. *J. Agric. Food Chem.* **2004**, *52*, 4338-4343.
33. Orsolic, N.; Knezevic, A.H.; Sver, L.; Terzic, S.; Basic, I. *J Ethnopharmacol.* **2004**, *94*, 307-315.
34. Rao, C.V.; Desai, D.; Simi, B.; Kulkarni, N.; Amin, S.; Reddy, B.S. *Cancer Res.* **1993**, *53*, 4182-4188.
35. Maggi-Capeyron, M.F.; Ceballos, P.; Cristol, J.P.; Delbosc, S.; Le Doucen, C.; Pons, M.; Leger, C.L.; Descomps, B. *J. Agric. Food Chem.* **2001**, *49,* 5646-52.
36. Bailly, F.; Cotelle, P. *Curr. Med. Chem.* **2005**, *12*, 1811-8.
37. Yeh, C.T.; Yen, G.C. *Free Radic. Res.* **2005**, *39*, 893-904.
38. Heber, D.; Bowerman, S. *J. Nutr.* **2001**, *131*, 3078S-81S.
39. Keum, Y.S.; Jeong, W.S.; Kong, A.N. *Mutat. Res.* **2004**, *555,* 191-202.
40. Wilkinson, J. 4th.; Clapper, M.L. *Proc. Soc. Exp. Biol. Med.* **1997**, *1216*, 192-200.
41. Chen, C.; Kong, A.N. *Trends Pharmacol. Sci.* **2005**, *26*, 318-326.
42. Coughtrie, M.W. *Pharmacogenomics J.* **2003**, *2*, 297-308.
43. Brix, L.A.; Barnett, A.C.; Duggleby, R.G..; Leggett, B.; McManus, M.E. *Biochemistry* **1999**, *38*, 10474-10479.
44. Littlewood, J.T.; Glover, V.; Sandler, M. *Br. J. Clin. Pharmacol.* **1985**, *19*, 275-278.
45. Anderson, R.J.; Kudlacek, P.E.; Clemens, D.L. *Chem. Biol. Interact.* **1998**, *109*, 53-67.
46. Weinshiboum, R.M. *Fed. Proc.* **1986**, *45*, 2223-2228.
47. Falany, J.L.; Greer, H.; Kovacs, T. Sorscher, E.J.; Falany, C.N. *Biochem. J.* **2002**, *364*, 115–120.
48. Ogawa, D.; Shikata, K.; Honke, K. et al. *J. Biol. Chem.* **2004**, 279, 2085–2090.
49. Hayes, J.D.; McMahon, M.; *Cancer Lett.* **2001**, *174*, 103-113.
50. Satoh, T.; Okamoto, S.I.; Cui, J.; Watanabe, Y.; Furuta, K. Suzuki, M.; Tohyama, K.; Lipton, S.A. *Proc. Natl. Acad. Sci. USA.* **2006**, *103*, 768-773.
51. Siess, M.H.; Le Bon, A.M.; Canivenc-Lavier, M.C.; Suschetet. M. *Biofactors* **2000**, 12, 193-199.
52. Yeh, C.T.;Yen, G.C. *J. Agric. Food Chem.* 2003, *51*, 1474-1479.

53. Yeh, C.T.; Shih, P.H.; Yen, G.C. *J. Agric. Food Chem.* **2004**, *52*, 4139-4143.
54. Yeh, C.T.; Yen, G.C. *Carcinogenesis* **2006**, *27*, 1008-1017.
55. Lafay, S.; Morand, C.; Manach, C.; Besson, C.; Scalbert, A. *Br. J. Nutr.* **2006**, *96*, 39-46.
56. Yeh, C.T.; Yen, G.C. *J. Nutr. Biochem.* **2006**, *17*, 561-569.
57. Humphrey, P.P.A.; Fenuik, W.; Perren, M.J.; Beresford, I.J.M.; Skingle, M.; Whalley, E.T. *Annu. NY Acad. Sci.* **1990**, *600*, 587-600.
58. Friberg, L.; Olesen, J.; Iversen, H.K.; Sperling, B. *Lancet* **1991**, *338*, 13-17.
59. Mine, K.; Okada, M.; Mishima, N.; Fujiwara, M.; Nagakawa, T. *Biol. Psychiatry* **1993**, *34*, 654-660.
60. Richard, K.; Hume, R.; Kaptein, E.; Stanley, E.L.; Visser, T.J.; Coughtrie, M.W. *J. Clin. Endocrinol. Metab.* **2001**, *86*, 2734-2742.
61. Maiti, S.; Grant, S.; Baker, S.M.; Karanth, S.; Pope, C.N.; Chen, G. *Biochem. Biophys. Res. Commun.* **2004**, *323*, 235-241.
62. Davies, E.; Tsang, C.W.; Ghazali, A.R.; Harris, R.M.; Waring, R.H. *Toxicol. In Vitro,* **2004**, *18*, 749-754.
63. Yeh, C.T.; Huang, S.M.; Yen, G.C. *J. Agric. Food Chem.* **2005**, *53*, 4766-4773.
64. Steele, V.E.; Kelloff, G.J.; Balentine, D.; Boone, C.W.; Mehta, R.; Baheri, D.; Sigman, C.C.; Zhu, S.; Sharma, S. *Carcinogenesis* **2000**, *21*, 63-67.
65. Dashwood, R.H.; Xu, M.; Hernaez, J.F.; Hasaniya, N.; Youn, K.; Razzuk, A. *Proc. Soc. Exp. Biol. Med.* **1999**, *220*, 239-243.
66. Song, L.L.; Kosmeder, J.W.; Lee, S.K.; Gerhauser, C.; Pezzuto, J.M. *Cancer Res.* **1999**, *59*, 578-585.
67. Tukey, R.H.; Strassburg, C.P.; Mackenzie, P.I. *Mol. Pharmacol.* **2002**, *62*, 446-450.
68. Ahmad, N.; Mukhtar, H. *Nutr. Rev.,* **1999**, *57*, 78-83.
69. Ow, Y.Y.; Stupans, I. *Curr. Drug Metab.* **2003**, 4, 241-248.
70. Duanmu, Z.; Kocarek, T.A.; *Drug Metab. Dispos.* **2001**, *29*, 1130-1135.
71. Kobayashi, M.; Yamamoto, M. *Antioxid. Redox Signal.* **2005**, *7*, 385-394.
72. Nguyen, T.; Sherratt, P.J.; Nioi, P.; Yang, C.S.; Pickett, C.B. *J. Biol. Chem.* **2005**, *280*, 32485-32492.
73. Hong, F.; Sekhar, K.R.; Freeman, M.L.; Liebler, D.C. *J. Biol. Chem.* **2005**, *280,* 31768-31775.
74. Motohashi, H.; Yamamoto, M. *Trends Mol. Med.* **2004**, *10*, 549-557.
75. Iida, K.; Itoh, K.; Kumagai, Y.; Oyasu, R.; Hattori, K.; Kawai, K.; Shimazui, T.; Akaza, H.; Yamamoto, M. *Cancer Res.* **2004**, *64*, 6424-6431.
76. Chan, K.; Han, X.D.; Kan, Y.W. *Proc. Natl. Acad. Sci. USA.* **2001**, *98*, 4611-4616.
77. Prochaska, H.J.; Santamaria, A.B.; Talalay, P. *Proc. Natl. Acad. Sci. USA.* **1992**, *89*, 2394-2398.

78. Krajka-Kuzniak, V.; Szaefer, H.; Baer-Dubowska, W. *Toxicol. Lett.* **2004**, *152*, 117-126.
79. Seger, R.; Krebs, E.G. *FASEB J.* **1995**, *9*, 726-735.
80. Obata, T.; Brown, G.E.; Yaffe, M.B. *Crit. Care Med.* **2000**, *28*, 67-77
81. Kong, A.N.; Owuor, E.; Yu, R.; Hebbar, V.; Chen, C.; Hu, R.; Mandlekar, S. *Drug Metab. Rev.* **2001**, *33*, 255-271.
82. Chen, C.; Yu, R.; Owuor, E.D.; Kong, A.N. *Arch. Pharm. Res.* **2000**, *23*, 605-612.
83. Yu, R.; Jiao, J.J.; Duh, J.L.; Gudehithlu, K.; Tan, T.H.; Kong, A.N. *Carcinogenesis* **1997**, *18*, 451-456.
84. Di Simplicio, P.; Rossi, R. *Biochim. Biophys. Acta.* **1994**, *1199*, 245-252.
85. Yen, G.C.; Yeh, C.T.; Chen, Y.J. *J. Agric. Food Chem.*, **2004**, *52*, 4121-4127.
86. Nishino, H.; Tokuda, H.; Satomi, Y. et al. *Asian Pac. J. Cancer Prev.*, **2000**, 1, 49-55.
87. Zhang, Y.; Gordon, G.B. *Mol. Cancer Ther.* **2004**, *3*, 885-893.
88. Shahrzad, S.; Aoyagi, K.; Winter, A.; Koyama, A.; Bitsch, I. *J. Nutr.* **2001**, *131*, 1207-1210.
89. Shahrzad, S.; Bitsch, I. *J. Chromatogr. B.* **2000**, *705*, 1033-1043.

Chapter 8

Transformation of Daidzein to Equol and Its Bioactivity

Mariko Uehara[1], Yoshiko Ishimi[2], Shin-ichi Katsumata[1], and Kazuharu Suzuki[1]

[1]Department of Nutritional Science, Faculty of Applied Bioscience, Tokyo University of Agriculture, 1–1–1 Sakuragaoka, Setagaya-ku, Tokyo 156–8502, Japan
[2]Project for Bio-index, National Institute of Health and Nutrition, Tokyo 162–8636, Japan

The structures of soybean isoflavones are similar to that of estrogen and have received attention as alternatives to hormone replacement therapy. Daidzein that is a major isoflavone found in soybean is metabolized to equol by intestinal microflora, and the metabolite exhibits a stronger estrogenic activity than daidzein. Recent studies suggest that the clinical effectiveness of isoflavones might be due to their ability to produce equol in the intestine. Although typical laboratory animal species consistently produce high levels of equol, only 30–50% of the human population can produce equol. Fructooligosaccharides, indigestible sugars, increased the bioavailability of daidzein and enhanced the transformation of daidzein to equol by stimulating intestinal microflora. In particular, equol may have indirect and direct effects on bone loss in rodents, osteoporotic models, and osteopenia. Possible bioactivity of equol based on its estrogenic properties with regard to bone metabolism and taking into consideration the status of intestinal microflora is discussed.

Several plant constituents have estrogen-like physiological effects. These constituents are called phytoestrogens (*1, 2*). Isoflavones (genistein and daidzein) in soybeans and lignans (enterolactone and enterodiol) that are formed in the intestine from their precursors in various grains, seeds, fruits, some vegetables, and tea are commonly known as phytoestrogens (*3, 4*). Dietary phytoestrogens may prevent sex hormone-related diseases, such as breast cancer and prostate cancer, through mechanisms that have not yet been adequately documented (*3, 5, 6*). On the other hand, they may also prevent postmenopausal osteoporosis through an estrogenic-like effect. For instance, the intake of soy products, genistein, daidzein, or other phytoestrogens has been found to prevent postovariectomized bone loss in rats and mice (*7, 8*), and post- or peri-menopausal women (*9-11*).

Almost all phytoestrogens, such as genistein and daidzein, in food exist as glycosides; thus, to facilitate the intestinal absorption and the physiological effect of these glycosides, it is necessary to hydrolyze the glycosidic bonds (*1-3*). These bonds are hydrolyzed by the glucosidases of intestinal bacteria such as *lactobacilli, bacteroides,* and *bifidobacteria* (*12, 13*). Recently it has been determined that the metabolites of isoflavones are more important than their precursors (14). The soybean isoflavone daidzein is metabolized to equol in the gastrointestinal tract by gut microflora (*14, 15*). Equol possesses a stronger affinity for estrogen receptors than daidzein (*16*) and has the strongest transcriptional activity among soybean isoflavonoids in yeast 2-hybrid systems (*17*). Furthermore, equol is a chiral molecule, which can exist as the enantiomers *R*(+)-equol and *S*(–)-equol. Setchell *et al.* (*18*) established *S*-equol as the exclusive product of human gut microfloral synthesis from daidzein and also showed that both enantiomers were bioavailable. Thus, equol is a biologically active metabolite of daidzein. Recent studies suggest that the clinical effectiveness of isoflavones might be due to their ability to produce equol in the intestine (*14*). Maximal responses to isoflavone intake are observed in individuals who are effective producers of equol. Although typical laboratory animal species consistently produce high levels of equol, only 30–50% of the population can produce equol (*14, 15, 19*). It is important to clarify the key factors involved in the transformation of daidzein into equol and the bioactivity of equol.

Equol Has a Direct Beneficial Effect on Bone Loss Due to Ovariectomy

To determine a direct effect of equol on bone loss, eight-week-old female mice were assigned to the following five groups: sham-operation (sham), ovariectomy (OVX), OVX + 0.1 mg/day equol (0.1 Eq), OVX + 0.5 mg/day equol (0.5 Eq), and OVX + 0.03 mg/day 17β-estradiol (E_2) (*20*). Equol and E_2 were subcutaneously administered by using a mini-osmotic pump. Four weeks after intervention, uterine weight was reduced by OVX and restored by E_2 administration. In contrast, equol at doses used in this study did not affect uterine atrophy in OVX mice. Bone mineral density (BMD) of the whole body measured by a PIXImus densitometer was reduced in OVX mice, while it was maintained by the administration of 0.5 mg/day equol as well as E_2. The BMD of the femur and lumbar spine was also reduced by OVX, and treatment with 0.5 mg/day equol prevented bone loss in dual-energy X-ray absorptiometry (DXA). In particular, the BMD of the proximal femur was the same as that of the proximal femurs of the mice in the sham group. E_2 prevented OVX-induced bone loss from all regions. These results suggest that equol inhibits bone loss apparently without estrogenic activity in the reproductive organs of OVX mice.

Antibiotics Decrease Equol Production from Daidzein

The efficacy of daidzein in inhibiting bone loss was examined. The effect of antibiotics on equol production and femoral BMD was determined in OVX mice that were administered daidzein (aglycone). Eight-week-old female mice were assigned to the following seven diet groups: sham, OVX, OVX + 0.3% kanamycin (KN), OVX + 0.1% daidzein (Dz), OVX + 0.04% KN + Dz (0.04 KN), OVX + 0.08% KN + Dz (0.08 KN), and OVX + 0.3% KN + Dz (0.3 KN). Four weeks after treatment, kanamycin decreased the equol concentration in the serum (Table I). Uterine weight was reduced by OVX and the reduced weight was unchanged by kanamycin or daidzein administration. Proximal and distal femur BMD was reduced in OVX mice, while it was maintained by the administration of 0.1% daidzein (Table I). However, kanamycin destroyed the preventive effect of daidzein on bone loss induced by OVX. These results suggest that it might be important to promote or activate intestinal microflora, which produce equol, to obtain the maximal effects of isoflavones on the prevention of bone loss in an estrogen-deficient status.

The Prebiotics Fructooligosaccharides Modify Isoflavone Metabolism

Fructooligosaccharides (FOS)—a mixture of indigestible and fermentable sugars—stimulate the growth of *bifidobacteria* in the intestine (*21-24*). It was postulated that dietary FOS may affect the bioavailability of isoflavone glycoside conjugates and therefore improve their absorption in the intestine. The kinetics of isoflavones in rats fed a 5% FOS-supplemented diet or a control diet were examined by measuring genistein and daidzein concentrations in blood collected from three different veins and by measuring urinary excretion at 24-

Table I. Serum Equol Concentration and Femoral Bone Mineral Density (BMD)

	Serum equol (nmol/L)	BMD (mg/cm^2)		
		Proximal	Middle	Distal
Sham	ND	39.5 ± 0.75 ^{a}	31.9 ± 0.46 ^{a}	38.6 ± 0.41 ^{a}
OVX	ND	36.9 ± 0.76 ^{b}	30.3 ± 0.59 $^{a, b}$	34.1 ± 1.05 $^{b, c}$
+KN0.3	ND	37.3 ± 1.05 ^{b}	31.2 ± 0.41 ^{a}	35.2 ± 0.45 $^{b, c}$
+Dz	3530.1 ± 790.0 ^{a}	39.8 ± 1.06 ^{a}	31.7 ± 0.54 ^{a}	36.7 ± 0.67 ^{a}
+Dz+KN0.04	111.7 ± 50.2 ^{b}	35.8 ± 0.64 ^{b}	30.5 ± 0.61 $^{a, b}$	35.3 ± 0.53 $^{b, c}$
+Dz+KN0.08	113.2 ± 22.2 ^{b}	35.1 ± 0.97 ^{b}	30.4 ± 1.00 $^{a, b}$	35.3 ± 0.69 $^{b, c}$
+Dz+KN0.3	92.4 ± 9.5 ^{b}	36.6 ± 0.59 ^{b}	28.8 ± 0.5 ^{b}	33.7 ± 0.69 ^{c}

Values are Means ± SEM, n=6.
sham: sham operated, OVX: ovariectomized, +KN0.3: OVX+0.3%kanamycin, +Dz: OVX+0.1% daidzein, OVX+Dz+0.04%kanamycin, +Dz+KN0.08: OVX+Dz+0.08%kanamycin, +Dz+KN0.3: OVX+Dz+0.3%ka
Means with different letters differ, $P < 0.05$.

and 48-h after a single intragastric administration of isoflavone glycoside conjugates (100mg mixture of daidzin, genistin and glycitin/kg body weight (BW)). The genistein concentration in the portal blood increased rapidly, reaching a peak in both the FOS-fed and control groups at 1 h after administration. The concentrations in the central venous blood were approximately half of those in the portal blood. In the FOS-fed group, both genistein and daidzein remained detectable in the tail venous blood after 24 to 48 h. The urinary excretion of both isoflavones at 24- and 48-h after administration was significantly higher in the FOS-fed group than in the control group. In a similar study, the equol concentration started to increase in the central venous blood at 12 h after the administration of isoflavone glycoside conjugates (100mg/kg BW) with FOS feeding and was significantly higher in the FOS-fed group than in the control group at 48 and 72 h (Table II). Thus, FOS modified the absorption and enterohepatic recirculation of isoflavones and enhanced equol production from daidzein.

Furthermore, in sham-operated and OVX mice, FOS increased the activity of β-glucosidase in cecum contents to hydrolyze the glycosidic bonds of isoflavone conjugates and stimulated the transformation of daidzein to equol in the administered isoflavone conjugates. This combination of isoflavones and FOS feeding inhibited bone loss induced by OVX. The other synergistic effects of dietary isoflavones and FOS supplementation on postgastrectomy (GX) osteopenia were also determined. Decreases in femoral trabecular BMD and bone breaking force by GX were inhibited in rats fed FOS with and without isoflavone diets, whereas supplementation with only isoflavone did not prevent bone loss in gastrectomized rats. Notably, a combination of isoflavones and FOS diet enhanced the production of equol from daidzein in gastrectomized rats.

Changes in Intestinal Microbiota on Administration of FOS and Isoflavones Determined by Terminal Restriction Fragment Length Polymorphism Analysis

To determine changes in intestinal microbiota produced by the administration of FOS and isoflavones, terminal restriction fragment length polymorphism (T-RFLP) analysis of amplified 16S rRNA genes was used (*25*). Fecal samples were collected between 0 and 24 h, 24 and 48 h, and 48 and 72 h after a single intragastric administration of isoflavone conjugates. Clone libraries of 16S rDNA were constructed and the predominant terminal restriction fragments were identified by comparing T-RFLP patterns in the fecal community with those of corresponding 16S rDNA clones. Sequence analysis indicated that despite isoflavone administration, FOS-fed rats were colonized mainly by members of *Bifidobacterium* and *Lactobacillus*. The results of real-

Table II. Time Course of Changes of Equol Concentrations in Central Venous Serum in Rats Fed the Control Diet of the 5% Fructooligosaccharide (FOS) Diet after a Single Dose of Isoflavone Glycoside Conjugates

	Serum equol (nmol/L)						
	1h	3h	6h	12h	24h	48h	72h
Control	271.8 ± 49.6	144.7 ± 17.5	282.2 ± 59.1	277.8 ± 75.5	1185.3 ± 280.8	861.5 ± 81.1	507.6 ± 0.4
FOS	198.9 ± 55.4	214.8 ± 41.5	371.4 ± 67.2	1048.5 ± 648.5	2361.0 ± 966.5	2172.6 ± 494.0 *	1211.0 ± 1.1 *

Values are Means ± SEM, n=5-7.

*Significantly different from the control at a particular time point, $P < 0.05$.

time polymerase chain reaction (PCR) indicated that *Lactobacillus* was classified as *L. murinus or L. animalis. Bifidobacterium* is one of species, which can produce equol. However, identifying equol-producing bacteria is complicated. It is known that equol is produced from daidzein through its intermediate metabolite dihydrodaidzein (*26*). Only one lactic acid bacterium (*Lactococcus garvieae*) has yet been identified that can produce equol directly from daidzein (*27*). From a human fecal sample, Decroos *et al.* have isolated a stable mixed microbial culture comprising four species (*Lactobacillus mucosae, Enterococcus faecium, Finegoldia magna,* and *Veillonella sp*) that is capable of transforming daidzein into equol and have examined the influence of some environmental conditions in the colon on equol production (*28*). They suggested that hydrogen in particular but also propionate and butyrate influenced equol production in a positive manner because hydrogen gas probably acts as an electron donor in the biotransformation, and short chain fatty acids are related to the production of hydrogen gas (*29*). FOS increase hydrogen gas, propionate, and butyrate *in vivo*, but in an *in vitro* study, FOS inhibited equol production. Further studies should be conducted to clarify this discrepancy between the results of the *in vivo* and *in vitro* studies.

Conclusions

It is important to promote or activate intestinal microbiota for the transformation of daidzein to equol. FOS might be candidate ingredients for enhancing equol production, but further studies should be conducted to clarify the discrepancy between the results of *in vivo* and *in vitro* studies. Furthermore, in a human study involving French postmenopausal women, FOS did not increase urinary equol production (*28*). However, racial differences might exist with regard to isoflavone metabolism. Further human studies involving Asian subjects are required.

Acknowledgment

We would like to thank Dr. Herman Adlercreutz, Dr. Atsutane Ohta and Dr. Kazuki Kanazawa for constructive discussions and valuable suggestions about phytoestrogens and FOS. Dr. Hidetoshi Kubota, Mr. Hiroya Endo and Ms. Yuko Baba from Meiji Seika Kaisha Ltd. are also acknowledged for the technical support of T-RFLP.

References

1. Murkies, A. L.; Wilcox, G.; Davis, S. R. *J. Clin. Endocrinol. Metab.* **1998**, *83*, 297-303.
2. Price, K. R.; Fenwick, G. R. *Food Addit. Contam.* **1985**, *2*, 73-106.
3. Adlercreutz, H.; Mazur, W. *Ann. Med.* **1997**, *29*, 95-120.
4. Mazur, W. E.; Wahala, K.; Rasku, S.; Salakka, A.; Hase, T.; Adlercreutz, H. *Br. J. Nutr.* **1998**, *79*, 37-45.
5. Adlercreutz, H. *Lancet Oncol.* **2002**, *3*, 364-373.
6. Gikas, P. D.; Mokbel, K. *Int. J. Ferti. Womens Med.* **2005**, *50*, 250-258.
7. Arjmandi, B. H.; Birnbaum, R.; Goyal, N. V.; Getlinger, M. J.; Juma, S.; Alekel, L.; Hasler, C. M.; Drum, M. L.; Hollis, B. W.; Kukreja, S. C. *Am. J. Clin. Nutr.* **1998**, *68*, 1364S-1368S.
8. Ishimi, Y.; Miyaura, C.; Ohmura, M.; Onoe, Y.; Sato, T.; Uchiyama, Y.; Ito, M.; Wang, X.; Suda, T.; Ikegami, S. *Endocrinology* **1999**, *140*, 1893-1900.
9. Greendale, G. A.; FitzGerald, G.; Huang, M. H.; Sternfeld, B.; Gold, E.; Seeman, T.; Sherman, S.; Sowers, M. *Am. J. Epidemiol.* **2002**, *155*, 746-754.
10. Mei, J.; Yeung, S. S.; Kung, A. W. *J. Clin. Endocrinol. Metab.* **2001**, *86*, 5217–5221.
11. Alekel, D. L.; Germain, A. S.; Peterson, C. T.; Hanson, K. B.; Stewart, J. W.; Toda, T. *Am. J. Clin. Nutr.* **2000**, *72*, 844-852.
12. Hawksworth, G.; Drasar, B. S.; Hill, M. J. *J. Med. Microbiol.* **1971**, *4*, 451-459.
13. Xu, X.; Harris, K. S.; Wang, H. J.; Murphy, P. A.; Hendrich, S. *J. Nutr.* **1995**, *125*, 2307-2315.
14. Setchell, K. D.; Brown, N. M.; Lydeking-Olsen, E. *J. Nutr.* **2002**, *132*, 3577-3584.
15. Atkinson, C.; Frankenfeld, C. L.; Lampe, J. W. *Exp. Biol. Med.* **2005**, *230*, 155-170.
16. Muthyala, R. S.; Ju, Y. H.; Sheng, S.; Williams, L. D.; Doerge, D. R.; Katzenellenbogen, B. S.; Helferich, W. G.; Katzenellenbogen, J. A. *Bioorg. Med. Chem.* **2004**, *12*, 1559-1567.
17. Morito, K.; Aomori, T.; Hirose, T.; Kinjo, J.; Hasegawa, J.; Ogawa, S.; Inoue, S.; Muramatsu, M.; Masamune, Y. *Biol. Pharm. Bull.* **2001**, *24*, 351-356.
18. Setchell, K. D.; Clerici, C.; Lephart, E. D.; Cole, S. J.; Heenan, C.; Castellani, D.; Wolfe, B. E.; Nechemias-Zimmer, L.; Brown, N. M.; Lund, T. D.; Handa, R. J.; Heubi, J. E. *Am. J. Clin. Nutr.* **2005**, *81*, 1072-1079.
19. Rowland, I. R.; Wiseman, H.; Sanders, T. A.; Adlercreutz, H.; Bowey, E. A. *Nutr. Cancer* **2000**, *36*, 27-32.
20. Fujioka, M.; Uehara, M.; Wu, J.; Adlercreutz , H.; Suzuki, K.; Kanazawa, K.; Takeda, K.; Yamada, K.; Ishimi ,Y.; *J. Nutr.* **2004**, *134*, 2623-2627.

21. Uehara, M.; Ohta, A.; Sakai, K.; Suzuki, K.; Watanabe, S.; Adlercreutz, H. *J Nutr.* **2001**, *131*, 787-795.
22. Ohta, A.; Uehara, M.; Sakai, K.; Takasaki, M.; Adlercreutz ,H.; Morohashi, T.; Ishimi ,Y. *J. Nutr.* **2002**, *132*, 2048-2054.
23. Uehara, M.; Ohta, A.; Ishimi, Y.; Morohashi, T.; Adlercreutz, H.; Watanabe, S.; Masuyama, R.; Suzuki, K. *J. Nutr.* **2002**, *132*, 616S-617S.
24. Hidaka, H.; Eida, T.; Takizawa, T.; Tokunaga, T.; Tashiro, Y. *Bifidobact. Microflora* **1986**, *5*, 37-50.
25. Nakanishi, Y.; Murashima, K.; Ohara, H.; Suzuki, T.; Hayashi, H.; Sakamoto, M.; Fukasawa, T.; Kubota, H.; Hosono, A.; Kono, T.; Kaminogawa, S.; Benno, Y. *Appl. Environ. Microbiol.* **2006**, *72*, 6271-6276.
26. Heinonen, S.; Wähälä, K.; Adlercreutz, H. *Anal. Biochem.* **1999**, *274*, 211-219.
27. Uchiyama, S.; Ueno, T.; Suzuki, T. *J. Intestinal Microbiol(Tokyo).* **2007**, *21*, 217-220.
28. Decroos, K.; Vanhemmens, S.; Cattoir, S.; Boon, N.; Verstraete, W. *Arch. Microbiol.* **2005**, *183*, 45-55.
29. Clavel, T.; Fallani, M.; Lepage, P.; Levenez, F.; Mathey, J.; Rochet, V.; Serezat, M.; Sutren, M.; Henderson, G.; Bennetau-Pelissero, C.; Tondu, F.; Blaut, M.; Dore, J.; Coxam, V. *J. Nutr.* **2005**, *135*, 2786-2792.

Chapter 9

Structures and Functionalities of Acylated Anthocyanins

Norihiko Terahara[1] and Toshiro Matsui[2]

[1]Department of Food Science for Health, Faculty of Health and Nutrition, Minami-Kyushu University, 5–1–2 Kirishima, Miyazaki 880–0032, Japan
[2]Department of Bioscience and Biotechnology, Division of Bioresource and Bioenvironmental Sciences, Faculty of Agriculture, Graduate School of Kyushu University, Fukuoka 812–8581, Japan

Anthocyanins are polyphenolic glycoside pigments widely distributed in the plant kingdom. Recently, along with health trend against aging, anthocyanins have attracted much attention as functional food factors that prevent life-style related diseases. To develop new stable anthocyanin materials, we elucidated a correlation between the structures and the functionalities of acylated anthocyanins with two or more aromatic acids in the plant materials such as a purple sweet potato storage root and a butterfly-pea flower. Of the functionalities, the antioxidative activity was evaluated on the basis of DPPH radical scavenging activity and lipid peroxidation inhibitory activity. It was demonstrated that aglycone and/or aromatic acids with free phenolic hydroxyl groups on neighboring positions and only aromatic acids at outer sides in polyacylated anthocyanin molecules (folded in solutions) were more effective for the antioxidative activity.

Anthocyanins (ANs) are polyphenolic glycoside pigments widely distributed in the plant kingdom. Over 600 kinds of anthocyanins have so far been found in nature (flowers, fruits, vegetables and so on) and pigmented cultured cells (*1-4*). Commonly, they are based on major six aglycons, pelargonidin, cyanidin (Cy), delphinidin (Dp), peonidin (Pn), petunidin and malvidin as demonstrated in Figure 1. They are glycosylated with monosaccharides like glucose, galactose, arabinose and xylose, or their combination di- or trisaccharides, so they are water-soluble. Cyanidin 3-glucoside occurs as most popular anthocyanin in plants (Figure 1). Also they are often acylated with aromatic acids (ARs) such as *p*-coumaric, caffeic, ferulic, sinapic and *p*-hydroxybenzoic acids and/or aliphatic acids such as malonic, acetic and succinic acids. They are called "acylated anthocyanins (AANs)", and constitute about half of all ANs found (*5*). For example, malonylshisonin from Shiso (*Perilla ocimoides*) leaves (*6*) and Heavenly Blue anthocyanin from flower of the morning glory (*Ipomoea tricolor*) (*7*) are typical AANs with acylating ARs (Figure 1). AANs with two or more ARs are more stable in weakly acidic or neutral aqueous solution than other ANs (*8, 9*). Hydrophobic stacking between the aglycone and aromatic rings of ARs protect from nucleophilic attack on C-ring C2-position of water molecules lead to the colorless pseudobases (hemiacetals, Figure 3b) (*10-13*).

Recently, ANs have been indicated to have antioxidative activity (AOA) (*14-17*), and many functionalities modulating physiological functions associated AOA or other mechanisms, such antihyperglycemic, visual function improvement, antimutagenic, antitumor, anti-virus, antihypertensive, anti-inflammation, reduction of hepatic insufficiency, and lowering of serum cholesterol, among others (*18-31*). Along with health trend, ANs attract attention as functional food factors to prevent the life-style related diseases.

Although AANs occur widely, their *in vivo* properties like functionalities, absorption, bioavailability, and utilization for foods have not progressed. To develop functional foods using AANs-containing materials, we examined the AAN structures-AOA relationship because AOA is related to biological functions and might be most suitable assessment for health benefits of AANs (Figure 1).

Materials and Methods

Plant Materials

To develop AAN-containing plant materials for functional foods, we extensively searched for many plant materials and examined their AAN qualities. Conclusively, we could select two materials, purple sweet potato (*Ipomoea batatas*) storage roots and butterfly-pea (*Clitoria ternatea*) flowers. Both are of tropical origins and their pigments are stable and have been traditionally used as food colorants (*32-34*).

Preparation of Plant Pigments

The plant materials were extracted with aqueous acetic or formic acid and purified with an adsorbed resin (Amberlite XAD-2000) column to give the purple

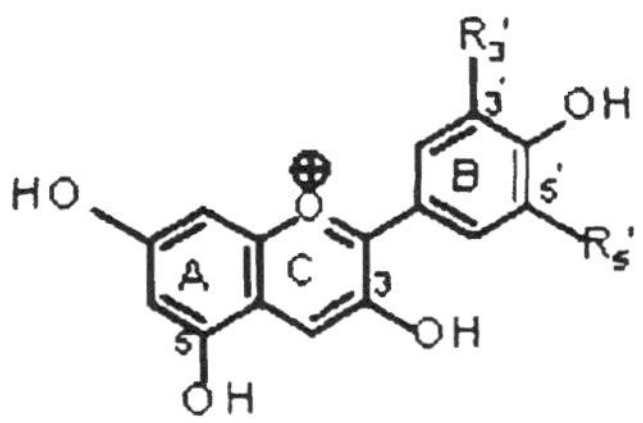

Aglycon	R_3	R_5
Pelargonidin	H	H
Cyanidin	OH	H
Peonidin	OMe	H
Delphinidin	OH	OH
Petunidin	OH	OMe
Malvidin	OMe	OMe

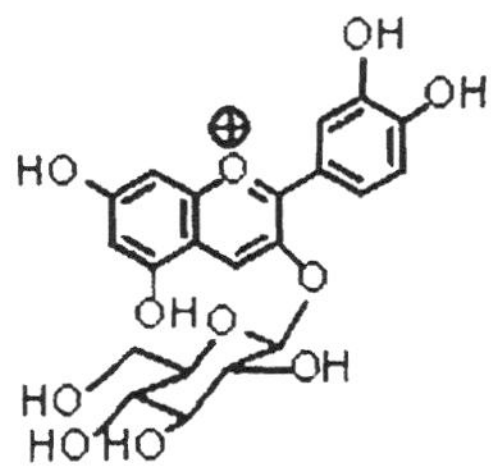

Cyanidin 3-glucoside

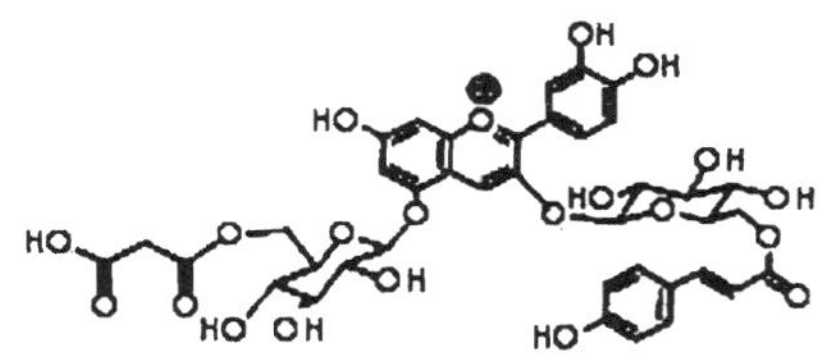

Malonylshisonin

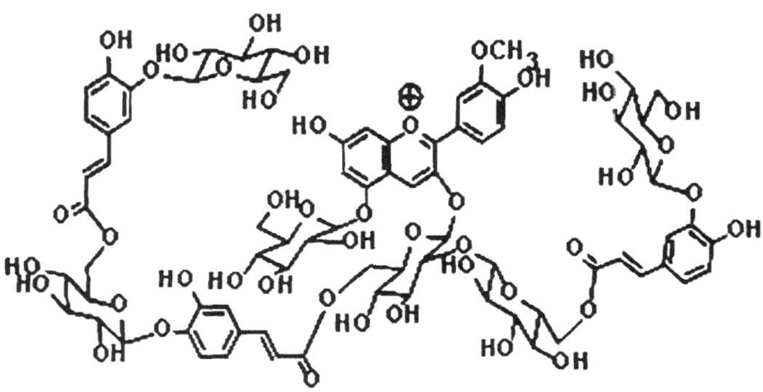

Heavenly Blue anthocyanin

Figure 1. Common anthocyanidins and anthocyanins (OMe, Methoxyl group).

sweet potato crude pigment powder (PSP, yield ca 0.8%) and the butterfly-pea flower crude pigment powder (BPP, yield ca 0.4%) according to the previously explained method (*35, 36*). They mainly contained respective AANs along with other polyphenols. Through further purification with a polyvinylpyrrolidone resin (Polyclar AT) column and preparative ODS-HPLC provided the trifluoroacetic acid salts of pure ANs, YGM-1-6 (Figure 2) from PSP and ternatin A-D (Figure 3) from BPP, respectively. The structures of isolated ANNs were confirmed by MS and NMR measurements in addition to chemical analyses (*35, 36*).

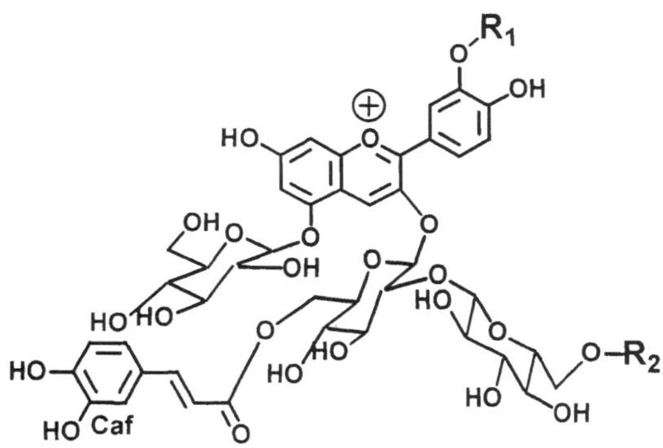

YGM	R_1	R_2
Cyanidin-based		
YGM-1a	H	Phb
YGM-1b	H	Caf
YGM-2	H	H
YGM-3	H	Fer
Peonidin-based		
YGM-4b	CH_3	Caf
YGM-5a	CH_3	Phb
YGM-5b	CH_3	H
YGM-6	CH_3	Fer

Figure 2. Purple sweet potato root anthocyanins (Phb, p-Hydroxybenzoic acid; Caf, Caffeic acid; Fer, Ferulic acid).

By alkaline deacylation, cyanidin and peonidin 3-sophoroside-5-glucosides (Cy3S5G and Pn3S5G) from PSP, and delphinidin 3, 3', 5'-triglucoside, also called deacylternatin (Da-T) from BPP, were prepared, respectively.

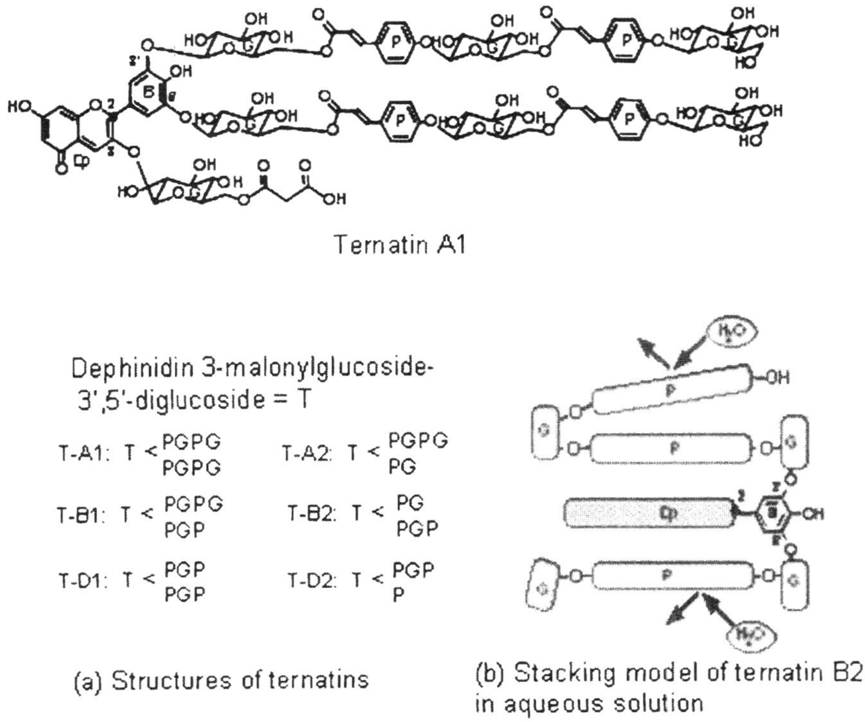

Figure 3. Butterfly-pea flowers anthocyanins (P, p-Coumaric acid; G, D-Glucose).

Evaluation of Antioxidative Activity (AOA)

We employed two *in vitro* methods to assess AOA of pigment samples. One was DPPH-radical scavenging activity assay based on 1,1-diphenyl-2-picrylhydrazyl (DPPH) fading with antioxidants at pH 7.4 according to Yamaguchi *et al.* (*37*) with some modification. The remaining DPPH quantity is in inverse proportion to antioxidative (radical scavenging) activity which is expressed as the DPPH-radical scavenging ratio (RS%). The other is lipid peroxidation inhibition assay base on ß-carotene bleaching inhibition with antioxidants at pH 7.0 according to Igarashi *et al.* (*38*) with some modifications. Linoleic acid was used as a lipid model and the peroxidation is initiated with lipoxygenase. Linoleate peroxides bleach coexisting ß-carotene without antioxidant. In this assay, remaining ß-carotene quantity is in proportion to the antioxidative activity which is designated as the lipid peroxidation inhibitory ratio (IR%).

Results and Discussion

Acylated Anthocyanins of Purple Sweet Potato and Butterfly-pea

Purple sweet potato root contains 7 or more YGMs (named for the ancestral variety name "Yamagawamurasaki") and butterfly-pea flower contains 8 or more ternatins and also minor preternatins (*34-36, 39-40*).

YGMs have the common chromophore structures, cyanidin or peonidin 3-caffeoylsophoroside-5-glucoside, and peonidin-based diacylated ANs are dominant in YGMs as shown in Figure 2 (*35*). The purple sweet potatoes have already been used to develop popular processed foods in Japan like juices, dressings, red vinegars, red spirits, red beer, ice cream, breads, and sweet stuffs, among others, but these are not functional foods. Like ternatin A1, all ternatins have a common skeleton, delphidin 3-malonylglucoside-3', 5'-diglucoside bearing with 3', 5'- side chains with *p*-coumarate-glucosyl units and presented as shown in Figure 3a (*36*). Ternatin A1 is the largest AAN ($C_{96}H_{107}O_{53}^{+}$, Mr=2108) in all ANs found by now (*42*). Blue color of ternatin solutions is exceptionally stable, which is convenient characteristic for food pigments. Since they associate in folding form through intramolecular hydrophobic-stacking in aqueous solutions, the form protects ternatins from water molecule attack leading to loss of color as shown in Figure 3b (*7*).

Antioxidative Activity of Crude Pigments

Both crude pigments PSP and BPP were examined in the antioxidative assays. These AOAs were paralleled in each assay, and PSP has somewhat stronger activity than BPP (data not shown). This is an inverse situation of their color stabilities (BPP ($t_{1/2}$=67days) > PSP ($t_{1/2}$=230 min), $t_{1/2}$=half life in pH 7.0 buffer solution at room temperature). Thus, both pigments were found to have moderate radical scavenging activities and lipid peroxidation inhibition as well as high stabilities.

Antioxidative Activity of Aromatic Acids and Deacylated Anthocyanins

In order to grasp how ARs exhibit their antioxidation property, we tested the activities of YGMs and ternatins. YGMs contain caffeic (Caf), ferulic (Fer) and *p*-hydroxybenzoic (Phb) acids, and ternatins contain *p*-coumaric acid (Pco). The result shows that the strongest compound is Caf, followed by Fer, Pco, and Phb (Figure 4a). In the acids, Caf only has a catechol (1,2-dihydroxybenzene) structure, which enhances the antioxidation by stabilizing the generated radicals with the many resonance structures after oxidation (Figure 4a) (*43*).

Hydoxycinnamic acids are stronger than hydroxybenzoic acids (Phb) in their antioxidant effect.

The AOAs of deacylated derivatives (Cy3S5G and Pn3S5G from YGMs and Da-T from ternatins) were examined. The results show that Cy3S5G is strongest of all as shown in Figure 4b. Only this cyanidin-based glycoside has a catechol structure in B-ring of the aglycon, so, like Caf, it is considered to stabilize the generated radicals more than Pn3S5G and Da-T that have only one phenolic hydroxyl group at position of B-ring.

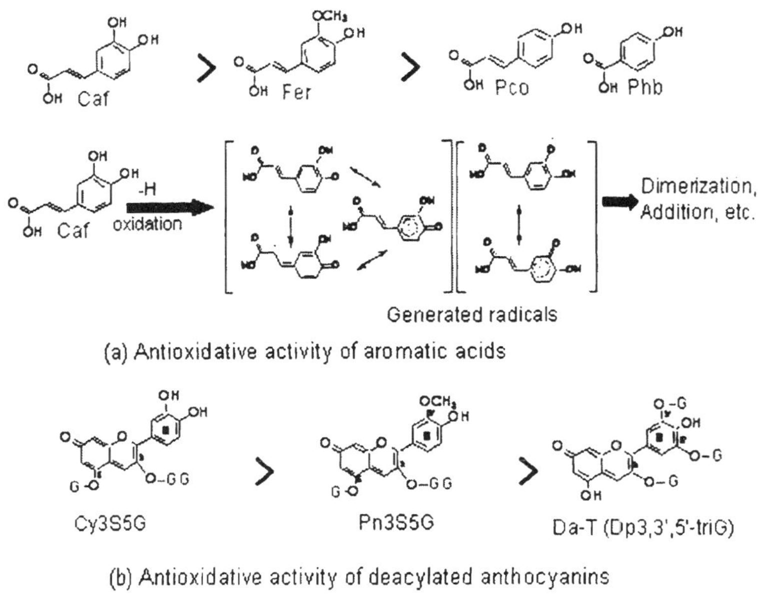

Figure 4. Antioxidative activity of aromatic acids and deacylated anthocyanins (G, D-Glucose).

YGMs' Structures and Antioxidative Activity

Figure 5 demonstrates the antioxidation test results of three YGMs (monoacylated YGM-5b (Pn 3-Caf-S-5-G), and diacylated YGM-3 (Cy 3-Caf-Fer-S-5-G) and YGM-6 (Pn 3-Caf-Fer-S-5-G)), the deacylated anthocyanins (Cy3S5G and Pn3S5G), and related acids (Caf and Fer). The order of activity was YGM-3 > YGM-6 > YGM-5b > Cy3S5G ≈ Caf > Pn3S5G > Fer. The results indicate that, (1) acylation with ARs enhances the AOAs of ANs, and AOA is proportional to the degree of acylation, (2) antioxidative intensities of YGMs are nearly equal to addition of those of corresponding DAs and acylating acids, (for example, intensity of AOA of YGM-3 ≈ Cy3S5G + Caf + Fer), and (3) cyanidin-based YGM-3 with a catechol moiety is stronger than peonidin-based YGM-6 and YGM-5b, as shown in the case of DAs (Figure 4b). Thus YGM-3 with both catechol type B-ring and caffeoyl group enhances AOA additively, and the AOA is the strongest of all antioxidants employed in this test.

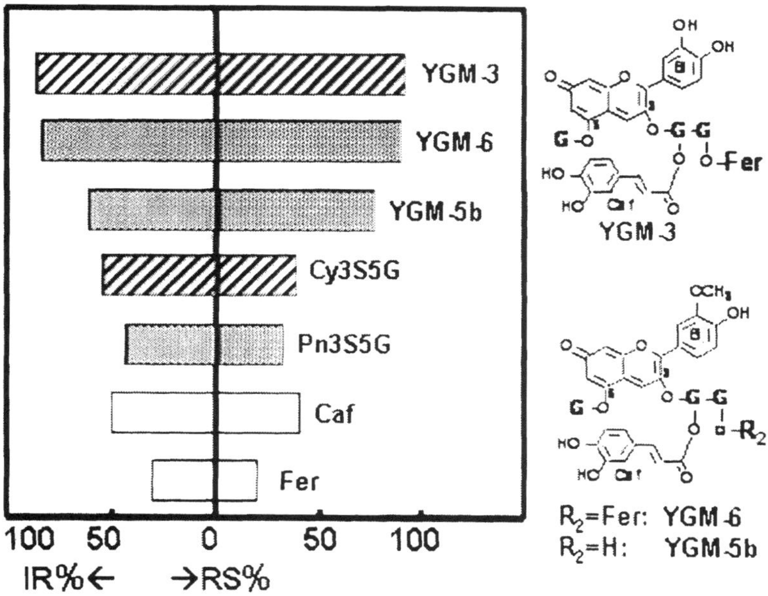

Figure 5. YGMs' structures and the antioxidative activity (IR%, lipid peroxidation inhibitory ratio; RS%, DPPH radical scavenging ratio; G, D-Glucose).

Ternatins' Structures and Antioxidative Activity

Figure 6 shows the antioxidative results of six ternatins, Da-T and Pco. Ternatin D (T-D) series have the highest activity, followed by B-series (T-B), A-series (T-A) and Da-T. The results show that, (1) only the terminal Pco(s) in 3', 5'-side chains enhance antioxidative activities, and the intensive degree is dependent on the terminal acid numbers (T-D series (two Pcos) > T-B series (one Pco) > T-A series (non Pco)), (2) antioxidative intensities of ternatins are nearly equal to additive intensities of Da-T and terminal Pco(s).

The terminal Pco(s)' enhancing effect is ascribable to ternatins' conformation in aqueous solution. As presented in color stability of ternatins, T-D1, for example, associates sterically not expanded but folding form through

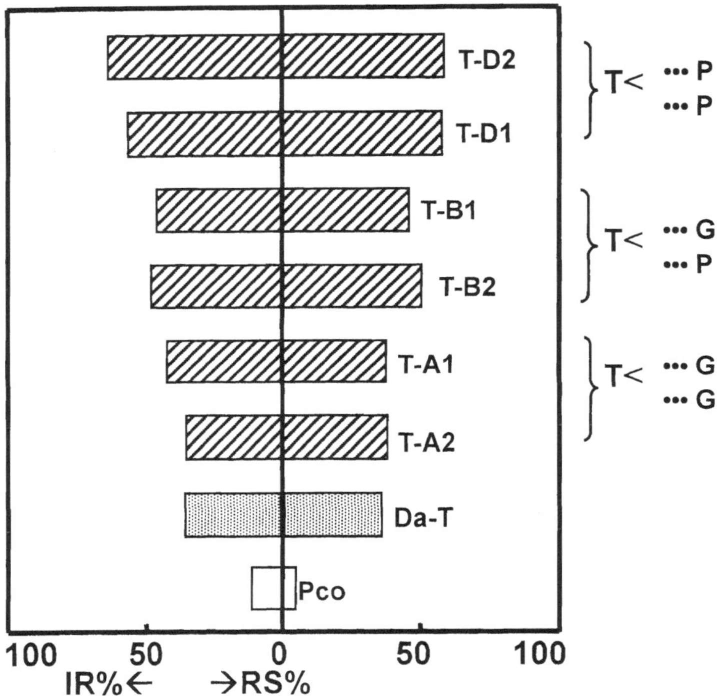

Figure 6. Ternatins' structures and the antioxidative activity (IR%, lipid peroxidation inhibitory ratio; RS%, DPPH-radical scavenging ratio; G, D-Glucose; P, p-coumaric acid)

intramolecular hydrophobic-stacking in a solution (Figure 7). The terminal Pco(s) occupy outside and the inner Pco(s) occupy inside of the folding pigment molecule. Therefore, the terminal Pco(s) only can enhance the AOA despite having multiple Pcos inner side of the folding molecule. This "burying effect" apparently decreases the AOA expected from AR number, and might be adaptive to other polyacylated ANs (*7, 8, 44- 47*).

Conclusions

The antioxidative activities of acylated anthocyanins, YGMs and ternatins may be summarized as follows.

1. Their AOAs are exerted by a radical scavenging and a lipid peroxidation inhibition mechanism.
2. Intensity of AOAs was nearly equal to the total of those of corresponding DAs and acylating acids, hence an additive and not synergistic effect.

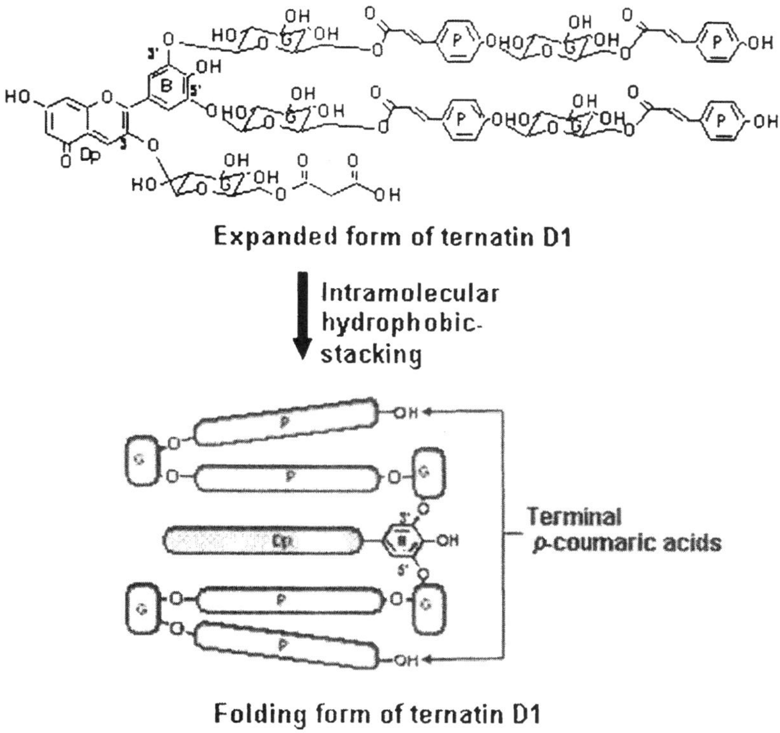

Figure 7. Conformation of ternatin D1 in aqueous solution (P, p-Coumaric acid; G, D-Glucose; Dp, delphinidin).

3. AOA-enhancing structural factors were as follows:
   a. AR acylation enhanced the activity more than corresponding deacylated (nonacylated) ANs.
   b. Catechol structures in B-ring of aglycons like cyanidin, and acylating AR like Caf powerfully enhanced the activities.
4. In case of polyacylated anthocyanins with three or more ARs, only terminal acids in side chains enhanced the AOA.

References

1. Strack, D.; Wray, V. In *The Flavonoids - Advances in Research since 1986*; Harborne, J. B., Eds.; Chapman and Hall, London, 1994.
2. Anderson, O. M.; Jordheim, M. In *Flavonoids: Chemistry, Biochemistry and Applications*; Andersen, O. M., Markham, K. R., Eds.; CRC Press: Boca Raton, FL, 2004.

3. Konczak, I.; Terahara, N.; Yoshimoto, M.; Nakatani, M.; Yoshinaga, M.; Yamakawa, O. *Trends Food Sci. Technol.* **2005**, *16*, 377-388.
4. Wu, X.; Beecher, G. R.; Holden, J. M.; Haytowitz, D. B.; Gebhardt, S. E.; Prior, R.L. *J. Agric. Food Chem,* **2006**, *54*, 4069-4075.
5. Mazza, G.; Miniati. E. *Anthocyanins in Fruits, Vegetables, and Grains*, CRC Press: Boca Raton, FL, 1993.
6. Yoshida, K.; Kondo, T.; Kameda, K.; Goto, T. *Agric. Biol. Chem.* **1990**, *54*, 1745-1751.
7. Kondo, T.; Kawai, T.; Tamura, H.; Goto, T. *Tetrahedron Lett.* **1987**, *28*, 2273-2276.
8. Goto, T. *Prog. Chem. Organ. Nat. Prod.* **1987**, *52*, 113-158.
9. Terahara, N. In *Proceedings of the 3rd International Congress & Symposium on Natural Colorants III*, Princeton, NJ, 1998.
10. Brouillard, R. In *Anthocyanins as Food Colors*; Markakis, P., Eds.; Academic Press: New York, NY, 1982, pp.1-40.
11. Goto, T.; Kondo, T. *Angew. Chem. Int. Ed. Engl.* **1991**, *30*, 17-33.
12. Figueiredo, P.; Elhabiri, M.; Toki, K.; Saito, N.; Dangles, O.; Brouillard, R. *Phytochemistry* **1996**, *41*, 301-308.
13. Yui, T.; Kawano, Y.; Terahara, N.; Ogawa, K. In *The 9th Blatislava Symposium on Saccharides, Smolenic Slovakia, 2000.*
14. Kahkonen, M. P.; Heinonen, M. *J. Agric. Food Chem.* **2003**, *51*, 628-633.
15. Wang, H.; Cao, G.; Prior, R. L. *J. Agric. Food Chem.* **1997**, *45*, 304-309.
16. Oki, T.; Masuda, M.; Furuta, S.; Nishiba, N.; Terahara, N.; Suda, I. *J. Food Sci.* **2002**, *67*, 1752–1756.
17. Philpott, M.; Gould, K. S.; Lim, C.; Ferguson, L. R. *J. Agric. Food Chem.* **2004**, *52*, 1511–1513.
18. Kong, J. M.; Chia, L. S.; Goh, N. K.; Chia, T. F.; Brouillard, R. *Phytochemistry* **2003**, *64*, 923-933.
19. Cho, J.; Kang, J. S.; Long, P. H.; Jing, J.; Back, Y.; Chung, K. S. *Arch. Pharm. Res.* **2003**, *26*, 821-825.
20. Knox, Y.; Hayashi, K.; Suzutani, T.; Ogasawara, M.; Yoshida, I.; Shiina, R.; Tsukui, A.; Terahara, N.; Azuma, M. *Acta virologica* **2001**, *45*, 209-215.
21. Hou, D. X. *Curr. Mol. Med.* **2003**, *3*, 149-159.
22. Hou, D. X.; Tong, X.; Terahara, N.; Luo, D.; Fujii, M. *Arch. Biochem. Biophys.* **2005**, *440*, 101–109.
23. Tsuda, T.; Horio, F.; Uchida, K.; Aoki, H.; Osawa, T. *J. Nutr.* **2003**, *133*, 2125-2130.
24. Serraino, I.; Dugo, L.; Dugo, P.; Mondello, L.; Mazzon, E.; Dugo, G.; Caputi, A. P.; Cuzzocrea, S. *Life Sci.* **2003**, *73*, 1097-1114.
25. Rossi, A.; Serraino, I.; Dugo, P.; Di Paola, R.; Mondello, L.; Genovese, T.; Morabito, D.; Dugo, G.; Sautebin, L.; Caputi, A. P.; Cuzzocrea, S. *Free Radical Res.* **2003**, *37*, 891-900.
26. Malik, M.; Zhao, C.; Schoene, N.; Guisti, M. M.; Moyer, M. P.; Magnuson, B. A. *Nutr. Cancer* **2003**, *46*, 186-196.
27. Galvano, F.; La Fauci, L.; Lazzarino, G.; Fogliano, V.; Ritieni, A.; Ciappellano, S.; Battistini, N. C.; Tavazzi, B.; Galvano, G. *J. Nutr. Biochem.* **2004**, *15*, 2-11.

28. Marko, D.; Puppel, N.; Tjaden, Z.; Jakobs, S.; Pahlke, G. *Mol. Nutr. Food Res.* **2004**, *48*, 318-325.
29. Matsui, T.; Ueda, T.; Oki, T.; Sugita, K.; Terahara, N.; Matsumoto, K. *J. Agric. Food Chem.* **2001**, *49*, 1948-1951.
30. Matsui, T.; Ueda, T.; Oki.; T, Sugita, K.; Terahara, N.; Matsumoto, K. *J. Agric. Food Chem.* **2001**, *49*, 1952-1956.
31. Matsui, T.; Ebuchi, S.; Kobayashi, M.; Fukui, K.; Sugita, K.; Terahara, N.; Matsumoto, K. *J. Agric. Food Chem.* **2002**, *50*, 7244-7248.
32. Bassa, I. A.; Francis, F. J. *J. Food Sci.* **1987**, *52*, 1753-1754.
33. Seaforth, C. E.; Williams, D. B. *Proc. Am. Soc. Hort. Sci.* **1966**, *88*, 418-485.
34. Shi, Z.; Bassa, I. A.; Gabriel, S.L.; Francis, F. J. *J. Food Sci.*, **1992**, *57*, 755-757.
35. Terahara, N.; Kato, Y.; Nakamura, M.; Maitani, T.; Yamaguchi, M.; Goda, Y. *Biosci. Biotech. Biochem.* **1999**, *63*, 1420-1424.
36. Terahara, N.; Saito, N.; Honda, T.; Toki, K.; Osajima, Y. *Phytochemistry* **1990**, *29*, 949-953.
37. Yamaguchi, T.; Takamura, H.; Matoba, T. ; Terao, J. *Biosci. Biotechnol. Biochem.*, **1998**, *62*, 1201-1204.
38. Igarashi, K.; Takanashi, K.; Makino, M.; Yasui, T. *Nippon Shokuhin Kogyo Gakkaishi.*, **1989**, *36*, 852-856.
39. Srivastava, B. K.; Pande, C. S. *Clitoria ternatea. Planta Med.*, **1977**, *32*, 138-140.
40. Odake, K.; Terahara, N.; Saito, N.; Toki, K.; Honda, T. *Phytochemistry* **1992**, *31*, 2127-2130.
41. Kondo, T.; Ueda, M.; Goto, T. *Tetrahedron* **1990**, *46*, 4749-4756.
42. Terahara, N.; Saito, N.; Honda, T.; Toki, K.; Osajima, Y. *Tetrahedron Lett.* **1990**, *31*, 2921-2924.
43. Bors, W.; Michel, C.; Stettmaier, K.; Lu, Y.; Foo, L. Y. *Biol. Res.* **2004**, *37*, 301-311.
44. Goto, T.; Kondo, T.; Kawai, T.; Tamura, H. *Tetrahedron Lett.* **1984**, *25*, 6021-6024.
45. Idaka, E.; Ohashi, Y.; Ogawa, T.; Kondo, T.; Goto, T. *Tetrahedron Lett.* **1987**, *28*, 1901-1904.
46. Kondo, T.; Yamashiki, J.; Kawahori, K.; Goto, T. *Tetrahedron Lett.* **1989**, *30*, 6055-6058.
47. Yoshida, K.; Kondo, T.; Kameda, K.; Kawakishi, S.; Lubag, A. J. M.; Mendoza, E. M. T.; Goto, T. *Tetrahedron Lett.* **1991**, *32*, 5575-5578.

Chapter 10

A Possible Mechanism That Flavonoids Exert Anticarcinogenesis with Activation of β-Glucuronidase in Cancerous Tissues

Naomi Oi, Takashi Hashimoto, and Kazuki Kanazawa

Department of Biosystems Science, Graduate School of Science and Technology, Kobe University, 1–1 Rokkodai, Nada, Kobe 657–8501, Japan

Flavonoids are immediately metabolized into their inactive forms as conjugates with glucuronic acid and/or sulfate during intestinal absorption. To exercise biological activity in our body, conjugated flavonoids should be de-conjugated into their aglycons. In the present study, the activity of β-glucuronidase, which can de-conjugate flavonoid glucuronides into their aglycons, was compared between normal and hepatocarcinogenic Fisher 344 rats induced by *N*-diethyl-nitrosamine and phenobarbital. In the liver of the hepatocarcinogenic rats, the β-glucronidase activity significantly increased compared to that in the normal rats. On the other hand, the activity in the kidney, lung, heart, thymus and plasma only slightly changed. These results suggest that β-glucuronidase is specifically activated in the inflammatory tissues such as cancerous tissues. This is a possible mechanism that flavonoids are able to exert biological activities in the carcinogenic tissue of our body.

Flavonoids occur abundantly in plant foods such as vegetables, fruits and tea. Flavonoids, *in vitro*, render strong biological effects including antioxidative, anti-inflammatory and anti-carcinogenic properties (*1-3*). Many epidemiological studies also suggest that the daily intake of flavonoids is inversely associated with the risk of certain cancers and coronary heart diseases (*4-6*). However, ingested flavonoids are immediately metabolized and conjugated with glucuronic acid and/or sulfate during intestinal absorption (*7, 8*). It is well-known that conjugated flavonoids have low biological activities compared with the aglycons due to their hydrophilicity and molecular size (*9*). Thus, the metabolism and bioavailability of flavonoids contradict the epidemiological studies (*4-8*).

On the other hand, β-glucuronidase, an exo-glycosidase, hydrolyzes glucuronide-conjugated flavonoids into their aglycons (*10*). Recent reports (*11, 12*) have indicated the activation of β-glucuronidase in several cancerous tissues. These reports suggest that flavonoid glucuronides would be de-conjugated into the active form in cancerous tissues and prevent cancer development. However, it is not known whether these events are specific to cancerous tissues and not to normal tissues.

In the present study, Fisher 344 rats were induced the early stage of carcinogenesis using a two-stage hepatocarcinogenesis model by *N*-diethylnitrosamine (DEN) and phenobarbital (PB). The β-glucuronidase activity in the liver and other tissues was compared between the normal and hepatocarcinogenic rats.

Materials and Methods

Animal Treatments

This study was approved by the Institutional Animal Care and Use Committee (Permission number: 17-03-02) and carried out according to the Guidelines of Animal Experimentation of Kobe University. Male Fisher 344 rats (5 weeks old; Clea Japan, Tokyo, Japan) were acclimatized for 1 week. They were housed in an animal facility maintained on a 12 h light/dark cycle at a constant temperature of 23 ± 1 °C. They were given free access to diet (Oriental MF diet, Oriental Yeast, Tokyo, Japan) and drinking water. As shown in Figure 1, 10 rats were divided into 2 groups; *i.e.*, the Control and DEN/PB groups. The DEN/PB group was intraperitoneally injected with DEN (100 mg/kg body weight) dissolved in saline once a week for 3 weeks. One week after the 3rd injection, the rats received 500 ppm of PB in their drinking water for 10 weeks, and then killed. The Control group was injected with saline instead of DEN, and given drinking water.

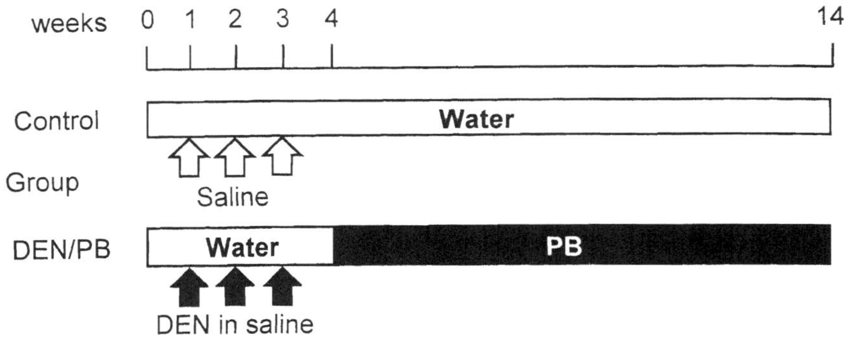

Figure 1. Animal experimental schedule. *Fisher 344 rats were intraperitoneally injected with DEN (100 mg/kg body weight), dissolved in saline followed by receiving PB (500 ppm), which is contained in drinking water after 1 week interval. The Control group was injected saline and given drinking water. At week 14, all rats were killed.*

Immunohistochemical Staining of Glutathione *S*-Transferase Placental Form (GST-P)

The liver fixed with paraformaldehyde was embedded in an OCT (optimal cutting temperature) compound (Sakura Finetech, Tokyo, Japan) and sectioned into 40 μm by acryostat. The sections were treated 3 times for 30 min with 0.6% hydrogen peroxide in 0.1 M PBS containing 0.1% Triton X-100 (PBST, pH 7.5) and then washed with PBST. Nonspecific protein binding was blocked by 5% normal goat serum (NGS; Chemicon International, Temecula, CA, USA) in PBST for 1h. The sections were incubated with rabbit anti-rat GST-P primary antibody (Medical and Biological Laboratories, Tokyo, Japan) diluted to 1:50 in PBST containing 5% NGS at 4°C overnight. They were incubated with goat anti-rabbit immunoglobulins conjugated peroxidase labelled-dextran polymer in Tris-HCl buffer (EnVision Plus, Dako, Kyoto, Japan) for 30 min, washed with 0.1 M Tris-HCl (pH 7.5) and then reacted with 0.05% 3,3'-diaminobenzidine (DAB), 0.01% hydrogen peroxide, and 0.08% ammonium nickel sulfate in 0.1M Tris-HCl (pH 7.5) for 4 min. The immunostained sections mounted on slides were dehydrated through a graded series of ethanol, cleared by xylene, and then coverslipped with an embedding compound. The DAB-stained GST-P positive foci were observed by light microscopy. Quantitative analysis of the GST-P positive foci was performed using NIH image version 1.61. The number and areas of GST-P positive foci >0.2 mm^2 in diameter were measured.

Measurement of β-Glucuronidase Activity

The β-glucuronidase activity was measured as described previously (*13*). Briefly, the liver, kidney, lung, heart and thymus were homogenized in 20 mM Tris-HCl (pH 7.4) at 4°C. The protein contents in the homogenates were determined according to the Lowry method (14). The homogenates or plasma were incubated with 50 μL of assay buffer (200 mM sodium acetate, pH 5.0 containing 10 mM EDTA, 0.01% bovine serum albumin, 0.1% Triton X-100 and 0.5 mM 4-mathylumbelliferyl-β-D-glucuronide (MUG) as a substrate) at 37°C. The enzymatic reaction was stopped by adding 150 μL of 200 mM sodium carbonate. The incubation mixture was subsequently added to 5 μL of 0.5 mM 9-chloromethylanthracecene (Tokyo Kasei Kogyo, Tokyo, Japan) as the internal standard, and centrifuged for 5 min at 15000 rpm. 4-Methylunbelliferone (MU), a hydrolyzed product, in the supernatants was analyzed by HPLC with a fluorescence detector (excitation at 355 nm, emission at 460 nm). The activity was expressed as MU nmols released from MUG.

Results and Discussion

The expression of GST-P, a liver-specific pre-cancer lesion, was measured by immunohistochemistry (Figure 2). The GST-P positive foci were not observed in the control group (Figure 2A). In contrast, the foci were significantly observed in the DEN/PB group; *i.e.*, approximately 622 foci per cm^2 were detected, and the area was approximately 2.11 mm^2/cm^2 in the tissues (Figure 2B). This result indicates that this experimental model obviously induced the early stage of hepatocarcinogenesis in the DEN/PB group rats.

To investigate whether the β-glucuronidase specifically increased in cancerous tissues, the activity of β-glucuronidase was compared between the livers of the Control and DEN/PB groups (Table I). The β-glucronidase in the GST-P-expressed liver (the DEN/PB group) significantly increased by 1.5-fold compared to the Control group; *i.e.*, 640 ± 105 (MU nmols/mg protein/h) in the DEN/PB group, and 430 ± 29 (MU nmols/mg protein/h) in the Control group. On the other hand, the enzymatic activity of the other tissues, kidney, lung, heart, thymus and plasma was no significant differences between both groups.

These results indicated that β-glucuronidase would be specifically activated in the early stage of carcinogenic tissues. There is a possibility that flavonoid aglycons increase by de-conjugation with activated β-glucuronidase in the carcinogenic tissues and prevent cancer development. To elucidate the contradiction between the metabolism and bioavailability of flavonoids and the epidemiologic studies, further studies are required; *e.g.*, a comparative study of

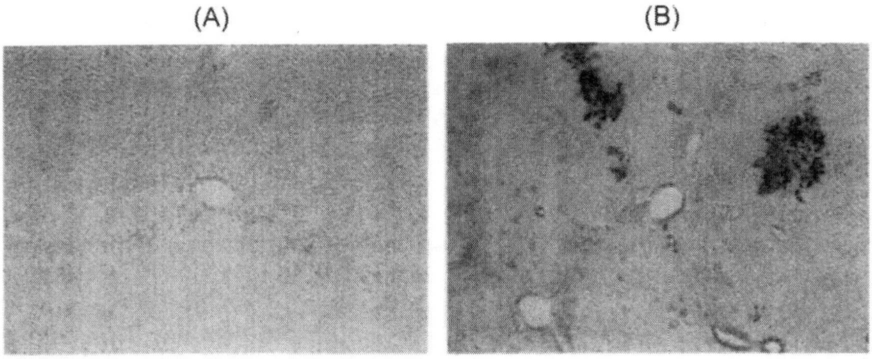

Figure 2. Expression of GST-P in liver. *Fisher 344 rats were treated according to the experimental schedule shown in Figure 1. The GST-P positive foci in the liver were stained with DAB as described in the Materials and Methods. The back blots were DAB-stained GST-P positive foci. The data are representative. Original magnification ×210.*

Table I. Comparison of β-Glucuronidase Activity in Tissues and Plasma

	β-Glucuronidase					
Group	Liver[a)]	Kidney[a)]	Lung	Heart[a)]	Thymus[a)]	Plasma[b)]
Control	430 ± 29	149 ± 14	118 ± 9	14.3 ± 1.1	141 ± 20	30.6 ± 5.8
DEN/P	640 ± 105*	158 ± 5	121 ± 7	15.8 ± 1.9	134 ± 9	31.3 ± 6.9

[a)] The values are mean±SD. (MU nmols/mg protein/h)
[b)] The values are mean±SD. (MU nmols/ml)
* Significantly different from the Control group ($P<0.05$) as determined by the Student's t test.

the concentration of the aglycons form in these tissues as well as the kinetic study of the activation of β-glucronidase, among others.

References

1. Nijveldt, R. J.; van Nood, E.; van Hoorn, D. E.; Boelens, P. G.; van Norren, K.; van Leeuwen, P. A. *Am. J. Clin. Nutr.* **2001**, *74*, 418-425.
2. Kimata, M.; Shichijo, M.; Miura, T.; Serizawa, I.; Inagaki, N.; Nagai, H. *Clin. Exp. Allergy* **2000**, *30*, 501-508.
3. Granado-Serrano, A. B.; Martin, M. A.; Bravo, L.; Goya, L.; Ramos, S. *J. Nutr.* **2006**, *136*, 2715-2721.
4. Arts, I. C.; Hollman, P. C. *Am. J. Clin. Nutr.* **2005**, *81*, 317S-325S.
5. Knekt, P.; Jarvinen, R.; Seppanen, R.; Hellovaara, M.; Teppo, L.; Pukkala, E.; Aromaa, A. *Am. J. Epidemiol.* **1997**, *146*, 223-230.
6. Knekt, P.; Kumpulainen, J.; Jarvinen, R.; Rissanen, H.; Heliovaara, M.; Reunanen, A.; Hakulinen, T.; Aromaa, A. *Am. J. Clin. Nutr.* **2002**, *76*, 560-568.
7. Murota, K.; Terao, J. *Arch. Biochem. Biophys.* **2003**, *417*, 12-17.
8. van der, Woude, H.; Boersma, M. G.; Vervoort, J.; Rietjens, I. M. *Chem. Res. Toxicol.* **2004**, *17*, 1520-1530.
9. Spencer, J. P.; Rice-Evans, C.; Williams, R. J. *J. Biol. Chem.* **2003**, *278*, 34783-34793.
10. O'Leary, K. A.; Day, A. J.; Needs, P. W.; Sly, W. S.; O'Brien, N. M.; Williamson, G. *FEBS Lett.* **2001**, *503*, 103-106.
11. Sperker, B.; Werner, U.; Murdter, T. E.; Tekkaya, C.; Fritz, P.; Wacke, R.; Adam, U.; Gerken, M.; Drewelow, B.; Kroemer, H. K. *Naunyn-Schmiedeberg's Arch. Pharmacol.* **2000**, *362*, 110-115.
12. Devasena, T.; Menon, V. P. *Phytother. Res.* **2003**, *17*, 1088-1091.
13. Sperker, B.; Schick, M.; Kroemer, H. K. *J. Chromatogr. B* **1996**, *685*, 181-184.
14. Lowry, O. H.; Rosebrough, N. J.; Farr, A. L.; Randall, R. J. *J. Biol. Chem.* **1951**, *193*, 265-275.

Cereal

Chapter 11

Phenolic Content and Antioxidant Activity of Whole-Wheat Grain and Its Components

Fereidoon Shahidi[1,2] and Chandrika Liyana-Pathirana[2]

Departments of [1]Biochemistry and [2]Biology, Memorial University of Newfoundland, St. John's, Newfoundland A1B 3X9, Canada

Grains, including wheat, are staple foods for many populations around the globe. In addition to providing carbohydrates and proteins, they also serve as a good source of phytochemicals including phenolics. Phenolics present are concentrated in the outermost layers of the grains and are present as free, esterified and insoluble-bound forms. Release of bound phenolics is important when evaluating the phenolic contents of grains. The phytochemicals present render a myriad of health benefits and may alleviate oxidative stress and other diseases in humans. Data provided support the general belief that whole grains are superior to processed products although the initial recommendation was primarily based on fiber related evidences.

Grains provide approximately two-thirds of the energy and protein intake in the world, especially in the developing countries (*1*). The most commonly consumed cereal grains are wheat, rice and maize while barley, sorghum, millet, oats, and rye are more common only in some countries than others depending on climate and cultural differences (*2*). According to Slavin (*3*) wheat accounts for one-third of the total worldwide grain production, while rice accounts for one-quarter. Cereal grains provide an excellent source of carbohydrate, dietary fiber and protein as well as a good source of B-group vitamins, vitamin E and a number of minerals such as iron, zinc, magnesium, and phosphorous. Moreover, whole grains contain a large number of identified phytochemicals such as phytoestrogens and phenolics, among others. All these components may act together in a synergistic fashion and may be protective against gastrointestinal cancers and cardiovascular diseases (*4*).

In general, grains are subjected to a number of processing steps such as milling, heat extrusion, cooking, and parboiling, among others (*1*). Moreover, commercial cereals are usually extruded, puffed, flaked, or altered in some manner to make consumer desirable products. During milling process, the bran and germ are separated from the starchy endosperm and the latter is ground to form flour. Nutrients and phytochemicals, in general, are asymmetrically distributed in the grain, with a higher concentration existing in the outer part of the grain (*5,6*). The refining of the grain results in reduced nutrient and phytochemical content (*5-7*). Eventually the refined grains have a higher starch content compared to whole grains.

Phenolic acids, particularly hydroxycinnamic acids (HCA), are found in wheat (*5*). Vanillic, *p*-coumaric, ferulic, and sinapic acids were predominant phenolic acids in whole grain of wheat and its processing by- products (*7*). The more refined the flour the lower would be its total phenolic content. Grains are thought to be rich sources of phenolic acids which are located in the bran layer particularly in the bound form compared to free and esterified forms (*7*).

Processing of Grains

In the past, grains have been consumed as 'whole grains' where the grains were crushed and milled to produce coarse flours. This process involved poor and incomplete separation of the bran and germ from the endosperm (*8*). However, as a consequence of the development of the roller mill separation there has been a dramatic increase in the consumption of refined grains especially in the Western industrialized countries (*8*). Most whole and refined grains are subjected to some type of processing to make a desirable product with optimum flavor, color, texture, and appearance that are shelf-stable (*3*). Whole grains are composed of three major parts namely the bran, the germ, and the endosperm. The relative amounts of these constituents differ from one species to another. Thus, bran contents of brown rice, corn, and wheat are approximately 30, 60, and 160 mg/g, respectively (*8*). Bran and germ, that contain many nutrients and non-nutrient inorganic and organic components, are lost during the refining process (*9*). According to the definition of American Association of Cereal

Chemists (*10*) whole grains shall consist of the intact, ground, cracked or flaked caryopsis whose principal anatomical constituents namely the endosperm, germ and bran are present in the same relative proportions as they exist in the intact caryopsis. In modern milling the individual components of the grain are initially separated followed by reconstituting to reform the wholegrain flour (*8*).

Cereals deliver many potentially anticarcinogenic agents predominantly existing in the aleurone layer of the grain. However, modern milling techniques usually eliminate the aleurone layer as this is tightly bound to the bran of the grain (*11*). In the refining process bran and germ are separated from the starchy endosperm thereby removing important disease preventing nutrients and phytochemicals such as lignans, tocotrienols, and phenolic compounds, and anti-nutrient constituents such as phytic acid, tannins, and enzyme inhibitors (*4*).

Slavin *et al.* (*12*) reported that milled wholegrains can be nutritionally superior to intact wholegrains for human consumption due to the removal of poorly digestable components during milling thereby enhancing bioavailability of nutrients. Refined grains that lack bran and germ have reduced nutrient content compared to wholegrains as the milling process results in the removal of dietary fiber, vitamins, minerals, lignans, phytoestrogens, phenolic compounds, and phytic acid to varying degrees (*13*). In general, the antioxidant potential and bioavailability of grain antioxidants depend on the species and varieties of grains, fractions of the grain (bran, germ, flour, wholegrain), and processing conditions (*5-7,14*).

The effects of primary processing, namely milling and pearling, on the antioxidant activity of wheat phenolics have been reported (*5,6*). According to Liyana-Pathirana *et al.* (*5*) wheat grains were "debranned" or "pearled" to various levels (10-50% in 10% increments) starting from the whole unprocessed grain. The processed grains and their by-products were collected separately at each level of pearling and their total phenolic content and antioxidant activity determined. The 10% by-product from pearling process possessed the highest total phenolic content (Table I). Subsequent removal of bran or external layers from the grain resulted in a decrease in total phenolic content. Among grain products, the unprocessed whole grain possessed the highest total phenolic content. A dilution of antioxidative constituents by starchy endosperm was monitored, as pearling removed the external layers.

We have also determined the effect of milling on the total phenolic content and antioxidant activity of wheat cultivars (*6*). Different milling fractions namely bran, flour, shorts, feed flour, and semolina were employed in the analyses. Among different milling fractions bran demonstrated the highest total phenolic content while endosperm possessed the lowest (Table II). The shorts and feed flour may contain different proportions of bran and germ and their total phenolic content may vary depending on the actual proportion of these fractions.

Table 1. Total Phenolic Content[a] of Whole Grains, Pearled Grains and Their By-products of Two Wheat Cultivars CWAD[b] and CWHRS[c]

Degree of pearling	*CWAD*		*CWRS*	
	Grain	*By-product*	*Grain*	*By-product*
0[d]	12.4	-	23.9	-
10	10.1	26.5	19.5	32.7
20	6.8	17.4	16.7	20.7
30	6.8	9.3	13.5	17.8
40	5.8	8.1	13.2	15.7
50	5.6	7.2	12.0	11.8

[a]μg ferulic acid equivalents/g defatted material; [b]Canadian Western Amber Durum; [c]Canadian Western Red Spring; [d]whole unprocessed grain

Table II. Total Phenolic Content[a] of Milling Fractions of Two Wheat Cultivars CWAD[b] and CWRS[c]

Milling fraction	*CWAD*	*CWRS*
Whole grains	769	1291
Bran	2279	3437
Flour	210	216
Shorts[d]	192	3146
Feed flour[e]	1404	2033
Semolina[e]	140	-

[a]μg ferulic acid equivalents/g defatted material; [b]Canadian Western Amber Durum; [c]Canadian Western Red Spring; [d] a mixture of bran, endosperm and germ; [e] a mixture of bran and low grade endosperm; [f]coarse flour prepared from durum wheat

Bound Phenolics and Effect of Hydrolysis on Their Solubility

Cereals are a rich source of low-molecular-weight compounds such as ferulic, caffeic, *p*-hydroxybenzoic, protocatechuic, *p*-coumaric, gentisic, sinapic, vanillic, and syringic acids, among others (*15,16*). Phenolic acids may form both ester and ether linkages owing to their bifunctional nature through reactions involving their carboxylic and hydroxyl groups, respectively. This allows phenolic acids to form cross-links with cell wall macromolecules (*17*). Bound phenolics may be released by alkali, acid or enzymatic treatment of samples prior to extraction (*18-23*). In barley, most phenolic acids exist in the bound form with other grain components such as starch, cellulose, β-glucan and pentosans (*17*). A similar observation was made for phenolic acids in ground rye grain that were released upon enzymatic hydrolysis (*24*). With respect to the histological distribution of arabinoxylans, their concentration increases from the centre to the periphery of the endosperm. Thus, they are mainly concentrated in the seed coats of cereals (*25*). The increased concentration of ferulates in the outer layers may be implicated in resistance to both insect and fungal pathogens. Thus, cross linking of phenolic compounds may provide a physical barrier to invasive disease development and consumption by insects (*22*).

Cereal grains are one of the most important food groups and their fibers are known to render health benefits that may be attributed to the nature of cell wall polymers and chemical architecture (*26*). Epidemiological data have shown that consumption of wheat bran is associated with a reduced risk of colorectal and gastric cancer (*27*). Cereals contain a high amount of HCA that may exert potential health benefits (*21*). In cereals HCA and their dimers exist mainly as esters bound to arabinoxylan. In many studies phenolic antioxidants of wheat have been tested in their free form only (*28-31*). In order to understand the total antioxidant activity of cereals, it is important to consider the contribution from their bound phenolics since under normal conditions phenolics occur mainly in the bound form (*18-23*). Ferulic acid is the major phenolic acid in many cereals and exists predominantly in the seed coat (*32*) while traces may be present in the starchy endosperm (*33*).

Liyana-Pathirana and Shahidi (*7*) reported the importance of insoluble-bound phenolics to antioxidant properties of wheat. The soluble-esters of phenolics and insoluble phenolics after alkaline hydrolysis of samples were extracted. The contribution of bound phenolics to total phenolic content was significantly higher than that of free and esterified fractions; phenolic constituents were primarily located in the external bran tissues (Figure 1). In order to take full advantage of potential health benefits of wheat it is essential that it is subjected to minimal or no processing that allows the retention of external bran layers.

The commonly existing *trans*-ferulic acid (4-hydroxy-3-methoxycinnamic acid) and *trans-p*-coumaric acid (4-hydroxycinnamic acid) are predominantly esterified to hemicellulose via covalent links to arabinofuranose in the heteroxylans (*34*). In general, digestion of cell wall materials is rather difficult and hence may survive gastrointestinal digestion to reach the colon. However, it has been shown that the colonic fermentation of such material may lead to the release of some of the bound phenolics and, hence, may exert their unique health benefits in the colon following absorption (*35*). In the gut, the action of microbial esterases may lead to the release of monomeric hydroxycinnamtes and their absorption into the circulatory system has been demonstrated in humans (*35*). Therefore, bound phenolic constituents may have a significant effect on

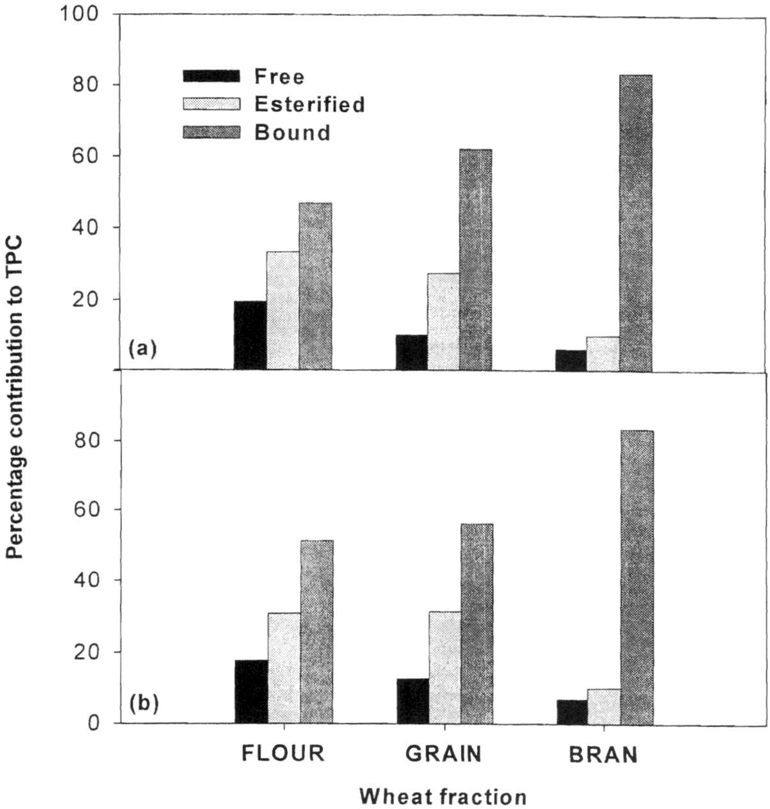

Figure 1. Percentage of contribution of free, esterified and bound phenolics to total phenolic content of (a) soft and (b) hard wheat

human health following digestion (*36*). Grains are known to possess unique phytochemicals, including ferulic and caffeic acids in a more concentrated form, that are complimentary to those present in fruits and vegetables (*37*). Liyana-Pathirana and Shahidi (*38*) subjected wheat samples to simulated gastrointestinal pH treatment prior to extraction of phenolics from the samples. The low pH improved the extractability of phenolics from wheat and hence suggested that simulated gastrointestinal conditions may solubilize bound phenolics. There was a significant increase in the content of total phenolics when the samples of both soft and hard wheat were subjected to simulated gastrointestinal pH treatment (Table III).

Table III. Total Phenolic Content of Soft and Hard Wheat As Such and After Subjected to Simulated Gastric Conditions

Milling fraction	*Soft wheat*		*Hard wheat*	
	Non-treated[a]	Treated[b]	Non-treated	Treated
Whole grain	62	294	81	252
Flour	22	74	23	118
Germ	356	850	251	699
Flour	181	959	243	574

[a]Samples subjected to pH treatment; [b]Samples as such

Antioxidant Activity of Whole Grains

Reactive oxygen species (ROS) and/or free radicals are generated as by-products or intermediates during normal metabolism (*39*). For the maintenance of normal physiological functions for human health a balance between formation and removal of ROS is essential. Oxidative stress results when the balance between the production of ROS overrides the antioxidant capacity of cells thereby causing damage to nucleic acids, proteins, and membrane lipids leading to several health problems including cancer and heart disease (*40,41*). Antioxidants are compounds that delay the onset or slow down the rate of oxidation of oxidizable substrates and reduce risks of these diseases or promote general human health (*3*). Antioxidants include many molecules such as minerals, vitamins, proteins, carbohydrates, and polyphenolics that are capable of donating an electron(s) or a hydrogen atom to a prooxidant (42). Avoidance of oxidative stress is important in protecting the body against the consequences of oxidative damage to biomolecules (*43*).

Recently, phenolic acids have gained attention due to their antioxidative, antiinflammatory, antimutagenic and anticancer properties as well as their ability to modulate some key enzymatic functions in the cell (*44*). The antioxidant activity of phenolic compounds depends largely on their bioavailability which is influenced by several factors. Thus, release of phenolics from the food matrix in the gut, stability of phenolics in the gut flora, their modification on the intestine and absorption into circulatory system through intestinal wall, and stability in the liver and accessibility to the tissue at the target site are some of the important factors affecting bioavailability of antioxidative components (*45,46*). Cereals and cereal-based foods may be an important vehicle to deliver nutraceutical components to the consumer as such foods are consumed widely and consistently as staples in many parts of the world (*47*).

Hydroxycinnamates have been shown to be absorbed in humans in the gastrointestinal tract (*48,49*). The most abundant phenolic acids in rye include ferulic acid, sinapic acid and the dimer 8-*O*-4-diferulic acid that exert antioxidant activity against oxidation of low density lipoprotein (LDL) cholesterol under *in vitro* conditions. Olthof *et al.* (*50*) have found that chlorogenic acid is partially absorbed in humans and may have biological effects in blood circulation while the fraction not absorbed may bring about biological effects in the colon. Caffeic acid has been shown to be a promising natural antioxidant that contributes to the antioxidant defense system under *in vivo* conditions using a rat model. Hence, caffeic acid may either contribute directly to the antioxidant system or it may impart a sparing effect on α-tocopherol in LDL cholesterol (*51*). Madhujith and Shahidi (*52*) determined the antioxidant and antiproliferative properties of six different barley cultivars. These authors have suggested that inclusion of whole barley into daily diet may render beneficial health benefits as phenolic extracts from whole grain barley demonstrated high antioxidant, antiradical, and antiproliferative potentials. In the gut the action of microbial esterases may release monomeric HCA and their absorption into the circulatory system has been demonstrated in both rats (*49*) and humans (*35*). Since absorption of dietary phenolics in human has been demonstrated, phenolic compounds of whole grains may result in increased antioxidant capacity of plasma after ingestion. Hence, phytochemicals may constitute an important source of dietary antioxidants responsible for a myriad of health benefits (*53*).

In general, the total antioxidant capacity of plant-derived materials cannot be evaluated by any single method due to the complex nature of phytochemicals (*54*). Therefore, two or more methods should always be employed in order to evaluate total antioxidant activity. Moreover, despite the use of several methods in testing antioxidant properties, these methods cannot be used as such to measure the antioxidant activity of plant extracts due to their often complex composition (*55*). Measurement of antioxidant activity in a biological system is of paramount importance since antioxidant properties allow the prevention of oxidative stress and related diseases. Hence, it is possible to reach an opinion about the intake of dietary antioxidants and to evaluate their contribution in the antioxidant status of humans (*42*). In the case of a food system, measurement of antioxidant activity allows one to evaluate the total antioxidant capacity of the system and its significance in enhancing the health of an individual (*56*).

Numerous *in vitro* assays have been used in the measurement of the antioxidant potential of food and biological materials. Many chemical assays have been based on the ability to scavenge different free radicals while others may eliminate any source of oxidative initiation such as inhibition of an enzyme, chelation of metal ions and reducing power, among others. There are some chemical assays that evaluate lipid peroxidation, in which a lipid or lipoprotein substrate is used and the degree of inhibition of oxidation of the substrate is measured (*57*).

Liyana-Pathirana *et al.* (*5*) determined antioxidant properties of wheat as affected by pearling on the basis of oxygen radical absorbance capacity (ORAC), inhibition of oxidation of low density lipoprotein (LDL) cholesterol and deoxyribonucleic acid (DNA), among others (Table IV). The antioxidant capacity of both pearled grains and by-products decreased significantly with the increased degree of pearling. Among grains unprocessed whole grains demonstrated the highest antioxidant potential. Moreover, the by-products always exhibited higher antioxidant capacity compared to their pearled counterparts. The effects of milling on antioxidant capacity of wheat were also determined (*6*). The milling products bran, flour, shorts, and feed flour were employed in the analyses. Bran tissues demonstrated the highest antioxidant activity while endosperm possessed the lowest (Table V).

Table IV. Antioxidant Capacity of the Pearled Grains and By-products of Wheat Cultivar CWRS[a] as Evaluated by Different Analytical Methods

	ORAC[b]		*Inhibition of LDL oxidation[c]*		*Inhibition of oxidation of DNA[d]*	
Pearling degree	*Pearled grains*	*By-product*	*Pearled grains*	*By-product*	*Pearled grains*	*By-product*
0[e]	95		866		351	
10	87	207	843	1587	312	819
20	79	189	727	1510	285	699
30	73	163	658	1359	248	583
40	58	138	588	1057	190	481
50	41	115	517	898	156	391

[a]Canadian Western Red Spring; [b]Oxygen radical absorbance capacity (μmol Trolox equivalents/g defatted sample; [c]μg protein retained/g defatted sample; [d]μg DNA retained/g defatted sample; [e]unprocessed whole grain

Liyana-Pathirana and Shahidi (*7*) determined the antioxidant capacity of bound phenolics compared to free and esterified phenolics. The total antioxidant capacity of free, esterified and bound phenolics of wheat was determined using Trolox equivalents antioxidant capacity (TEAC) assay and expressed as μmol Trolox equivalents/g of defatted wheat. The percentage contribution of free, esterified and bound phenolics of wheat extracts to total antioxidant capacity is shown in Figure 2. The total antioxidant capacity of wheat fractions of both wheat samples revealed the significance of bound phenolics as >80% of total antioxidant capacity was due to bound phenolics. The contribution of free phenolics to total antioxidant capacity was less than 2% in all wheat fractions examined while the contribution of esterified phenolics was significantly higher compared to that of free phenolics, but significantly less than that of bound phenolics.

Table V. Antioxidant Capacity of Milling Fractions of Whaet Cultivar CWRS[a] as Evaluated by Different Analytical Methods

Milling fraction	*ORAC*[b]	*Inhibition of LDL oxidation*[c]	*Inhibition of DNA oxidation*[d]
Whole grains	100	834	395
Bran	301	1603	786
Flour	45	401	172
Shorts	246	1426	743
Feed flour	199	1123	608

[a]Canadian Western Red Spring; [b]Oxygen radical absorbance capacity (μmol Trolox equivalents/g defatted sample; [c]μg protein retained/g defatted sample; [d]μg DNA retained/g defatted sample

Whole Grains and Disease Prevention

Regular consumption of fruits, vegetables, and whole grains is known to reduce the risk of a number of chronic diseases. Hence, dietary modification by increasing the daily intake of plant foods may have a significant impact on preventing chronic diseases (*37*). Whole grains, in particular, provide a wide range of nutrients and biologically active constituents which reduce the incidence of various diseases (*12*). In this regard, wheat is a staple food for a majority of world's population and serves as a source of potentially health enhancing components such as dietary fiber, phenolics, tocopherols and carotenoids (*58*) if consumed as whole grains. Plant-derived phenolic compounds have received a great deal of interest recently due to their suggested health benefits (*59*). Plant foods, particularly whole grain cereals, have been associated with decreased risk of a number of diseases such as obesity, diabetes, coronary heart disease and some cancers (*4*). According to Slavin *et al.* (*60*) the proposed mechanism for the protective effects of wholegrains include the effects of phenolic compounds existing in the bran fraction. Epidemiological studies have shown an inverse relationship between consumption of wholegrains and incidence of chronic diseases such as diabetes (*61,62*), coronary heart disease (*63,64*), colon dysfunction (*65*), and some types of cancer (*27,63,66-69*). The health benefits have been attributed to the fiber and phytochemical contents of whole grains (*70*).

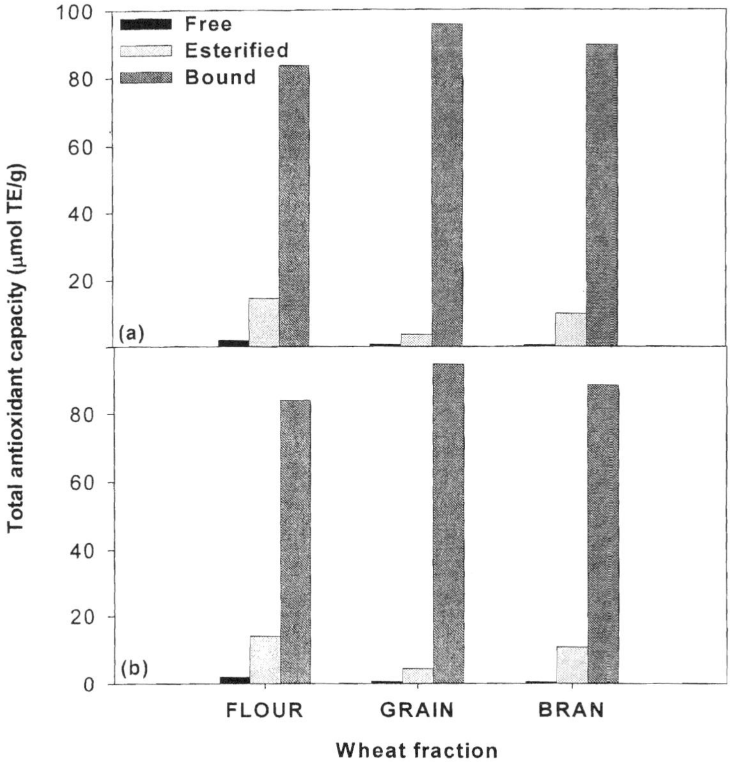

Figure 2. Total antioxidant capacity (μmol Trolox equivalents/g defatted sample) of free, esterified and bound phenolics of (a) hard and (b) soft wheat.

Rimm *et al.* (*71*) reported that the consumption of dietary fiber derived from cereals substantially reduces the risk of coronary heart diseases (CHD). According to Pietinen *et al.* (*72*) there is a significant inverse association between cereal fiber intake and mortality due to CHD. Andlauer and Fürst (*11*) reported that high intakes of whole grains are associated with a reduced risk of cancer particularly colorectal and gastric cancers of the alimentary tract. These authors discussed the anticarcinogenic properties of cereal fiber and antioxidant and demonstrated that wheat bran decreased fecal bile acid concentration while bran from oat and corn exhibited an opposite effect. Fecal bile acid concentration is one of the risk factors for colon carcinogenesis both in humans and animals (*11*).

Higher intake of whole grain foods is known to lower the risk of ischemic stroke among women, independent of known cardiovascular disease (CVD) risk factors (*73*). According to Fung *et al.* (*74*) a diet high in whole grains is associated with a reduced risk of type 2 diabetes in men. The health benefits have been attributed to cereal fiber and hence the authors suggested that efforts should always be made to replace refined grains with whole grain foods. Jensen *et al.* (*75*) found an inverse relationship between the intake of whole grains and bran with coronary heart disease risk in a prospective cohort study of male health professionals. Moreover, increased consumption of wholegrains improved insulin sensitivity and lowered serum insulin concentrations in humans (*63,76*). Anderson and Hanna (*76*) also stated that high intake of whole grains reduces the risk of coronary artery disease, in addition to reduced insulin resistance. According to Jang *et al.* (*77*) increased consumption of whole grains substantially reduces the risk factors of CAD and diabetes in CAD patients.

The exact mechanisms that explain how whole grains could contribute to disease prevention are not well known. However, the positive effects of whole grains may be due to their effects on gastrointestinal tract, antioxidant protection and intake of phytoestrogens (*4*). Whole grain foods also contain phytochemicals such as phenolic compounds that play important roles in disease prevention with their synergistic action with vitamins and minerals (*4*). The health benefits associated with high fiber diet may come from the other components and not just from fiber itself (*60*). Thus, phytochemicals such as phytoestrogens, antioxidants, and phenolic compounds which together with vitamins and minerals such as vitamin E and selenium, respectively, that play an important role in disease prevention (*4*).

References

1. Pedersen, B.; Knudsen, K.E.; Eggum, B.O. *World Rev. Nutr. Diet* **1989**, *60*, 1-91.
2. Southgate D. A. T. In *Human Nutrition and Dietetics*. J. S. Garrow, W. P. T. James, A. Ralph (eds), London, Churchill Livingstone, 2000, pp. 333-347.
3. Slavin, J.L. *Asia Pacific J. Clin. Nutr.* **2000**, *9*(Suppl.), S23-S27.
4. Slavin, J.L.; Jacobs, D.; Marquart, L.; Wiemer, K. *J. Am. Diet. Assoc.* **2001**, *101*, 780-785.
5. Liyana-Pathirana, C. M.; Dexter, J.; Shahidi, F. *J. Agric. Food Chem.* **2006**, *54*, 6177-6184.
6. Liyana-Pathirana, C. M.; Shahidi, F. *J. Cereal Sci.* **2007**, *45*, 238-247.
7. Liyana-Pathirana, C. M.; Shahidi, F. *J. Agric. Food Chem.* **2006**, *54*, 1256-1264.

8. Seal, C. J. *Proc. Nutr. Soc.* **2006**, *65*, 24-34.
9. Smith, A. T.; Kuznesof, S.; Richardson, D. P.; Seal, C. J. *Proc. Nutr. Soc.* **2003**, *62*, 455-467.
10. American Association of Cereal Chemists, Definition/Reports. http://www.aaccnet.org/definitions/wholegrain.asp (accessed August 2005)
11. Andlauer, W.; Fürst, P. *Cereal Foods World* **1999**, *44*(2), 76-78.
12. Slavin, J. L.; Jacobs, D.; Marquart, L. *Crit. Rev. Food Sci. Nutr.* **2000**, *40*, 309-326.
13. Slavin, J. *Nutr. Res. Rev.* **2004**, *17*, 99-110.
14. Zieliński, H.; Kozłowska, H. *J. Agric. Food Chem.* **2000**, *48*, 2008-2016.
15. Emmons, C. L.; Peterson, D. M.; Paul, G. L. *J. Agric. Food Chem.* **1999**, *47*, 4894-4898.
16. Kăhkŏnen, M. P.; Hopia, A. I.; Vuorela, H. J.; Rauha, J-P.; Pihlaja, K.; Kujala, T. S.; Heinonen, M. *J. Agric. Food Chem.* **1999**, *47*, 3954-3962.
17. Yu, J.; Vasanthan, T.; Tamelli, F. *J. Agric. Food Chem.* **2001**, *49*, 4352-4360.
18. Sosulski, F.; Krzysztof, K.; Lawrence, H. *J. Agric. Food Chem.* **1982**, *30*, 337-340.
19. Krygier, K.; Sosulski, F.; Hodge, L. *J. Agric. Food Chem.* **1982**, *30*, 330-334.
20. Krygier, K.; Sosulski, F.; Hogge, L. *J. Agric. Food Chem.* **1982**, *30*, 334-336.
21. Andreasen, M. F.; Landbo, A-K.; Christensen, L. P.; Hansen, A.; Meyer, A. S. *J. Agric. Food Chem.* **2001**, *49*, 4090-4096.
22. Zupfer, J. M.; Churchill, K. E.; Rasmusson, D. C.; Fulcher, R. G. *J. Agric. Food Chem.* **1998**, *46*, 1350-1354.
23. Bartolome, B.; Gomez-Cordoves, C. *J. Sci. Food Agric.* **1999**, *79*, 435-439.
24. Andreasen, M. F.; Christensen, L. P.; Meyer, A. S.; Hansen, A. *J. Sci. Food Agric.* **1999**, *79*, 411-413.
25. Lempereur, I.; Rouau, X.; Abecassis, J. *J. Cereal Sci.* **1997**, 25, 103-110.
26. Bunzel, M.; Ralph, J.; Kim, H., Lu, F.; Ralph, S. A.; Marita, J. M.; Hatfield, R. D.; Steinhart, H. *J. Agric. Food Chem.* **2003**, *51*, 1427-1434.
27. Jacobs, D. R.; Slavin, J.; Marquart, L. *Nutr. Cancer* **1995**, *24,* 221-299.
28. Yu, L.; Haley, S.; Perret, J.; Harris, M.; Wilson, J.; Qian, M. *J. Agric. Food Chem.* **2002**, *50*, 1619-1624
29. Yu, L.; Perret, J.; Harris, M.; Wilson, J.; Haley, S. *J. Agric. Food Chem.* **2003**, *51*, 1566-1570.
30. Zhou, K.; Laux, J. J.; Yu, L. *J. Agric. Food Chem.* **2004**, *52*, 1118-1123.
31. Zhou, K.; Yu, L. J. Agric. Food Chem. **2004**, *52*, 1112-1117.
32. Watanabe, M.; Ohshita, Y.; Tsushida, T. *J. Agric. Food Chem.* **1997**, *45*, 1039-1044.
33. Pussayanawin, V.; Wetzel, D. L.; Fulcher, R. G. *J. Agric. Food Chem.* **1988**, *36*, 515-520.

34. Mueller-Harvey, I.; Harley, R. D.; Harris, P. J.; Curzon, E. H. *Carbohyd. Res.* **1986**, *148*, 71-85.
35. Kroon, P. A.; Faulds, C. B.; Ryden, P.; Robertson, J. A.; Williamson, G. () *J. Agric. Food Chem.* **1997**, *45*, 661-667.
36. Miller, H. E.; Rigelhof, F.; Marquart, L.; Prakash, A.; Kanter, M. *J. Am. Coll. Nutr.* **2000**, *19*, 312S-319S.
37. Dewanto, V.; Wu, X.; Liu, R. H. *J. Agric. Food Chem.* **2002**, *50*, 4959-4964.
38. Liyana-Pathirana C. M.; Shahidi, F. *J. Agric. Food Chem.* **2005**, *53*, 2433-2440.
39. Halliwell, B.; Gutteridge, J. M. C. *Free Radicals in Biology and Medicine.* Clarendon Press, Oxford, UK., 1989.
40. Halliwell, B.; Gutteridge, J. M. C.; Cross, C. E. *J. Lab. Clin. Med.* **1992**, *119*, 598-620.
41. Halliwell, B. *Free Radical Res.* **1996**, *25*, 57-74.
42. Halliwell, B.; Murcia, M. A.; Chirico, S.; Aruoma, O. I. *Crit. Rev. Food Sci. Nutr.* **1995**, *35*, 7-20.
43. Sies, H.; Stahl, W.; Sevanian, A. *J. Nutr.* **2005**, *135*, 969-972.
44. Goupy, P.; Hugues, M.; Boïvin, P.; Amiot, M. J. *J. Sci. Food Agric.* **1999**, *79*, 1625-1634.
45. Donovan, J. L.; Bell, J. R.; Kasim-Karakas, S.; German, J. B.; Walzem, R. L.; Hansen, R. J.; Waterhouse, A. L. *J. Nutr.* **1999**, *129*, 1662-1668.
46. Neilsen, S. E.; Young, J. F.; Daneshvar, B.; Lauridsen, S. T.; Knuthsen, P.; Sandström, B.; Dragsted, L. O. *Brit. J. Nutr.* **1999**, *81*, 447-455.
47. Awika, J. M.; Dykes, L.; Gu, L.; Rooney, L. W. Prior, R. L. *J. Agric. Food Chem.* **2003**, *51*, 5516-5521.
48. Kroon, P. A.; Williamson, G. *J. Sci. Food Agric.* **1999**, *79*, 355-361.
49. Ghesson, A.; Provan, G. J.; Russell, W. R.; Scobbie, L.; Richardson, A. J.; Stewart, C. *J. Sci. Food Agric.* **1999**, *79*, 373-378.
50. Olthof, M. R.; Hollman, P. C. H; Katan, M. B. *J. Nutr.* **2000**, *131*, 66-71
51. Nardini, M.; Natella, F.; Gentili, V.; DiFelice, M.; Scaccini, C. *Arch. Biochem. Biophys.* **1997**, *342*, 157-160.
52. Madhujith, T.; Shahidi, F. *J. Agric. Food Chem.* **2007**, *55*, 5018-5024.
53. Prior, R. L.; Cao, G. *Horticult. Sci.* **2000**, *35*, 588-592.
54. Chu, Y.; Chang, C.; Hsu, H. *J. Sci. Food Agric.* **2000**, *80*, 561-566.
55. Yamaguchi, T.; Takamura, H.; Matoba, T. Terao, *J. Biosci. Biotechnol. Biochem.* **1998**, *62*, 1201-1204.
56. Fogliano, V.; Verde, V.; Randazzo, G., Ritieni, A. *J. Agric. Food Chem.* **1999**, *47*, 1035-1040.
57. Sanchez-Moreno, C.; Larrauri, J. A. *Food Sci. Technol. Int.* **1998**, *4*, 391-400.
58. Abdel-Aal, E-S. M.; Hucl, P. *J. Agric. Food Chem.* **2003**, *51*, 2174-2180.

59. Kris-Etherton, P. M.; Hecker, K. D.; Bonanome, A.; Coval, S. M.; Binkoski, A. E.; Hilpert, K. F.; Griel, A. E.; Etherton, A. E. *Am. J. Med.* **2002**, *113*, 71S-88S.
60. Slavin, J. L.; Martini, M. C.; Jacobs, D. R. Jr.; Marquart, L. *Am. J. Clin. Nutr.* **1999**, *70*(Suppl.), 459S-463S.
61. Meyer, K. A.; Kushi, L. H.; Jacobs, D. R. Jr.; Slavin, J.; Sellers, T. A.; Folsom, A. R. *Am. J. Clin. Nutr.* **2000**, *71*, 921-930.
62. Fung, T. T.; Hu, F. B.; Pereira, M. A.; Liu, S.; Stampfer, M. J.; Colditz, G. A.; Willett, W. C. *Am. J. Clin. Nutr.* **2002**, *76*, 535-540.
63. Jacobs, D.R. Jr.; Meyer, K.A.; Kushi, L.H.; Folsom, A.R. *Am. J. Clin. Nutr.* **1998**, *68*, 48-257.
64. Liu, S.; Stampfer, M. J.; Hu, F. B.; Giovanucci, E.; Rimm, E.; Manson, J. E.; Hennekens, C. H.; Willett, W. C. *Am. J. Clin. Nutr.* **1999**, *70*, 412-419.
65. Schweizer, T. E.; Wursch, P. *Eur. J. Clin. Nutr.* **1991**, *59*, 989-995.
66. Smigel K. *J. Natl.Cancer Inst.* **1992**, *84*(2), 80-81.
67. Jacobs D. R. Jr.; Meyer K. A.; Kushi L. H.; Folsom A. R. () *Am. J. Public Health* **1999**, *89*, 322-329.
68. Jacobs, D. R.; Meyer H. E.; Solvoll K. *Eur. J. Clin. Nutr.* **2001**, *55*, 137-143.
69. Nicodemus, K.K.; Jacobs, D.R.; Folsom, A.R. *Cancer Causes Control* **2001**, *12*, 917-925.
70. Chatenoud, L.; Tavani, A.; Vecchia, C.A.; Jacobs, D.R.; Negri, E. *Int. J. Cancer* **1998**, *77*, 24-28.
71. Rimm, E.B.; Ascherio, A.; Giovanucci, E.; Spiegelman, D.; Stampfer, M.J.; Willett, W.C. *JAMA* **1996**, *275*, 447-541.
72. Pietinen, P.; Rimm, E.B.; Korhonen, P.; Hartman, A.M.; Willett, W.C.; Albanes, D.; Vertamo, J. *Circulation* **1996**, *94*, 2720-2727.
73. Liu, S.; Manson, J.E.; Stampfer, M.J.; Rexrode, K.M.; Hu, F.B.; Rimm, E.B.; Willett, W.C. *JAMA* **2000**, *284*(12), 1534-1540.
74. Fung, T. T.; Hu, F. B.; Pereira, M. A.; Liu, S.; Stampfer, M. J.; Colditz, G. A.; Willet, W. C. *Am. J. Clin. Nutr.* **2002**, *76*, 535-540.
75. Jensen, M.K.; Koh-Banerjee, P.; Hu, F.B.; Franz, M.; Sampson, L.; Grønbæk, M.; Rimm, E.B. *Am. J. Clin. Nutr.* **2004**, *80*, 1492-1499.
76. Anderson, J.W.; Hanna ,T.J. *J. Nutr.* **1999**, *29*, 1457s-1466s.
77. Jang Y, Lee JH, Kim OY, Park HY, Lee SY. *Arterioscler. Thromb. Vasc. Biol.* **2001**, *21*, 2065-2071.

Chapter 12

All Natural Whole-Wheat Functional Foods for Health Promotion and Disease Prevention

Liangli Yu and Margaret Slavin

Department of Nutrition and Food Science, University of Maryland, College Park, MD 20742

With an aging society, much research has been devoted to decreasing the risk and severity of chronic diseases, along with improving life quality. It has long been accepted that diet plays a role in the prevention or development of chronic diseases. Modifications in food and ingredient technologies may enhance the health beneficial effects of certain foods beyond that of basic nutrition, possibly enhancing their ability to prevent chronic diseases, and thus making them functional foods. This contribution discusses the innovation of post-harvest treatments on health beneficial components in wheat-based food ingredients, along with the effect of storage conditions on antioxidant availability in wheat-based food ingredients. It also devotes attention to studying fruit seed flours for their potential as antioxidant-rich nutraceutical ingredients and as natural shelf-life enhancers. Additionally, all-natural whole wheat foods containing fruit seed flours are reported. An example for developing value-added consumer-desired functional foods is also discussed.

Functional foods are foods that provide health benefits beyond basic nutritional needs (*1*). In other words, humans consume food in order to provide our bodies with the essential nutrients needed to survive, whereas a functional food provides biologically active components that, while not necessary for immediate survival, may help to reduce the risk of chronic disease and promote overall health.

The potential benefits from functional foods can be viewed from the perspectives of both health and economics. The increase in chronic diseases over the past several decades has had tremendous impact on our overall health and the cost of health care. Heart disease is the leading cause of death for both men and women in the United States, followed by cancer and stroke (*2*). Prevention of these and other aging-related diseases and health problems could save billions of dollars in health care costs, not to mention the drastic reduction in personal and societal suffering.

The development of a functional food is a long and complex process, which requires multidisciplinary approaches (Figure 1). For instance, plant scientists are needed to breed crops with desirable traits for functional foods. Biological techniques must be employed in screening ingredients for bioactivity, while chemical techniques are needed to study effects of processing methods, storage conditions, and other possible sources of variation in availability of the bioactive components in the final functional food product. Studies involving nutrition, biochemistry, and pharmacology must be employed in order to assess the safety and efficacy of the functional food ingredient. Food scientists must incorporate the ingredient into a food for consumption, without causing detrimental effects to the food's palatability and the functional ingredient's efficacy. Various measures of quality control involving these fields must be employed to ensure a consistent product. The resultant product must finally be introduced for sale, with consideration of proper packaging, labeling, and distribution to the intended market.

Wheat is an important agricultural commodity and one of the most widely consumed food ingredients in the United States and in the Western diet in general. The development of wheat-based functional foods that promote health are likely to be more easily accepted than a food that is not typically consumed, such as soy or fish oil. Wheat grain contains natural antioxidants, concentrated in the aleurone fraction of the bran (*3,4,5*). This fraction is removed in the refining process used to produce white flours but is retained in whole wheat flours. Furthermore, wheat phenolic acids, a group of phenolic antioxidants, are present in an insoluble bound form, which may not be released from the food matrix for absorption. Post-harvest treatment, ingredient storage approaches, and food processing conditions may alter the availability of wheat antioxidants including the phenolic acids.

Fruit seeds are by-products from fruit processing—they are sometimes pressed to produce fruit seed oils. Fruit seed flours are then the by-products of

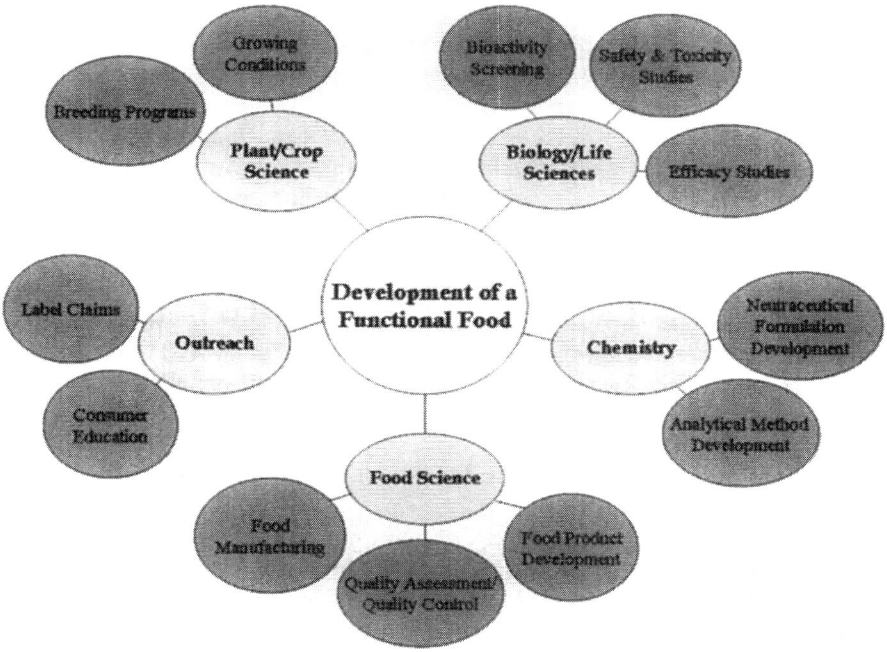

Figure 1. Development of a functional food following the identification of a possible bioactive component

fruit seed oil production and are used in low-value animal feed or treated as waste. Characterizing the nutraceutical properties of these fruit seed flours and demonstrating their potential utilization in functional foods may add value to the fruit production and processing industries as well as the fruit seed oil producers, while also improving human health.

This chapter discusses the development of all-natural functional foods rich in natural antioxidants by using whole-wheat flour and selected fruit seed flours. Also included is the brief summary of our recent research on novel "green processing procedures" to enhance the availability of natural antioxidants in wheat-based food ingredients.

Materials and Methods

Akron, Jagalene and Trego wheat samples were provided by Dr. Scott Haley, Department of Soil and Crop Science, Colorado State University (Fort Collins, CO). All varieties were from the 2004 growing season and were ground and separated into flour and bran using a Quadromat Junior experimental mill. The solid residues leftover after cold-pressing seed oil production of selected fruit seeds—black raspberry, red raspberry, blueberry, cranberry, pinot noir grape and chardonnay grape—were obtained from Badger Oil Co. (Spooner, WI). Black raspberry seed oil was also obtained from Badger Oil Co., while additional chardonnay grape and black raspberry seed flours were obtained from Botanical Oil Innovations, Inc. (Spooner, WI) for the antimicrobial and prevention of lipid peroxidation studies. The enzymes Viscozyme L, Pectinex 3XL, Flavourzyme 500L and Celluclast 1.5L were from Novozymes Corp. (Bagsvaerd, Denmark) and obtained through Sigma-Aldrich (Milwaukee, WI). Porcine liver esterase was also obtained through Sigma-Aldrich, while Ultraflo L was obtained from Novozymes North America (Franklinton, NC). All chemical reagents and solvents used were of the highest quality commercially available.

Effects of Post-harvest Enzymatic Treatment on Antioxidant Availability in Wheat Bran

The bran samples of Akron and Jagalene hard red winter wheat varieties were treated with the selected enzymes following a laboratory solid-state enzymatic reaction procedure (*6,7*). Briefly, wheat bran was treated with each enzyme at 6 doses ranging from 0 to 221.6 U/g wheat bran for 72 hours at ambient temperature. These enzymatically treated wheat brans were extracted with 100% ethanol and tested for their scavenging capacities against $ABTS^{\bullet+}$, $DPPH^{\bullet}$, and peroxyl radicals (ORAC). A known amount of the ethanol extract for each treated bran sample was used to prepare the dimethyl sulfoxide (DMSO) solution for hydroxyl radical scavenging capacity (HOSC) and total phenolic content (TPC) assays. The enzymatically treated wheat bran samples were also subjected to phenolic acid composition determination.

Effects of Ingredient Storage Conditions on Antioxidant Availability in Wheat Bran

Grain, bran and 40-mesh bran of Akron and Trego wheat varieties, were exposed to temperatures of 25, 60, or 100 °C for 9 days. Samples were taken at 0, 1, 2, 3, 5, and 9 days, ground to a fine powder and extracted with ethanol (*8*). These extracts were then tested for their peroxyl (ORAC), $ABTS^{\bullet+}$, $DPPH^{\bullet}$, and superoxide radical anion ($O_2^{\bullet-}$) scavenging capacities, as well as their total phenolic contents as discussed below. The wheat samples were also examined for their phenolic acid compositions.

Nutraceutical Properties of the Selected Fruit Seed Flours

Black raspberry, red raspberry, blueberry, cranberry, pinot noir grape, and chardonnay grape seed flours were extracted in 50% acetone at ambient temperature. The 50% acetone extracts were examined for their TPC, Fe^{2+} chelating activity, and peroxyl (ORAC) and DPPH• scavenging capacities following laboratory protocols. After removing solvent from a known amount of 50% acetone solution, the residues were quantitatively redissolved in DMSO and tested for their potential anti-proliferative capacity against HT-29 human colon cancer cells. Additionally, these seed flours were extracted with acidic methanol containing 2% 12 M HCl (v/v). After removing solvent, the residue was redissolved in DMSO for the total anthocyanin content assay (*9*).

These fruit seed flours were also examined for their total oil contents and fatty acid compositions in the oils. The oils were extracted from the fruit seed flours using a Soxhlet apparatus with petroleum ether as the solvent. After determining the total oil contents, these oils were converted to the fatty acid methyl esters and subjected to GC analysis for their fatty acid compositions.

Preservative Properties of the Selected Fruit Seed Flours

The fruit seed flours were extracted with 100% ethanol via a Soxhlet apparatus for 3 h. The ethanol extracts were evaluated for their abilities to inhibit total lipid peroxidation in fish oils using a Rancimat instrument, model 743 (Metrohm Ltd., Herisau, Switzerland), and their capacity to preserve long chain n-3 fatty acids, EPA and DHA, by GC analysis of fatty acid compositions (*10,11*). These extracts were also tested for their DPPH• scavenging capacity, ORAC value, and total phenolic contents. Additionally, antibacterial activity of the extracts was tested against *Escherichia coli* ATCC 25922 and *Listeria monocytogenes* ATCC 19114 (*11*).

Antioxidant Activity Assays

Oxygen Radical Absorbance Capacity (ORAC)

ORAC values were measured using fluorescein as a fluorescent probe on a Victor[3] multilabel plate reader (Perkin-Elmer, Turku, Finland) according to the laboratory protocol described by Moore *et al.* (*12*). The final reaction mixtures contained 8.16×10^{-8} M fluorescein, 0.36 M AAPH, as well as extract, solvent for the blank, or standard in various concentrations. Trolox was used as the antioxidant standard, and fluorescence was read every minute with excitation and emission wavelengths of 485 and 535 nm, respectively.

Hydroxyl Radical ($HO^{\bullet}$) Scavenging Capacity (HOSC)

HOSC was measured according to a laboratory protocol described by Moore *et al.* (*13*) using a Victor^3 multilabel plate reader (Perkin-Elmer, Turku, Finland). Final reaction mixtures contained 30 μL of sample, solvent for the blank, or standard in various concentrations along with 170 μL of 9.28×10^{-8} M fluorescein in 75 mM phosphate buffer, 40 μL of 0.1990 M H_2O_2, and 60 μL of 3.43 mM $FeCl_3$. Trolox was used as the standard, and fluorescence was read every minute for 3 h with excitation and emission wavelengths of 485 and 535 nm, respectively.

Fe^{2+} Chelating Capacity

Fe^{2+} chelating capacity was determined using the 2,2-bipyridyl competition assay described previously (*4*). The final reaction mixture contained 30 μL of 1.8 mM $FeSO_4$, 400 μL standard or sample solution, 200 μL of pH 7.4 Tris-HCl buffer, 50 μL of 0.1% 2,2'-bipyridyl in 0.2 M HCl, and 200 μL of 7% β-cyclodextrin in 50% acetone. EDTA was used as the standard, and absorbance was read at 522 nm.

$ABTS^{\bullet+}$ Scavenging Activity

Radical cation $ABTS^{\bullet+}$ scavenging capacity was measured according to a laboratory procedure described by Miller and Rice-Evans (*14*). Radicals were generated by reacting ABTS with manganese dioxide at ambient temperature for 30 min. The absorbance of $ABTS^{\bullet+}$ was adjusted to 0.7 at 734 nm. Extracts were mixed with $ABTS^{\bullet+}$ solution and absorbance at 734 nm was read at 1 or 1.5 min reaction time. $ABTS^{\bullet+}$ scavenging capacity was calculated according to a standard curve prepared with trolox.

$DPPH^{\bullet}$ Scavenging Activity

$DPPH^{\bullet}$ scavenging activity was measured according to previously published protocols (*15-18*). In all cases, the extracts of interest were mixed with a freshly prepared $DPPH^{\bullet}$ solution, and absorbance was read at either 515 or 517 nm at various intervals up to 40 minutes.

Superoxide Anion Radical ($O_2^{\bullet-}$) Scavenging Capacity

$O_2^{\bullet-}$ scavenging activity was evaluated using an E-109X-Band electron spin resonance (ESR) spectrometer (Varian, Inc., Palo Alto, CA) according to a previously published protocol (*3*). $O_2^{\bullet-}$ was generated using the xanthine/ xanthine oxidase system. The final reaction mixture was 100 µL and contained 2 mM xanthine, 200 mM BMPO (5-tert-butoxycarbonyl-5-methyl-1-pyrroline *N*-oxide), 0.2 mM DTPA (diethylenetriaminepentaacetate), 0.08 U/mL xanthine oxidase, and 20 mg wheat equivalents/mL sample extracts. ESR spectra were recorded at 2 minutes of reaction with 10 mW incident microwave power and 100 kHz field modulation of 1 G at room temperature.

Inhibition of Lipid Peroxidation in Fish Oil

Oxidative stability index (OSI) was measured on a Rancimat instrument, model 743 (Metrohm Ltd., Herisau, Switzerland), as reported previously (*10,11*). Briefly, 6 mL of menhaden oil samples containing two different levels of each extract were heated to 80 °C with an air flow rate of 7 L/h. Extension of the induction time was calculated according to that of fish oil sample with no antioxidant (the negative control) and used to compare the capacity of the tested fruit seed flour extracts for their inhibitory capacity against lipid peroxidation in fish oil.

Total Phenolic Contents (TPC)

The TPC for wheat and fruit seed flour samples was determined using the Folin-Ciocalteu (FC) reagent as described by Yu and others (*19*). Each final reaction mixture contained 50 µL of extract or standard, 250 µL FC reagent, 750 µL of 20% sodium carbonate, and 3 mL of pure water. Absorbance was read at 765 nm after 2 h of reaction at ambient temperature. Gallic acid was used as the standard.

Total Anthocyanin Contents (TAC)

The TAC was measured only for the fruit seed flour samples. A pH differential method was used (*9*). It exploits the structural transformations of anthocyanin molecules, and therefore light absorbance capacities, at different pHs.

GC Analysis of Fatty Acid Compositions

Fatty acid compositions in the oils extracted from the fruit seed flours, as well as fresh fish oil and the oxidized fish oils with or without the fruit seed flour extracts were analyzed according to the laboratory protocol explained by Parker *et al.* (*10*). Fatty acid methyl esters were prepared by reacting 1 mg of oil with 0.2 mL NaOH-MeOH (0.5M) for 5 min and 0.5 mL of 4% HCl-MeOH (w/w) for 5 min at ambient temperature. GC analysis was performed using a Supelco 2380 (Bellefonte, PA) column, 30 m × 0.25 mm i.d. with a 0.20 μm film thickness, on a Shimadzu GC-2010 with a FID detector and a Shimadzu AOC-20i autosampler (Shimadzu, Columbia, MD) following a laboratory protocol explained by Parry *et al.* (*9*). Helium was used as the carrier gas at a flow rate of 0.8 mL/min. Column temperature was programmed as 142 °C to 184 °C at a rate of 6 °C/min, held for 3 min, and to 244 °C at a rate of 6 °C/min. Fatty acid identification was carried out by comparing GC retention time of each peak with those of the commercial standards.

Phenolic Acid Composition Determination

Phenolic acid composition was measured according to a previously published laboratory protocol (*12,17*). Acetone/methanol/water (7/7/6, v/v/v) was used to extract soluble free and conjugated phenolic acids and to separate them from the insoluble bound phenolic acids. The soluble free and conjugated phenolic acids were separated by their solubility in ethyl acetate/ethyl ether (1/1, v/v) at pH 1-2. The soluble conjugated phenolic acids were converted to their free forms by acidic hydrolysis. Insoluble bound phenolic acids remaining in the solid residue after the acetone/methanol/water extraction were hydrolyzed with NaOH, and re-extracted in ethyl acetate/ethyl ether (1/1, v/v) at pH 2. The phenolic acids were quantified via reverse-phase HPLC using a C18 column (250 mm × 4.6 mm) according to a laboratory protocol described elsewhere (*17*).

Antiproliferative Ability Assay

The HT-29 human colon cancer cell proliferation tests were the same as those described by Parry and others (*9*). In short, cells were plated in 12-well plates, incubated for 24 hours in control media at 37 °C at 5% CO_2. They were then treated with media containing either 3 or 6 mg flour equivalents/mL culture media or a control media containing the same volume of DMSO as that in the extracts. Live cells were counted daily for 4 days.

Results and Discussion

Consumer desire is always one of the most important driving forces for food innovation and development. Current consumer purchasing trends in "natural", safe, and health beneficial food products stimulate the research and development of functional foods that are capable of reducing the risk of chronic diseases or managing certain health problems through innovation of "green processing technologies", novel "natural nutraceutical food ingredients", eliminating or reducing the level of synthetic food preservatives such as BHA, and replacing them with natural alternatives. In 2005, the Institute of Food Technologists Expert Report on Functional Foods proposed 7 steps to follow in the development of a functional food for sale in the marketplace, beginning with the identification of a relationship between the food component(s) and their demonstrated health benefit. Subsequently, efficacious and safe doses must be established, followed by development of a suitable food vehicle, communication of the benefits to consumers, and in-market surveillance to further confirm efficacy and safety (*1*). As depicted in Figure 1, many of the steps in functional food development may and do occur simultaneously. The need for input from multiple scientific fields is highlighted. The results presented here evoke discussion of the possibility of enhancing the availability of nutraceutical components in food products through developing novel nutraceutical ingredients, post-harvest treatments, ingredient storage practices, and food preparation conditions.

Effects of Post-harvest Enzymatic Treatment on Antioxidant Availability in Wheat Bran

Understanding the factors that affect the bioavailability of antioxidants and subsequently using that knowledge to enhance bioavailability have been and remain as some of the most difficult challenges in developing a whole wheat functional food. Most of the phenolic acids in wheat, known for their antioxidant qualities, are bound to the cell walls of the bran, thus preventing them from being absorbed (*20*). Previous studies have utilized enzymes to free phenolic acids from the cell wall matrix, but those studies used aqueous reaction systems. While successful in freeing phenolics, the use of aqueous enzymatic reactions is not feasible in a commercial setting due to cost, amount of waste generated, and the large amount of effort needed to isolate the product from its aqueous phase.

Recently, we treated Akron and Jagalene wheat bran samples with five commercial food grade enzyme preparations (Viscozyme L, Pectinex 3XL, Flavourzyme 500L, Ultraflo L, and Celluclast 1.5L) and one purified non-food-grade enzyme (porcine liver esterase) in a solid-state reaction. Enzyme doses of 0, 2.26, 4.52, 9.04, 18.09, and 221.6 units per gram of bran were used, and the antioxidant properties of the enzymatically treated bran samples were evaluated. As shown in Table I, all tested enzymes were able to dose-dependently enhance the releasable or available hydroxyl radical scavenging capacity (HOSC) of Akron wheat bran. Similar effects were observed for Jagalene wheat bran samples under the experimental conditions. Ultraflo L was the most efficient enzyme in increasing HOSC values for both wheat varieties, with a 4.5-fold increase for Akron at 9 U/g and a 2.76-fold increase for Jagalene at 18 U/g.

Table I. Hydroxyl Radical Scavenging Capacity (HOSC) of Akron wheat Bran Samples Treated with Various Levels of Enzymes. Values are reported in μmol trolox equivalents/gram bran. NM = not measured.

	Enzyme Concentration (Units/g bran)					
	0	2.26	4.52	9.04	18.09	221.6
Ultraflo L	16.43	59.29	59.81	73.31	55.69	NM
Viscozyme L	16.43	18.22	23.96	29.30	31.26	NM
Esterase	16.43	21.83	27.53	31.06	25.83	19.68
Celluclast 1.5L	16.43	16.34	20.83	24.32	16.39	58.28
Pectinex 3XL	16.43	13.75	14.97	18.46	18.08	24.92
Flavourzyme 500L	16.43	26.06	17.09	42.68	46.89	65.80

These enzymes also dose-dependently increased the available ORAC values and DPPH$^{\bullet}$ scavenging capacities, and had no effect or increased ABTS$^{\bullet+}$ scavenging capacities in bran samples of both wheat varieties (*7*). Ultraflo L was the most efficient enzyme in enhancing these free radical scavenging activities in wheat bran samples under the experimental conditions. The ORAC value increased by as much as 4.3-fold in Akron wheat bran using 221 U/g Celluclast 1.5L and 3.5-fold in Jagalene wheat bran using 9 U/g Ultraflo L. Porcine esterase actually decreased the ABTS$^{\bullet+}$ scavenging capacity at lower doses and showed no effect at higher doses for both wheat varieties.

This study also showed that solid-state enzymatic treatment was able to increase the measurable TPC values in bran samples of both wheat varieties under the experimental conditions (Table II) (*7*). Ultraflo L was most effective in releasing phenolics per unit of enzyme, but increasing enzyme levels above 9 U/g did not further enhance the release of phenolics in both wheat bran varieties. Treatment of both varieties with Flavourzyme 500L at 221 U/g produced the highest TPC values with 5.6 and 5.1-fold increases for Akron and Jagalene, respectively. Additionally, this study demonstrated that solid-state enzymatic treatments increased the availability of soluble phenolic acids in wheat bran samples (*7*). These data indicate the potential to enhance the availability of wheat antioxidants through post-harvest treatments.

Table II. Total phenolic contents (TPC) for Akron wheat bran samples treated with various levels of enzymes. Values are measured in mg gallic acid equivalents/gram dry weight. NM = not measured.

	Enzyme Concentration (Units/g bran)					
Enzyme	*0*	*2.26*	*4.52*	*9.04*	*18.09*	*221.6*
Ultraflo L	0.185	0.599	0.692	0.838	0.811	NM
Viscozyme L	0.185	0.213	0.167	0.292	0.355	NM
Esterase	0.185	0.065	0.145	0.152	0.147	0.132
Celluclast 1.5L	0.185	0.173	0.180	0.204	0.173	0.370
Pectinex 3XL	0.185	0.151	0.159	0.225	0.314	0.569
Flavourzyme 500L	0.185	0.268	0.213	0.372	0.425	1.055

Effect of Storage Conditions on Antioxidant Availability in Wheat-based Food Ingredients

The grain, bran, and 40-mesh bran samples of Akron and Trego wheat varieties were stored at 25, 60, and 100 °C for 0, 1, 2, 3, 5, and 9 days, extracted with 100% ethanol, and evaluated for their antioxidant properties (*8*). The results showed that wheat grain was most resistant to loss of antioxidant capacity from thermal treatments in accelerated storage testing, as compared to the bran or 40-mesh bran forms. Whereas the ground bran held at 100 °C lost almost all of its $ABTS^{\bullet+}$ scavenging capacity, the wheat grain remained at about its original level of $ABTS^{\bullet+}$ scavenging capacity (*8*). Similar results, though not as dramatic, were seen in ORAC value and $DPPH^{\bullet}$ scavenging capacity estimations. Figure 2 shows the effects of thermal treatment during 9 days of storage on the ORAC value of wheat grain, bran, and 40-mesh bran extracts. $O_2^{\bullet-}$ scavenging activity tests with ESR did not show significant changes in the grain, bran and ground bran samples, though the data were not quantitative. Earlier in 2004, a study showed that more antioxidants could be extracted from smaller wheat particles, pointing to the possibility of using finely ground wheat as a better source of available antioxidants (*5*). Taken together, these results suggest that reduction of bran particle size in processing may enhance the release of wheat antioxidants but also decrease their thermal and storage stability. This may indicate that grain is a better storage form of wheat for preserving wheat antioxidants. This information is important for optimizing the storage conditions to allow for the greatest available level of antioxidant capacity in wheat-based food ingredients and food products. Further research is necessary to find the optimal balance between preservation and extractability of the antioxidants.

Nutraceutical Properties of the Selected Fruit Seed Flours

Fruit seeds are often a by-product of juice and wine production, leaving megatons of seeds to be thrown out every year in the United States and around the world. The seeds, however, may be processed into specialty oils with varying flavors, colors and beneficial properties and the seed flours. A study in our laboratory with cold-pressed black raspberry seed oil and flour revealed that both materials contained strong scavenging capacities against $DPPH^{\bullet}$ and $ABTS^{\bullet+}$ radicals (*21*). Additionally, the black raspberry seed flour contained a significant level of oil rich in α-linolenic acid, the essential n-3 fatty acid. This study suggested there is potential to develop nutraceutical ingredients from the fruit seed flours for improving human health while adding value to the fruit production and processing industries.

To develop novel value-added nutraceutical uses for fruit seed flours in functional food products, characterization of their health beneficial properties and demonstration of their utilization in model food products are required. Recently, cold-pressed black raspberry, red raspberry, blueberry, cranberry, pinot noir grape and chardonnay grape seed flours were investigated for their antioxidant capacities and antiproliferative properties against human HT-29 colon cancer cells (*9*). The results revealed that all the tested fruit seed flours are rich sources of natural antioxidants. These antioxidants may directly react with and quench $DPPH^{\bullet}$ and oxygen radicals (ORAC), form chelating complex with Fe^{2+}, and inhibit HT-29 cell proliferation. The fruit seed flour extracts also contained significant levels of TPC and total anthocyanin contents (TAC) (*9*). Their ORAC values are shown and compared on a per dry weight basis in Figure 3. The chardonnay seed extract exhibited the highest ORAC values, strongest $DPPH^{\bullet}$ radical scavenging capacity, and greatest TPC among all tested fruit seed flours. Also reported were the TAC values, oil contents, and fatty acid composition of the oils (*9*).

Additionally, cranberry, black raspberry and chardonnay seed flour extracts were evaluated for their ability to inhibit HT-29 human colon cancer cell growth (Figure 4). All three extracts exhibited anti-proliferative effects on HT-29 cells, while chardonnay showed the strongest effect by completely killing all cells within 24 hours of treatment. The black raspberry and cranberry anti-proliferative effects were dose dependent.

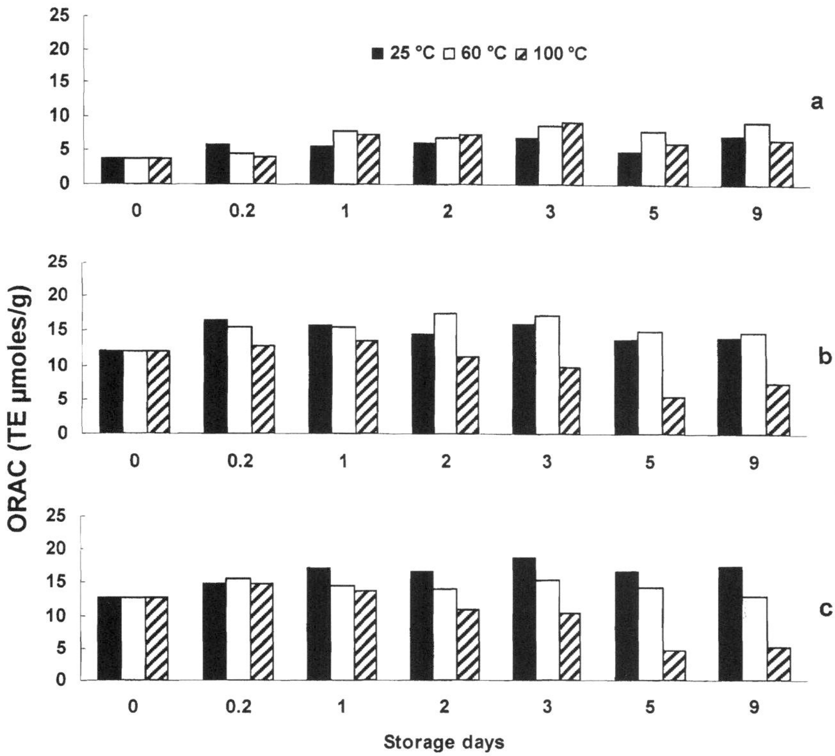

Figure 2. Oxygen Radical Absorbance Capacity (ORAC) under accelerated storage testing of Akron (a) wheat, (b) bran, and (c) 40-mesh bran. Samples were held at 25, 60 and 100 °C for up to 9 days, extracted in ethanol and then evaluated for their ORAC capacity. TE stands for trolox equivalents. A greater ORAC value is associated with a stronger antioxidant activity.

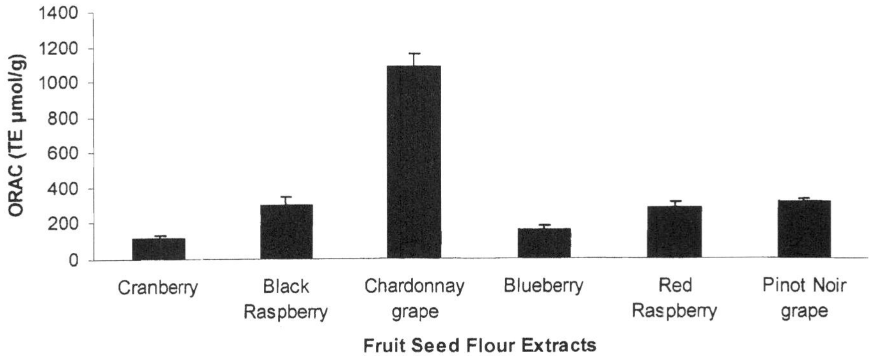

Figure 3. ORAC values of selected fruit seed flour extracts. Values are expressed on a dry weight basis.

Food Preservative Properties of the Selected Fruit Seed Flours

It is known that phenolics may have antioxidant and antimicrobial properties. Ethanol extracts of the seed flours were prepared and tested for their ability to extend the stability of menhaden oil under accelerated testing conditions and to preserve longer chain n-3 fatty acids (EPA and DHA), along with their capacities to react with DPPH$^{\bullet}$ and peroxyl (ORAC) radicals. The extracts were capable of dose-dependently extending the induction time of the fish oil under the experimental conditions (Figure 5). The chardonnay grape seed flour extract at its highest dose of 16.7 mg flour equivalent/mL was able to extend that length of induction time by about 1.5 times that of the positive control, mixed tocopherols.

Effects of the fruit seed flour extracts on the growth of *Escherichia coli* and *Listeria monocytogenes*, common food-borne pathogens, were also measured. All fruit seed flour extracts tested (black currant, chardonnay grape, pinot noir grape, black raspberry, and cranberry) had bactericidal effects on *E. coli* at 35 °C, leaving no detectable viable *E. coli* (*11*). The extracts had no effect on *L. monocytogenes* at 35 °C, but did exhibit bactericidal effects at 4 °C. Chardonnay and pinot noir grape seed extracts left no detectable *E. coli* alive, while black currant, black raspberry and cranberry seed extracts left 0.03, 1.43 and 2.86% cells surviving, respectively. No bactericidal effects were shown by the control, mixed tocopherols (*11*). These data showed that these fruit seed flours contain natural food preservatives that exhibit both antioxidant and antibacterial activities, suggesting their utilization in food formulae to replace or decrease the amount of synthetic food preservatives.

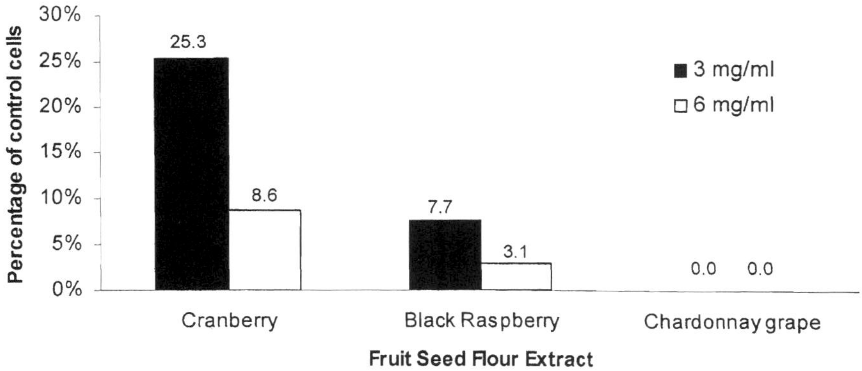

Figure 4. Antiproliferative effects of selected fruit seed flour extracts on HT-29 human colon cancer cells after 4 days of treatment. Cells were exposed to 50% acetone extracts of the fruit seed flours in final concentrations of 3 and 6 mg seed flour equivalent per mL initial culture media. The control was treated with culture media containing the same concentration of DMSO as in the treatment culture media. Treatment with the chardonnay grape seed flour extract eliminated all living cells after 24 hours treatment in both doses.

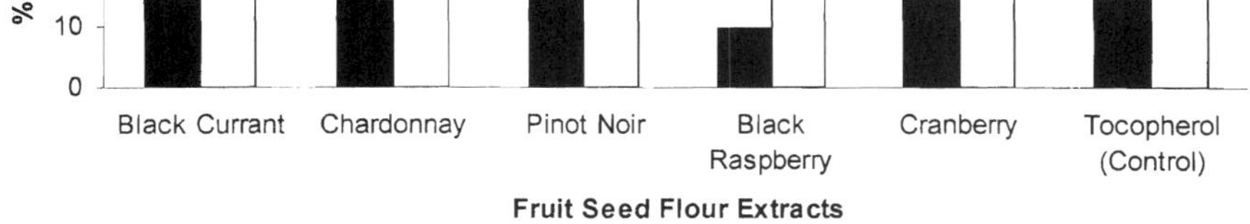

Figure 5. Oxidative stability index extension of menhaden fish oil by fruit seed flour extracts. The solid column represents the low dose of extracts, and the open column represents the high dose, about 7 and 16 mg seed flour equivalents per mL fish oil, respectively. Ethanol extracts of the seed flours were used in this study.

Additionally, these fruit seed flours contain a significant level of oils which may add essential fatty acids in desired ratios. These fruit seeds had different color and flavor, which may contribute to the overall sensory properties of the final food products. It needs to be noted that much further testing is required to determine their efficacy in food matrices and their safety and availability prior to their use in commercial food products.

All-natural Whole Wheat Functional Food Products

Muffins and breads containing different levels of wheat bran and the fruit seed flours were prepared. When evaluated by in-house sensory tests, these model foods had no difference in their sensory properties as compared to commercial conventional whole wheat muffins and bread with similar formulae. Figure 6 shows the all-natural whole-wheat muffins rich in antioxidants. These muffins contained no synthetic colorants, flavoring agent, or food preservatives. This preliminary study demonstrates the food application of the fruit seed flours, and warrants additional pilot human studies to evaluate and confirm the health beneficial effects of these functional foods.

Conclusions

We have shown that small changes in storage and processing methods have the potential to alter overall antioxidant capacity of whole wheat food ingredients and food products. Further research is necessary to optimize the antioxidant availability thorugh improving the post-harvest treatments, ingredient storage, and the processing techniques, and also to elucidate the efficacy of these ingredients and/or final functional food products in animal studies and human trials. Only after these studies, it may be possible to draw conclusion about the association between the enhanced antioxidant capacity and a desired health benefit.

Because many factors can affect antioxidant contents of a final food product, each individual antioxidant-rich functional food—whole wheat, fruit seed flour or otherwise—should ideally be evaluated individually for its safety and efficacy in animal or human trials. The large sample size necessary for these types of studies and consequently their high cost make them unrealistic in many cases. As a result, much of the evidence remains promising, but at the level of laboratory research. Nonetheless, the potential is present for such minor changes—on the part of food manufacturers and consumers—to have an additive effect and greatly enhance human health and well-being.

Figure 6. Various all-natural, functional whole wheat muffin formulations.

References

1. Clydesdale, F. *Food Technol.* **2004**, *58*(12), 35-40.
2. Centers for Disease Control and Prevention. Available at: http://www.cdc.gov/nccdphp/overview.htm
3. Zhou, K.; Yin, J. J.; Yu, L. *J. Agric. Food Chem.* **2005**, *53*, 3916-3922.
4. Zhou, K.; Yu, L. *J. Agric. Food Chem.* **2004**, *52*, 1112-1117.
5. Zhou, K.; Laux, J.; Yu, L. *J. Agric. Food Chem.* **2004**, *52*, 1118-1123.
6. Yu, L.; Perret, J. *J. Agric. Food Chem.* **2003**, *51*, 492-495.
7. Moore, J.; Cheng, Z.; Su, L.; Yu, L. *J. Agric. Food Chem.* **2006**, *54*, 9032-9045.
8. Cheng, Z.; Su, L.; Moore, J.; Zhou, K.; Luther, M.; Yin, J.-J.; Yu, L. *J. Agric. Food Chem.* **2006**, *54*, 5623-5629.
9. Parry, J.; Su, L.; Moore, J.; Cheng, Z.; Luther, M.; Rao, J. N.; Wang, J.-Y.; Yu, L. *J. Agric. Food Chem.* **2006**, *54*, 3773-3778.
10. Parker, T. D.; Adams, D. A.; Zhou, K.; Yu, L. *J. Food Sci.* **2003**, *68*, 1240-1243.

11. Luther, M.; Parry, J.; Moore, J.; Meng, J.; Zhang, Y.; Cheng, Z.; Yu, L. *Food Chem.* **2007**, *104*, 1065-1073.
12. Moore, J.; Hao, Z.; Zhou, K.; Luther, M.; Costa, J.; Yu, L. *J. Agric. Food Chem.* **2005**, *53*, 6649-6657.
13. Moore, J.; Yin, J.; Yu, L. *J. Agric. Food Chem.* **2006**, *54*, 617-626.
14. Miller, N.; Rice-Evans, C. *Free Rad. Res.* **1997**, *26*, 195-199.
15. Yu, L.; Adams, D.; Gabel, M.; *J. Agric. Food Chem.* **2002**, *50*, 4135-4140.
16. Yu, L.; Perret, J.; Davy, D.; Wilson, J.; Melby, C. L. *J. Food Sci.* **2002**, *67*, 2600-2603.
17. Zhou, K.; Su, L.; Yu, L. *J. Agric. Food Chem.* **2004**, *52*, 6108-6114.
18. Yu, L.; Scanlin, L.; Wilson, J.; Schmidt, G. *J. Food Sci.* **2002**, *67*, 582-585.
19. Yu, L.; Haley, S.; Perret, J.; Harris, M. *Food Chem.* **2002**, *78*, 457-461.
20. Scalbert, A.; Williamson, G. *J. Nutr.* **2000**, *130*, 2073S-2085S.
21. Parry, J.; Yu, L. *J. Food Sci.* **2004**, *69*, 189-193.

Chapter 13

Total Phenolic Content and Antioxidant Activity of Cereals

Ting Sun[1] and Chi-Tang Ho[2]

[1]Department of Food Science, Louisiana State University, Baton Rouge, LA 70803
[2]Department of Food Science, Rutgers, The State University of New Jersey, New Brunswick, NJ 08903

Cereals are good sources of various antioxidants that can prevent cardiovascular diseases and certain types of cancer. For barley, buckwheat, corn, millet, oat, rice, rye, sorghum and wheat, the total phenolics content of the selected cereals have been summarized in the literature. Based on the available data of 2,2′-azinobis-(3-ethylbenzothiazoline-6-sulfonic acid) (ABTS) and 2,2′-diphenyl-1-picrylhydrazyl (DPPH) free radical scavenging activity in the literature, the antioxidant activity of sorghum (not including white sorghum) is higher than the other cereals, rye has the lowest antioxidant activity, and the other cereals are in the medium antioxidant activity group. The total phenolics content and antioxidant activity of cereals are significantly correlated.

Cereals are important staple foods. According to the annual global production of major cereal grains in 2005, the mostly consumed cereals with production amount from high to low was corn, wheat, rice, barley, sorghum, millet, oat, rye and buckwheat. Corn, wheat and rice accounted for 87% of all grain production (*1*). Cereal grains supply most of their food energy as starch. Whole grains are good sources of dietary fiber and essential fatty acids. Cereals also contain phenolic acids, flavonoids, vitamins and minerals. Contribution of the identified antioxidants in grains to health benefit has not been fully explored (*2*). The major health benefits of cereals include antibacterial, antiviral, anti-inflammatory, and antiallergic effect. Cereals can also slow digestibility, lower cholesterol, prevent cardiovascular disease and reduce colon cancer (*3,4*). It is generally suggested that the health benefit of grains are mainly due to their antioxidants. The objective of this review is to summarize and compare the total phenolics content and antioxidant activity of the commonly consumed cereals.

Phenolic Acids of Cereals

Ferulic acid is the most abundant phenolic acid in cereals. Wheat contains 0.8-2 g/kg dry weight of ferulic acid, which is 90% of total phenols present. The other major phenolic acids in cereals are caffeic acid, vanillic acid, *p*-coumaric acid, *p*-hydroxybenzoic acid, protocatechuic acid, sinapic acid and syringic acid (*5*). The total amount of phenolic acid in cereals is as high as 500 mg/kg of groat (*5*). Phenolic acids exist mainly as benzoic and cinnamic acid derivatives in cereals. Cereal grains contain several parts: hull, bran, endosperm and germ. The distribution of phenolic acids in cereal groat is not even, generally, the outer layer, such as bran, contains higher amounts of phenolic acids than the center of grain. It is found that *p*-coumaric acid in barley increases greatly from the center to the outer layer of grain.

Free and soluble conjugated phenolic acids in cereal flour are very low and their total amount ranges between 2.3 (wheat) and 16.5 (corn) μg/kg (*6*). No single method can extract all the cereal antioxidants. Methanol, ethanol, and acetone are commonly used to extract free phenolic acid in cereals, and small amounts of HCl or acetic acid can be added to release the soluble conjugated phenolic acids. Bound phenolic acids are not available for extraction by these solvents. Most phenolic acids in cereals exist as insoluble bound forms associated with cell wall polysaccharides. To hydrolyze the bound phenolic acids from cell walls, alkali hydrolysis and enzyme hydrolysis are used. There is an increase of phenolic acid content by 14.2 mg ferulic acid equivalents/g crude extract after hydrolysis of whole wheat (*7*). Phenolic acids are the common antioxidants in cereals and have been thought to contribute significantly to the total antioxidant properties of cereals.

Total Phenolic Content of Cereals

The total phenolics content of cereals is often determined using Folin-Ciocalteu reagent, which includes phenolic acids, flavonoids, ascorbic acid, among others. In most studies, cereal groat is extracted with methanol or acetone without complete hydrolysis of the bound phenolics, which are the data reported in Table I. Data from the literature were used to calculate the same unit of μmol/g dry weight. The total phenolics content is reported using different standards, such as gallic acid, ferulic acid or catechin equivalents. Based on the data in Table I, relative total phenolics content index (%) is developed to facilitate the comparison of the total phenolics content of cereals expressed as different standards. For the data expressed as each standard, the highest value of cereal was considered as 100%, and the values of the other cereals are expressed as a percentage of it. Then, for each cereal, the mean value of the percentage for different standards is calculated as this cereal's relative total phenolics content index (%). Grain color of sorghum includes red, black, brown and white, and the data of each color of sorghum are reported individually. As shown in Figure 1, the rank of the relative total phenolics content index of cereal groat from high to low is: brown sorghum, red sorghum, black sorghum, wheat, barley, oat, buckwheat, rice, corn, millet, white sorghum, and rye. The high total phenolics content group includes brown, red and black sorghum. The low total phenolics content group includes rye and white sorghum, and the other cereals belong to the medium total phenolics content group.

Antioxidant Activity of Cereals

More than 15 methods are available for determination of antioxidant activity of foods (*8*). However, there are no officially standardized methods established yet for analyzing the antioxidant activity of foods. Hydrogen atom transfer (HAT) reactions include oxygen radical absorbance capacity (ORAC), total radical-trapping antioxidant parameter (TRAP), and inhibition of autoxidation of induced low-density lipoprotein (LDL). In most HAT-based methods, antioxidant and substrate compete for thermally generated peroxyl radicals through the decomposition of azo compounds. Electron transfer-based methods include 2,2′-azinobis-(3-ethylbenzothiazoline-6-sulfonic acid) (ABTS), ferric reducing antioxidant power (FRAP), 2′-diphenyl-1-picrylhydrazyl (DPPH) and cupric reducing antioxidant capacity (CUPRAC) methods. These methods measure the ability to reduce an antioxidant by the color change of the oxidant. However, there is no simple universal method by which antioxidant activity can be measured accurately and quantitatively (*8*). Therefore, data generated with several methods are used in this article. To facilitate the comparison, data from different methods are standardized (see below for more detailed information).

As little research about antioxidant activity of millet is found, the data of antioxidant activity of millet is not reported in this review. In most research on antioxidant activity of cereals, antioxidants are extracted with methanol, ethanol or acetone, without hydrolysis to release the conjugated phenolic acid; the bound phenolic acids were not extracted by these solvents. Thus the obtained antioxidant activity of cereals is smaller compared to sample with bound phenolics hydrolyzed during extraction. The antioxidant activity data of eight

Table I. Total Phenolic Content (μmol/g dry weight) of Some Cereal Groats Extracted Without Hydrolyzing the Insoluble Bound Phenolic Acids.

		Catechin equivalent		*Gallic acid equivalent*		*Ferulic acid equivalent*	
Barley		4.3	(*9*)*			10.2	(*10*)
Buckwheat		13.1	(*9*)	17.9	(*11*)		
Corn				1.6	(*12*)	2.9	(*13*)
Millet		11.2	(*14*)				
Oat		2.1	(*9*)	1.6	(*15*)	6.7	(*10*)
				1.0	(*16*)		
				1.5	(*17*)		
Rice				10.1	(*18*)		
Rye		2.9	(*19*)				
Sorghum	Black			39.2	(*20*)		
				40.6	(*21*)		
	Brown			117.6	(*20*)		
				15.3	(*21*)		
	Red	129.1	(*14*)	32.7	(*21*)		
				32.3	(*21*)		
	White	4.6	(*14*)	6.5	(*20*)		
				5.9	(*21*)		
				3.0	(*22*)		
Wheat				4.6	(*23*)	12.0	(*24*)
				1.2	(*19*)	10.6	(*25*)

*References are given in parentheses.

cereals are collected from the references, ABTS and DPPH methods are the two most commonly used methods to analyze the antioxidant activity of cereals. Similar to the total phenolic content index, the data in Table II are expressed as a percentage of the highest value of a cereal (100%) determined by each method and the mean value of the percentages from different methods is calculated as this cereal's relative antioxidant activity index (%). As shown in Figure 2, the antioxidant activity of cereals can be divided into high, medium and low groups. The high group includes brown, black and red sorghum. Rye has low antioxidant activity, and the other cereals are among the medium antioxidant activity group. With the increase of total phenolics content, antioxidant activity of cereal groat was also increased. Total phenolics content and antioxidant activity of cereal groat is significantly correlated ($n = 11$, $R = 0.84$, $p < 0.01$). Red, brown and black sorghum have both high total phenolics content and antioxidant activity, rye has both the lowest total phenolics content and antioxidant activity, and the other cereals belong to the group with medium total phenolics content and antioxidant activity.

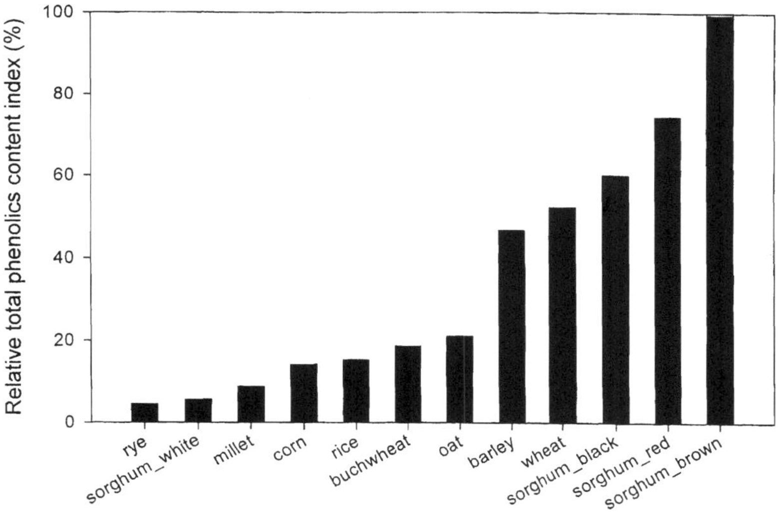

Figure 1. Relative total phenolics content index of cereal groat extracted without hydrolyzing the insoluble bound phenolic acids.

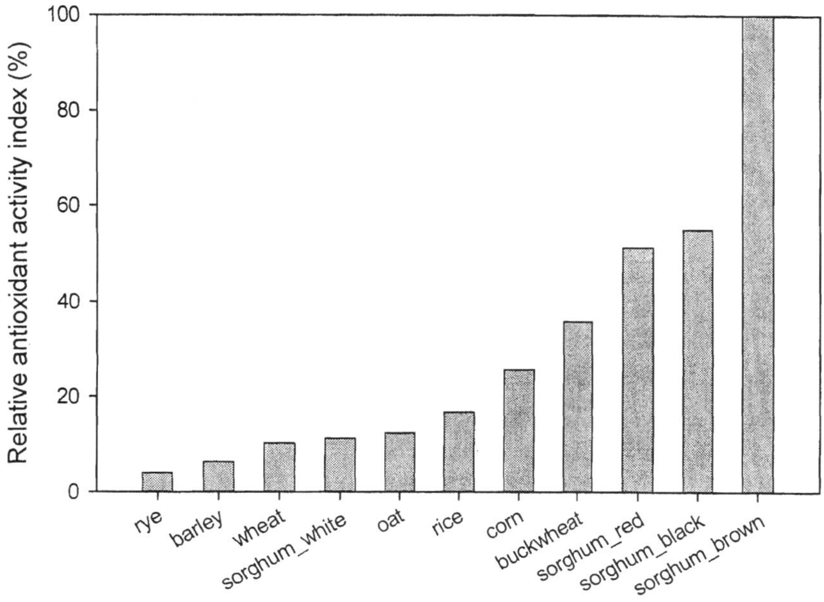

Figure 2. Relative antioxidant activity index of cereal groats extracted without hydrolyzing the insoluble bound phenolic acids.

Table II. Antioxidant Activity of Some Cereals (μmol Trolox equivalent/g dry weight) Extracted without Hydrolyzing the Insoluble Bound Phenolic Acids Determined with ABTS and DPPH Methods.

		ABTS		*DPPH*	
Barley		10.1	(*9*)*	1.1	(*26*)
		11.3	(*2*)		
		14.6	(*10*)		
Buckwheat		23.2	(*9*)	53.2	(*27*)
Corn		18.9	(*28*)	21.2	(*29*)
		48.1	(*30*)		
Oat		29.4	(*9*)	0.9	(*31*)
		2.8	(*32*)	27.0	(*29*)
		4.3	(*10*)		
Rice		22.2	(*30*)	13.4	(*29*)
Rye		4.2	(*9*)		
Sorghum	Black	57.1	(*20*)	41.2	(*20*)
		110.2	(*21*)	50.1	(*21*)
		44.1	(*33*)		
	Brown	226.0	(*20*)	202.3	(*20*)
		32.2	(*21*)	15.2	(*21*)
		55.1	(*32*)		
	Red	53.2	(*20*)	28.1	(*20*)
		112.3	(*21*)	40.3	(*21*)
		56.0	(*33*)		
	White	6.2	(*20*)	6.0	(*20*)
		10.3	(*21*)	5.1	(*21*)
		39.1	(*33*)		
Wheat		9.5	(*30*)	29.6	(*29*)
		1.7	(*23*)	7.4	(*25*)
		2.4	(*32*)		
		0.4	(*24*)		

*References are given in parentheses.

More work on total phenolics content and antioxidant activity of cereals is needed. It is necessary to extract cereals with hydrolysis of the bound phenolics before determining their total phenolics content and antioxidant activity, as this will more accurately reflect the real antioxidant activity of cereals in the human body after digestion. It is also important to determine antioxidant activity of cereals using other methods in addition to ABTS and DPPH method to reflect the antioxidant activity of different polarities of antioxidants in cereals.

References

1. Wikipedia, Cereal, http://en.wikipedia.org/wiki/Cereal.
2. Madhujith, T.; Izydorczyk, M.; Shahidi, F. *J. Agric. Food Chem.* **2006**, *54*, 3283-3289.
3. Hill, M. J.; Path, F. R. *Nutri. Res.* **1998**, *18*, 653-659.
4. Zielinski, H. *Polish Journal of Food and Nutrition Sciences* **2002**, *11*, 3-9.
5. Andlauer, W.; Furst, P. *Cereal Foods World* **1998**, *43*, 356-360.
6. Sosulski, F.; Krygier, K.; Hogge, L. *J. Agric. Food Chem.* **1982**, *30*, 337-340.
7. Liyanapathirana, C.; Shahidi, F. *Biofactors* **2004**, *21*, 325-328.
8. Huang, D. J.; Ou, B. X.; Prior, R. L. *J. Agric. Food Chem.* **2005**, *53*, 1841-1856.
9. Zielinski, H.; Kozlowska, H. *J. Agric. Food Chem.* **2000**, *48*, 2008-2016.
10. Zdunczyk, Z.; Flis, M.; Zielinski, H.; Wroblewska, M.; Antoszkiewicz, Z.; Juskiewicz, J. *J. Agric. Food Chem.* **2006**, *54*, 4168-4175.
11. Quettier-Deleu, C.; Gressier, B.; Vasseur, J.; Dine, T.; Brunet, C.; Luyckx, M.; Cazin, M.; Cazin, J. C.; Bailleul, F.; Trotin, F. *J. Ethnopharmacol.* **2000**, *72*, 35-42.
12. Dewanto, V.; Wu, X. Z.; Liu, R. H. *J. Agric. Food Chem.* **2002**, *50*, 4959-4964.
13. Camire, M. E.; Dougherty, M. P.; Briggs, J. L. *Cereal Chem.* **2005**, *82*, 666-670.
14. Towo, E. E.; Svanberg, U.; Ndossi, G. D. *J. Sci. Food Agric.* **2003**, *83*, 980-986.
15. Emmons, C. L.; Peterson, D. M. *Cereal Chem.* **1999**, *76*, 902-906.
16. Emmons, C. L.; Peterson, D. M.; Paul, G. L. *J. Agric. Food Chem.* **1999**, *47*, 4894-4898.
17. Peterson, D. M.; Emmons, C. L.; Hibbs, A. H. *J. Cereal Sci.* **2001**, *33*, 97-103.
18. Goffman, F. D.; Bergman, C. J. *J. Sci. Food Agric.* **2004**, *84*, 1235-1240.
19. Kahkonen, M. P.; Hopia, A. I.; Vuorela, H. J.; Rauha, J. P.; Pihlaja, K.; Kujala, T. S.; Heinonen, M. *J. Agric. Food Chem.* **1999**, *47*, 3954-3962.

20. Awika, J. M.; Rooney, L. W.; Wu, X. L.; Prior, R. L.; Cisneros-Zevallos, L. *J. Agric. Food Chem.* **2003**, *51*, 6657-6662.
21. Dykes, L.; Rooney, L. W.; Waniska, R. D.; Rooney, W. L. *J. Agric. Food Chem.* **2005**, *53*, 6813-6818.
22. Kamath, V. G.; Chandrashekar, A.; Rajini, P. S. *Journal of Cereal Science* **2004**, *40*, 283-288.
23. Yu, L.; Haley, S.; Perret, J.; Harris, M.; Wilson, J.; Qian, M. *J. Agric. Food Chem.* **2002**, *50*, 1619-1624.
24. Liyana-Pathirana, C. M.; Shahidi, F. *J. Sci. Food Agric.* **2006**, *86*, 477-485.
25. Mpofu, A.; Sapirstein, H. D.; Beta, T. *J. Agric. Food Chem.* **2006**, *54*, 1265-1270.
26. Bonoli, M.; Verardo, V.; Marconi, E.; Caboni, M. F. *J. Agric. Food Chem.* **2004**, *52*, 5195-5200.
27. Przybylski, R.; Lee, Y. C.; Eskin, N. A. M. *J. Am. Oil Chem. Soc.* **1998**, *75*, 1595-1601.
28. Wettasinghe, M.; Bolling, B.; Plhak, L.; Parkin, K. *J. Food Sci.* **2002**, *67*, 2583-2588.
29. Miller, H. E.; Rigelhof, F.; Marquart, L.; Prakash, A.; Kanter, M. *J. Am. Coll. Nutr.* **2000**, *19*, 312S-319S.
30. Nagah, A. M.; Seal, C. J. *J. Sci. Food Agric.* **2005**, *85*, 1177-1185.
31. Sun, T.; Xu, Z.; Godber, J. S.; Prinyawiwatkul, W. *Cereal Chem.* **2006**. *83*, 451-454.
32. Yu, L.; Perret, J.; Davy, B.; Wilson, J.; Melby, C. L. *J. Food Sci.* **2002**, *67*, 2600-2603.
33. Dicko, M. H.; Gruppen, H.; Traore, A. S.; Van Berkel, W. J. H.; Voragen, A. G. J. *J. Agric. Food Chem.* **2005**, *53*, 2581-2588.

Antioxidants

Chapter 14

Evaluation of Antioxidant Activity of Curcumin-Free Turmeric (*Curcuma longa* L.) Oil and Identification of Its Antioxidant Constituents

Yongxiang Yu[1], Feng Chen[1], Xi Wang[2], Jeff Adelberg[3], Felix H. Barron[1], Yenhui Chen[1], and Hau Yin Chung[4]

Departments of [1]Food Science and Human Nutrition, [2]Genetics and Biochemistry, and [3]Horticulture, Clemson University, Clemson, SC 29634
[4]Food and Nutritional Sciences Programme, Department of Biology, The Chinese University of Hong Kong, Hong Kong SAR, China

Antioxidant capacity of curcumin-free turmeric (*Curcuma longa* L.) oil was evaluated by two different *in vitro* assays: the DPPH radical scavenging assay and reducing power assay. Results showed that the turmeric oil (TO) possessed strong free radical scavenging activity and reducing power when compared to standard antioxidants such as butylated hydroxytoluene (BHT) and α-tocopherol (VitE). An aliquot of 20 µL/mL TO showed 91% free radical scavenging activity in the DPPH* assay, which was comparable to 10 mM BHT (86%) and 10 mM VitE (96%) under the same conditions. In the reducing power assay, the absorbance at 700 nm of 20 µl/mL TO was 1.085, which was comparable to 10 mM BHT (1.164). Higher concentration of TO at 100 µL/mL reached an absorbance of 1.537, which had no significant difference with 10 mM VitE (1.530). Among the complex constituents in the crude TO, ar-turmerone, turmerone, curlone and α-terpineol that were found to be the major components responsible for the detected antioxidant activities were isolated and identified using various chromatographic techniques including silica gel open column chromatography, normal phase HPLC, and GC-MS. These results showed that TO and some of its inherent bioactive components could potentially serve as alternative natural antioxidants.

There is an increasing interest in antioxidant activities of phytochemicals in diets due to an accumulated evidence that natural antioxidants can protect human bodies against excessive reactive oxygen species (ROS), which are considered to be the harmful by-products generated during normal cell aerobic respiration. Intake of exogenous antioxidants may help to maintain an adequate antioxidant status and the normal physiological function of a living system (*1*). Also, antioxidant is a crucial factor influencing the final food quality in food processing. Therefore, natural and/or synthetic antioxidants are often added as food additives to improve or preserve food quality related to texture, color, flavor and nutritional values, as well as the shelf life of food. Although some synthetic antioxidants such as butylated hydroxyanisole (BHA) and butylated hydroxytoluene (BHT) are very effective and commonly used in the food industry, they are restricted to a certain degree in commercial applications because of possible damaging effect to human health (*2*). For this reason, it is important to explore naturally occurring non- or less-toxic antioxidants that can be used to prevent food oxidative deterioration. In addition, natural antioxidants also have important usage as nutraceutical or cosmetic ingredients.

Turmeric (*Curcuma longa* L.) is a member of the *Zingiberaceae* family, which includes another well known plant, ginger (*Zingiber officinale* Rosc.). Turmeric has been used as an important food ingredient in India for thousands of years because of its special aromatic flavor and attractive color. Govindarajan (3) reported that the dried rhizome of turmeric contained 3-5% essential oil and 0.02-2.0% yellow curcuminoids. Steam distillation of turmeric powder is the most commonly used method to extract the turmeric oil which generally has a pale yellow color and a peppery and aromatic odor. The major components of the oil include α-phellandrene, 1,8-cineol, zingiberene, ar-curcumene, turmerone, β-sesquiphellandrene, curlone and dehydrozingerone (*3*). Turmeric is also used as a traditional Indian medicine in the Ayurvedic system due to its medicinal properties that were highly esteemed by the Indian people. Many researches have shown that curcumin and curcuminoids have various biological activities, such as antioxidant (*4-7*), anticancer (*8-9*), and anti-arthritis (*10*) activities. Despite numerous investigations on biological activities of curcumin, few have involved the curcumin-free turmeric oil that is considered a by-product with little commercial importance (*11,12*). Therefore, there are strong economic reasons to explore new functionalities of other components in the curcumin-free turmeric oil.

In this study, we investigated the antioxidant capacity of curcumin-free turmeric oil using two complimentary *in vitro* assays: 2,2-diphenyl-1-picrylhydrazyl (DPPH) radical scavenging assay and reducing power assay. The antioxidant activity of curcumin-free TO was compared with that of the commercial standard antioxidants, BHT and VitE. Moreover, some TO components showing strong antioxidant activity were further separated from the crude TO by various chromatographic techniques such as open-column silica gel chromatography and normal phase HPLC, and identified by GC-MS.

Materials and Methods

Materials and Chemicals

The crude turmeric oil (TO) was purchased from Aromaland Company (Santa Fe, NM). Curcumin standard, 2,2-diphenyl-1-picrylhydrazyl (DPPH), BHT, α–tocopherol and α-terpineol were obtained from Sigma Chemical Co. (St. Louis, MO). Potassium ferricyanide and ferric chloride were obtained from J. T. Baker Chemical Co. (Phillipsburg. NJ). Trichloroacetic acid and all solvents in HPLC grade were purchased from Fisher Scientific (Suwanee, GA).

HPLC Analyses of Curcumin and the Crude TO

A Shimadzu LC-10AT HPLC system (Kyoto, Japan) was equipped with a Waters (Milford, MA) C_{18} reverse phase column (Xterra^{TM}, 4.6 x 150 mm, 5 μm) and a SPD-M10A photodiode-array (PDA) detector set at 420 nm. UV spectra were recorded in a region of 200-800 nm. Mobile phase was consisted of solvents: (A) water (0.25% acetic acid) and (B) acetonitrile, and run in a gradient profile in: 0-17 min, 40-60% B; 17-32 min, 60-100% B; 32-38 min, 100% B; 38-40 min, 100-40% B; 41 min, stop. Total flow rate of the mobile phase was 0.6 ml/min. Injection volume was 20 μL. Curcumin standard (dissolved in ethanol, 0.15 mg/mL) and TO (dissolved in DCM/methanol (1:2, v/v), 100 μL/mL) were analyzed by HPLC using the aforementioned method.

Fractionation and Identification of Antioxidants from Crude TO

Silica Gel Open Column Chromatography

A glass open column (25 x 2.5 cm) was packed with silica gel (70-230 mesh, 60Å). An aliquot of 1.5 mL crude TO was loaded and eluted by multistep solvent gradients as follows: hexane, dichloromethane (DCM), ethyl acetate, acetonitrile, and methanol. Flow rate of the mobile phase was 4 mL/min. Each collected fraction was 4 mL. After the open-column separation, those TO fractions possessing strong DPPH* scavenging activities were pooled for further analyses.

Normal Phase HPLC Analysis

A Spherisorb silica column (250 x 4.6 mm, 5 μm; Waters, Milford, MA) was installed on a Shimadzu LC-10AT HPLC system (Kyoto, Japan) and equilibrated with DCM. To separate the eluted components in the DCM fraction obtained from the open column chromatography, an aliquot of 20 μL of the pooled fractions were further separated by the HPLC normal phase column and eluted at a flow rate of 1 mL/min by isocratic DCM. For separation of the components in the ethyl acetate fraction, 50 μL of the pooled fractions were separated by the same column and eluted by the same flow rate by linear gradient of methanol from 0 to 10% at 5-15 min to wash out polar compounds. The eluant was collected into 1 mL fractions after passing through the PDA detector for further analysis.

GC-MS Identification

A Shimadzu GC-MS system consisting of a GC-17A with a QP5050 Mass Spectrometer (Kyoto, Japan) equipped with a DB-5 capillary column (60 × 0.25 mm, thickness 0.25 μm; J&W Scientific, Folsom, CA) for all volatile chemicals' quantitative and qualitative analyses. The oven temperature was programmed from 60 to 280°C at a ramp rate of 8°C/min and held at 280°C for 30 min. The injector and ion source temperatures were set at 180 and 290°C, respectively. The detector voltage was 70 eV, and the scanning mass range was *m/z* 43-350. Helium was used as the carrier gas at a column flow rate of 1.2 mL/min. The sample injection volume was 2 μL with a split ratio of 1:1. Identification of compounds was based on comparison of their mass spectra and retention indices (RIs) with those of the authentic standards. Chemical RIs were calculated using a series of *n*-alkanes (C_8-C_{26}). If standard compounds were not available, each unknown compound was tentatively identified by comparing the mass spectrum with that in the Wiley and NIST mass spectral databases and the previously published RIs obtained under the same conditions (*13-15*).

Antioxidative Capacity

DPPH Scavenging Assay

DPPH scavenging assay was adopted using the method described by Yamaguchi *et al.* (*16*) with minor modifications. An aliquot of 0.4 mL of sample dissolved in DCM was mixed with 0.4 mL of 0.25 mM DPPH solution dissolved in DCM. The mixture was shaken vigorously and left in the dark at room temperature for 30 min. A control consisted of 0.4 mL solvent instead of sample. Because the crude TO had a light yellowish color, 0.4 mL TO (at each concentration) was added to 0.4 mL solvent as a blank to eliminate the color interference from TO (see Equation 1). The absorbance of the resulting solution was measured spectrophotometrically at 517 nm. The scavenging effect of DPPH was calculated by the following equation:

$$\text{Scavenging effect (\%)} = \left(1 - \frac{\text{absorbance of sample at 517 nm} - \text{blank}}{\text{absorbance of control at 517 nm}}\right) \times 100 \qquad (1)$$

Reducing Power Assay

The reducing power of TO was determined by the method of Yen and Chen (*17*) and Chung *et al.* (*18*) with minor modifications. Turmeric oil was dissolved in acetone to prepare solutions in different concentrations. An aliquot of 0.5 mL of the sample was mixed with 1 mL of 1% potassium ferricyanide [$K_3Fe(CN)_6$], then the mixture was incubated at 50°C for 20 min, followed by the addition of 1 mL of trichloroaceteic acid (10%) to the mixture and centrifugation for 10 min. The upper layer of the solution (1 mL) was mixed with 1 mL distilled water and 0.2 mL $FeCl_3$ (0.1%). The absorbance of the mixture was measured at 700 nm. Higher absorbance of the reaction mixture indicated a higher reducing power.

Triplicates were performed for each concentration of the tested samples and standards in these two methods. The experiments were repeated three times on different days.

Statistical Analysis

The data of the antioxidant activities of TO, BHT and VitE were subjected to the analysis of variance (ANOVA) and the least significant difference (LSD at $p<0.05$). Analyses were performed using the statistical software SAS 9.1 operated on the Windows system (SAS Institute Inc., Cary, NC).

Results and Discussion

Confirming No Curcumin and Curcuminoids in Crude TO

Standards of curcumin and its derivatives were analyzed by the HPLC resulting in three peaks (Figure 1), including [1] bisdemethoxycurcumin; [2] demethoxycurcumin; and [3] curcumin. These chemicals showed strong absorbance wavelength at 420-430 nm. Comparing the HPLC profile of the standards with that of TO, it was confirmed that our sample was the curcumin-free turmeric oil.

Determination of the Antioxidant Activity of Crude TO

The antioxidant activity of crude TO was initially measured by the DPPH* scavenging assay and reducing power assay, and compared with that of 10 mM

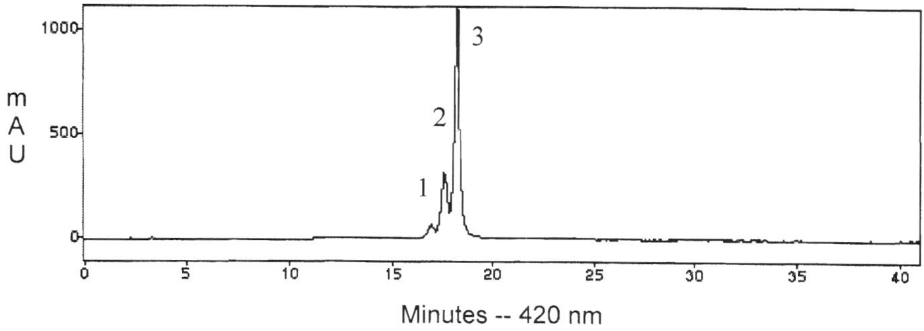

Figure 1. HPLC chromatogram of curcumin and curcuminoids and their absorbance at 420nm. 1. bisdemethoxycurcumin; 2. demethoxycurcumin; 3. curcumin.

BHT and 10 mM VitE. Crude TO was prepared in a serial concentrations (1, 5, 10, 20, 40, 70, 100 μL/mL) and tested by the two aforementioned assays. Results shown in Figure 2A indicated that the DPPH scavenging activity of the crude TO increased with the increasing TO concentration. When 0.25 mM DPPH solution was used, it could be saturated by the crude TO at the concentration of 20 μL/mL and above, and resulted in a scavenging activity of approximate 90%, which was comparable to the antioxidant ability of 10 mM BHT and 10 mM VitE at the same condition.

Figure 2B shows the reducing powers of TO, BHT, and VitE. Similar to the DPPH assay, the reducing power of crude TO also increased with the increasing concentration. The spectrometric absorbance of the reducing power of 20 μL/mL TO was 1.085, which was comparable to that of 10 mM BHT (1.164), but lower than that of 10 mM VitE (1.530) (p<0.001). When the concentration of TO increased to 100 μL/mL, its absorbance increased to 1.537, which had a close reducing power to 10 mM VitE without significant difference (p<0.001). These results demonstrated that the TO was electron donors, and could react with free radicals and convert them to more stable products (*17*).

Separation and Identification of Antioxidants in Crude TO

Since the tested TO had shown strong antioxidant activities in two assays, its inherent antioxidants were then subjected to a series of sequential separations with silica gel open column chromatography and normal phase HPLC, and finally identified by GC-MS.

When the crude TO was separated by the silica gel open column chromatography using a stepwise solvent elution method with different solvents such as hexane, DCM, ethyl acetate, acetonitrile and methanol, a total of 235 fractions were collected. As shown in Figure 3, there were two strong

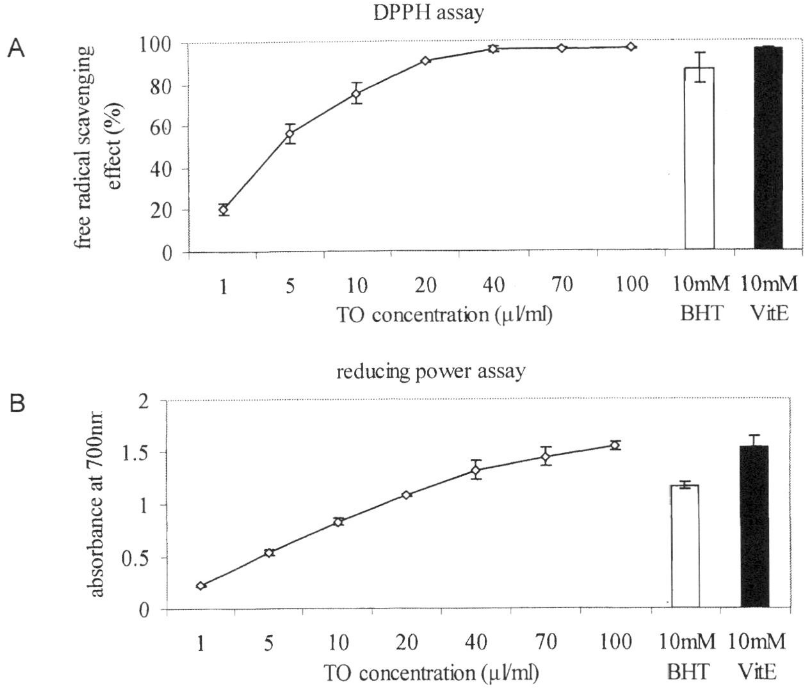

Figure 2. Antioxidative capacity of crude TO. A: DPPH assay, 10 mM BHT and 10 mM VitE were used as standards; B: Reducing power assay, 10 mM BHT and 10 mM VitE were used as standards.

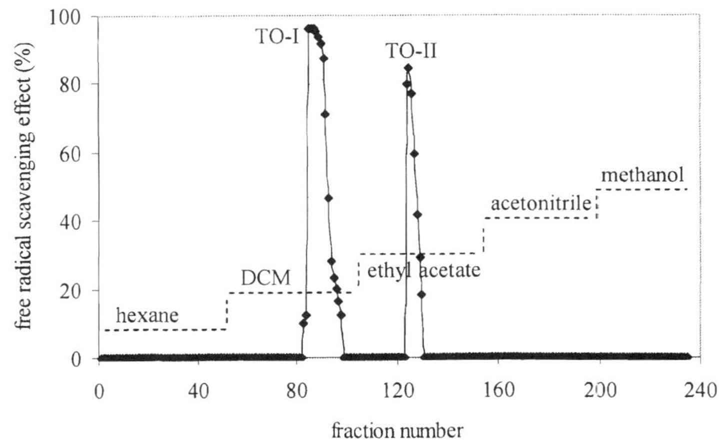

Figure 3. Free radical scavenging activity of fractions of crude TO separated by silica gel open column chromatography.

antioxidant peaks corresponding to the DPPH assay, namely, TO-I and TO-II, respectively. The TO-I contained the fractions 85-95 that were eluted by DCM; while the TO-II comprised the fractions 123-129 eluted by ethyl acetate. The strength of antioxidant activities of both TO-I and TO-II were similar, with the antioxidant activity values close to 85% in the DPPH test. To further separate antioxidants in these fractions, all fractions within the TO-I and the TO-II were pooled respectively and separated by the Spherisorb silica HPLC column. HPLC separation chromatograms and online antioxidant determination by the DPPH assay are shown in Figure 4. Major components of TO-I were eluted by DCM. TO-II was eluted by DCM and methanol. In Figure 4A, there were 3 peaks, named TO-I-1, TO-I-2 and TO-I-3, respectively. However, only the fraction TO-I-3 showed a high antioxidant activity (90%). Figure 4B showed two chromatographic peaks, denoted as TO-II-1 and TO-II-2. The fraction TO-II-2 had a light yellow color and possessed a medium free radical scavenging activity (60%). Since GC-MS analysis indicated that the TO-II-2 fraction contained more than 10 compounds (Figure 5C), it was further concentrated and separated by DCM/methanol (v/v 94:6) using the same column, which resulted into two peaks (Figure 4C). However, only the second peak (TO-II-2-b) showed the antioxidant activity. Therefore, both the fraction TO-I-3 and the fraction TO-II-2-b were selected for further chemical identification by GC-MS.

Identification of Antioxidants in TO-I and TO-II Separated by Spherisorb Silica HPLC

Previous research (*19*) found that crude TO could contain nearly 100 compounds, including major volatile chemicals such as ar-turmerone, turmerone, β-sesquiphellandrene and curcumene, among others. In our sample, the major components were ar-curcumene (16.4%), zingiberene (16.5%), β-sesquiphellandrene (14.4%), ar-turmerone (16.8%) and turmerone (16.5%) (Figure 5A, Table I). Our finding was similar to the previous report. Further chemical separation and identification of the compounds in the fractions revealed that there were only 3 compounds in the fraction TO-I-3 and more compounds in the fraction TO-II-2-b. As profiled in Figure 5, ar-turmerone, turmerone, and curlone were concentrated in the fraction TO-I-3 (Figure 5B), while α-terpineol was enriched in the fraction TO-II-2-b (Figure 5D). All these chemicals were identified after comparing their RIs and mass spectra with those in an essential oil library (*15*) and the Wiley and NIST mass spectral databases, as well as with standard compounds under the same experimental conditions. Figure 6 shows the structures of the four identified compounds. These results confirmed the previous suspect that the antioxidant activity of turmeric oil may be also attributed to the chemicals such as ar-turmerone, turmerone, and some other oxygenated compounds (*20*). Although the mixture of ar-turmerone, turmerone and curlone showed high antioxidant activity in our preliminary

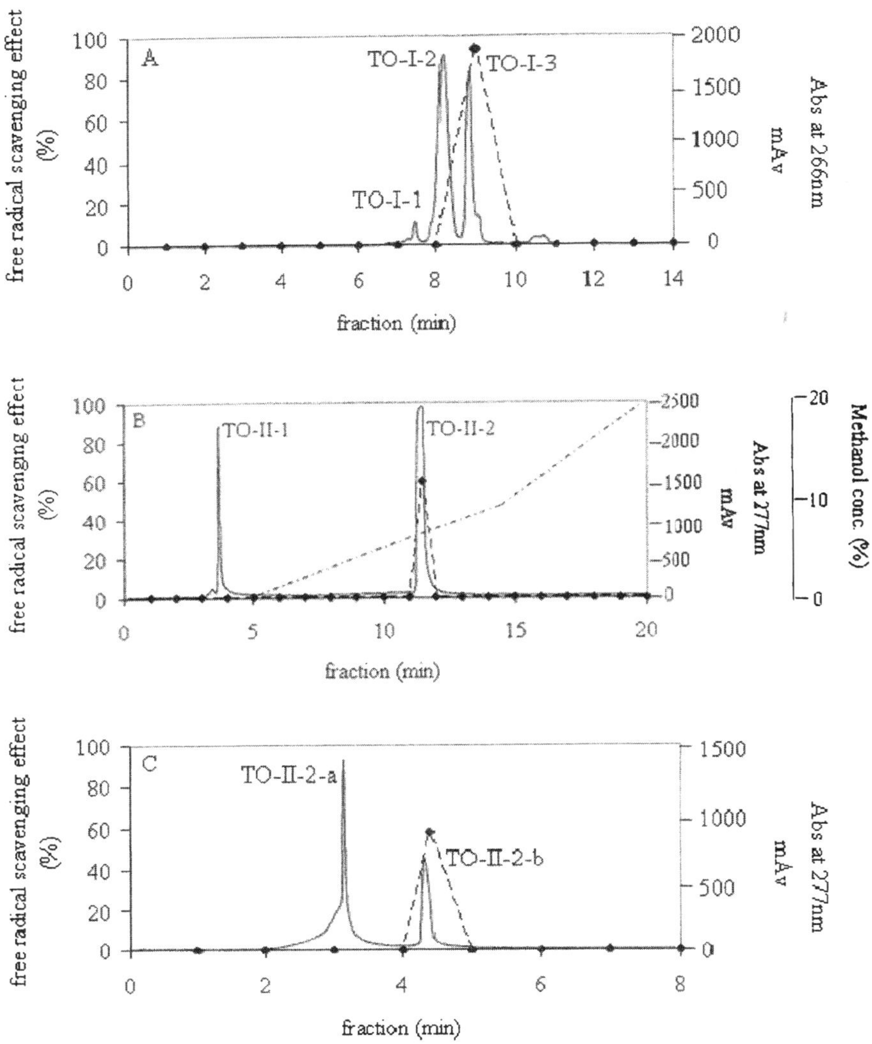

Figure 4. Chromatograms of TO-I and TO-II and their free radical scavenging activities. A: TO-I separated by Spherisorb silica HPLC, isocratic method, mobile phase: DCM; B: TO-II separated by Spherisorb silica HPLC, gradient method, mobile phase: a. DCM. b. methanol; C: TO-II-2 separated by Spherisorb silica HPLC, gradient method, mobile phase: DCM /methanol (94:6, v/v).

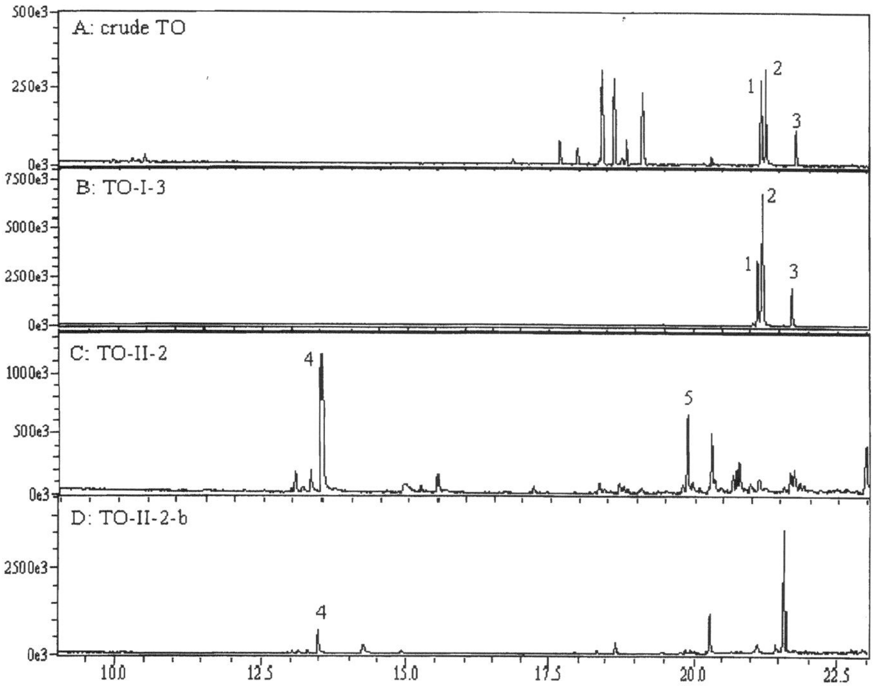

Figure 5. Gas chromatographic profiles of crude TO and its fractions. 1. ar-turmerone; 2. turmerone; 3. curlone; 4. α-terpineol; 5. turmerol.

Table I. Composition of Crude TO.

Retention time (min)	*Compound*	*RI*[a]	*Composition (%)*	*Identification method*
17.650	*Trans*-caryophyllene	1444	4.58	MS, RI
17.958	Farnesene	1463	2.93	MS, RI
18.417	ar-curcumene	1491	16.44	MS, RI
18.617	Zingiberene	1505	16.51	MS, RI
18.842	β-bisabolene	1518	4.82	MS, RI
19.117	β-sesquiphellandrene	1537	14.35	MS, RI
21.183	ar-turmerone	1675	16.84	MS, RI
21.267	Turmerone	1680	16.45	MS, RI
21.767	Curlone	1715	7.08	MS, RI

[a] RI was calculated using a series of *n*-alkanes (C_8-C_{26}).
MS: Mass spectra.

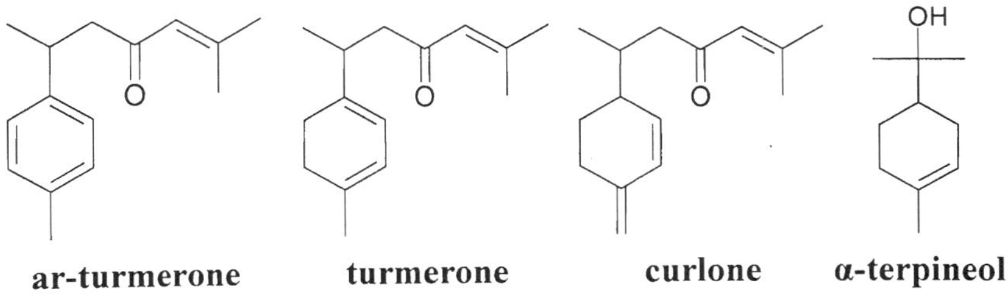

Figure 6. Structure of turmeric components with antioxidant activities.

screening, their individual concentration-activity relationship could not be established due to the lack of available standards.

The fraction TO-II-2 was light yellow when concentrated. GC-MS analysis showed that the fraction contained more than 10 compounds that included turmerol and α-terpineol as the major constituents (Figure 5C). The fraction TO-II-2-b became colorless when the naturally yellow turmerol was removed. The concentrated fraction TO-II-2-b possessed 58% free radical scavenging activity when 0.1 mM DPPH solution was used. Although α-terpineol (α,α,4-trimethyl-3-cyclohexene-1-methanol) showed a moderate free radical scavenging activity during the concentration range of 10 to 100 μL/mL (Figure 7), it showed very weak reducing power.

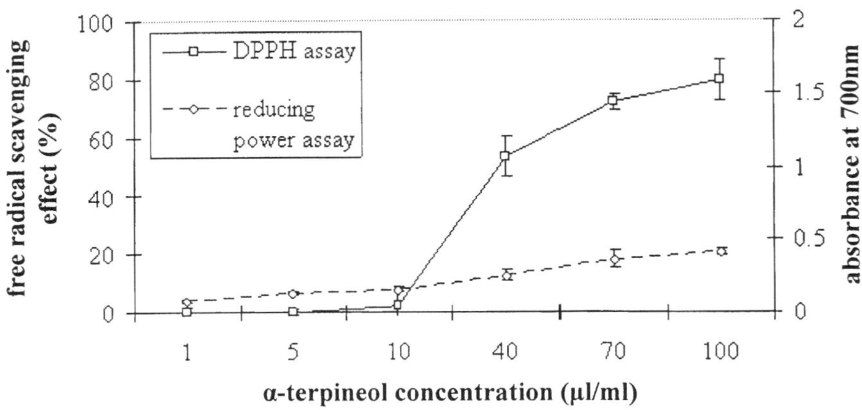

Figure 7. Free radical scavenging assay and reducing power assay of α-terpineol.

Conclusions

Many studies have addressed the benefits of using turmeric because of its inherent bioactive chemicals such as curcumin and curcuminoids. However, there is a lack of information of the potential values of curcumin-free turmeric oil. Our present study revealed that ar-turmerone, turmerone, curlone and α-terpineol in the curcumin-free TO also possessed some strong antioxidant activities. Considering their other biological activities such as antibacterial (*13*), antifungal (*21*), antivenom (*22*) and insect repellent (*23*) activities, the curcumin-free TO that was considered a by-product without commercial value might be able to be converted to a value-added product, for example, using TO as an alternative natural antioxidative food additive.

References

1. Mimić-Oka, J.; Simić, D.V.; Simić, T.P. *Med. Biol.* **1990**, *6*, 11-22.
2. Ito, N.; Fukushima, S.; Tsuda, H.; Shirai, T.; Hagiwara, A.; Imaida, K. In *Food Toxicology: Real or Imaginary Problems?* Gibson, G.G. and Walker, R. London and Philadelphia: Taylor & Francis. **1985**, pp. 181-198.
3. Govindarajan, V.S. *Crit. Rev. Food Sci. Nutr.* **1980**, *12*, 199-301.
4. Daniel, S.; Limson, J.L.; Dairam, A.; Watkins, G.M.; Daya, S. *J. Inorg. Biochem.* **2004**, *98*, 266-275.
5. Ruby, A.J.; Kuttan, G.; Babu, K.D.; Rajasekharan, K.N.; Kuttan, R. *Cancer Lett.* **1995**, *94*, 79-83.
6. Chatterjee, S.; Desai, S.R.P.; Thomas, P. *Food Res. Int.* **1999**, *32*, 487-490.
7. Rukkumani, R.; Aruna, K.; Varma, P. S.; Rajasekaran, K.N.; Menon, V.P.M. *J. Pharm. Pharmaceut. Sci.* **2004**, *7*, 274-283.
8. Nagai, S.; Kurimoto, M.; Washiyama, K.; Hirashima, Y., Kumanishi, T.; Endo, S. *J. Neuro-Oncol.* **2005**, *74*, 105-111.
9. Verma, S.P.; Salamone, E.; Goldin, B. *Biochem. Biophy. Res. Commun.* **1997**, *233*, 692-696.
10. Funk, J.L.; Oyarzo, J. N.; Frye, J.B.; Chen, G.; Lantz, R.C.; Jolad, S.D.; Sólyom, A.M.; Timmermann, B.N. *J. Nat. Prod.,* **2006**, *69*, 351-355.
11. Saju, K.A.; Venugopal, M.N.; Mathew, M. J. *Curr. Sci.*, **1998**, *75*, 660-662.
12. Jayaprakasha, G.K.; Rao, L.J.M.; Sakariah, K.K. *Trends Food Sci. Technol.*, **2005**, *16*, 533-548.
13. Negi, P.S.; Jayaprakasha, G.K.; Rao, L.J.M.; Sakariah, K. K. *J. Agric. Food Chem.,* **1999**, *47*, 4297-4300.
14. Chatterjee, S.; Variyar, P.S.; Gholap, A.S.; Padwal-Desai, S.R.; Bongirwar, D.R. *Food Res. Int.,* **2000**, *33*, 103-106.

15. Adams, R.P. *Identification of essential oil components by gas chromatography / quadrupole mass spectroscopy.* Carol Stream, IL Allured Publishing Corporation. 2001
16. Yamaguchi, T.; Takamura, H.; Matoba, T.; Terao, J. *Biosci. Biotechnol. Biochem.*, **1998**, *62*, 1201-1204.
17. Yen, G. C.; Chen, H. Y. *J. Agric. Food Chem.*, **1995**, *43*, 27-32.
18. Chung, Y. C., Chang, C. T., Chao, W. W., Lin, C. F., & Chou, S. T. Antioxidative activity and safety of the 50% ethanolic extract from red bean fermented by *Bacillus subtilis* IMR-NK1. *J. Agric. Food Chem.*, *50*, 2454-2458.
19. Raina, V.K.; Srivastava, S.K.; Jain, N.; Ahmad, A.; Syamasundar, K.V.; Aggarwal, K.K. *Flav. Fragrance J.*, **2002**, *17*, 99-102.
20. Jayaprakasha, G.K.; Jena, B.S.; Negi, P.S.; Sakariah, K.K. *Verlag der Zeitschrift für Naturforschung*, **2002**, *57c*, 828-835.
21. Jayaprakasha, G.K.; Negi, P.S.; Anandharamakrishnan, C.; Sakariah, K.K. *Verlag der Zeitschrift für Naturforschung*, **2001**, *56c*, 40-44.
22. Ferreira, L. A. F.; Henriques, O.B.; Andreoni, A.A.S.; Vital, G.R.F.; Campos, M.M.C.; Habermehl, G.G.; Moraes, V.L.G.D; *Toxicon,* **1992**, *30*, 1211-1218.
23. Su, H.C.F.; Horvat, R.; Jilani, G. *J. Agric. Food Chem.* **1982**, *30*, 290-292.

Chapter 15

Identification and Evalutation of Antioxidant Phenolic Compounds in Parsley (*Petroselinum crispum var. neapolitanum*) and Radish (*Raphanus sativus* L.) Sprout

Hyun-Jin Kim[1,2], Feng Chen[2], Ju-Hee Choi[3], Xi Wang[4], and Yueming Jiang[5]

[1]Department of Food Science and Human Nutrition, Clemson University, Clemson, SC 29634
[2]Food Convergence Technology Division, Korea Food Research Institute, Bundang, Sungnam, Gyeonggi 463746, Korea
[3]Department of Food Engineering, Mokpo National University, Chungkey, Muangun, Chonnam 534729, Korea
[4]Department of Genetics and Biochemistry, Clemson University, Clemson, SC 29634
[5]South China Botanical Gardens, Chinese Academy of Sciences, Guangzhou 510650, People's Republic of China

Free phenolic antioxidant compounds (i.e., caffeic acid in parsley and ferulic acid, methyl ferulate, isoferulic acid, sinapic acid, and methyl sinapate in radish sprout) in parsley and radish sprout were separated and identified by chromatographic techniques. Their antioxidant capacities were determined by the DPPH radical scavenging and the β-carotene-linoleic acid assays. Caffeic acid had a lower EC_{50} value than two antioxidant standards, BHT and α-tocopherol in the DPPH$^{\cdot}$ scavenging assay, but its inhibitory efficacy against the discoloration of β-carotene was much weaker than that of the standards and other phenolics, except isoferulic acid, in the β-carotene-linoleic acid assay. All phenolic constituents in the radish sprout had lower antioxidant activities than the standard antioxidants in both assays, but the crude methanolic extract of radish sprout showed a stronger free radical scavenging capacity than the parsley extract in the DPPH$^{\cdot}$ assay. It was also found that the methylation of hydroxyl group on phenol ring decreased the free radical scavenging activity.

Oxidative and reductive reactions are common and essential processes in biological systems. However, problems may arise when the electron flow becomes uncoupled leading to excessive generation of free radicals known as reactive oxygen species (ROS) such as superoxide ($O_2^{\cdot -}$), peroxyl ($ROO^{\cdot}$), alkoxyl ($RO^{\cdot}$), hydroxyl ($HO^{\cdot}$), and reactive nitrogen species such as nitric oxide ($NO^{\cdot}$) (*1-2*). These aforementioned radicals can cause oxidative damage to different cellular marcromolecules including lipids, proteins, and DNA. The damage is considered to be closely associated with various diseases and the aging process in humans (*3-7*).

To protect against free radicals, the human body has two different endogenous antioxidant systems: enzymatic system and nonenzymatic system. The former includes superoxide dismutase, glutathione peroxidase, catalase, and D-T diaphorase, among others, which metabolize superoxide, hydrogen peroxide, and lipid peroxides to prevent production of the toxic radicals. The latter consists of plasma proteins, urate, glutathione, dihydrolipoic acid, histidine-containing peptides, and melatonin, among others, which nonspecifically scavenge free radicals (*1,8*). Although endogenous antioxidant systems in most cases work well in human body, some ROS/RNS still escape and cause various damages that may lead to different diseases such as cancer and cardiovascular disease because of various continuous environmental and physical inducers such as UV radiation, smoke, carcinogens, toxins, excessive exercise and injury (*9-10*). To scavenge the excessive ROS/RNS, exogenous dietary antioxidants seem to be necessary as complements to the endogenous antioxidants. During the past decades many dietary antioxidants have been investigated, including one of the most widely studied group, phenolic compounds (*1, 11-15*).

Phenolic compounds are widely distributed in the plant kingdom and are mostly derived from phenylalanine via the phenylpropanoid pathway. These compounds naturally function as secondary metabolites to protect plants against biotic and abiotic stresses (*16-18*). Recently, some phenolic compounds have been intentionally incorporated into our diets because of their various health benefits such as anticancer, antioxidant, and anti-inflammatory activities (*8, 13-15*). Although the scientific basis of the relationship between the epidemiological efficacy and the free radical scavenging ability of phenolic compounds is still in debate, many investigators believe that the beneficial health effects of dietary phenolic compounds are due to their strong antioxidant activities (*14*). Such antioxidant activities are mainly due to the hydroxyl groups on the phenol ring, but the hydroxyl groups of phenolic compounds are sometimes glycosylated with various sugars such as glucose, galactose, rhamnose, xylose, and arabionose in plants (*12-13*). The glycosylation is reported to decrease the original antioxidant activities of the phenolic compounds (*13*). These glycosylated phenolic compounds can be hydrolyzed to the aglycone, or free phenolic compounds, by enzymes in the intestine for their absorption and bioavailability. After absorption, these free phenolic compounds are conjugated through glucuronidation, sulfation, and methylation, and then act as antioxidants in the human body (*1,12,14*).

In recent years there has been an increasing public interest in natural antioxidants, which has led to an extensive search for effective natural antioxidants in our diets. However, in spite of the intensive research on fruits, vegetables, herbs and spices, attempts to identify antioxidant free phenolic compounds in parsley (*Petroselinum neapolitanum*) and radish (*Raphanus sativus* L.) sprout are rather limited (*19-20*). Therefore, the aim of this research was to investigate the antioxidant values of the natural antioxidants inherent in these two common plants. Antioxidant free phenolic compounds in parsley and radish sprouts were separated and identified using Sephadex LH-20 chromatography, HPLC, and GC/MS. In addition, their antioxidant activities were evaluated and compared with the antioxidant standards (i.e., butylated hydroxytoluene (BHT) and α-tocopherol) by the DPPH˙ scavenging and the β-carotene-linoleic acid methods.

Materials and Methods

Chemicals and Plants

2,2-Diphenyl-1-picrylhydrazyl (DPPH˙), α-tocopherol, butylated hydroxytoluene (BHT), caffeic acid, gallic acid, and Folin-Ciocalteu's reagent were purchased from Sigma Chemical Co. (St. Louis, MO). Sinapic acid and β-carotene were obtained from Fluka Chemical Co (Milwakee, WI). Ferulic acid, linoleic acid, bis-(trimethylsilyl) trifluoroacetamide (BSTFA), isoferulic acid, and methyl ferulate were purchased from Aldrich Chemical Co. (Milwakee, WI), TCI (Portland, OR), Supelco (Bellefonte, PA), Acros (NJ), and Alfa Aesar

(Ward Hill, MA), respectively. Tween 20 and all HPLC solvents in analytical grade were from Fisher Scientific (Suwanee, GA).

In this study, parsley and radish sprouts were grown in 0.4 L plastic pots containing a commercial potting mixture (Fafard 3-B Mix, Fafard Inc., Anderson, S.C.) in the greenhouse located at the Clemson University's main campus at Clemson, South Carolina, and watered every other day. Parsley plants were fertilized once a week at irrigation with 1g/L of a 20 N-4.4 P-16.7 K water solution fertilizer (Peters 20-10-20 Peat-Lite Special, Scotts-Sierra Horticultural Products Co., Marysville, Ohio) during the cultivation, but radish sprouts were not fertilized. Greenhouse cooling/heating set points were 27/25 °C under natural light condition.

Extraction and Separation of Antioxidant Phenolic Compounds from Parsley and Radish Sprout

After harvesting, parsley (55-day-old) and radish sprouts (7-day-old) were ground in liquid nitrogen into fine powders. One hundred grams of both plant powders were mixed with 100 mL of 80% aqueous methanol to extract the phenolic compounds. The mixture was shaken at room temperature for 12 hr and then centrifuged at 2000 g for 20 min. After centrifugation, each methanol supernatant of parsley and radish sprouts was used for determination of crude phenolic compounds.

Solvent Separation and Sephadex LH-20 Open Column Chromatography

After concentrating the methanolic extracts using a rotary evaporator at 35°C, suspicious interfering compounds were successively removed by hexane, chloroform, dichloromethane, and ethyl acetate. The samples were then separated by Sephadex LH-20 open column chromatography. Each sample (6 mL) was loaded on the top of the column (60 cm × 2.5 cm) which was packed with Sephadex LH-20 resin and equilibrated with water, and then linearly eluted with gradually increasing methanol from 0 to 30% at 4 ml/min to collect fractions in size of 8 mL.

HPLC Separation

Further separation of the targeted antioxidant phenolic compounds was performed on a Shimadzu LC-10AT HPLC system (Kyoto, Japan), which consisted of a YMC™ ODS-AQ reverse-phase C_{18} column (250 mm × 4.6 mm, 5 μm; Waters, Milford, MA) and a Shimadzu SPD-M10V diode array detector (DAD). The column was initially equilibrated with HPLC water containing 0.05% trifluoroacetic acid (TFA). Fifty microliters of each sample that were separated from the Sephadex LH-20 open column chromatography was injected and eluted with HPLC water containing 0.05% TFA (solvent A) and acetonitrile (solvent B) at a flow rate of 0.8 mL/min for parsley or 1 mL/min for radish sprout. The absorbance of the eluant was scanned from 200 to 500 nm by DAD.

Acidic Hydrolysis of Phenolic Conjugates and Derivatization of Free Phenolics

To obtain free phenolic compounds, phenolic glycosides were separated by C_{18}-HPLC and hydrolyzed by hydrogen chloride (HCl). The sample (1 mL) was mixed with 4 mL of 2 N HCl in 50% aqueous methanol and incubated at 90°C for 1 hr. After incubation, methanol was purged off by high purity nitrogen gas. The hydrolyzed free phenolic compounds were extracted with ethyl acetate. Free phenolic compounds were then derivatized with bis-(trimethylsilyl) trifluoroacetamide (BSTFA) prior to the GC/MS analysis. The volatile TMS-phenolics were prepared as follows. One milliliter of ethyl acetate containing free phenolic compounds was dried under nitrogen gas and then the dried sample was reacted with 50 μL of BSTFA in the capped vial at 70°C for 1 hr.

Identification of the Antioxidant Free Phenolic Compounds

A DB-5 capillary column (60 m × 0.25 mm, thickness 0.25 μm; J&W Scientific, Folsom, CA) was installed in a Shimadzu GC-17A instrument that was also connected to a QP 5050 Mass Spectrometer (MS) detector (Kyoto, Japan). The GC oven temperature was programmed from 60 to 280°C at a rate of 10°C/min and held at 280°C for 5 min. The injector and ion source temperatures were 220 and 290°C, respectively. The detector voltage was set at 70 eV and the MS spectra were obtained in the mass range of m/z 43-350. Helium was used as the carrier gas at a flow rate of 1.1 mL/min. One microliter of each derivatized sample was injected into the column in a split mode at 1:5. Identification of compounds was based on comparison with mass spectra and retention index (RI) of the authentic standards. Tentative identification of methyl sinapate was based on comparison with the mass spectra of Wiley and NIST mass spectral databases and our own RI database.

Determination of Total Amount of Phenolic Compounds

The total phenolic contents in parsley and radish sprouts were determined using Folin-Ciocalteu's reagent by the method of Singleton and Rossi (*21*). Fifty microliters of the crude methanolic extract were mixed with 450 μL of distilled water and 250 μL of 2 N Folin-Ciocalteu reagent. The mixture was incubated with 1.25 mL of 20% Na_2CO_3 at 25°C for 20 min and then centrifuged at 2000 g for 10 min. The absorbance of the supernatant was spectrophotometrically measured at 735 nm. The standard curve was prepared using gallic acid (GA).

Determination of Antioxidant Activity

The antioxidant capacities of parsley and radish sprouts were assessed by the $DPPH^{\cdot}$ scavenging assay and the β-carotene-linoleic acid assay. Butylated hydroxytoluene (BHT) and α-tocopherol were used as standards.

DPPH˙ Scavenging Assay

Scavenging activities on $DPPH^{\cdot}$ by the crude extracts of parsley and radish sprout and pure chemicals were determined according to the method of Yamaguchi *et al.* (*22*) with slight modification. The reaction mixture containing 0.1 mL of sample, 0.3 mL of 0.1 M Tris-HCl (pH 7.4), 0.1 mL of methanol, and 0.5 mL of 0.3 mM $DPPH^{\cdot}$ was vigorously shaken and incubated in the darkness at room temperature for 10 min. After incubation, the absorbance of the reaction mixture was spectrophotometrically measured at 517 nm and the scavenging activity of $DPPH^{\cdot}$ was calculated by the following formula:

$$\text{Scavenging activity (\%)} = \left(1 - \frac{\text{absorbance of sample at 517 nm}}{\text{absorbance of control at 517 nm}}\right) \times 100 \quad (1)$$

β-Carotene-Linoleic Acid Assay

Antioxidant capacities of the crude extracts of parsley and radish sprouts and reference chemicals were determined by measuring the discoloration of β-carotene due to its reaction with radicals which were formed by linoleic acid oxidation in an emulsion (*23*). Five milliliters of β-carotene solution that was prepared by dissolving 10 mg of β-carotene in 50 mL of chloroform were mixed with 50 mg of linoleic acid and 500 mg of Tween 20. The chloroform in the mixture solution was evaporated under nitrogen and then 100 mL of distilled water were added to the mixture. Sample (0.1 mL) was mixed with 1 mL of the emulsion and incubated at 50°C for 30 min and the reaction mixture without sample was used as a blank. The absorbance of the reaction mixture was spectrophotometrically measured at 470 nm and the antioxidant activity was calculated by the following formula:

$$\text{Antioxidant activity } (\%) = \left(1 - \frac{S_0 - S_{30}}{B_0 - B_{30}}\right) \times 100 \qquad (2)$$

Where S_0 and B_0 are the absorbance values of the sample and the blank, respectively, before incubation, while S_{30} and B_{30} are the absorbance of the sample and blank, respectively, after incubation at 50°C for 30 min.

EC_{50} value used to evaluate antioxidant capacities of the crude extracts, identified phenolic compounds, and standards were the effective concentration at which DPPH radicals were scavenged and β-carotene was bleached by 50%, respectively.

Statistical Analysis

The experimental data of the total amount of phenolic compounds and the determined antioxidant capacities were tested in triplicate and subjected to the analysis of variance (ANOVA) and the least significant difference (LSD at $p < 0.05$) by the SAS software for Windows V8 (SAS 9.1, SAS Inst. Inc., Cary, NC). To calculate the EC_{50} values of the phenolic compounds, the data on the antioxidant activities were subjected to ANOVA and analyzed with nonlinear regressions (SAS 9.1).

Results and Discussion

Total Amounts of Phenolic Compounds of Parsley and Radish Sprout and Their Antioxidant Activities

Total amount of phenolic compounds in parsley was equivalent to 26 mg gallic acid equivalent/g fresh tissue, which is 30% higher than that of the radish sprouts in 20 mg gallic acid equivalent/g fresh tissue (Table I). However, when the antioxidant capacities of both methanolic extracts were determined by two different methods: the DPPH$^{\cdot}$ scavenging assay and the β-carotene-linoleic acid system, the EC_{50} value of the radish sprout extract in the former test was 9 mg/mL which was much lower than that of parsley (60 mg/mL), while both extracts had similar inhibitory capacities against the β-carotene bleaching in the latter system with close EC_{50} values at 2.3 mg/mL (Table I).

These results indicated that the antioxidants in two plant extracts, when tested by different antioxidant methods, had been reflected by multiple facets. Firstly, the antioxidants in parsley had stronger capacity to inhibit the coupled oxidation of β-carotene and linoleic acid than to scavenge the $DPPH^{\cdot}$ and the radish sprouts extract possessed potent antioxidant capacities in both assays. Secondly, the result implied that no single bioassay was capable of providing a comprehensive picture of the antioxidant profile of a plant extract owing to its complex constituents. Finally, there might exist additive, synergistic and/or antagonistic effects among the extracted chemicals. In this context, there was a necessity to investigate and clarify the paradox. Therefore, antioxidant phenolic compounds in the both crude methanolic extracts were separated and characterized by the following chromatographic methods.

Table I. Total amounts of Phenolic Compounds and Antioxidant Capacities of the MethanolicExtracts of Parsley and Radish Sprout

Materials	*Total phenolic compounds (mg GA equivalent/g tissue)*	*Antioxidant capacity (EC_{50}, mg/mL)*	
		Scavenging activity	*Antioxidant activity* [a]
Parsley	25.91 ± 0.83	59.85 ± 1.99	2.26 ± 0.06
Radish sprout	19.66± 0.92	8.67 ± 0.66	2.28 ± 0.05

[a] antioxidant activity was determined using the β-carotene-linoleic acid assay.

Separation of Antioxidant Phenolic Compounds in Methanolic Extracts of Parsley and Radish Sprout

To remove non-phenolic compounds in the parsley and radish sprout extracts, both extracts were sequentially separated by the following solvents: hexane, chloroform, dichloromethane, and ethyl acetate. Through this process, it was found that most antioxidant phenolic compounds of parsley and radish sprouts still remained in the aqueous methanol layer (data not shown). This also suggested that the remaining antioxidant phenolic compounds had strong hydrophilic properties. After solvent separation, each aqueous methanol layer was further separated by a Sephadex LH-20 open column with the aid of an on-line DPPH˙ assay. As shown in Figure 1, the parsley extract had one active peak (PI) (Figure 1A), while radish sprouts extract exhibited two active peaks (RI and RII) (Figure 1B). All these antioxidant fractions showed strong free radical scavenging activities at about 90%. Moreover, the sharp chromatographic peak of PI implied it might contain less number of phenolic compounds, while the broad chromatographic and antioxidant peaks for the radish sprout extract revealed more co-eluted phenolics. To further separate the phenolics in the collected fractions of PI, RI, and RII, the reverse phase C_{18}-HPLC with the aid of on-line DPPH˙ assay was used. Four different antioxidant-active peaks (I, II, III, and IV) were separated from fraction RII, while only one active peak was found in fractions PI and RI (Figure 2). Although PI and RI were well separated, the intensities of phenolic compounds in these two fractions were smaller than the intensities of other peaks in fraction RII. However, all RII active peaks seemed to still contain some unresolved compounds due to the low resolution of HPLC.

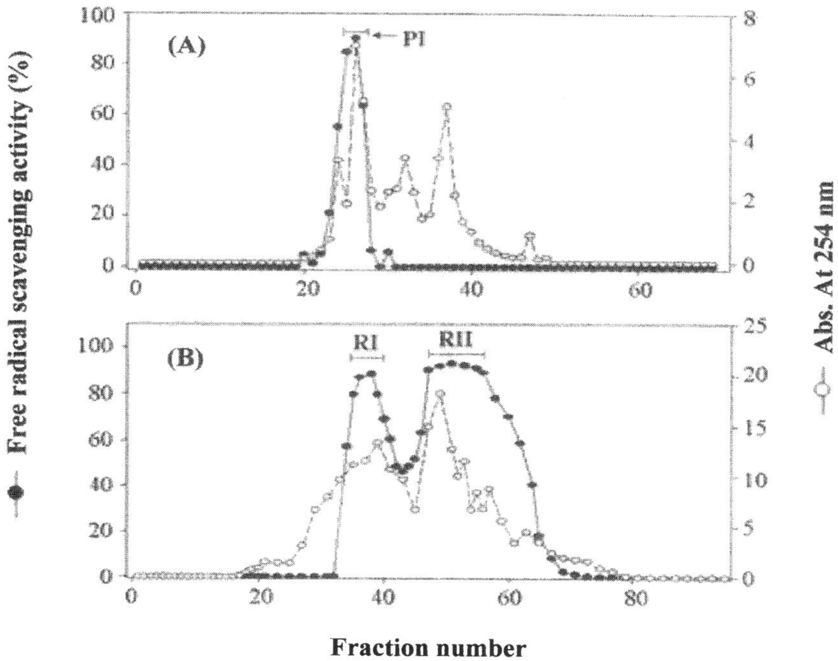

Figure 1. Sephadex LH-20 open column chromatograms of the crude methanolic extracts of parsley (A) and radish sprout (B) and free radical scavenging activities of collected fractions (PI, RI and RII).

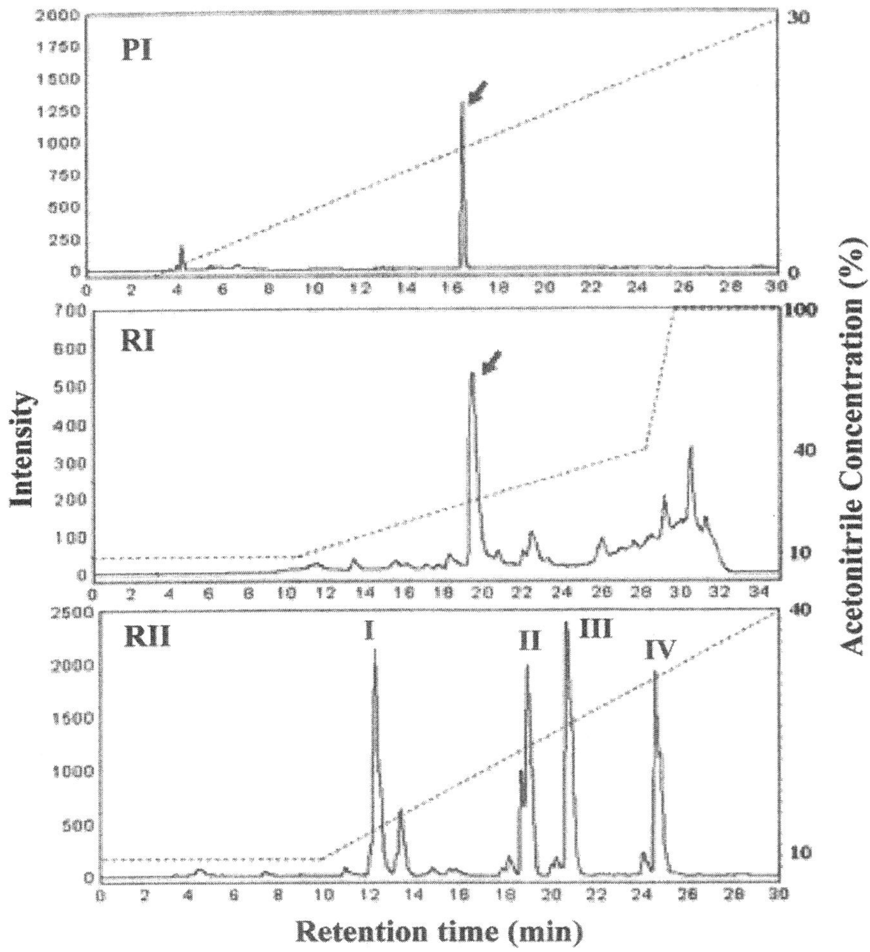

Figure 2. HPLC chromatograms of the antioxidants (PI, RI, and RII) that were partially pre-purified by Sephadex LH-20 open column chromatography. Samples were eluted with HPLC water containing 0.05% TFA and acetonitrile at a flow rate of 0.8 mL/min for PI or 1 mL/min for RI and RII. The arrowed peaks in PI and RI, and numbered peaks in RII indicate the peaks having free radical scavenging activity.

Identification of Antioxidant Free Phenolic Compounds in Fractions PI, RI, and RII (I, II, III, and IV)

All collected active peaks from HPLC were hydrolyzed with 2 N HCl in 50% methanol to break down the glycosidic conjugation between the phenolic compounds and sugars. Then the recovered free phenolic compounds were derivatized by BSTFA to make volatile phenolics-TMS derivatives to facilitate the GC/MS analysis (Figure 3). Caffeic acid in PI and five other compounds (ferulic acid, isoferulic acid, methyl ferulate, sinapic acid, and methyl sinapate) in RII were successfully separated and identified. However, no GC peaks in fraction RI could be identified by matching our available authentic phenolic compounds, hence we did not pursue this fraction any further. Nevertheless, a major phenolic compound in parsley, apigenin, (*24*) could not be recovered by our method. Additionally, it was found, based on the chromatographic peak area of each identified compound, that the amount of caffeic acid in PI (Figure 3) was much lower than those of other identified compounds in RII.

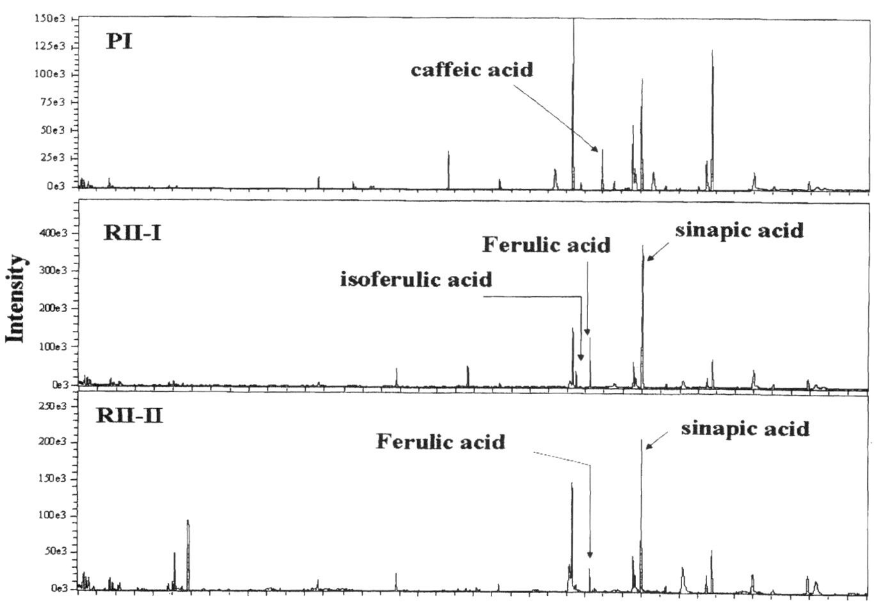

Figure 3. GC-MS chromatograms of antioxidant free phenolic compounds separated in parsley and radish sprout. Phenolic compounds were hydrolyzed and derivatized by BSTFA prior to GC-MS analysis.

Antioxidant Capacities of the Free Phenolics in Fractions PI and RII

Antioxidant capacities of the identified free phenolic compounds were in turn investigated by the DPPH˙ scavenging assay and β-carotene-linoleic acid assay. Their EC_{50} values were also compared with standard antioxidant controls (TableII). In the DPPH˙ scavenging assay, pure caffeic acid was the most potent antioxidant among the tested chemicals, with an EC_{50} value at 0.25 mM. In contrast, sinapic acid was the strongest radical scavenger among the four available phenolic compounds (i.e., ferulic acid, methyl ferulate, isoferulic acid, and sinapic acid) found in radish sprouts, while isoferulic acid was the least active antioxidant. The free radical scavenging activity of sinapic acid was also comparable to those of the two standard antioxidants and its EC_{50} value (0.34 mM) was 2-7 times lower than its counterparts in radish sprouts. Nevertheless, all the identified compounds and two standard antioxidants, except the unavailable methyl sinapate, possessed potent free radical scavenging activities and their capacities decreased in the following order: caffeic acid > α-tocopherol ≥ BHT ≈ sinapic acid > ferulic acid > methyl ferulate > isoferulic acid. However, despite the strongest free radical scavenging activity of caffeic acid in our DPPH˙ test, its lower concentration in the parsley crude extract resulted in its low scavenging activity compared to that of the radish sprouts extract (Table I), though the stronger activity exhibited by the radish sprouts extract might be attributable to the possible additive and/or synergistic effects among its phenolics (Figure 3).

Further investigation of the complementary antioxidant activity of these phenolic compounds was assessed by the β-carotene-linoleic acid assay. This assay is widely used because β-carotene is extremely sensitive to free radical-mediated oxidation. β-carotene in this emulsified model undergoes rapid discoloration in the absence of antioxidants. All phenolic compounds identified from parsley and radish sprout had much lower activities than those of BHT and α-tocopherol (Table II). Also, compared to the EC_{50} values determined by the DPPH˙ scavenging assay, most natural plant phenolics listed in Table II, except methyl ferualte, had higher EC_{50} values in the β-carotene-linoleic acid assay. Particularly, caffeic acid and sinapic acid showed remarkably increased EC_{50} values by approximately six and three fold, respectively, in the β-carotene-linoleic acid compared to values determined by the DPPH˙ scavenging assay, which meant caffeic acid and sinapic acid had lower antioxidant activities in the emulsified β-carotene-linoleic acid system. Caffeic acid had a higher EC_{50} value than each other compound identified, except isoferulic acid, in the radish sprouts. However, ferulic acid, isoferulic acid, and methyl ferulate only marginally changed their EC_{50} values in the DPPH˙ scavenging assay and β-carotene-linoleic acid assay. In the β-carotene-linoleic acid assay the antioxidant activities of all identified compounds and two standard antioxidants decreased in the following order: α-tocopherol > BHT >> methyl ferulate > ferulic acid > sinapic acid > caffeic acid > isoferulic acid. These results also indicated that, in the parsley extract, there were some other antioxidants that had inhibitory activity against β-carotene and lioneic acid oxidation because parsley extract and radish sprouts extract had very close activities, which could not be explained by the antioxidant capacity of caffeic acid alone. Unfortunately, these other chemicals were not successfully identified in this study.

Table II. Antioxidant Capacities of FreePhenolic Compounds Identified in Parsley and Radish Sprout

Materials	*identified antioxidants*	*antioxidant capacity (EC_{50}, mM)*	
		scavenging activity	*antioxidant activity*
Parsley	Caffeic acid	0.247 ± 0.019	1.385 ± 0.030
Radish sprout	Ferulic acid	0.635 ± 0.003	0.683 ± 0.008
	Methyl ferulate	0.764 ± 0.016	0.600 ± 0.011
	Isoferulic acid	2.335 ± 0.108	2.390 ± 0.059
	Sinapic acid	0.344 ± 0.019	1.001 ± 0.071
Standard antioxidant	BHT	0.342 ± 0.036	0.011 ± 0.000
	α-tocopherol	0.318 ± 0.026	0.004 ± 0.000

Due to the fact that all phenolic compounds (Figure 4) identified in both plants were derived from caffeic acid, their chemical structures were very similar, except for their number or the position of methylations and the additional methoxy group of sinapate. This provides us an opportunity to investigate the structure-activity relationship (SAR) of all the compounds identified. After the methylation of ortho-hydroxyl group of caffeic acid to form ferulic acid, the EC_{50} value of caffeic acid increased by 2.6 times to that of ferulic acid in the free radical scavenging test. This demonstrated that the hydroxyl group had played a vital role in hydrogen-donation. Further methylation of carboxyl group of the ferulic acid made methyl ferulate having an EC_{50} value 20% higher than that of ferulic acid in the same assay. This indicated that the carboxyl group might partially contribute its hydrogen and/or methylation of the carboxyl group has shifted the electrons to and strengthened the O-H bond resulting in less dissociation. In addition, the scavenging capacity of isoferulic aicd, an isomer of ferulic acid, was 3.7 times lower than that of ferulic acid. This indicates that the position of methylation of the hydroxyl group in the phenol ring is markedly important in influencing the free radical scavenging activity of an antioxidant compound. Such observation is in agreement with previous reports for the functions of the hydroxyl group in the phenol ring in free radical scavenging (*25-26*). Based on these results, it is assumed that the EC_{50} value of the methyl sinapate, an unavailable compound for this study, might have had a higher value than that of the sinapic acid that has a value of 0.34 mM for free radical scavenging activity.

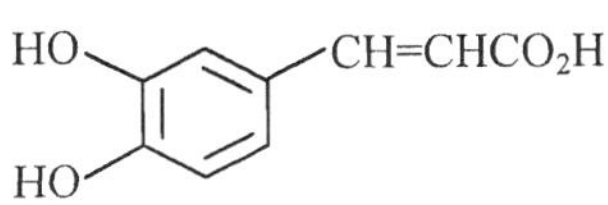

Caffeic acid

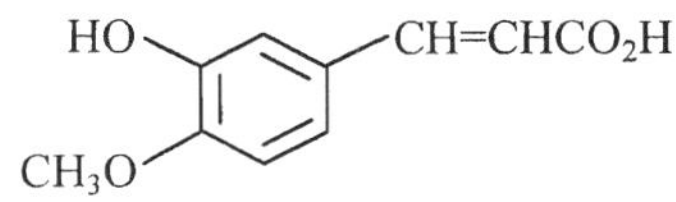

Isoferulic acid

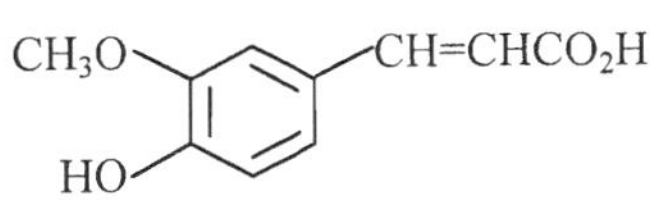

Ferulic acid

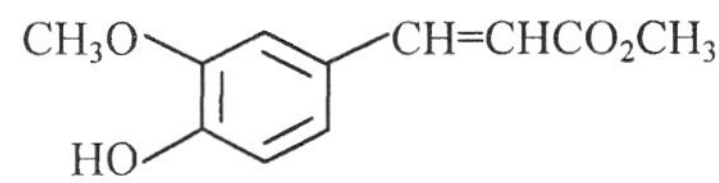

Methyl ferulate

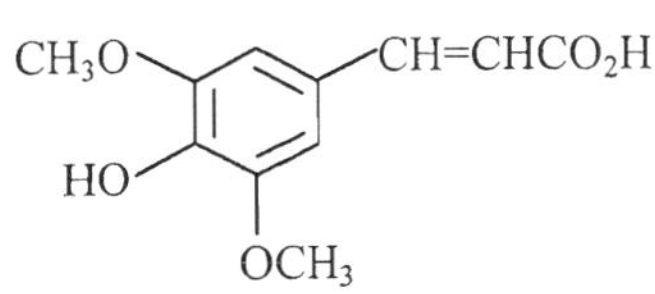

Sinapic acid

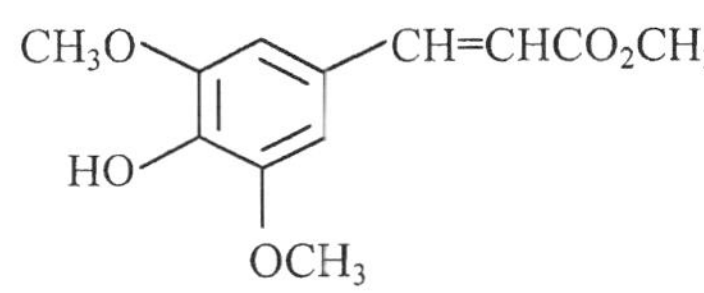

Methyl sinapate

Figure 4. Structures of the antioxidant free phenolic compounds identified in parsley and radish sprout.

Conclusions

Many studies have been performed to search for effective natural antioxidants in vegetables, fruits, and herbs, but little effort has been made for edible parsley and radish sprouts. We have now confirmed that parsley and radish sprouts have antioxidant activities due to their inherent phenolic constituents. The antioxidant phenolic compounds in the free form were separated and identified by combined methods including the chromatographic methods and two *in vitro* antioxidant assays. Caffeic acid from parsley and five other free phenolic compounds (ferulic acid, methyl ferulate, isoferulic acid, sinapic acid, and methyl sinapate) in radish sprouts were identified. Caffeic acid showed the strongest free radical scavenging activity in the DPPH˙ test, but had low antioxidant capacity in the β-carotene-linoleic acid assay. In addition, the radish sprouts extract possessed a strong radical scavenging activity and a strong inhibitory capacity against β-carotene bleaching. In the structure-activity relationship study, we found methylation of the hydroxyl group on the phenol ring could decrease the free radical scavenging activity of a compound, though more research is needed to reveal and validate the exact mechanisms involved. Nevertheless, this study suggests that both plants, especially radish sprout, may serve as a useful dietary antioxidant source.

References

1. Pietta, P. G. *J. Nat. Prod.* **2000**, *63*, 1035-1042.
2. Chidambara Murthy, K. N.; Jayaprakasha, G.; Singh, R. P. *J. Agric. Food Chem.* **2002**, *50*, 4791-4795.
3. Govindarajan, R.; Vijayakumar, M.; Pushpangadan, P. *J. Ethnopharmacol.* **2005**, *99*, 165-178.
4. Murakami, S.; Murakami, H. *Neurobiol. Aging* **2005**, *26*, 899-905.
5. Sastre, J.; Martín, J. A.; Gómez-Cabrera, M.-C.; Pereda, J.; Borrás, C.; Pallardó, F. V.; Viña, J. *Free Radical Biol. Med.* **2005**, *38*, 575-582.
6. Sastre, J.; Pallardó, F. V.; Viña, J. *Free Radical Biol. Med.* **2003**, *35*, 1-8.
7. Barja, G. *Trends Neurosci.* **2004**, *27*, 595-600.
8. Middleton, E.; Kandaswami, C.; Theoharides, T. C. *Pharmacol. Rev.* **2000**, *52*, 673-751.
9. Ichihashi, M.; Ueda, M.; Budiyanto, A.; Bito, T.; Oka, M.; Fukunaga, M.; Tsuru, K.; Horikawa, T. *Toxicol.* **2003**, *189*, 21-39.
10. Halliwell, B. *Lancet*, **1994**, *344*, 721-724.
11. Vitaglione, P.; Morisco, F.; Caporaso, N. *Crit. Rev. Food Sci. Nutr.* **2004**, *44*, 575-586.
12. Karakaya, S. *Crit. Rev. Food Sci. Nutr.* **2004**, *44*, 453-464.
13. Rice-Evans, C. A.; Miller, N. J.; Paganga, G. *Trends Plant Sci.* **1997**, *2*, 152-159.
14. Yang, C. S.; Landau, J. M.; Huang, M.-T.; Newmark, H. L. *Amnu. Rev. Nutr.* **2001**, *21*, 381-406.
15. Havsteen, B. H.. *Pharmacol. Therapeutics* **2002**, *96*, 67-202.
16. Dixon, R. A.; Paiva, N. L. *Plant Cell* **1995**, *7*, 1085-1097.
17. Bennett, R. N.; Wallsgrove, R. M. *New Phytol.* **1994,** *127*, 617-633.
18. Kessler, A.; Baldwin, I. T. *Annu. Rev. Plant Biol.* **2002**, *53*, 299-328.
19. Zheng, W.; Wang, S. Y. *J. Agric. Food Chem.* **2001**, *49*, 5165-5170.

20. Takaya, Y.; Kondo, Y.; Furukawa, T.; Niwa, M. *J. Agric. Food Chem.* **2003**, *51*, 8061-8066.
21. Singleton, V. L.; Rossi, I. A. *Am. J. Enol. Viticult.* **1965**, *16*, 144-158.
22. Yamaguchi, T.; Takamura, H.; Matoba, T.; Terao, J. *Biosci. Biotechnol. Biochem.* **1998**, *62*, 1201-1204.
23. Kumazawa, S.; Taniguchi, M.; Suzuki, Y.; Shimura, M.; Kwon, M.-S.; Nakayama, T. *J. Agric. Food Chem.* **2002**, *50*, 373-377.
24. Sakakibara, H.; Honda, Y.; Nakagawa, S.; Ashida, H.; Kanazawa, K. *J. Agric. Food Chem.* **2003**, *51*, 571-581.
25. Seeram, N. P.; Nair, M. G. *J. Agric. Food Chem.* **2002**, *50*, 5308-5312.
26. Wang, H.; Nair, M. G.; Strashurg, G. M.; Booren, A. M.; Gray, J. I. *J. Nat. Prod.* **1999**, *62*, 86-88.

Chapter 16

Antioxidant Constituents in Tree Nuts: Health Implications and Aflatoxin Inhibition

Russell J. Molyneux[1], Noreen Mahoney[1], Jong H. Kim[1], Bruce C. Campbell[1], and Ann E. Hagerman[2]

[1]Western Regional Research Center, Agricultural Research Service, U.S. Department of Agriculture, Albany, CA 94710
[2]Department of Chemistry and Biochemistry, Miami University, Oxford, OH 45056

Epidemiological studies have indicated that antioxidant compounds in fruits and vegetables may have particular health benefits to consumers. Different classes of phenolics are often encountered in foods and these exert beneficial health effects related to their antioxidant activity, chemopreventive effect and anticarcinogenic potential, governed by the structural characteristics of the bioactive components present. Tree nuts in general have high levels of antioxidant constituents which, especially in walnuts, confer resistance to formation of aflatoxins on *Aspergillus flavus* infection. It is postulated that aflatoxigenesis is due to oxidative stress on the fungus and that compounds capable of relieving oxidative stress should be capable of reducing aflatoxin. Similarly, antioxidants in the diet are believed to mediate the effects of oxidative damage.

Tree nuts, especially almonds, pistachios, and walnuts are an extremely valuable agricultural commodity in both domestic and international trade. In the United States virtually all of these crops are produced in California, having a combined value to the State of almost $3.5 billion in 2005, with 40-60% being exported, depending on the nut variety (*1,2*). As a major agricultural crop, tree nuts are unusual in that they are not a primary nutrient source. In many countries, nuts are consumed as a snack food, eaten out of hand, and as such undergo fairly minimal processing. For other uses, such as incorporation into baked goods or preparation of nut butters and marzipan, they may be simply blanched or ground. There is therefore little opportunity to influence either positive or negative quality characteristics and these are primarily dependant on nut species or cultivar and their handling during harvesting, drying and packing. Control of quality factors must therefore be focused on conditions during the growing season and at harvest time.

Healthy Foods and Antioxidants

The relationship of foods with health is exceptionally complex but it can be divided into two major factors, nutritional quality and safety. Nutritional quality factors are intrinsic nutritional components, health-promoting constituents, and organoleptic properties such as appearance, flavor, aroma and texture. Within certain limits these should be maximized in order to appeal to the consumer. For example, even a pleasant, attractive odor constituent can be overpowering at too high a level and become repellent. In general, nutritional quality factors are accessible by the consumer, either through product labeling as to the quantity of calories, vitamins, minerals, etc. in each serving, or through individual perception as to desirability of the food. Safety quality factors are toxins and contaminants, which must be minimized or even eliminated, if possible. The consumer is generally not aware of the presence of such compounds and is therefore completely reliant on regulatory organizations, at either producer or governmental levels, for safety assurance. Furthermore, the information is not easily obtained since tolerance and measured levels are not marked on the product.

It is obviously preferable for enhancement of nutritional quality factors and elimination of noxious compounds to be achieved through intrinsic means rather than through additives or processing methods. This is particularly true for tree nuts which are regarded as "natural" foods and where fortification or detoxification would be regarded by the consumer as unacceptable. As will be seen, the natural antioxidants present in almonds, pistachios and especially walnuts, are not only associated with human health benefits but can control mycotoxins formed by fungal attack on the nuts.

Antioxidants in Tree Nuts

Consumption of tree nuts is associated with a healthy diet and the Food and Drug Administration (FDA) has allowed a qualified health claim for a relationship between the consumption of nuts and reduced risk of coronary heart disease (*3*). Nuts are a good source of protein, fiber and dietary fats, having only low levels of undesirable saturated fats but high levels of unsaturated fats. Monounsaturated fatty acids (oleic) predominate in almonds and pistachios and polyunsaturated fatty acids (linoleic and linolenic) in walnuts (*4*). A meta-analysis of five controlled dietary clinical intervention trials in humans with walnuts demonstrated a consistent decrease in serum cholesterol levels and reduced risk of heart disease (*5*). Similar effects have been found with other nuts, including pistachios and almonds (*6*).

The majority of health benefits appear to derive from antioxidants present in tree nuts. It has been shown in randomized intervention trials that protection against oxidative stress related diseases is not provided by supplementation with antioxidants such as ascorbic acid, α-tocopherol and β-carotene. A different class of antioxidants must therefore be responsible and the most likely candidate is polyphenolic compounds (*7*). The inhibitory effect of both a polyphenol-rich walnut extract and ellagic acid on *in vitro* oxidation of human plasma and low-density lipoproteins has been suggested to be responsible for the cardioprotective effect of walnuts (*8*). A survey of hydrophilic and lipophilic antioxidant capacities in commonly consumed foods in the U.S. showed that nuts were high in total antioxidant capacity (*9*). However, the correlation between total phenolics content and hydrophilic antioxidant capacities for all foods analyzed was not strong, indicating a significant molecular structural component in relation to antioxidant activity.

Antioxidants in tree nuts fall into three major structural classes: hydrolysable tannins, phenolic acids, and flavonoids. Hydrolysable tannins can be exceptionally complex but in their simplest form they consist of a carbohydrate core (usually glucose) esterified with gallic acid and/or hexahydroxydiphenic acid moieties (*10, 11*). Further esterification through the phenolic acid groups leads to depside linkages. In the case of glucose-based tannins, the initial biosynthetic product is pentagalloyl glucose, with subsequent transformations involving hydrolysis of some of the galloyl ester groups or oxidative coupling to give the hexahydroxydiphenic esters (*12*). This is exemplified by the structure of tellimagrandin (Figure 1), a major tannin constituent of walnuts, from which at least 16 hydrolysable tannins of varying degrees of complexity have been isolated and structurally identified (*13*).

In contrast to walnuts, in which antioxidant compounds are restricted to hydrolysable tannins and very small quantities of the tannin-derived hydrolysis products, gallic and ellagic acids, pistachios and almonds have a much greater variety of phenolic antioxidants. In pistachios, these include anacardic acids, typical of the family Anacardiaceae to which *Pistacia* species belong (unpublished results), caffeic acid, and chlorogenic acid, its quinate ester. In addition, pistachios also have hydrolysable tannins, but unlike walnuts these are quinic acid rather than glucose-based (unpublished results). The stereochemistry of quinic acid is such that dimerization of gallic acid moieties cannot occur and

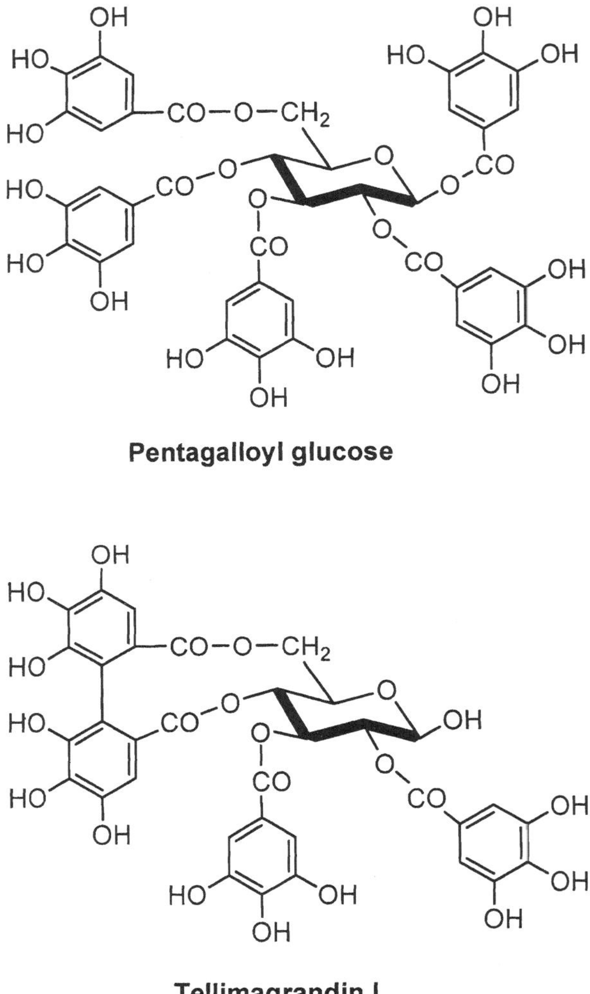

Figure 1. Structures of hydrolysable tannins.

there are thus no hexahydroxydiphenic acid structures involved in these tannins. In almonds, the phytochemicals consist of simple phenolic acids, including 4-hydroxybenzoic, protocatechuic, and vanillic acids, and flavonoids, such as catechins and isorhamnetin-, kaempferol- and quercetin-3-*O*-glucosides, with the largest proportion located in the seed coat (*14*).

Mycotoxins in Tree Nuts

The health benefits of eating tree nuts could potentially be offset by their ability to become infected by various fungi, either spoilage organisms that affect their appearance, or toxigenic fungi (primarily *Aspergillus* species) that produce mycotoxins such as the nephrotoxic ochratoxins or hepatocarcinogenic aflatoxins (Figure 2). In more developed countries, the risk from the fungal toxins to the ordinary consumer is low due to careful sorting of the product and the relatively minor contribution of nuts to the diet. Furthermore, there is a strong association of hepatocarcinogenicity due to aflatoxins with infection with hepatitis B (*15*). However, in countries where the fungi affect subsistence crops and hepatitis is endemic, as in many parts of Africa, the threat is much greater and even acute toxicity can be observed on occasion (*16*).

A primary concern for countries that produce tree nuts is the strict regulations imposed on mycotoxin levels in tree nuts. Of greatest concern are those for aflatoxins in tree nuts imported into the European Union, with tolerance levels of 4 ppb total aflatoxins and 2 ppb aflatoxin B_1 (*17*). In 2005, there were 827 rapid alerts or notifications for mycotoxins in tree nuts, of which 94% were for aflatoxins and 5% for ochratoxins (*18*). For aflatoxins, the preponderance of the aflatoxin notifications (46%) were for pistachios from Iran but for the U.S. there were 28 notifications for almonds and 13 for pistachios, in addition to one notification for ochratoxins in pistachios. Interestingly, there were no notifications for walnut imports. Although the U.S. numbers may seem relatively small, they represent considerable economic loss to the producer and exporter, with costs associated with reprocessing or return of the shipment and increased surveillance of subsequent imports. Tree nuts produced in the U.S. are an exceedingly valuable commodity with a total income to the state of California of $3.42 billion in 2005, of which $2.00 billion accrued from exports. This results in considerable incentive to find natural methods to limit or eliminate aflatoxin formation in tree nuts.

Resistance of Tree Nut Species to Aflatoxin Formation

The RASFF (Rapid Alert System for Food and Feed) notification system (*18*), together with anecdotal information, indicates that there is differential resistance among tree nut species to aflatoxigenesis, with walnuts being the least

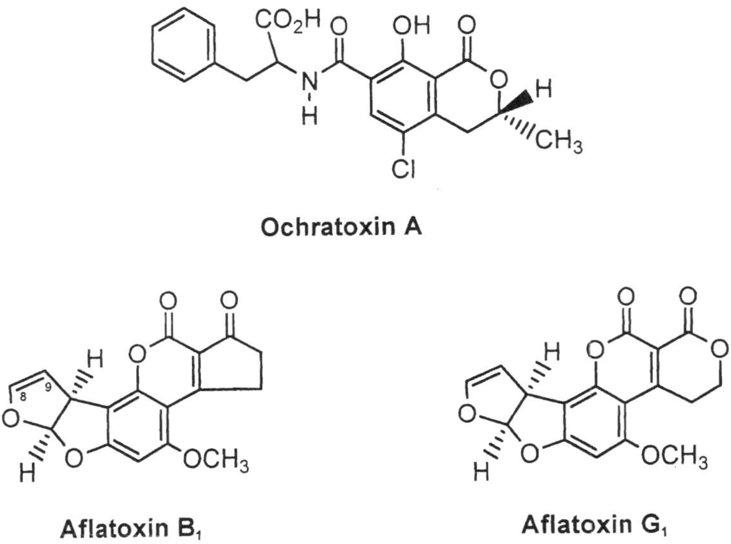

Figure 2. Structures of ochratoxin and aflatoxins B1 and G1; aflatoxins B2 and G2 are saturated at the 8,9-double bond.

likely to become contaminated. Although insect attack is known to be a vector for *Aspergillus* spores to enter the nut, there does not appear to be a physical explanation for these differences, since walnuts are just as susceptible to codling moth and navel orangeworm infestation as almonds and pistachios. It seems likely therefore that the resistance is primarily a chemical effect, either through suppression of fungal growth or of aflatoxin biosynthesis. Nut kernels were therefore tested *in vitro* for their ability to support growth of *A. flavus* and aflatoxin production. A selection of 34 almond cultivars or selections produced aflatoxin in the range of 20-192 μg/plate whereas 26 walnut varieties produced 0-28 μg/plate; the single commercial pistachio cultivar in California, 'Kerman', was intermediate in level with 40 μg/plate (*19*). The most potent inhibitor of aflatoxigenesis was the 'Tulare' walnut variety, with no detectable aflatoxin produced, and further experiments were therefore conducted using this variety.

Separation of the seed coat (pellicle) from the kernel itself showed on testing that the antiaflatoxigenic factor(s) were located in the former, with no inhibition produced by the kernel material, which even enhanced aflatoxin production at higher concentrations in the media, probably due to increased nutrient availability. Bioassay-directed fractionation of the seed coat constituents, by extraction with solvents of increasing polarity, established that the activity was due to the hydrolysable tannins, which are essentially the only non-structural components of the pellicle (*11*, *13*). Tellimagrandin (Figure 1) is representative of walnut hydrolysable tannins, but the variety and complexity of the tannins present as a group is exceptional and it is impossible to isolate and evaluate each individual component. However, a measure of the overall quantity of tannin in any particular sample can be obtained by hydrolysis with methanolic HCl, giving methyl gallate and ellagic acid, both of which can be determined by HPLC (*20*). When walnut varieties were compared against each other and versus other nut species by this procedure, 'Tulare' was found to have the highest levels of gallic acid, with therefore the greatest amount of hydrolysable tannin (*21*). Higher gallic acid levels showed an inverse correlation with ability of walnut kernels to produce aflatoxins *in vitro*, showing that this compound and its parent tannins were responsible for antiaflatoxigenic activity (Table I). One the other hand, there was no clear correlation with ellagic acid content and suppression of aflatoxigenesis. Analysis of walnut pellicle for 'Tulare' and 'Chico' cultivars throughout the growing season for two consecutive years showed that gallic acid levels were consistently higher for 'Tulare' and showed less of a decline with maturity of the nut (*21*).

Antiaflatoxigenic Activity of Phenolic Antioxidants

It has been suggested that aflatoxin formation is promoted by oxidative stress (*22*) and eugenol, a propenylphenol, has been shown to substantially suppress aflatoxigenesis through inhibition of lipid peroxidation (*23*). If this is a general mechanism, then any antioxidant should lower aflatoxin levels relative to a control sample with no antioxidant present. When a selection of phenolic natural products present in tree nuts were tested *in vitro*, this proved to be the case (unpublished results). Pentagalloyl glucose (*24*) was particularly active with an inhibitory activity relative to control of 99.8%. Hydrolysable tannins in general were exceptional potent, reducing aflatoxin levels by >98%, as did the pistachio constituent, caffeic acid. However, the quinic acid ester of the latter, chlorogenic acid, was less effective with an inhibition percentage of 88.5%. It is noteworthy that caffeic acid is a potent inhibitor of 5- and 12-lipoxygenases (IC_{50} 3.7 and 5.1 μM, respectively) whereas chlorogenic acid is non-inhibitory even at 100 μM. *A. flavus* is known to be capable of surviving in tannin vats and possesses a tannase capable of hydrolyzing tannins to their constituent parts (*25*), the core carbohydrate (glucose or quinic acid), gallic acid, and ellagic acid which derives from hexahydroxydiphenic acid through spontaneous lactonization. Surprisingly, quinic acid was an excellent aflatoxigenesis inhibitor, even though it is non-phenolic, whereas gallic and ellagic acids had lower inhibition levels, typical of other phenolic acids (59.5-85.6%). Nevertheless, all of the tested compounds inhibited aflatoxin formation to a significant extent, demonstrating support for the hypothesis that oxidative stress plays a role in aflatoxigenesis and that natural antioxidants are capable of suppressing the formation of the toxins.

Table I. Correlation of Aflatoxin Production with Gallic and Ellagic Acid Content of Tree Nut Seed Coat

Nut Species/Variety	*Aflatoxin (µg/plate)*	*Gallic acid (% dw)*	*Ellagic acid (% dw)*
English walnut (*Juglans regia*):			
Tulare	0	3.2	14.0
Chico	56	1.8	11.0
Black walnut (*Juglans hindsii*):			
Rawlins	44	1.0	3.1
Pistachio (*Pistacia vera*):			
Kerman	40	0.5	n.d.
Almond (*Prunus amygdalus*):			
Nonpareil	172	trace (<0.01)	n.d.
Mission	202	trace (<0.01)	n.d.

Phenolic Antioxidants as Preservatives

Even foods that are commonly regarded as healthy will not be eaten if they are distasteful or unattractive to the consumer. This is especially true of tree nuts, which can develop rancid off-flavors due to their high content of mono- or polyunsaturated fatty acids. In prepared foods such oxidative deterioration is often prevented by the addition of synthetic antioxidants such as butylated hydroxyanisole (BHA) and butylated hydroxytoluene (BHT) but this is not an option with nuts which are most commonly eaten in an unprocessed or minimally processed form. The presence of natural antioxidants is therefore the only protection against lipid oxidation. Fortunately, almonds, pistachios and walnuts all possess natural phenolics, especially phenolic acids, flavonoids and hydrolysable tannins. However, these compounds are located primarily in the seed coat rather than the kernel itself so that blanching could remove much of the protective effect. The lipophilic, hydrophilic and total antioxidant capacities, together with total phenolic levels, have been determined for fruits, vegetables, cereals and nuts commonly consumed in the United States (*9*). Among tree nuts the total antioxidant capacity (TAC) was in the order pecans > walnuts > hazelnuts > pistachios > almonds, with other species being considerably lower. The TAC correlated well with total phenolic content, measured as gallic acid equivalents per gram. The demonstrated activity of sucrose gallate esters as radical scavengers in lipid peroxidation (*26*) suggest that appropriate levels of structurally related hydrolysable tannins should prevent or delay the development of rancidity.

Phenolic Antioxidants and Disease Prevention

There is a considerable evidence that various disease states, especially those associated with aging such as heart disease, neurological degeneration and cancer may be a consequence of oxidative stress (*27-29*). It is a reasonable corollary that foods high in antioxidants will have a salutary effect and epidemiological evidence has demonstrated that diets consisting of a considerable proportion of fruits and vegetables, including nuts, are conducive to maintenance of good health (*30-32*).

In most tree nuts a considerable proportion of the total antioxidant capacity accrues from the hydrolysable tannins. The biological and pharmacological activities of this class of compounds have been comprehensively reviewed by Haslam (*33*), and include bacteriocidal, anthelmintic, and antihepatoxic properties, and suppression of replication of human immunodeficiency virus (HIV) and herpes simplex virus (HSV). In addition to their potent antioxidant activity towards cellular prooxidant states, hydrolysable tannins can also function by complexation of transition metal ions and through association with proteins and peptides. Complexation of hydrolysable tannins with proteins is responsible for their astringent taste properties but mild astringency in medicinal plants has been shown to be beneficial for treatment of digestive disorders. It is probable that a combination of these molecular functions is involved in mitigating many diseases and additional research is needed to unravel these effects as well as to determine effects of specific structural features and rates of absorption and metabolism.

Implications

Hydrolysable tannins and other phenolic antioxidants naturally present in tree nuts can serve three specific functions in relation to their health benefits as a food. They can prevent the formation of deleterious mycotoxins, reduce potential off-flavors, and prevent disease related to oxidative stress. It appears that there are no major contraindications to their presence and research needs to be concentrated on determining the optimal levels and specific structural features best suited to such properties and to ensure that breeding programs for new varieties incorporate them as a factor.

References

1. National Agriculture Statistics Service, USDA. 2006. California Fruit and Nut Review (10/18/06).

http://www.nass.usda.gov/Statistics_by_State/California/Publications/recent_reports.asp

2. Foreign Agricultural Service, USDA . 2006. FAS Online. Horticultural and Tropical Crops Division. http://www.fas.usda.gov/htp/horticulture/nuts.html
3. U.S. Food and Drug Administration, Center for Food Safety and Applied Nutrition, Office of Nutritional Products, Labeling and Dietary Supplements (July 14, 2003) "Qualified Health Claims: Letter of Enforcement Discretion - Nuts and Coronary Heart Disease (Docket No 02P-0505)". http://www.cfsan.fda.gov/~dms/qhcnuts2.html
4. U.S. Department of Agriculture, Agricultural Research Service. 2006. USDA National Nutrient Database for Standard Reference, Release 19. http://www.nal.usda.gov/fnic/foodcomp/search/
5. Feldman, E. B. *J. Nutr.* **2002**, *132*, 1062S-1101S.
6. Kris-Etherton, P. M.; Zhao, G.; Binkoski, A. E.; Coval, S. M.; Etherton, T. D. *Nutr. Rev.* **2001**, *59*, 103-111.
7. Halvorsen, B. L.; Carlsen, M. H.; Phillips, K. M.; Bøhn, S. K.; Holte, K.; Jacobs Jr., D. R.; Blomhoff, R. *Am. J. Clin. Nutr.* **2006**, *84*, 95-135.
8. Anderson, K. J.; Teuber, S. S.; Gobeille, A.; Cremin, P.; Waterhouse, A. L.; Steinberg, F. M. *J. Nutr.* **2001**, *131*, 2837-2842.
9. Wu, X.; Beecher, G. R.; Holden, J. M.; Haytowitz, D. B.; Gebhardt, S. E.; Prior, R. L. *J. Agric. Food Chem.* **2004**, *52,* 4026 -4037.
10. Hagerman, A. E. *Tannin Handbook*; http://www.users.muohio.edu/hagermae/tannin.pdf.
11. Jurd, L. *J. Am. Chem. Soc.* **1958**, *80*, 2249-2252.
12. Quideau, S.; Feldman, K. S. *Chem. Rev.* **1996**, *96*, 475-503.
13. Fukuda, T.; Ito, H.; Yoshida, T. *Phytochemistry* **2003**, *63*, 795-801.
14. Milbury, P. E.; Chen, C.-Y.; Dolnikowski, G. G.; Blumberg, J. B. *J. Agric. Food Chem.* **2006**, *54*, 5027-5033.
15. Henry, S. H.; Bosch, F. X.; Bowers, J. C. *Adv. Exp. Med. Biol.* **2002**, *504*, 229-33.
16. Lewis, L.; Onsongo, M.; Njapau, H.; Schurz-Rogers, H.; Luber, G.; Kieszak, S., Nyamongo, J., Backer, L., Dahiye, A. M., Misore, A., DeCock, K.; Rubin, C.; and the Kenya Aflatoxicosis Investigation Group. 2005. *Environ. Health Persp.* **2005**, *113*, 1763-1767.
17. Commission of the European Community. 1998. Commission Directive 98/53/EC of 16 July 1998 laying down the sampling methods and the methods of analysis for the official control of the levels of certain contaminants in foodstuffs. *Off. J. Eur. Commun. Leg.* **1998**, *L201*, 93.
18. European Commission. The Rapid Alert System for Food and Feed (RASFF) Annual Report 2005. http://ec.europa.eu/food/food/rapidalert/report2005_en.pdf

19. Mahoney, N., Molyneux, R. J., McKenna, J., Leslie, C. A., McGranahan, G. *J. Food Sci.* **2003**, *68*, 619-622.
20. Lei, Z.; Jervis, J.; Helm, R. F. *J. Agric. Food Chem.* **2001**, *49*, 1165-1168.
21. Mahoney, N.; Molyneux, R. J. *J. Agric. Food Chem.* **2004**, *52*, 1882-1889.
22. Jayashree, T.; Subramanyam, C. *Free Radical Biol. Med.* **2000**, *29*, 981-985.
23. Jayashree, T.; Subramanyam, C. *Lett. Appl. Microbiol.* **1999**, *28*, 179-183.
24. Chen, Y.; Hagerman, A. E. *J. Agric. Food Chem.* **2004**, *52*, 4008-4011.
25. Yamada, H.; Adachi, O.; Watanabe, M.; Sato, N. *Agric. Biol. Chem.* **1968**, *32*, 1070-1078.
26. Dufour, C.; da Silva, E.; Potier, P.; Queneau, Y.; Dangles, O. *J. Agric. Food Chem.* **2002**, *50*, 3425-3430.
27. Diaz, M. N.; Frei, B.; Keaney, Jr., J. F. *New Engl. J. Med.* **1997**, *337*, 408-416.
28. Christen, Y. *Am. J. Clin. Nutr.* **2000**, *71*, 621S-629S.
29. Ames, B. N.; Gold, L. S.; Willet, W. C. *Proc. Natl. Acad. Sci. U.S.A.* **1995**, *92*, 5258-5265.
30. Block, G.; Patterson, B.; Suber, A. *Nutr. Cancer* **1992**, *18*, 1-29.
31. Ness, A. R.; Powles, J. W. *Int. J. Epidemiol.* **1997**, *26*, 1-13.
32. Young, I. S.; Woodside, J. V. *J. Clin. Pathol.* **2001**, *54*, 176-186.
33. Haslam, E. *J. Nat. Prod.* **1996**, 59, 205-215.

Chapter 17

Natural Bioactive Antioxidants for the Enrichment of Seafood

I. Medina[1], M. Cascante[2], J. L. Torres[3], and M. Pazos[1]

[1]IIM-CSIC, Eduardo Cabello 6, 36028-Vigo, Spain
[2]University of Barcelona, Martí I Franquès 1, 08028-Barcelona, Spain
[3]IIQAB-CSIC, Jordi Girona 18–26, 08034-Barcelona, Spain

Exploring new, effective natural antioxidants for the food industry and consumers demanding natural ingredients in foodstuffs has intensified in recent years. This has a special relevance for fatty and semi-fatty fish and fish products having high amounts of n-3 polyunsaturated fatty acids (PUFA), prone to undergo oxidation and rancidity. This work proposes the use of flavonoids obtained from different agrofood by-products as natural and bioactive antioxidants in fish products. Different parameters were found relevant. Properties such as the influence of the molecular structure on the redox and chelating capacities, the distribution of the antioxidant in the oxidation sites, and the incidence of the antioxidant on the endogenous redox balance of fish muscle will be discussed. Besides increasing the shelf-life of functional seafood preparations, the addition of some of these compounds may be beneficial for the prevention of cancer due to the induction of apoptosis in different cancer cells. Thus, the product based on fish muscle supplemented with bioactive antioxidants appears to be an interesting and stable functional food offering the combined action of n-3 PUFA and natural polyphenols.

Fish products have an essential role in many traditional occidental and oriental diets due to their composition and the high number of fish species. Fish and seafood contribute significantly to "healthy diets" by their high content on n-3 polyunsaturated fatty acids n-3 PUFA, and other important components as high quality proteins, vitamins or minerals (*1*). However, fish products are very prone to degradation (*2*). Different physical and biochemical changes lead to deterioration and loss of freshness and make difficult processing and transformation. Among these changes, lipid oxidation is one of the most important. It conduces to rancid off-flavors and reduces the shelf–life of fish products especially during storage (*3*).

Fish rancidity can not be totally avoided, but some procedures can minimise the rate of lipid oxidation. The use of antioxidants as food additives has increased as an effective methodology for controlling rancidity and its deleterious consequences (*4*). However, the effectiveness of these antioxidants for inhibiting lipid oxidation of a complex matrix like fish muscle is difficult to predict. The different antioxidants show different efficacy depending on the type of fat or food and even depending on the processing or manipulation (5). Antioxidants in fish can be scavengers of free radicals involved or generated in the oxidative reactions (*5*), can reduce the pro-oxidants present in fish muscle such as hemoglobin (*6*) or can reinforce the action of the endogenous antioxidant system of fish tissues (*7*).

Synthetic phenolics as butylated hydroxytoluene (BHT), butylated hydroxyanisole (BHA) and tert-butyl hydoquinone (TBHQ) are usually employed for minimizing rancidity of foodstuffs. However, their use is lately questioned (*8*), and the current legislation and the restrictions and preferences of consumers limit their use in foodstuffs. In the last years, the inclusion of plant extracts in food as preservatives is a common practice in the food industry. Natural polyphenols are valuable alternatives to synthetic phenolics. They can effectively scavenge radicals resulting in the application of bioactive antioxidant compounds in foods prone to rancidity development (*9*).

Natural Flavonoids: Procyanidins

Procyanidins, oligomeric catechins (Figure 1), are common in vegetable and forest by-products. They have demonstrated to be potent free radical scavengers increasingly appreciated as chemopreventive agents against health conditions (*10-12*). We have investigated the ability of procyanidins for inhibiting oxidation of chilled horse mackerel muscle stressing the factors involved. The compounds tested here had similar molecular structures with different polymerisation (number of phenolic residues) and galloylation degrees (gallate content). Additionally since procyanidins will be, at least in part, bioavailable in the colon after ingestion of the supplemented seafood (*13*), their effect on colon cancer cells has been evaluated.

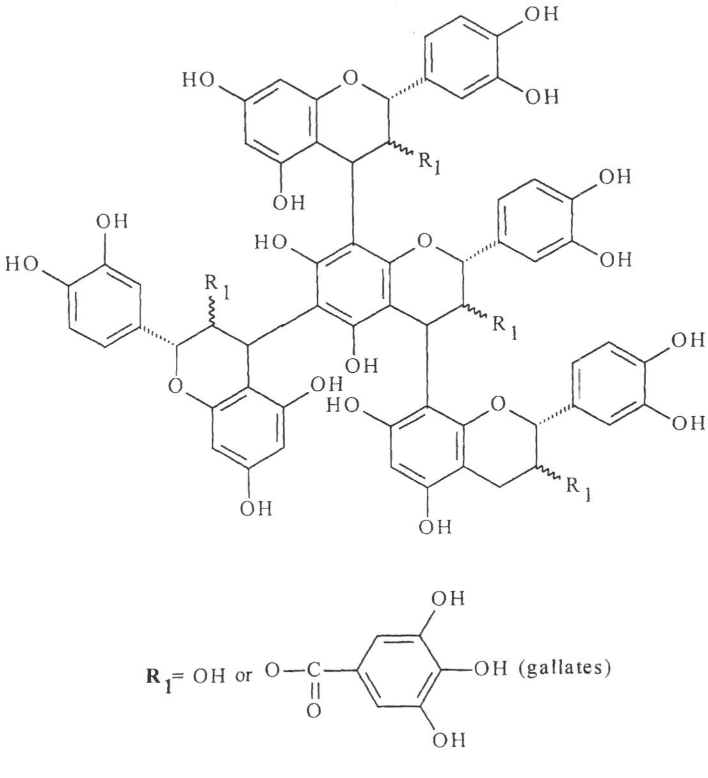

Figure 1. General structure of procyanidins.

Mechanism Involved in the Antioxidant Role of Procyanidins in Fish Muscle

We have tested the activity of different procyanidins extracted from grape (g) and pine (p) by-products according to Torres *et al.* (*14*) in minced horse mackerel muscle stored at 4°C (Table 1). Procyanidins effectively inhibited lipid oxidation of the fish homogenate by retarding the generation of peroxides and thiobarbituric acid reactive substances (TBARS) (Figure 2).

Table I. Polymerization and Galloylation Degrees of the Most Effective Procyanidins in Fish Muscle

Procyanidin source	*Polymerization*[a]	*Galloylation*[b]
OW grape	1.7	15
OW pine	2.1	0
IV Grape	2.2	25
IV Pine	2.9	0

[a] mean number of phenolic residues; [b] mean percentage of gallates

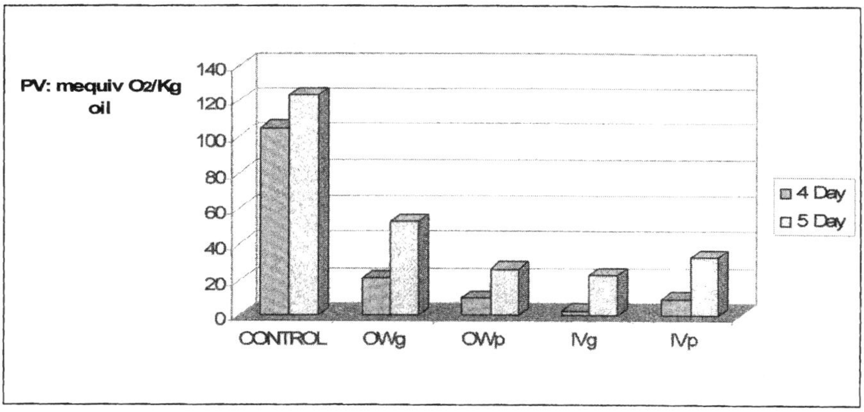

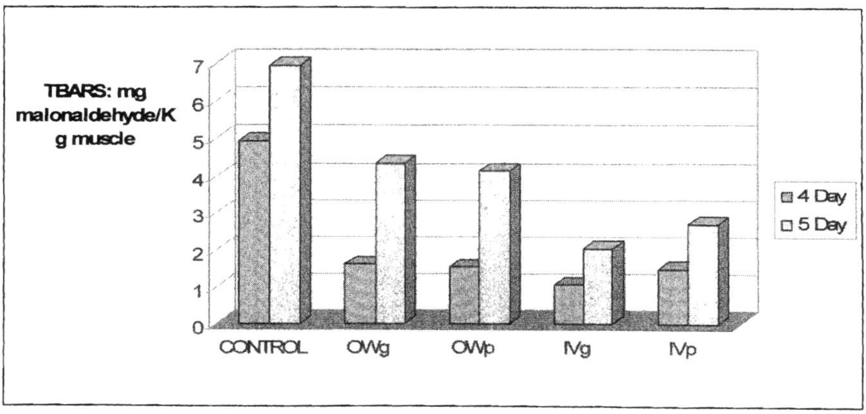

Figure 2. Peroxide value and TBARS formation during chilled storage of minced fish muscle supplemented with grape (g) and pine (p) procyanidin.

The tested compounds significantly retarded the induction periods of oxidation and the amount of lipid oxidation by-products formed. An increment on galloylation degrees of procyanidins increased the antioxidant efficacy. With regards to polymerization, there was an optimum number of phenolic residues, around 2.2-2.4 in which the best efficiency was achieved. The fraction showing the highest antioxidant activity on fish muscle was IVg: 25% of galloylation and 2.2 degree of polymerization. IVg was also able to significantly preserve the amounts of eicosapentaenoic acid (EPA) and docosahexaenoic acid (DHA) during chilled storage. This fraction can donate 5.8 electrons/mol and shows a chelating capacity of 94% compared to that of ethylenediaminetetraacetic (EDTA). It is also a polar fraction showing a partitioning coefficient of 24 % oil/water. By correlating these data with the antioxidant activity found in chilled fish muscle, the capacity of procyanidins for donating electrons seemed to play the most significant role for retarding the development of rancidity in fish muscle. Instead, the properties related with the ability for chelating metals were not correlated with the inhibitory activities. The more polar procyanidins, IV and OW, were also most active for inhibiting lipid oxidation. In addition to their reducing capacities, they may establish hydrophobic and hydrophilic interactions depending on the medium and reach a higher incorporation into the oxidative sensitive-sites like fish membranes.

Since phenolic antioxidants can reinforce the action of other antioxidants, we have also examined the effect of procyanidins IVg on the endogenous α-tocopherol of fish muscle. *In vivo*, fish contains an antioxidant system that stabilizes its high content of unsaturated lipids and involves α-tocopherol, ubiquinone, carotenoids, glutathione and ascorbate (*5*). In post mortem conditions, endogenous antioxidants are consumed sequentially. Procyanidins IVg were able to preserve α-tocopherol present in fish muscle from oxidation (Figure 3). Such preservation was strongly correlated with the inhibition of rancidity achieved by IVg on the fish homogenate. To test a possible antioxidant synergism between procyanidins and tocopherol, we have generated the α-tocopheroxyl radical by reaction of α-tocopherol with 1,1-diphenyl-2-picrylhydrazyl (DPPH) radical. The radical was unambiguously monitored by electron spin resonance (ESR) spectroscopy. ESR data showed a decrease of tocopheroxy radical in the presence of procyanidins IVg. Therefore, procyanidins can regenerate α-tocopherol from α-tocopheroxy and to provide a synergistic effect for stabilizing fish muscle.

The results of this work reveal the significance of the capacity for donating electrons and the effect on the endogenous tocopherol to explain the ability of phenolic antioxidants supplemented for retarding and inhibiting lipid oxidation of fish muscle.

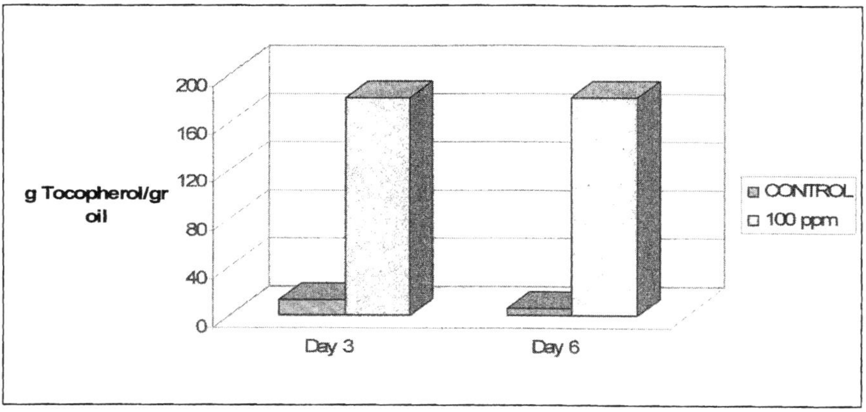

Figure 3. Preservation of initial endogenous α-tocopherol in minced fish muscle supplemented with 100 ppm of grape procyanidins IVg after 3 and 6 weeks of storage at 4 °C.

Functional Properties of Grape Procyanidins

The possible functional properties of the procyanidins IVg in the colon, their effect on the cell viability, cell cycle and apoptosis in colon cancer cells was investigated. We have used HT29 epithelial colon carcinoma immortalized cell lines and two other cell lines as control models of non-cancer colon cells. The results demonstrated that the decrease in cell number was dose-dependent and the polyphenols affected very little the cell viability in all three cell lines. There was an effect on the cell cycle in cancer HT29 cells but the cell cycle distribution in non-cancer cell lines was not affected by the procyanidin mixture. The procyanidins induced the appearance of 16% of apoptotic HT29 cells. Interestingly the pro-apoptotic effect was selective for cancer cells. These results show that the supplementation of grape procyanidins stabilizes fatty fish muscle against oxidation and the product can be an interesting functional food offering the combined action of fish and natural polyphenols

References

1. Ackman, R. G. In *Marine biogenic lipids, fats and oils*. CRC Press: Boca Raton, FL, 1989, pp. 103-137.
2. Liston, J. In *Chemistry and biochemistry of marine food products*. AVI Publishing Co.: Westport, CT, 1982, pp. 27-37.

3. Flick, G. F.; Martin, J. In *Advances in seafood biochemistry*. Technomic Publishing Co.: Lancaster, PA, 1992, pp. 99-122.
4. Madrid, A.; Cenzano, J. M. In *Los aditivos en los alimentos según la Unión europea y la Legislación española*. AMV Ed.; Madrid, 2000, 13-78.
5. Frankel, E. N. *Lipid Oxidation*. The Oily Press, Scotland, 1998.
6. Richards, M. P.; Kellerher, S.; Hultin, H.O. *J. Agric. Food. Chem.* **1998**, *46*, 4363-4371.
7. Pazos, M.; González, M. J.; Gallardo, J. M.; Torres, J. L.; Medina, I. *Eur. Food Res. Technol.* **2005**, *220*, 514-519.
8. Parke, D. V.; Lewis D. F. *Food Addit. Contam.* **1992**, *5*, 561-577.
9. Shahidi, F; Naczk, M. In *Food Phenolics: sources, chemistry, effects, applications*. Technomic Publishing Company; Lancaster, PA, 1995, pp. 1-5.
10. Packer, L.; Rimbach, G.; Virgili, F. *Free Radical Biol. Med.* **1999**, *27*, 704-724.
11. Lu, J. B.; Ho, C. T.; Ghai, G.; Chen, K. Y. *Cancer Res.* **2000**, *60*, 6465-6471.
12. Hayakawa, S.; Saeki, K.; Sazuka, M.; Suzuki, Y.; Shoji, Y.; Ohta, T.; Kaji, K.; You, A.; Isemura, M. *Biochem. Biophys. Res. Commun.* **2001**, *285*, 1102-1106.
13. Gorelik, S.; Kanner, J. *J. Agric. Food Chem.* **2001**, *49*, 5945-5950.
14. Torres, J. L.; Varela, B.; García, M. T.; Carilla, J.; Matito, C.; Centelles, J. J.; Cascante, C.; Sort, X.; Bobet, R. *J. Agric. Food Chem.* **2002**, *50*, 7548-7555.

Chapter 18

Antioxidant Activity of Volatile Extracts Isolated from Various Herbs and Spices

Cheong Kim and Kwang-Geun Lee

Department of Food Science and Technology, Dongguk University, Seoul 100–715, Korea

Herbs and spices have been used for providing humans with tastes in foods. In addition to taste, their beneficial health effects have also been widely attracted by food scientists. However, studies on natural antioxidants have mainly focused on molecules such as flavonoids and polyphenolic compounds whose formula weight (FW) is over 400 Da. In this study, we dealt with relatively lower FW molecules, lower than 400 Da, such as volatile compounds and their mixtures. Aroma extracts of various plants—15 herbs and 4 spices—were prepared by two extraction methods, namely distillation under reduced pressure followed by liquid-liquid extraction (DRP-LLE) and solvent assisted flavor extraction (SAFE). The extracts were examined for antioxidant activity by four different assays such as aldehyde-carboxylic acid assay, lipid/malonaldehyde (MA) assay, blood plasma/MA assay, and conjugated diene assay. The antioxidant activity was compared with that of known antioxidant, vitamin E or butylated hydroxytoluene (BHT).

Since synthetic antioxidants such as butylated hydroxyanisole and butylated hydroxytoluene have been reported to be toxic to experimental animals (*1*), antioxidants naturally occurring in plants are beginning to be considered as possible safe alternatives to synthetic antioxidants (*2*). Natural antioxidants are generally concentrated in seeds, beans and the nuts of plants (*3*). Naturally occurring antioxidants such as vitamin C and vitamin E as well as phenolic compounds, found in tea and red wine, possess the ability to prevent oxidative damage. Oxidative damage is associated with the onset of many diseases including cancer, cardiovascular disease, cataract, atherosclerosis, diabetes, arthritis, immune deficiency diseases, aging, and brain dysfunction (*4-6*).

However, studies on natural antioxidants have mainly focused on molecules such as flavonoids and polyphenolic compounds whose formula weight (FW) is over 400 Da. Investigation of beneficial health effects of relatively low FW compounds (LFWC) including various volatiles has been scarce. So far a lot of naturally occurring volatile compounds were found in plants as well as in cooked foods and they were considered primary only as flavors and fragrances (*7*). The leaves and flowers of plants containing numerous aroma chemicals have been used widely in folk medicine and in modern aromatherapies due to their potent beneficial health effects (*8*).

There would be many reasons why the health effects of LFWC has not been studied. First of all, it is hard to find robust assays to evaluate health effects of LFWC. Dut to high volatility of LFWC the assay should be carefully chosen for accurate measurement of beneficial health effects such as antioxidant activity. In our current study four different assays were introduced for measuring antioxidant activity of various volatile extracts. Aroma extracts of various plants—15 herbs and 4 spices—were prepared by two extraction methods, namely distillation under reduced pressure followed by liquid-liquid extraction (DRP-LLE) and solvent assisted flavor extraction (SAFE). The extracts were examined for antioxidant activity by four different assays such as aldehyde-carboxylic acid assay, lipid/malonaldehyde (MA) assay, blood plasma/MA assay, and conjugated diene assay.

Experimental

Materials

Green pepper (*Capsicum annuum* L.), welsh onion (*Allium fistulosum* L.), onion (*Allium cepa* L.), and garlic (*Allium scorodoprasm* L.) were grown and harvested at Dongguk University Farm located in Goyang, Korea, in 2004. The other plants such as Angelica tenuissimae roots (*Angelica tenuissima* Nakai), peppermint leaves (*Mentha arvensis* var. piperascens), pine needles (*Pinus sylvestris* L.), and Sweetflag leaves (*Acorus gramineus* Rhizoma), thyme leaves (*Thymus vulgaris* L.), basil leaves (*Ocimum basilicum* L.), rosemary leaves (*Rosemarynus officinalis* L.), chamomile flowerheads (*Chamaemelum nobile* L.), lavender flowerheads (*Lavandula angustifolia* P. Miller), cinnamon bark (*Cinnamomum cassia* Blume) and dried clove buds [*Syzygium aromaticum* (L.) Merr. et Perry] were purchased from a local market and identified by Prof. Byung-Soo Koo, College of Oriental Medicine, Dongguk University, Kyeongju, Korea. Eucalyptus leaves (*Eucalyptus polyanthemos* Schauer, *Eucalyptus globules, Eucalyptus perriniana*) were purchased from Faylor's Eucalyptus Farms (Temecula, CA). All chemicals used in this study were purchased from reliable commercial sources.

Distillation under Reduced Pressure Followed by Liquid-liquid Extraction (DRP-LLE)

Fresh plants were placed in a round-bottom flask with 1 L deionized water and then steam distilled at 55 °C for 3 h under reduced pressure (95 mm Hg). The distillate (200 mL) was extracted with 50 mL dichloromethane using a liquid-liquid continuous extractor for 6 h. After the extract was dried over anhydrous sodium sulfate, the solvent was removed by a distillation with a Vigreux column. The distillation was stopped when the volume of extract was reduced to approximately 1 mL, and then the solvent was further removed under a purified nitrogen stream. The sample was stored at 5 °C until antioxidative tests were carried out.

Solvent Assisted Flavor Evaporation (SAFE)

Plant was grinded or homogenized and extracted with 200 mL of dichloromethane for 12 h at room temperature. The filtered extract was added to the dropping funnel of the SAFE apparatus, which was heated to 40°C with a circulating water bath. The distillation flask (500 mL) was heated to 40°C in a water bath. The receiving flask for the distillate as well as the safety-cooling trap of the SAFE apparatus were cooled with liquid nitrogen. The SAFE apparatus was connected to a high vacuum pump (<0.01 Pa) and then the extract in the dropping funnel was added in small aliquots into the distillation flask over a period of 20 min. The frozen distillate was thawed at room temperature. The extract was dried over anhydrous sodium sulfate and concentrated to about 0.6 mL using a Vigreux column (15 x 1 cm) and water bath at 40-50 °C.

Aldehyde/Carboxylic Acid Assay (ACA)

The aldehyde/carboxylic acid assay (ACA) was developed for determining the long-term antioxidant potential of a chemical or a group of chemicals (*9*). This method is based on the autoxidation of aldehydes to carboxylic acids with active oxygen species such as a hydroxy radical (*10*). Fatty aldehydes are readily converted to the corresponding fatty acids in an oxygen-rich dichloromethane solution through a radical-type reaction (*11*). According ACA, the lack of antioxidative properties of a test solution will result in aldehyde conversion via oxidation to its corresponding carboxylic acid through a radical-type reaction in an oxygen rich environment. The extent of oxidation was measured by monitoring changes in the amount of aldehyde remaining using gas chromatography. Although ACA requires prolonged time periods, it offers a better measure of the oxidation process that occurs in foods and beverages.

Lipid/MA assay (LMA) and blood plasma/MA assay (BMA)

A capillary gas chromatography (GC) method for specific determination of MA has been developed (*12*). In this method, the MA formed from lipids upon oxidation is derivatized to 1-methylpyrazole (1-MP), which is subsequently determined by a GC with a nitrogen-phosphorus detector (NPD) (*13*). The lipids used in this assay were cod liver oil and blood plasma. In particular, the MA determined by the blood plasma/MA assay (BMA) may serve as an effective biomarker for monitoring the extent of lipid oxidation in biological systems. The BMA assay mimics the antioxidant activities of a chemical or mixture of chemicals on *in vivo* lipid peroxidation. Methods involved in LMA are useful for a rapid analysis of samples, but occur under artificially strong oxidative conditions. Therefore, these assays may not accurately represent oxidative processes which are often associated with food systems. Generally, in food systems, oxidation processes occur slowly over a period of 40 days.

Conjugated Diene Assay (CDA)

Measurements of conjugated diene hydroperoxides in methyl linoleate (MeLo) were carried out according to the procedures previously described (*14*). Various concentrations of each volatile extract dissolved in dichloromethane were added to methyl linoleate (1 g), and dichloromethane was removed under pure nitrogen purging (20 seconds). Oxidation of methyl linoleate in screw-capped amber vials (15 mL) was carried out at 40°C in a shaker water bath. An aliquot of sample (10 mg) was taken at 24-hour intervals and dissolved in 5 mL of 2,2,4-trimethylpentane (isooctane) for spectrometric measurement (HP 8452A diode array UV spectrophotometer, Palo Alto, CA) of conjugated diene absorption at 234 nm. The spectrophotometer was set to zero with 2,2,4-trimethylpentane. All analyses were carried out in triplicate. Results were calculated as millimoles MeLo hydroperoxides per kilogram of methyl linoleate with a molar absorptivity of 26,000 (*15*).

Results and Discussion

Antioxidant Activity of Various Herbs Measured by Aldehyde/Carboxylic Acid Assay (ACA)

The aldehyde/carboxylic acid assay (ACA) is a relatively simple assay compared to typical assays for antioxidant activity of an individual chemical or a group of chemicals. Volatile chemicals dissolved in an organic solvent (non- or less-polar phase) can be easily measured for their antioxidant potential by this assay. In addition, this assay is appropriate to assess long-term antioxidant potential of natural antioxidants because aldehyde oxidation is monitored over 40 days. This assay has been used to examine antioxidant activity of natural aroma extracts isolated from plant sources (*16, 17*).

Figure 1 shows the amounts of hexanal remaining in samples over a period of 40 days. GC peak area ratios were calculated by dividing the GC peak area of hexanal by the GC peak area of the internal standard, undecane. Antioxidant activities of known antioxidants—BHT and α-tocopherol—were used to confirm the testing system. All extracts exhibited dose-dependent inhibitory activity at concentrations of 50-500 μg/mL and antioxidant activity. After 30 days, control samples exhibited over 95% oxidation of hexanal to hexanoic acid. In control samples, the rate of hexanoic acid formation was slow during the first 10 days but increased rapidly after 10 days. Hexanal was completely oxidized to hexanoic acid after 40 days.

At all concentrations of thyme, almost 100% hexanal remained in the samples, indicating that the thyme extract possesses potent antioxidant activity against hexanal oxidation. The percentage of hexanal remaining in solutions treated with different amounts of volatile extracts, α-tocopherol, and BHT is shown in Figure 1. The inhibitory effect of the extracts on the formation of hexanoic acid was in the following descending order: thyme > basil = clove bud = pine needle > sweet flag leaf = peppermint leaf > *Eucalyptus polyanthermos* > Angelica tenuisimae root > rosemary = *Eucalyptus globules* > *Eucalyptus perriniana* > chamomile > aloe, lavender, cinnamon. At a concentration of 500 μg/mL, thyme, basil, clove bud, pine needle, and *Eucalyptus polyanthermos*

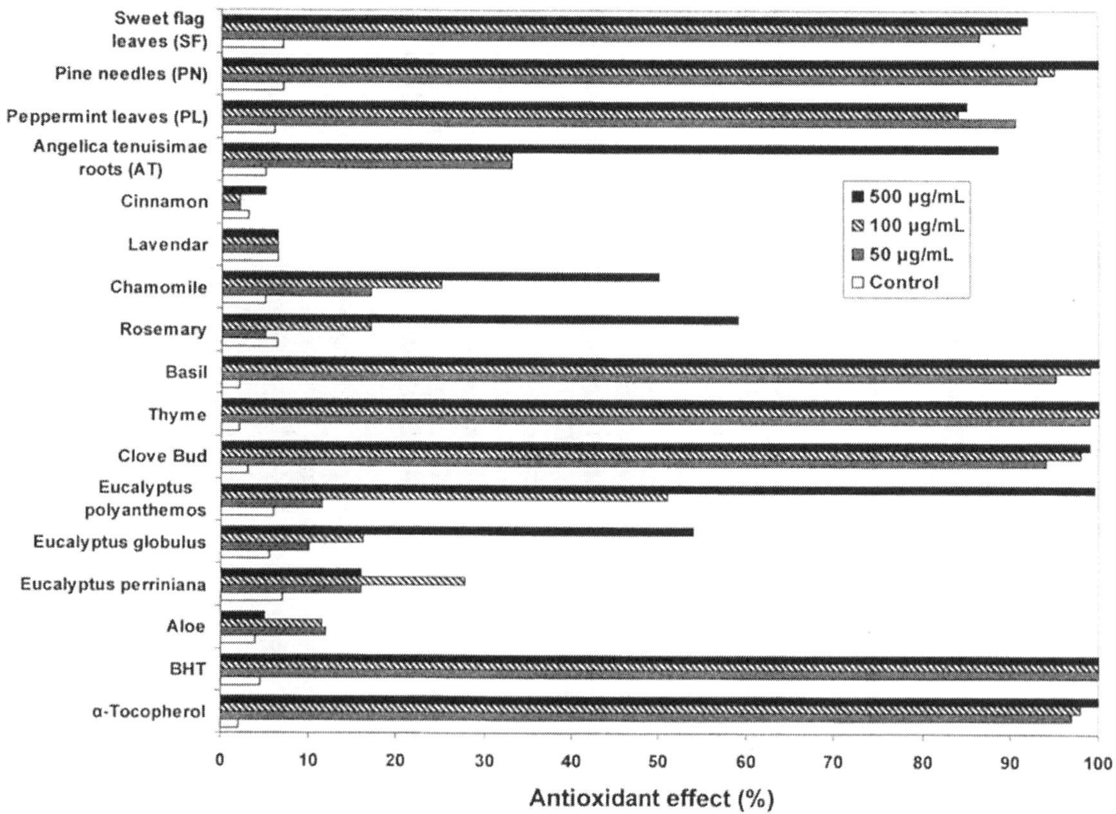

Figure 1. Antioxidant activity of various herbs measured by Aldehyde/carboxylic acid assay (ACA).

extract inhibited hexanal oxidation by 100%, for 40 days. For the volatile extract of thyme, basil, clove bud, and pine needle, the percentages of hexanal remaining were 95-100% at concentrations from 50 to 500 μg/mL. Their inhibitory effect were comparable to that of BHT and α-tocopherol which showed over 99% at concentrations varying from 50 to 500 μg/mL.

Antioxidant Activities of Volatile Extracts in the Lipid/MA Assay (LMA)

Since thiobarbituric acid (TBA) assay has been criticized for its low specificity and overestimation of malonaldehyde (MA), more accurate methods for quantifying MA concentrations have been investigated (*18*). The lipid/MA assay (LMA), a capillary GC method for specific determination of MA, has been selected for examining antioxidant properties of a chemical or mixture of chemicals (*19-21*). Figure 2 shows the results of the LMA in the presence of various herb extracts along with those of α-tocopherol and BHT. The results were similar to those obtained in the ACA. The inhibitory effect of the extracts on formation of MA was in the following descending order clove bud > *Eucalyptus polyanthermos* > *Eucalyptus globules* > *Eucalyptus perriniana* > aloe. Clove bud extract and *Eucalyptus polyanthemos* extract inhibited MA formation by 93 and 86%, respectively, at a level of 160 μg/mL. These antioxidative activities were comparable to those of α-tocopherol and BHT which inhibited MA formation by 96 and 97% at the same concentration of 160 μg/mL.

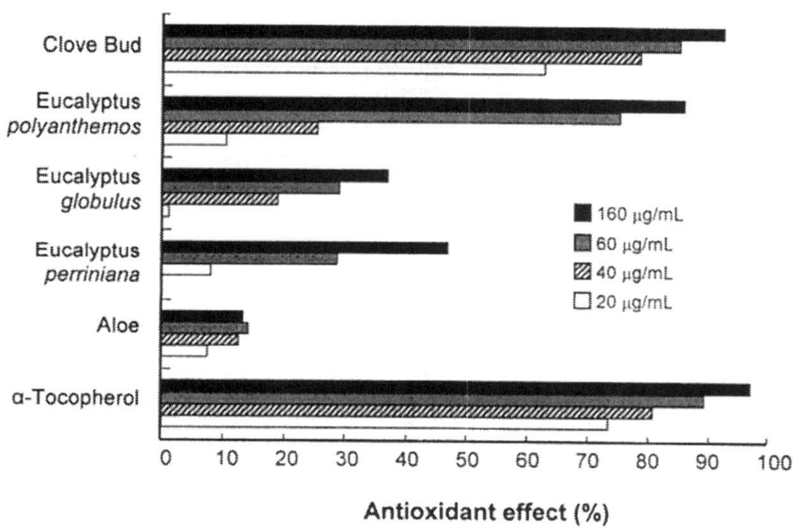

Figure 2. Antioxidant activity of various herbs measured by Lipid/MA assay (LMA). The values of concentration are in μg/mL.

According to the previous reports the antioxidative activities of clove buds and eucalyptus leaves are due in part to the contributions of aroma chemicals such as eugenol, eugenyl acetate, and thymol (*16*).

Antioxidant Activities of Volatile Extracts in the Blood Plasma/MA Assay (BMA)

Human blood plasma is a pale straw-colored fluid and contains 90-92% water and 6-8% protein in the solid matter of plasma. The majority of the plasma proteins are conjugated proteins such as glycoproteins and lipoproteins. Lipids in blood plasma are transported in the form of a macromolecular lipid-protein complex and termed "lipoproteins" due to their hydrophobic nature. The mechanisms by which oxidative modification of blood plasma lipids contribute to the progression of atherosclerosis have not been elucidated so far. One hypothesis is that reactive aldehydes generated endogenously during the process of lipid peroxidation are associated with the oxidative modification of lipids (*22*).

Figure 3 shows the inhibitory effect of aroma extracts (clove buds and eucalyptus leaves) along with those of α-tocopherol and BHT toward MA formation in horse blood plasma upon oxidation. All samples of the extracts, α-tocopherol, and BHT exhibited dose-dependent inhibitory activity. The amount of MA formed from 50 μL of blood plasma in a control sample was 770 ± 37 nmol under the same conditions used in this study. The aroma extract of clove buds exhibited a level of antioxidant activity comparable to that of α-tocopherol or BHT. The extracts of eucalyptus and clove inhibited MA formation by 23 and 48%, respectively, at a level of 400 μg/mL. On the other hand, α-tocopherol and BHT inhibited MA formation by 52 and 70%, respectively, at the same level.

Antioxidant Activities of Volatile Extracts in the Conjugated Diene Assay (CDA)

In the conjugated diene assay (CDA), the amount of total conjugated diene hydroperoxides produced from polyunsaturated lipids such as methyl linoleate (MeLo) upon oxidation, was determined quantitatively by the absorption maximum at 234 nm (*23*). This method has been applied to measure antioxidant activity of plant and wine extracts (*24, 25*), flavonoids (*26*), and tocopherols (*27*). In this assay, no initiator or metal-catalyst such as ferrous chloride or hydrogen peroxide was added for initiating and accelerating autoxidation, because the main oxidation phase is not an initiation phase but a propagation phase (*26*).

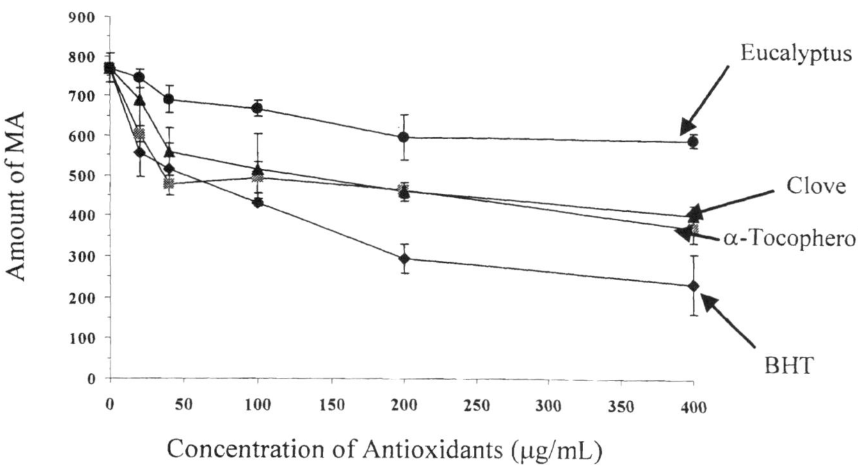

Figure 3. Inhibitory effects (%) of aroma extract of clove buds, eucalyptus leaves, α-tocopherol, and BHT toward MA formation from horse blood plasma upon oxidation at various levels.

Figure 4 illustrates the effect of volatile extracts on the oxidation of methyl linoleate. In CDA six different herbs were measured for their antioxidant activity. The formation of conjugated diene hydroperoxides in oxidizing methyl linoleate was most effectively inhibited by thyme. The volatile extracts of thyme and basil inhibited hydroperoxide formation in methyl linoleate dose dependently, at concentrations from 20 to 200 μg/mL. Inhibitory effects (%) of various concentrations of volatile extracts, α-tocopherol, and BHT on the formation of conjugated diene hydroperoxides from MeLo are shown in Figure 4.

The effectiveness of the various volatile extracts on MeLo oxidation was in the following descending order: thyme > basil > rosemary, chamomile > lavender, cinnamon. The trend of inhibition of MeLo hydroperoxides was quite similar to that of ACA. The inhibitory effects of thyme, basil, rosemary, chamomile, lavender, and cinnamon were 99, 68, 38, 31, 5, and 2%, respectively, at a concentration of 200 μg/mL. The inhibitory effect of thyme was similar to that of α-tocopherol or BHT at concentrations from 50 to 200 μg/mL. At concentrations from 20 to 200 μg/mL, BHT and α-tocopherol inhibited MeLo oxidation by 100 and 96-99%, respectively.

The strong antioxidant activity of the volatile extract of thyme is mostly due to their main volatile components, thymol and carvacrol (28). The antioxidant activities of thymol and carvacrol have been reported using various testing systems (*29-31*). In the case of the volatile extract of basil, eugenol, reported as a major volatile component, probably contributes significantly to the strong antioxidant activity of the extract (*28*). Eugenol showed high antioxidant activity in the aldehyde/carboxylic acid assay (*16*).

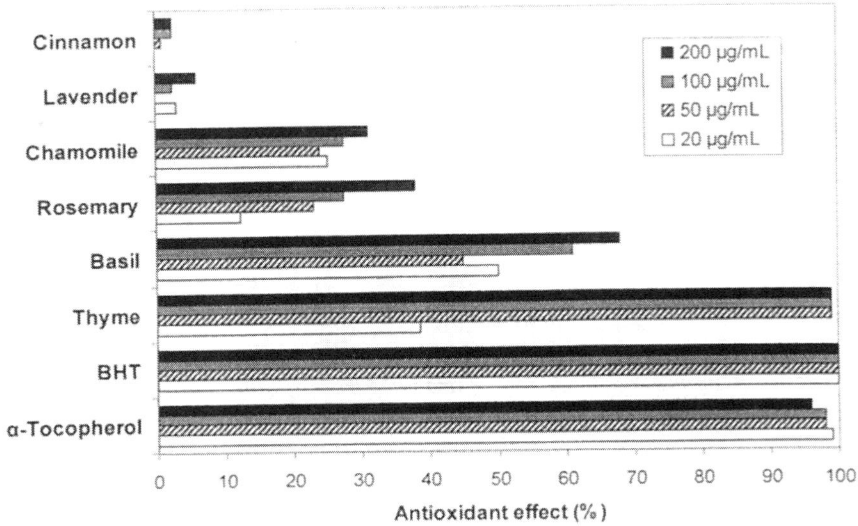

Figure 4. Antioxidant activity of various herbs measured by Conjugated diene assay (CDA).

Antioxidant Activity of Various Spices

To measure antioxidant activities of the aroma extracts of spices in this study, the ACA and LMA were used. Figure 5 shows the remaining hexanal in samples in the ACA. ACA can be used to measure the antioxidant potential of a chemical in an organic, non-aqueous phase (*32*). All extracts exhibited a dose-dependent inhibitory activity. The concentration of hexanoic acid content increased as the hexanal content decreased. At extract concentrations of 10, 50, 100, 500 μg/mL, green pepper inhibited hexanal oxidation by 10, 15, 46, and 51%, respectively. Welsh onion extract inhibited hexanoic acid formation by 9, 17, 29, and 95% in 10, 50, 100, 500 μg/mL, respectively. Onion extract inhibited hexanal oxidation by 3, 13, 51, and 96% for 30 days at concentrations of 10, 50, 100, 500 μg/mL, respectively. At extract concentrations of 10, 50, 100, 500 μg/mL, garlic inhibited hexanal oxidation by 1, 11, 97, and 98%, respectively. Antioxidant activity of welsh onion extract was higher than others at extract concentrations of 50 μg/mL. At extract concentrations of 100 and 500 μg/mL, garlic showed higher activity than others. The antioxidative activities of green pepper, welsh onion, onion, and garlic extracts were comparable to those of α-tocopherol, which inhibited hexanal oxidation by 50 75, 90, and 99% at the same concentrations of 10, 50, 100, and 500 μg/mL.

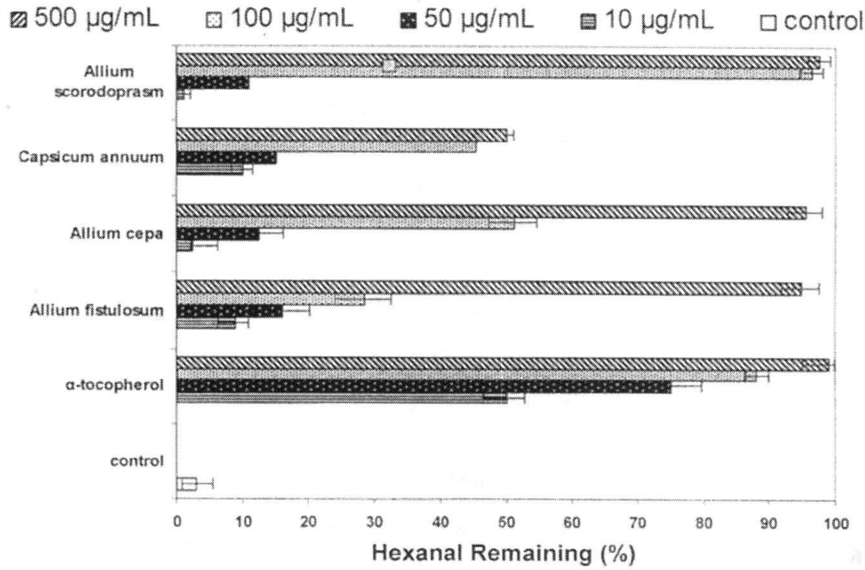

Figure 5. Percentage of hexanal remaining in samples containing different amounts of aroma extract of Capsicum annuum L., Allium fistulosum L., Allium cepa L., and Allium scorodoprasm L., and α-tocopherol.

Figure 6 shows the results of the LMA in the presence of green pepper, welsh onion, onion, and garlic extracts, along with those of α-tocopherol. The results were similar to those obtained in the ACA. Green pepper extract inhibited MA formation by 27, 45, and 49%, respectively, at 20, 100, and 500 μg/mL. At extract concentrations of 20, 100, 500 μg/mL, welsh onion inhibited MA formation by 34, 45, and 58%, respectively. Onion extract inhibited MA formation by 38, 58, and 62%, respectively, at concentrations of 20, 100, 500μg/mL. At extract concentrations of 20, 100, 500 μg/mL, garlic inhibited MA formation by 47, 62, and 58%, respectively. The antioxidative activities of green pepper, welsh onion, onion, and garlic extracts were comparable to those of α-tocopherol, which inhibited MA formation by 54, 68, and 93% at the same concentrations of 20, 100, 500 μg/mL.

To date there are few reports to elucidate volatile antioxidants in the four plants of this study. According to the previous report (*33*), glycosides such as trans-*p*-sinapoyl-β-D-glucopyranoside and querrcetin 3-O-α-L-rhamno pyranoside in hot pepper extract had strong antioxidant activity in two assays: the β-carotene-fractions linoleic acid system and the 1,1-diphenyl-2-picrylhydrazyl discoloration test. Aqueous extract of welsh onion in the range 0.05-1.0 μg/mL showed a potent concentration-dependent reduction of xanthine oxidase activity, and oxidative damage of albumin was inhibited by 0.1-2.0 μg/mL (*34*).

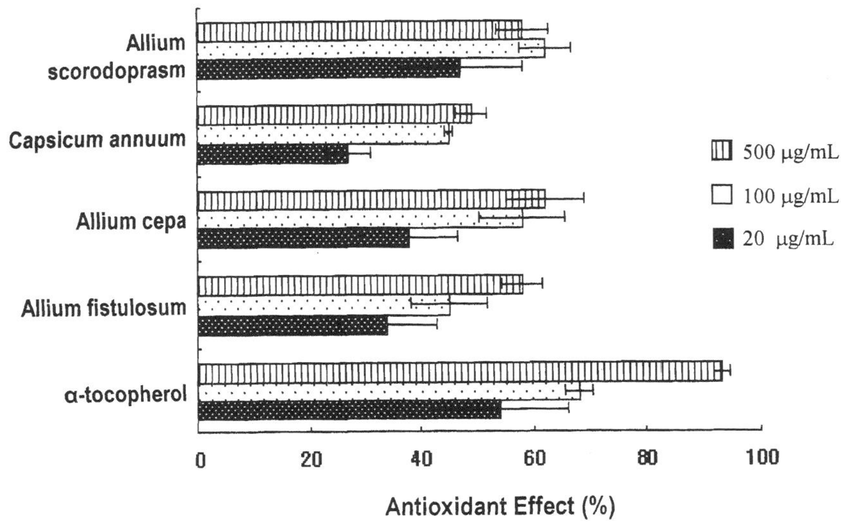

Figure 6. Inhibitory effects (%) of aroma extract of Capsicum annuum L., Allium fistulosum L., Allium cepa L., Allium scorodoprasm L., and α-tocopherol toward malonaldehyde formation from cod liver oil.

Conclusions

Natural antioxidants are becoming increasingly important in food and medicine, because many diseases such as cancer, cardiovascular disease, cataract, atherosclerosis, diabetes, arthritis, immune deficiency diseases, aging, and brain dysfunction, still remain difficult to treat. Searching for new natural antioxidants would be one of important sources to cure the diseases and also would be very useful to maintain food freshness, flavor, taste, and color by preventing oxidation deterioration. Aroma extracts isolated plants, primarily considered only as flavors and fragrances are now considered as one of natural antioxidants. The leaves and flowers of plants containing numerous aroma chemicals have been used widely in folk medicine and in modern aromatherapies.

Aroma extracts are composed of a complex mixture of chemicals in which not all compounds can be expected to have an antioxidant potential. It is nearly impossible to test the possible synergism between the more than 80 compounds observed in the gas chromatograms of aroma extracts. Since tremendous numbers of these aroma compounds are present in plants, so their combined activities may be comparable to those of known antioxidants, BHT and α-tocopherol. Since humans are exposed to reactive oxygen species 24 hours a day, it is extremely important that antioxidant supplements be taken constantly and consistently. Exposure and ingestion of these aroma extracts in this study may help to prevent *in vivo* oxidative damage, such as lipid peroxidation, which is associated with many diseases, including cancer, arteriosclerosis, diabetes, and immune deficiency.

References

1. Ford, S.M.; Hook, J.B.; Bond, J.T. *Food Cosmet. Toxicol.* **1980**, *18*, 21-26.
2. Frankel, E.N.; Huang, S.-W.; Aeschbach, R.; Prior, E. *J. Agric. Food Chem,* **1996**, *44*, 131-135.
3. Namiki, M.. *Crit. Rev. Food Sci. Nutr.* **1990**, *29*, 273-300.
4. Henle, E. S.; Luo, Y.; Gassmann, W.; Linn, S. *J. Biol. Chem.*, **1996**, *271*, 21177-21186.
5. Beckman, K. B.; Ames, B. N. *Physiol. Rev.* **1998**, *78*, 547-581.
6. Pietta, P.; Simonetti, P.; Mauri, P. *J. Agric. Food Chem.* **1998**, *46*, 4487-4490.
7. Bemelmans, J. M. H. In *Isolation, separation and identification of volatile compounds in aroma research.*; Maarse, H.; Belz, R., Ed.; Academie-Verlag: Berlin, 1978, pp. 4-49.
8. Vickers, A. *Massage and aromatherapy* Chapman & Hall: London, 1996, pp. 127-167.
9. Macku, C.; Shibamoto, T. *J. Agric. Food Chem.* **1991**, *39*, 1990-1993.
10. Horner, L. In *Autoxidation and Antioxidants;* Lundberg, W. O., Ed.; John Wiley & Sons: New York, NY, 1961, pp. 197-202.
11. Nonhebel, D. C.; Tedder, J. M.; Walton, J. C. *Radicals.* Cambridge niversity Press: London, 1979, pp. 155-157.
12. Umano, K.; Dennis, K. J.; Shibamoto, T. *Lipids* **1988**, *23*, 811-814.
13. Miyake, T.; Shibamoto, T. *J. Agric. Food Chem.* **1998**, *46*, 3694-3697.
14. Heinonen, I. M.; Lehtonen, P. J.; Hopia, A. I. *J. Agric. Food Chem.* **1998**, *46*, 25-31.
15. Chan, H. W.; Levett, G. *Lipids* **1977**, *12*, 99-104.
16. Lee, K. G.; Shibamoto, T. *Food Chem.* **2001**, *74*, 443-448.
17. Lee, K. G.; Shibamoto, T. *J. Sci. Food Agric.* **2001**, *81*, 1573-1579.
18. Esterbauer, H.; Cheeseman, K. H. *Meth. Enzymol.* **1990**, *186*, 407-431.
19. Wong, J. W.; Hashimoto, K.; Shibamoto, T. *J. Agric. Food Chem.* **1995**, *43*, 2707-2712.

20. Eiserich, J. P.; Wong, J. W.; Shibamoto, T. *J. Agric. Food Chem.* **1995**, *43*, 647-650.
21. Lee, K. G.; Shibamoto, T. *J. Agric. Food Chem.* **2000**, *48*, 4290-4293.
22. Esterbauer, H., Schaur, R.J. and Zollner, H. *Free Radical Biol. Med.* **1991**, *11*, 81-128.
23. Frankel, E. N. In *Lipid Oxidation*; Oily Press: Dundee, Scotland, 1998, pp. 33-35.
24. Kahkonen, M. P.; Hopia, A. I.; Heinonen, M. *J. Agric. Food Chem.* **2001**, *49*, 4076-4082.
25. Kahkonen, M. P.; Hopia, A. I.; Vuorela, H. J.; Rauha, J. P.; Pihlaja, K.; Kujala, T. S.; Heinonen, M. *J. Agric. Food Chem.* **1999**, *47*, 3954-3962.
26. Pekkarinen, S. S.; Heinonen, I. M.; Hopia, A. I. *J. Sci. Food Agric.* **1999**, *79*, 499-506.
27. Hopia, A. I.; Huang, S. W.; Schwarz, K.; German, J. B.; Frankel, E. N. *J. Agric. Food Chem.* **1996**, *44*, 2030-2036.
28. Teissedre, P. L.; Waterhouse, A. L. *J. Agric. Food Chem.* **2000**, *48*, 3801-3805.
29. Lacroix, M.; Smoragiewicz, W.; Pazdernik, L.; Kone, M. I.; Krzystyniak, K. *Food Res. Int.* **1997**, *30*, 457-462.
30. Schwarz, K.; Ernst, H.; Ternes, W. *J. Sci. Food Agric.* **1996**, *70*, 217-223.
31. Yanishlieva, N. V.; Marinova, E. M.; Gordon, M. H.; Raneva, V. G. *Food Chem.* **1999**, *64*, 59-66.
32. Ka MH; Choi EH; Chun HS; Lee KG. *J. Agric. Food Chem.* **2005**, *53*, 4124-4129.
33. Materska M; Perucka I. *J. Agric. Food Chem.* **2005**, *53*, 1750-1756.
34. Wang B. S.; Lin S. S.; Hsiao W. C.; Fan J. J.; Fuh L. F.; Duh P. D. *Food Chem.* **2006**, *98*, 149-157.

Chapter 19

Antioxidative Stress Peptides

Y. Mine and S. Katayama

Department of Food Science, University of Guelph, Guelph, Ontario N1G 2W1, Canada

Oxidative stress is a state characterized by an excess of reactive oxygen species (ROS) in the body, which creates a potentially unstable cellular environment linked to tissue damage, degenerative disease, and accelerated aging. Many environmental factors are implicated in the generation of ROS. Specific dietary antioxidants have been considered to modulate oxidative stress and suppress gut inflammation and carcinogenesis. Recently, we have developed phosphopeptides (PPPs) from hen egg yolk phosvitin and found PPPs act as inhibitors of lipid peroxidation and radical scavengers. In the present study, the protective effects of PPPs against hydrogen peroxide (H_2O_2)-induced oxidative stress in an *in vitro* assay was evaluated using human intestinal epithelial cells, Caco-2. The effects of PPPs on intracellular glutathione (GSH) levels and several antioxidative enzymes were also investigated. The results suggest that PPPs could suppress oxidative stress-induced gut inflammatory disorders and may potentially provide new bioactive peptides against tissue oxidative stress.

Oxidative stress is a highly oxidized environment within cells due to increased levels of free radicals. Free radicals can be defined as a highly reactive molecule with one or more unpaired electrons in their outer orbit. The unbalanced molecule tries to steal electrons from another molecule to make itself more stable, and this starts a chain reaction that can harm the cells of the body. Radicals derived from oxygen (reactive oxygen species; ROS) represent the most important class of radical species generated in living systems (*1*). ROS are divided into free oxygen radicals and non-radical ROS. Free oxygen radicals include superoxide (O_2^-), hydroxyl- (•OH), nitric oxide (•NO), alkoxyl- (RO•), or peroxyl- (ROO•) radicals. Non-radical ROS include hydrogen peroxide (H_2O_2), organic hydroperoxides (ROOH) and hypochloride (HOCl). It has been estimated that the average person has around 10,000-20,000 free radicals each in cell each day (*2*). ROS are produced by neutrophils and macrophage during inflammation and are by-products of mitochondria-catalyzed electron transport reactions and other mechanisms (*3, 4*). Many environmental factors are also implicated in the generation of ROS, including exposure to alcohol, cigarette smoke, atmosphere pollutants, UV light, toxins, and over-exercise (*5*). In addition, free radicals and lipid hydroperoxides may either preexist in the diet or arise from polyunsaturated fatty acids (*6-8*).

ROS are generally recognized for playing a dual role as both deleterious or beneficial species (*2*). Beneficial effects occur at low/moderate concentrations. Under physiological conditions, ROS are key components of cellular metabolism and play multiple positive roles in energy production, phagocytosis, cell growth, and cellular signaling regulation. In contrast, the harmful effect of free radicals is termed "oxidative stress". This occurs when there is an overproduction of ROS, most frequently either by excessive stimulation of NAD(P)H by cytokines, or by the mitochondrial electron transport chain and xanthine oxidase. The resultant excess ROS can damage the cellular lipids, membranes, proteins, and DNA, and consequently lead to loss of their biological function. Oxidative stress has been associated with the development of a number of human diseases, including cardiovascular disease, cancer, neurological disorders, diabetes, inflammatory, and other diseases, as well as the ageing process (*9-13*).

Recent studies (*9*) have revealed that ROS play an important role in several aspects of intracellular signaling regulation. Probably the most significant effect of ROS on signaling pathways has been observed in the mitogen-activated protein kinase (MAPK) pathways (*14*). The MAPK pathways involve activation of nuclear transcription factors. These factors control the expression of protective genes that repair damaged DNA, power the immune system, arrest the proliferation of damaged cells, and induce apoptosis. For example, the nuclear transcription factor NF-κB is involved in inflammatory responses and AP-1 is important for cell growth and differentiation.

The human body is equipped with an antioxidative defense system that mitigates the damaging effects of oxidative stress on cells (*15*). These involve enzymatic antioxidant defenses (superoxide dismutase, SOD; glutathione peroxidase, GPx; catalase), non-enzymatic antioxidants (glutathione, GSH; thioredoxin, TRX), and dietary antioxidants (vitamins, minerals, carotenoids, and polyphenols). These materials function individually and cooperatively as a firm antioxidant network. The activities and the intracellular levels of these antioxidants are very important aspect for the survival of living organisms and health.

Food Factors for Antioxidative Stress

Recently, there has been a considerable interest in search for natural antioxidants. These compounds have been discovered in various food sources, particularly in plant materials. Some of these compounds show antioxidant activities not only in vitro but also in vivo. Naturally occurring antioxidants have a potential for use as components in functional foods or nutraceuticals.

Vitamins

Vitamins are known to provide energy, combat stress, and maintain normal health. Most vitamins are not directly produced by the body and must be obtained from the diet, such as fruits, vegetables, grains, and cereals. Vitamin C (L-ascorbic acid) is a water-soluble antioxidant and physiologically scavenges ROS and free radicals, in particular the hydroxyl radicals (*16*). Vitamin C also can regenerate other low-molecular-weight antioxidants, such as α-tocopherol, GSH, urate, and β-carotene, from their respective radical species (*17*). The benefit of vitamin C intake was reported from clinical trials of stomach cancer incidence and cardiovascular disease (*18, 19*). Vitamin E is a lipophilic vitamin that exists in eight different forms, and α-tocopherol is a powerful vitamin E activity (*20*). The main activity of α-tocopherol is protection against lipid peroxidation (*21, 22*).

Carotenoids

Carotenoids are fat-soluble pigments synthesized by plants and microorganisms and are present in many foods, particularly fruits, vegetables, and fish. Various studies have indicated that carotenoids, in particular lycopene and β-carotene, deactivate free radicals and may prevent cancer, heart disease, stroke, and other diseases. The antioxidant activities of carotenoids are associated with their radical scavenging properties and their exceptional singlet oxygen quenching abilities. More recent studies (*23, 24*) have indicated that carotenoids could regulate transcription factors associated with various cell functions, such as cell proliferation, differentiation, and apotosis. Carotenoids are also capable of regulating transcriptional control of the expression of antioxidative enzymes. The transcription factor Nrf2, which binds to ARE (regulatory regions of antioxidant genes), appears to be essential for the induction of antioxidative enzymes. Several studies have shown that antioxidants present in diet may work as anti-cancer agents by activating this transcription system (*25*). Animal studies have indicated that some carotenoids induce antioxidative enzymes. Breinholt et al. (*26*) have shown that the administrated lycopene to rats induces several antioxidant enzymes, such as

NAD(P)H quinone oxidoreductase 1 (NQO1), GST, GPx, glutathione reductase (GR), and SOD, in liver. In another study, Bhuvaneswari et al. (*27*) reported that the administration of lycopene suppressed 7,12-dimethylbenz[*a*]anthracene (DMBA)-induced oral carcinogenesis by an increase in the levels of GSH, GPx, GST, and GR. Induction of antioxidative enzymes appears to be an effective means for achieving protection against a variety of carcinogens in humans.

Flavonoids

Flavonoids are a class of water-soluble pigments found in many plants and can be categorized according to their chemical structure (the position of the substituents and the number of hydroxyl groups) (*28-30*). The most actively studied flavonoids are red grape resveratrol, green tea catechins, onion quercetin, and soy daidzein. These compounds serve as powerful antioxidants and play important roles in maintaining health, including the treatment and prevention of cancer, cardiovascular disease and other pathological disorders. Flavonoids can act as scavengers of reactive radical species and thereby become effective inhibitors of lipid peroxidation. The phenoxy radical intermediates are relatively stable, so flavonoids may function as terminators of radical reactions.

Recent experimental data indicate that several flavonoids can stimulate the transcription of antioxidant and detoxification defense systems through ARE elements. For example, resveratrol is capable of enhancing endogenous antioxidants and phase II enzymes (enzymes involved in detoxification of electrophilic compounds), leading to increased resistance to oxidative and electrophilic injury (*31-33*). Green tea polyphenol extracts also stimulate the transcription of phase II enzymes by ARE activation probably utilizing the MAPK signaling pathway (*34*). The efficacy differs according to the polyphenol structure, which may be related to the different capacity to act as ligands for receptors unknown so far. It has been demonstrated that flavonoids having slight structural differences exhibited a markedly different efficiency in inducing phase II enzymes, suggesting that their beneficial effects arise from the presence of hydroxyl groups and the 2,3-double bond in the C ring (*35, 36*). Among five different green tea catechins, epigallocatechin-3-gallate and epicatechin-3-gallate showed the strongest activity in inducing gene expression such as phase II detoxifying enzymes in human hepatoma HepG2 cells, indicating the efficacy of the 3-gallate group (*37*).

Antioxidative Stress Activity of Phosvitin Phosphopeptides

Bioactive peptides are specific protein fragments that have a positive impact on body's function or condition and consequently influence health (*38*). These peptides are inactive within the sequence of the precursor protein and can be released by proteolytic enzymes during gastrointestinal digestion or during fermentation with microorganisms (*39*). Today, hen egg proteins are considered as an important source of bioactive peptides, and an increasing number of bioactive peptides have been identified in hen egg protein hydrolysates. So far, various biological activities, including antimicrobial activities (*40*), immunomodulatory (*41*), anti-tumor (*42*), and anti-hypertensive activities (*43, 44*), and disease prevention and treatment (*45, 46*), has been reported.

Hen egg yolk phosvitin is a highly phosphorylated protein with a molecular weight (MW) of 35000 Da that comprises 10% phosphorus and 6.5% carbohydrates (*47*). It contains 123 serine (Ser) residues accounting for 57.5% of the total amino acid residues, and most of these Ser are monoesterified with phosphate (*48*). We prepared oligophosphopeptides (PPPs) with MWs of 1000-3000 Da from egg yolk phosvitin by partial alkaline dephosphorylation and following tryptic hydrolysis (*49*). PPPs with 35% phosphate retention enhanced calcium- and iron-binding ability and inhibited the formation of phosphate precipitate, suggesting that PPPs provide a new nutraceutical by increasing calcium and iron uptake in the intestinal tract (*49-51*). We further demonstrated that PPPs showed free radical scavenging and antioxidant activities against lipid peroxidation. Many studies strongly indicate that a variety of dietary antioxidants exhibit protective effects against oxidative injury. We therefore evaluated the protective effects of PPPs derived from hen egg yolk against oxidative stress in an in vitro assay using human intestinal epithelial cells.

Protective effects of PPPs on gut inflammatory disorders in Caco-2

Gastrointestinal epithelial cells are exposed to oxidative stress that results in inflammation of the gut mucosa through the production of proinflammatory cytokines such as IL-8 and TNF-α (*52–54*). Especially, IL-8 production activates inflammatory cells (neutrophils), which extend the tissue damage (*55, 56*). Thus, a reduction in IL-8 production may provide antioxidative protection against stress activity in gastrointestinal epithelial cells. In this study, Caco-2 cell lines were used as a model of the intestinal epithelium. The Caco-2 monolayer cells cultured with 5% FBS-DMEM were treated with various concentrations of phosvitin, partially dephosphorylated phosvitin (DPV) and its tryptic digests, PPPs (35% phosphorus retained) for 2 h at 37°C. The cells were then incubated with H_2O_2 (1 mM) for 6 h, and the secretion of IL-8 was determined by ELISA. DPV and PPP inhibited IL-8 secretion in a concentration-dependent manner, whereas phosvitin did not significantly reduce the secretion (Figure 1) (*57*). The inhibitory effect of PPP on IL-8 production was almost identical to that of GSH, a powerful antioxidant. Thus, we focused on the protective effects of PPPs against oxidative stress in intestinal epithelial cells.

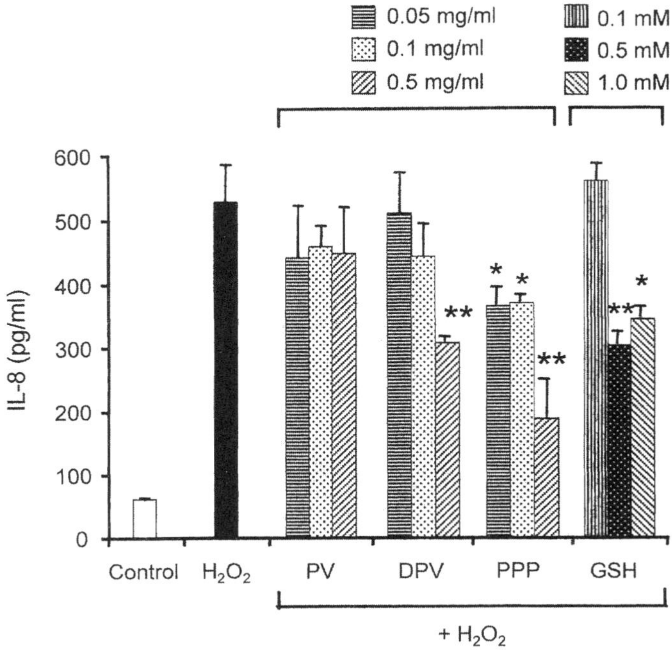

*Figure 1. Effects of phosvitin and its derivatives on IL-8 secretion in H_2O_2-treated Caco-2 cells. Cells were cultured with 5% FBS-DMEM, and treated with various concentrations of phosvitin and its derivatives (0-0.5 mg/mL) and GSH (0-1.0 mM) at 37℃ for 2 h, and then incubated with H_2O_2 (1 mM) for 6 h. PV, DPV, and PPP denotes phosvitin, partially dephosphorylated phosvitin, and phosvitin oligophosphopeptides, respectively. * $P < 0.05$ and ** $P < 0.001$, compared with cells treated with H_2O_2.*

PPPs were further separated into three fractions (PPP-1, PPP-2, and PPP-3) on Mono Q HR 5/5 anion-exchange column (Bio-Ral Laboratories Ltd) according to different contents of phosphorus and serine (*57*). The phosphorus contents of PPP-1, PPP-2, PPP-3 were 0, 7.2, and 18.9%, and the Ser contents were 9.5, 18.4, 52.2%, respectively (*57*). The inhibitory effects of the three PPP fractions on IL-8 secretion in H_2O_2-treated Caco-2 cells were then investigated. All PPP fractions decreased the IL-8 secretion in a concentration-dependent manner, with inhibition increasing with the amount of phosphorus (Figure 2). It seems that the phosphorus moieties in PPP fractions have pivotal effects; however, phosphoserine itself did not markedly decrease the secretion. Phosvitin, containing 10% phosphorus, did not significantly inhibit IL-8 secretion, and the effect was dramatically enhanced by enzymatic hydrolysis of phosvitin to oligophoshopeptides.

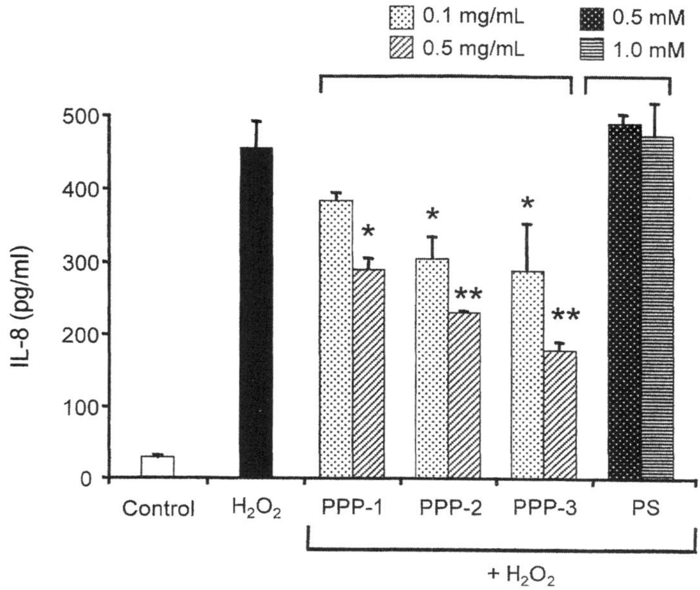

*Figure 2. Effects of PPP fractions on IL-8 level in H_2O_2 treated Caco-2 cells. PS denotes phosphoserine. * $P < 0.05$ and ** $P < 0.001$, compared with group with H_2O_2 alone.*

The concentration of intracellular malondialdehyde (MDA) concentration was measured as a second marker of the H_2O_2-induced tissue oxidative stress in the cultured cells. The polyunsaturated fatty acids of cellular membranes are prone to peroxidation by free radicals and oxygen species, which are subsequently degraded to MDA (*58*). Treatment of Caco-2 cells with 1 mM H_2O_2 for 6 h resulted in a significant increase in the MDA concentration; however, pretreatment with PPP-2 and PPP-3 suppressed MDA production (Figure 3) (*57*). In contrast, PPP-1 exerted no significant protective effect. This suggests that only those PPP fractions containing phosphorus protect H_2O_2-induced lipid peroxidation from Caco-2 cells.

Upregulation of GSH synthesis-related enzyme and antioxidant enzymes by PPPs

We have already confirmed that phosvitin and PPPs possess a strong radical scavenging activity for DPPH radical, indicating that H_2O_2-scavenging activity of PPPs might be responsible for the suppression of IL-8 secretion and MDA production in H_2O_2-treated Caco-2 cells. However, phosvitin did not exhibit any

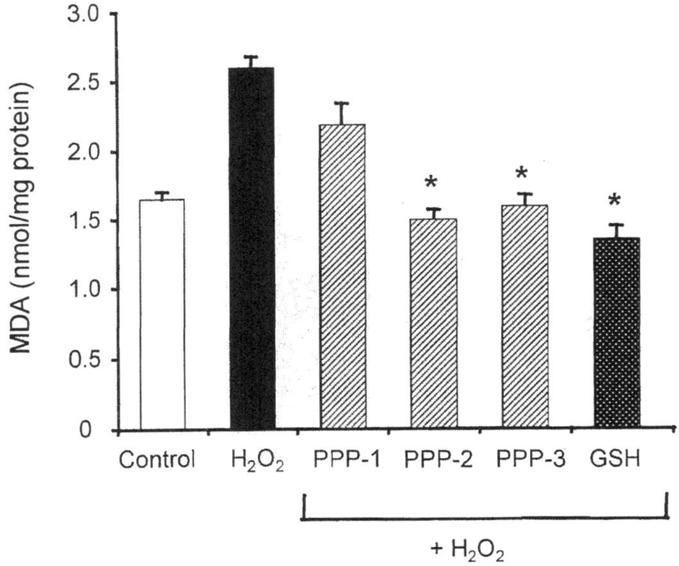

*Figure 3. Effects of PPP fractions on malondialdehyde (MDA) levels in H_2O_2-treated Caco-2 cells. Cells were treated with PPP fractions (0.5 mg/mL) and GSH (1.0 mM) at 37 °C for 2 h, and then incubated with 1 mM H_2O_2 for 6 h. * $P < 0.05$, compared with group with H_2O_2 alone.*

significant antioxidative activity against H_2O_2-induced oxidative stress. Thus, we hypothesized that other defense systems such as intracellular antioxidants and antioxidative enzymes might contribute to the mechanism underlying antioxidative stress activity of PPPs. We thus investigated whether PPP-1 and PPP-3 can enhance the intracellular antioxidative defense systems, including antioxidants (GSH) and several antioxidant enzymes (GR, GST, and catalase).

GSH is the main non-enzyme antioxidant in cells, and it protects against oxidative damage in systems in which it scavenges radicals, eliminates lipid peroxidation products, preserves the tiol-disulfide status of proteins, and repairs oxidative damage (*59*). The intracellular redox homeostasis capacity is substantiated primarily by GSH and TRX. It is usually assumed that GSH depletion reflects oxidative damage. The GSH level was lower in H_2O_2-treated Caco-2 cells than in control cells; however, pretreatments with PPP-1 and PPP-3 for 2 h restored GSH levels (Figure 4). In addition, increase in the GSH level was also observed when the cells were pretreated with PPP-3 for 2 h without H_2O_2 treatment. Thus, this indicates that PPPs can maintain high levels of GSH regardless of oxidative stress. It seems likely that the biosynthesis system of GSH in the cells is activated by PPPs. The levels of GSH are regulated by γ-glutamylcysteine synthetase (γ-GCS) activity and heavy subunit of γ-GCS (γ-

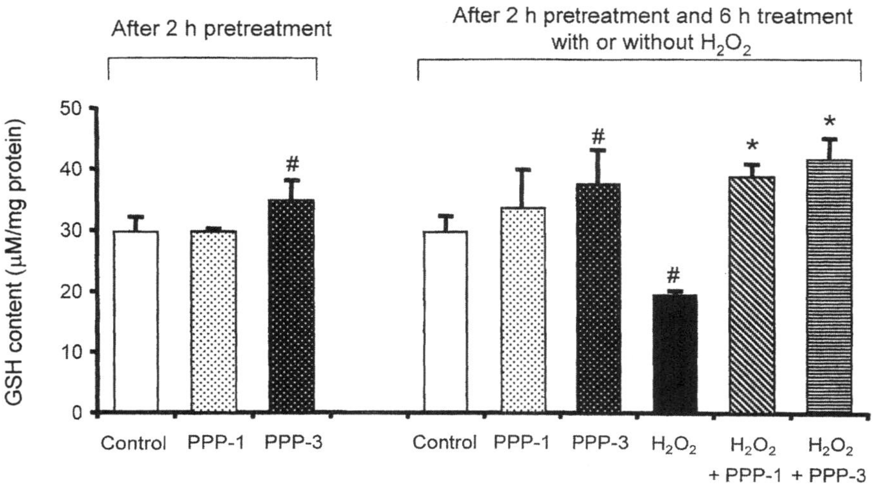

*Figure 4. Effects of PPP fractions on GSH content in H_2O_2-treated Caco-2 cells. # $P < 0.05$, compared with cells treated with control cells. Cells were treated with PPP-1 and PPP-3 (0.5 mg/mL) at 37 ℃ for 2 h, and then incubated with H_2O_2 (1 mM) for 6 h. * $P < 0.05$, compared with cells treated with H_2O_2 alone.*

GCS-HS) is a key domain. PPP-3 significantly increased γ-GCS-HS mRNA expression and γ-GCS activity in H_2O_2-treated Caco-2 cells, compared to GAPDH gene expression (Figure 5A) (*60*). There were no significant different levels of γ-GCS-HS mRNA expression between the control and PPPs without H_2O_2 treatment. γ-GCS activity was significantly elevated by PPP-3 treatment in Caco-2 cells, compared to the values in cells treated with H_2O_2 alone (Figure 5B). Thus, our data suggest that the increase in GSH levels is mainly due to transcriptional upregulation of γ-GCS-HS mRNA by PPP-3 under oxidative stress. These results support the idea that PPP-3 can protect intestinal epithelial cells from H_2O_2-induced oxidative stress by activation of GSH biosynthesis.

A variety of antioxidant enzymes and antioxidants are also associated with the defense systems against oxidative injury (*61, 62*). For example, harmful H_2O_2 can be decomposed by catalase or GPx. In the reaction catalyzed by GPx, GSH is transformed to the oxidized form (GSSG), which can be subsequently recycled into GSH by GR. The lipid hydroperoxides were primarily detoxified by GST. Therefore, we next investigated whether the activities of antioxidant enzymes such as GR, GST, and catalase could be elevated by treatment of PPPs. Treatment of Caco-2 cells with 1 mM H_2O_2 for 6 h in the absence of PPPs decreased the activities of GR and catalase (Figure 6). Pretreatment with PPP-3 for 2 h prior to the induction of oxidative stress enhanced the catalase activity;

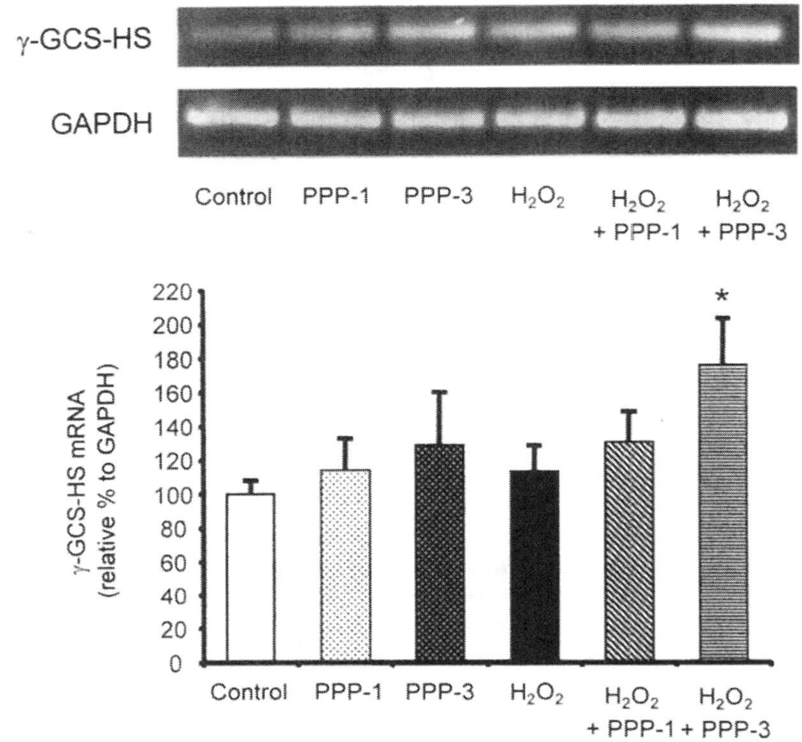

*Figure 5A. Effects of PPP-1 and PPP-3 on γ-GCS-HS mRNA expression in H_2O_2-treated Caco-2 cells. Total RNA was isolated from control cells and cells exposed with H_2O_2 (1 mM) for 1 h followed by pretreatment with PPP-1 and PPP-3 (0.5 mg/mL) for 2 h. * $P < 0.05$, compared with cells treated with H_2O_2 alone.*

whereas PPP-3 significantly increased the activities of GR, GST, and catalase in H_2O_2-treated Caco-2 cells. In contrast, PPP-1 showed no significant increase. These results suggest that PPP-3 can exert a protective effect against H_2O_2-induced oxidative stress by enhancing the activities of antioxidant enzymes.

In view of their molecular structures, PPP-3 has a large phosphorus moiety in contrast with PPP-1 that is without phosphorus. The present data clearly demonstrated that PPP-3 exhibited more significant increase in activities of GSH-synthesis enzyme and antioxidative enzymes than PPP-1. Therefore, it seems that upregulating effects of PPP-3 on intracellular enzyme activities are mainly involved in phosphorus moieties in oligophosphopeptides; however, phosphoserine did not induce the antioxidative response. Therefore, it is

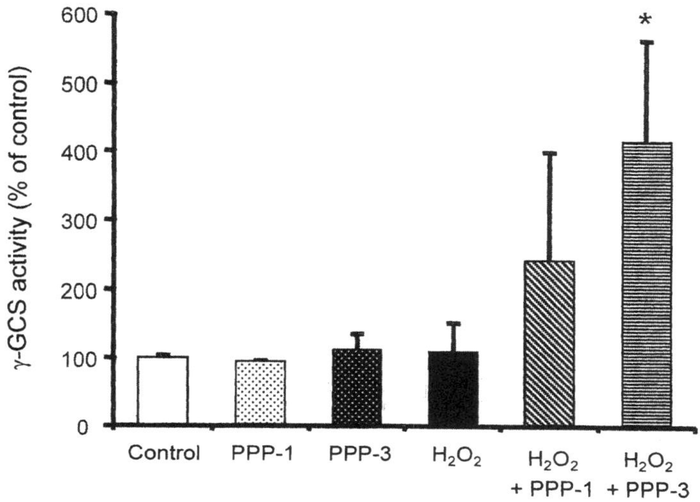

*Figure 5B. Effect of PPP-1 and PPP-3 on γ-GCS activity in H_2O_2-treated Caco-2 cells. Cells were treated with PPP-1 and PPP-3 (0.5 mg/mL) at 37 ℃ for 2 h prior to incubation with 1 mM H_2O_2 for 6 h. * $P < 0.05$, compared with cells treated with H_2O_2 alone.*

reasonable that the important key factors are peptides molecular size and amino acid composition as well as phosphorus.

PPPs exhibit antioxidative effects on stress by several mechanisms (Figure 7). These mechanisms include: (i) regulation of cell redox homeostasis by induction of GSH biosynthesis systems and the recycling pathway (implicated in the maintenance of a high GSH/GSSG ratio); (ii) enhancement of intracellular detoxifying systems by upregulation of catalase and GST (leading to scavenge radicals and inhibit lipid peroxidation). Induction of theses antioxidant and detoxifying enzymes may be involved in the ARE activation utilizing MAPK signaling pathway. Further studies on the effects of PPPs on transcription factor, such as Nrf2, are necessary.

Conclusions

Oxidative stress and radical attack are known to be major factors associated with the initiation and propagation of the development of various diseases. Dietary factors that are thought to exert protective effects include vitamins, polyphenols, and carotenoids. Considering the recently growing attraction of oxidative stress, focusing on the antioxidative effect of bioactive peptides will

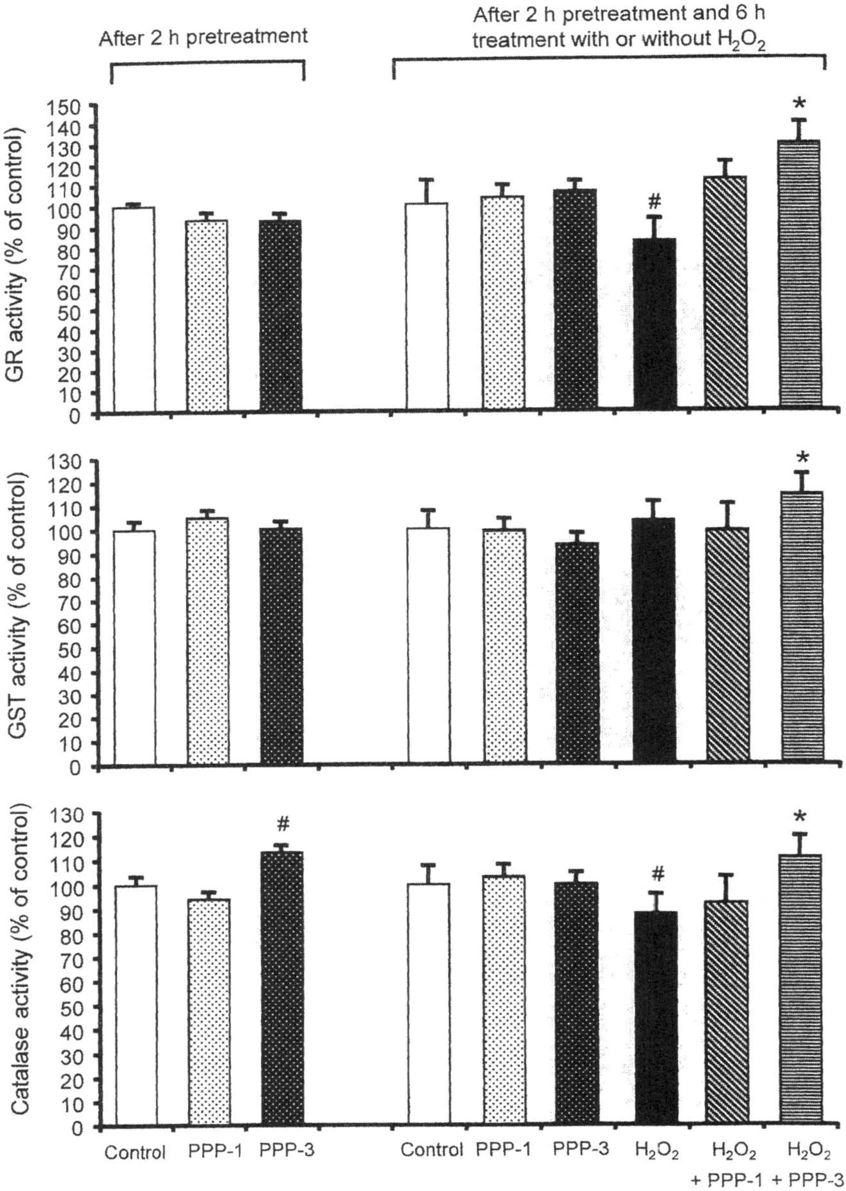

*Figure 6. Effect of PPP-1 and PPP-3 on antioxidative enzyme activities in H_2O_2-treated Caco-2 cells. The cells were treated with PPP-1 and PPP-3 (0.5 mg/mL) at 37 ℃ for 2 h, and then exposed with or without 1 mM H_2O_2 for 6 h. # $P < 0.05$, compared with cells treated with control cells. * $P < 0.05$, compared with cells treated with H_2O_2 alone.*

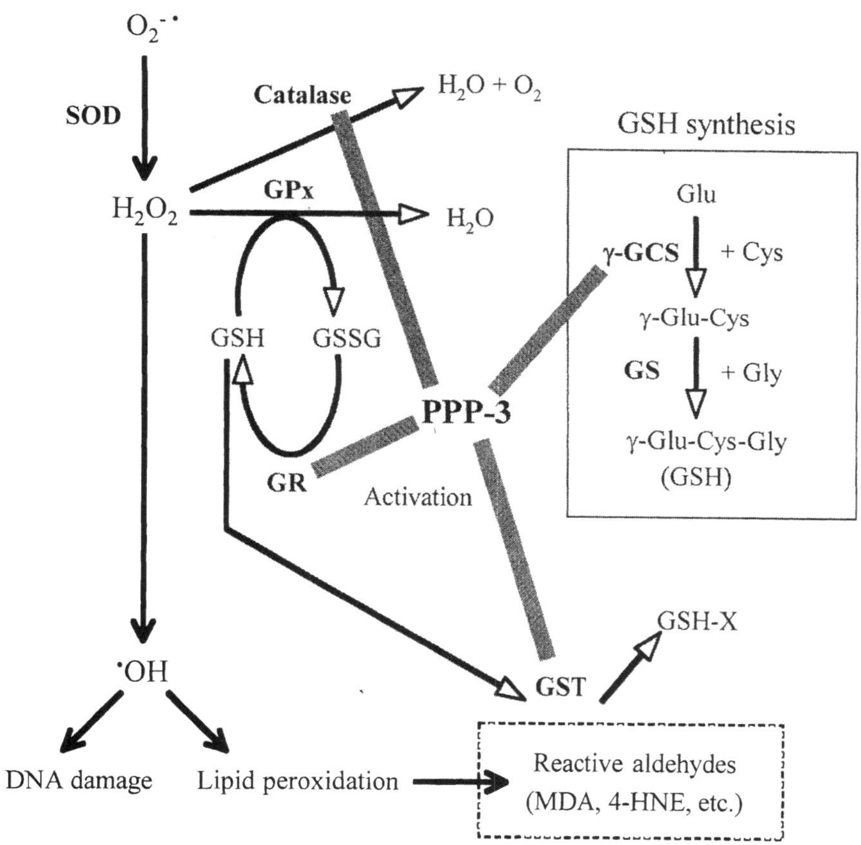

Figure 7. Hypothetical mechanism of protective effect of PPP-3 against H_2O_2-induced oxidative stress. PPP-3 enhances the activation of GR and γ-GCS, key enzymes in GSH biosynthesis and thereby maintains at high level of GSH in the cells. PPP-3 also increases catalase and GST activities, so that H_2O_2 and reactive aldehydes derived from lipid peroxidation are detoxified. GS and 4-HNE denote GSH synthetase and 4-hydroxy-2-nonenal, respectively.

provide new nutraceuticals and functional food ingredients. We here reported that oligophosphopeptides from hen egg yolk phosvitin exert antioxidant activity against oxidative stress by activating antioxidant defense systems in the intestinal epithelial cells.

Further work on the antioxidant effects *in vivo* is necessary. Generally, the beneficial role of antioxidant supplementation to healthy individuals remains controversial, thereby it needs to be evaluated in patients with stress. In addition, further work on the potential pro-oxidant effects of antioxidants as well as the beneficial role also need to be carried out. Recent experiments (*63-66*) using cells have shown not only loss of antioxidant effectiveness but also pro-oxidant effects of carotenoids at high carotenoid concentrations. It is possible that these effects are associated with carotenoid aggregation; however, reasons for this activity reduction have yet been fully unravelled. Another study (*67*) reported that at high pressures of oxygen, carotenoids (generated through the hydrogen abstraction) can react with oxygen to generate a carotenods-peroxyl radical, which become pro-oxidants by promoting oxidation of unsaturated lipid. Thus, it should be noted that there are studies exploring optimal dose of antioxidants that render in maximum effectiveness.

References

1. Miller, D. M.; Buettner, G. R.; Aust, S. D. *Free Radic. Biol. Med.* **1990**, *8*, 95-108.
2. Valko, M.; Izakovic, M.; Mazur, M.; Rhodes, C. J.; Telser, J. *Mol. Cell Biochem.* **2004**, *266*, 37-56.
3. Cadenas, E. *Annu. Rev. Biochem.* **1989**, *58*, 79-110.
4. Valko, M.; Rhodes, C. J.; Moncol, J.; Izakovic, M.; Mazur, M. *Chem-Biol. Interact.* **2006**, *160*, 1-40.
5. Bunker, V. W. *Med. Lab. Sci.* **1992**, *49*, 299-312.
6. Kaneda, T.; Miyazawa, T. *World Rev. Nutr. Diet.* **1987**, *50*, 186-214.
7. Kubow, S. *Free Radic. Biol. Med.* **1992**, *12*, 63-81.
8. Aw, T. Y. *Free Radic. Res.* **1998**, *28*, 637-646.
9. Droge, W. *Physiol. Rev.* **2002**, *82*, 47-95.
10. Dalle-Donne, I.; Rossi, R.; Colombo, R.; Giustarini, D.; Milzani, A. *Clin. Chem.* **2006**, *52*, 601-623.
11. Jenner, P. *Ann. Neurol.* **2003**, *53 Suppl 3*, S26-36; discussion S36-28.
12. Sayre, L. M.; Smith, M. A.; Perry, G. *Curr. Med. Chem.* **2001**, *8*, 721-738.
13. Dhalla, N. S.; Temsah, R. M.; Netticadan, T. *J. Hypertens.* **2000**, *18*, 655-673.
14. Sun, Y.; Oberley, L. W. *Free Radic. Biol. Med.* **1996**, *21*, 335-348.
15. Kehrer, J. P. *Crit. Rev. Toxicol.* **1993**, *23*, 21-48.
16. Frei, B.; England, L.; Ames, B. N. *Proc. Natl. Acad. Sci. U. S. A.* **1989**, *86*, 6377-6381.

17. Halliwell, B. *Free Radic. Res.* **1996**, *25*, 439-454.
18. Jacobs, E. J.; Connell, C. J.; McCullough, M. L.; Chao, A.; Jonas, C. R.; Rodriguez, C.; Calle, E. E.; Thun, M. J. *Cancer Epidemiol. Biomarkers Prev.* **2002**, *11*, 35-41.
19. Collins, A. R.; Gedik, C. M.; Olmedilla, B.; Southon, S.; Bellizzi, M. *FASEB J.* **1998**, *12*, 1397-1400.
20. Buettner, G. R. *Arch. Biochem. Biophys.* **1993**, *300*, 535-543.
21. Richter, C. *Chem. Phys. Lipids.* **1987**, *44*, 175-189.
22. Gutteridge, J. M. C. *Clin. Chem.* **1995**, *41*, 1819-1828.
23. Niles, R. M. *Mutat. Res-Fund. Mol. M.* **2004**, *555*, 81-96.
24. Sharoni, Y.; Danilenko, M.; Dubi, N.; Ben-Dor, A.; Levy, J. *Arch. Biochem. Biophys.* **2004**, *430*, 89-96.
25. Kwak, M. K.; Egner, P. A.; Dolan, P. M.; Ramos-Gomez, M.; Groopman, J. D.; Itoh, K.; Yamamoto, M.; Kensler, T. W. *Mutat. Res-Fund. Mol. M.* **2001**, *480*, 305-315.
26. Breinholt, V.; Lauridsen, S. T.; Daneshvar, B.; Jakobsen, J. *Cancer Lett.* **2000**, *154*, 201-210.
27. Bhuvaneswari, V.; Velmurugan, B.; Balasenthil, S.; Ramachandran, C. R.; Nagini, S. *Fitoterapia.* **2001**, *72*, 865-874.
28. RiceEvans, C. A.; Miller, N. J.; Paganga, G. *Free Radic. Biol. Med.* **1996**, *21*, 417-417.
29. Rice-Evans, C. *Curr. Med. Chem.* **2001**, *8*, 797-807.
30. Schroeter, H.; Boyd, C.; Spencer, J. P. E.; Williams, R. J.; Cadenas, E.; Rice-Evans, C. *Neurobiol. Aging.* **2002**, *23*, 861-880.
31. Romero-Perez, A. I.; Ibern-Gomez, M.; Lamuela-Raventos, R. M.; de la Torre-Boronat, M. C. *J. Agric. Food Chem.* **1999**, *47*, 1533-1536.
32. Jang, M.; Cai, L.; Udeani, G. O.; Slowing, K. V.; Thomas, C. F.; Beecher, C. W.; Fong, H. H.; Farnsworth, N. R.; Kinghorn, A. D.; Mehta, R. G.; Moon, R. C.; Pezzuto, J. M. *Science.* **1997**, *275*, 218-220.
33. Cao, Z. X.; Li, Y. B. *Eur. J. Pharmacol.* **2004**, *489*, 39-48.
34. Yu, R.; Jiao, J. J.; Duh, J. L.; Gudehithlu, K.; Tan, T. H.; Kong, A. N. *Carcinogenesis.* **1997**, *18*, 451-456.
35. van Zanden, J. J.; Geraets, L.; Wortelboer, H. M.; van Bladeren, P. J.; Rietjens, I. M. C. M.; Cnubben, N. H. P. *Biochem. Pharmacol.* **2004**, *67*, 1607-1617.
36. Wang, D. J.; Jensen, R. H.; Williams, K. E.; Pallavicini, M. G. *Proteomics.* **2004**, *4*, 2175-2183.
37. Chen, C.; Yu, R.; Owuor, E. D.; Kong, A. N. T. *Arch. Pharm. Res.* **2000**, *23*, 605-612.
38. Kitts, D. D.; Weiler, K. *Curr. Pharm. Des.* **2003**, *9*, 1309-1323.
39. Meisel, H. *Biopolymers.* **1997**, *43*, 119-128.
40. Mine, Y.; Ma, F.; Lauriau, S. *J. Agric. Food Chem.* **2004**, *52*, 1088-1094.
41. Hirota, Y.; Yang, M. P.; Araki, S.; Yoshihara, K.; Furusawa, S.; Yasuda, M.; Mohamed, A.; Matsumoto, Y.; Onodera, T. *J. Vet. Med. Sci.* **1995**, *57*, 825-829.

42. Watanabe, K.; Tsuge, Y.; Shimoyamada, M.; Ogama, N.; Ebina, T. *J. Agric. Food Chem.* **1998**, *46*, 3033-3038.
43. Matoba, N.; Usui, H.; Fujita, H.; Yoshikawa, M. *FEBS Lett.* **1999**, *452*, 181-184.
44. Fujita, H.; Sasaki, R.; Yoshikawa, M. *Biosci. Biotechnol. Biochem.* **1995**, *59*, 2344-2345.
45. Kassaify, Z. G.; Li, E. W.; Mine, Y. *J. Agric. Food Chem.* **2005**, *53*, 4607-4614.
46. Kobayashi, K.; Hattori, M.; Hara-Kudo, Y.; Okubo, T.; Yamamoto, S.; Takita, T.; Sugita-Konishi, Y. *J. Agric. Food Chem.* **2004**, *52*, 5740-5746.
47. Taborsky, G.; Mok, C. C. *J. Biol. Chem.* **1967**, *242*, 1495-1501.
48. Allerton, S. E.; Perlmann, G. E. *J. Biol. Chem.* **1965**, *240*, 3892-3898.
49. Jiang, B.; Mine, Y. *J. Agric. Food Chem.* **2000**, *48*, 990-994.
50. Jiang, B.; Mine, Y. *Biosci. Biotechnol. Biochem.* **2001**, *65*, 1187-1190.
51. Feng, F. Q.; Mine, Y. *Int. J. Food Sci. Tech.* **2006**, *41*, 455-458.
52. Yamamoto, K.; Kushima, R.; Kisaki, O.; Fujiyama, Y.; Okabe, H. *Inflammation.* **2003**, *27*, 123-128.
53. Shimada, T.; Watanabe, N.; Hiraishi, H.; Terano, A. *Dig. Dis. Sci.* **1999**, *44*, 266-273.
54. Verhasselt, V.; Goldman, M.; Willems, F. *Eur. J. Immunol.* **1998**, *28*, 3886-3890.
55. Zwahlen, R.; Walz, A.; Rot, A. *Int. Rev. Exp. Pathol.* **1993**, *34 Pt B*, 27-42.
56. Bickel, M. *J Periodontol.* **1993**, *64*, 456-460.
57. Katayama, S., Xu' X,. Fan M.Z., and Mine' Y. *J. Agric. Food Chem.*, **2006**, *54*, 773-778
58. Halliwell, B.; Chirico, S. *Am. J. Clin. Nutr.* **1993**, *57*, 715S-724S; discussion 724S-725S.
59. Shan, X. Q.; Aw, T. Y.; Jones, D. P. *Pharmacol. Ther.* **1990**, *47*, 61-71.
60. Katayama, S., Ishikawa, S., Ming M.Z., and Mine, Y. *J. Agric. Food Chem.*, **2007**, *55*, 2829-2835
61. Cao, Z.; Li, Y. *Biochem. Biophys. Res. Commun.* **2002**, *292*, 50-57.
62. Yang, Y.; Cheng, J. Z.; Singhal, S. S.; Saini, M.; Pandya, U.; Awasthi, S.; Awasthi, Y. C. *J. Biol. Chem.* **2001**, *276*, 19220-19230.
63. Lowe, G. M.; Booth, L. A.; Young, A. J.; Bilton, R. F. *Free Radic. Res.* **1999**, *30*, 141-151.
64. Woods, J. A.; Bilton, R. F.; Young, A. J. *FEBS Lett.* **1999**, *449*, 255-258.
65. Palozza, P.; Calviello, G.; Serini, S.; Maggiano, N.; Lanza, P.; Ranelletti, F. O.; Bartoli, G. M. *Free Radic. Biol. Med.* **2001**, *30*, 1000-1007.
66. Eichler, O.; Sies, H.; Stahl, W. *Photochem. Photobiol.* **2002**, *75*, 503-506.
67. Kennedy, T. A.; Liebler, D. C. *J. Biol. Chem.* **1992**, *267*, 4658-4663.

Chapter 20

Antioxidant and Antiinflammatory Activities of Licorice Root (*Glycyrrhiza uralensis*): Aroma Extract

Aki Tanaka and Takayuki Shibamoto

Department of Environmental Toxicology, University of California, Davis, CA 95616

The volatile fraction obtained from a steam distillate of Licorice was examined for antioxidant activity by a hexanal/hexanoic acid assay. Antioxidant activity of the residual aqueous fraction was also tested by a malonadehyde/gas chromatography (MA/GC) assay. The volatile fraction inhibited hexanal oxidation by 99% over 40 days at a level of 50 μg/mL. The GC/MS analysis of the volatile fraction yielded hexanol (52.75 mg/mL) and hexanal (31.75 mg/mL) as major components. The volatile fraction exhibited strong anti-inflammatory activity when assayed by ELSA using lipoxygenase. The steam distillate also showed strong dose dependent anti-inflammatory activity. However, the aqueous solution with the volatiles removed did not show any anti-inflammatory activity.

Many traditional Chinese herbs, including licorice root, have been used to treat various diseases since the beginning of recorded history. Today, over seven thousand herbs have been employed by 80% of the world's population (*1, 2*).

Licorice root, specifically the *Glycyrrhiza* species of the Fabaceae family (*3*), has been one of the most important medicinal herbs used in China, and one that has been in use there for over 6,000 years (*4*). Licorice root is one of the richest sources of biologically active compounds (*5*). For example, some flavonoids isolated from licorice root extract have been shown to exhibit strong antioxidant activity toward lard oxidation (*6*). Glycyrrhizin and glycyrrhetinic acid, which are the main components of licorice root, have been clinically used in the treatment of hyperlipemia, artherosclerosis, and allergic inflammation (*2*).

Previous studies have focused on the non-volatile licorice root constituents, such as flavonoids with sugar moieties. However, plants also contain numerous aroma chemicals, which have been widely used in both folk medicine and aroma therapies (*7*), suggesting that they have some beneficial health effects in addition to their pleasant odor. Aroma chemicals are highly-volatile low-molecular-weight compounds and have been isolated and identified in natural plants. Aroma chemicals have been studied mainly from the aspects of flavor and fragrance chemistry. However, medicinal activities of aroma chemicals have been discovered lately. For example, the antioxidant activities of aroma extracts obtained from spices and herbs have been reported (*8*),

Essential oils, which are comprised mainly of aroma chemicals and which are isolated from medicinal plants such as chamomile, clove, and eucalyptus, have anti-microbial and antioxidant properties (*9, 10*). They have also been widely used for aromatherapy (*11*). Essential oils are prepared from natural plants, herbs, by steam distillation followed by solvent extraction. In the present study, aroma extract and residual aqueous solution isolated from licorice root by a water distillation were examined for antioxidant and anti-inflammatory activities to investigate the possible additional benefits of licorice consumption.

Materials and Methods

Chemicals and Materials

α-Tocopherol was bought from Sigma Chemical Co. (St. Louis. MO). Glycyrrhizin was purchased from Wako Pure Chemical Industries (Osaka, Japan). Buffer solutions (pH 4.00, 7.00, and 10.00), dimethyl sulfoxide (DMSO), and 10% trichloroacetic acid solution (TCA) were from Fisher Co. (Pittsburgh, PA).

Licorice root (*Glycyrrhiza uralensis*) was collected and dried naturally by Dr. John F. Louie (Sacramento, CA) in Heilongjiang Province, Northwestern

China. Lipoxygenase inhibitor screening assay (LISA) kit was purchased from Cayman Chemical Co. (Ann Arbor, MI).

Sample Preparation for Volatile Extract and Aqueous Solution

Licorice root (50 g) was mixed with 1 L deionized water in a 2 L two necked-flask and allowed to stand for 1 h. The water in the flask was boiled using a mantle heater and 500 mL of water distillate was collected in a beaker. The distillate was extracted with 125 mL of dichloromethane for 6 h using a liquid-liquid continuous extractor. The extract was dried over anhydrous sodium sulfate for 12 h. After removal of the sodium sulfate, the extract was condensed to approximately 3 mL in volume using a rotary evaporator. The extract was further condensed to exactly 0.2 mL under a purified nitrogen stream (aroma extract sample) and then stored at – 20 °C until used for further experiments. Residual aqueous solution (aqueous solution sample) in the extractor was transferred into a flask and stored at 5 °C until used for further experiments.

Antioxidant Test of Aroma extract by Aldehyde/Carboxylic Acid Assay

Antioxidant activity of the aroma extract was tested using its inhibitory effect on the oxidation of hexanal to hexanoic acid (*12*). Various amounts of the extract were added to a 2 mL dichlomethane solution of hexanal (3 mg/mL) containing 0.2 mg/mL of undecane as a gas chromatographic (GC) internal standard. The oxidation of the sample solution was initiated by heating at 60 °C for 10 min in a sealed vial and then storing at room temperature. A blank sample was prepared following the same procedure without a test sample. Butylated hydroxytoluene (BHT) and α-tocopherol were used as a positive control. The amount of hexanal was measured after 45 days.

Fractionation of Aqueous Solution Sample

A residual aqueous solution from the dichloromethane extraction was condensed to a paste (dark brown color) using a rotary evaporator. The paste sample was placed in a glass column (40 cm × 4.5 cm i.d.) packed with Amberlite XAD-2 resin (Aldrich Chemical Co., Milwaukee, WI). The sample was eluted sequentially with a 1 L each water/methanol solution—100/0 (Fraction I); 95/5 (Fraction II); 80/20 (Fraction III); 50/50 (Fraction IV); 20/80 (Fraction V); and 0:100 (Fraction VI); and a final fraction of 1 L of acetone (Fraction VII). The seven fractions obtained were condensed to 0.3 mL using first a rotary evaporator and subsequently a purified nitrogen stream to 0.1 mL.

Anti-inflammatory Test of Aroma Extract and Fractions from Aqueous Solution

Anti-inflammatory tests were conducted using the lipoxygenase inhibitor screening assay (LISA) kit (*13, 14*).

The solutions provided commercially in the LISA Kit were 0.1 M Tris-HCl assay buffer (pH 7.4), developing agents 1 and 2 (chromogen), soybean-enzyme 15-lipoxygenase (15-LOX) standard, arachidonic acid, and KOH. An assay buffer was diluted tenfold with HPLC-grade water before use. Chromogen, which was used within 1 h, was prepared by mixing equal amounts of developing agents 1 and 2. A blank well was prepared by adding an assay buffer solution (100 μL) to a well plate supplied from the LISA Kit. A positive control well was made by mixing a 10 μL of 15-LOX solution and a 990 μL of assay buffer. A substrate solution was prepared by mixing 25 μL of arachidonic acid and 25 μL of KOH in ethanol. After the substrate solution was vortexed, it was diluted with 950 μL of HPLC-grade water. The substrate solution was used within 30 min to prevent degradation. The solution prepared was stored at 0 °C until used.

A15-LOX solution (10 μL), a testing sample (10 μL), and assay buffer (980 μL) were placed in the testing well. The reaction was initiated by adding a 10 μL of substrate solution to a positive control well and a testing sample well. All testing wells, were covered and placed on a shaker (Bellco Biotechnology, Vineland, NJ) for 5 min. Chromogen (100 μL) was added to the reaction wells to stop the enzyme catalysis and prevent further development of the reaction. The level of hydroperoxide in the samples was measured using a microplate reader (Molecular Devices Corporation, Sunnyvale, CA). The entire assay was performed in duplicates.

The three concentrations (62, 125, 250 μg/mL) of glycyrrhizin (standard anti-inflammatory chemical), the volatile extract, and the fractions from the aqueous solution were tested. Concentration levels of 31, 62, and 125 μg/mL were used for Fractions I and II because their original paste sample was highly viscous. The volatile extract and glycyrrhizin were diluted with DMSO. Fractions I, II, and III were diluted with HPLC-grade water. Fractions IV, V, and VI were diluted with methanol. Fraction VII was diluted with acetone.

Analysis of Glycyrrhizin in Fractions from Aqueous Sample

Glycyrrhizin in the aqueous Fractions was analyzed by an Agilent 1100 model HPLC system equipped with a 150 mm × 4.6 mm i.d.) Alltima C-18 5μ column (Alltech, Deerfield, IL) and a multiple wavelength detector. Mobile phase A was 10 mM citric acid and mobile phase B was methanol. The gradient mode was initially set at A/B ratio of 85/15 from 0 to 5 min, then linearly increased to 60/40 at 40 to 85 min. The flow rate was 1.0 ml/min. The detector was set at 254; injection volume was 5 μL.

Results and Discussion

Figure 1 shows the results of antioxidant tests on the volatile extract. The autoxidation of hexanal to hexanoic acid was monitored for 45 days. BHT (50 μg/mL), a synthetic standard antioxidant, inhibited hexanal oxidation by 100% throughout the 45 day period. α-Tocopherol, a natural antioxidant, inhibited hexanal oxidation by 95% during same period of time. All volatile extract samples, with the exception of 10 μg/mL volatile extract, inhibited hexanal oxidation by over 80% for 45 days.

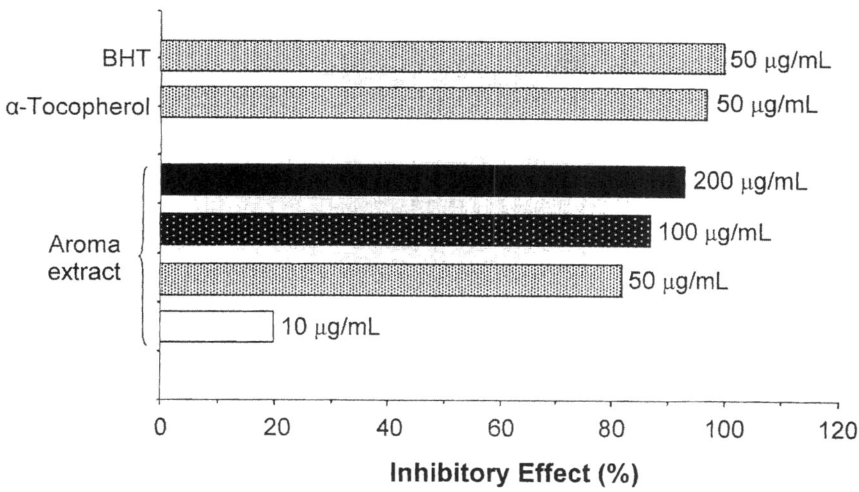

Figure 1. Antioxidant activity of aroma extracts and standard natural antioxidants

Figure 2 shows the inhibitory effect of aroma extracts at representative levels monitored at 5 day interval for 45 days. Hexanal converted to hexanoic acid completely after 45 days. Aroma extract inhibited hexanal oxidation by 50% over 25 days even at a low level of 10 mg/mL. The inhibitory activity of aroma extract at 200 μg/mL was comparable to that of 50 μg/mL level of α-tocopherol.

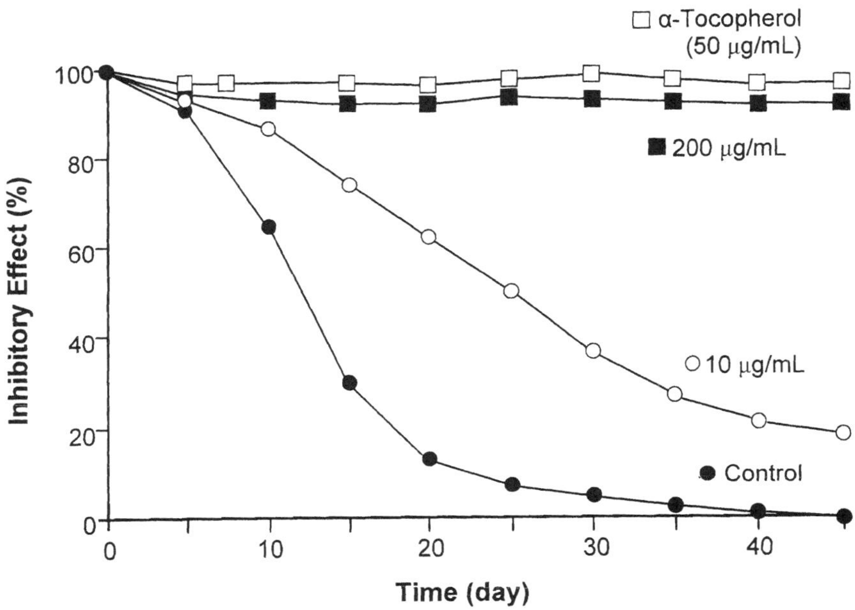

Figure 2. Inhibitory effect of aroma extracts at representative levels

Some antioxidant activity derived from licorice root steam distillate has been previously reported (*15*). However, the antioxidant activity of the volatile extract of licorice has not received much attention yet. Under the conditions of the present study, the volatile extract exhibited strong antioxidant activity, suggesting that licorice root contains certain volatile compounds possessing antioxidant potential.

Figure 3 shows the results of anti-inflammatory tests on fractions from an aqueous sample and glycyrrhizin. Glycyrrhizin exhibited 100% anti-inflammatory activity at the level of 125 μg/mL. Its inhibitory activity was 75% and 66% at the levels of 62 μg/mL and 31 μg/mL, respectively. Fraction VII showed the highest anti-inflammatory activities, which were over 75% at the three levels (62, 125, and 250 μg/mL) tested. Fractions V and VI exhibited moderate anti-inflammatory activities. They inhibited lipoxygenase formation by nearly 50% at the levels of 62, 125, and 250 μg/mL. Fraction IV showed slight anti-inflammatory activities—about 20% at the three levels tested. Dose related activity was not observed in the Fractions, even though it was clearly observed in glycyrrhizin. Fractions I, II, and III exhibited slight pro-inflammatory activities. These results may be due to the presence of various chemicals with anti- and pro-inflammatory activities in the fractions from aqueous licorice samples. The presence of glycyrrhizin may play an important role in the anti-inflammatory activity of such fractions.

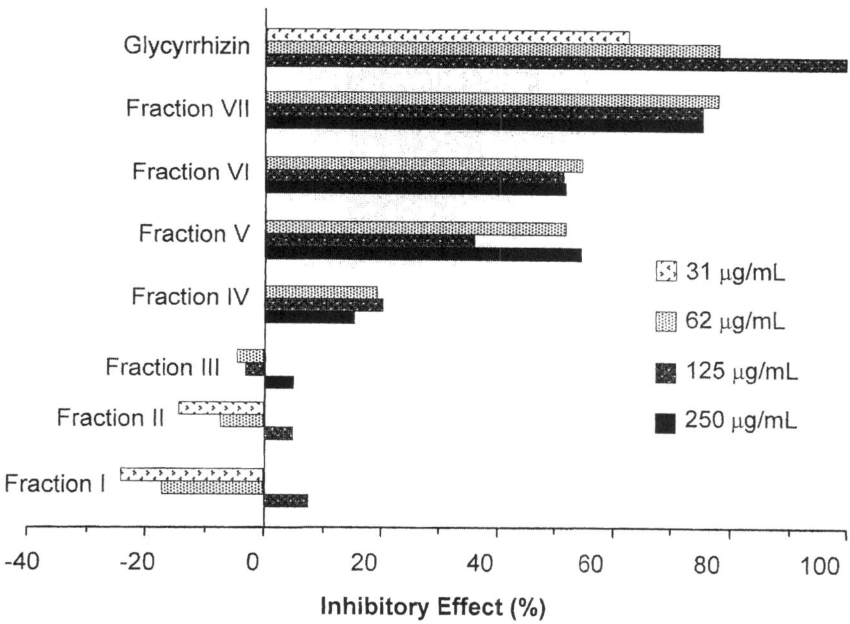

Figure 3. Anti-inflammatory activity of fractions from an aqueous sample and glycyrrhizin

Figure 4 shows the results of glycyrrhizin analysis in each fraction along with its structure. Fraction VII, which showed the highest anti-inflammatory activity, contained the greatest amount of glycyrrhizin (0.86 mg/mL). Fraction V and VI also contained glycyrrhizin (0.065 and 0.033 mg/mL, respectively). Glycyrrhizin was not detected in Fractions I, II, III, and IV, which—except for Fraction IV—showed some pro-inflammatory activities. These results indicate that glycyrrhizin contributed significantly to the anti-inflammatory activity of Fractions.

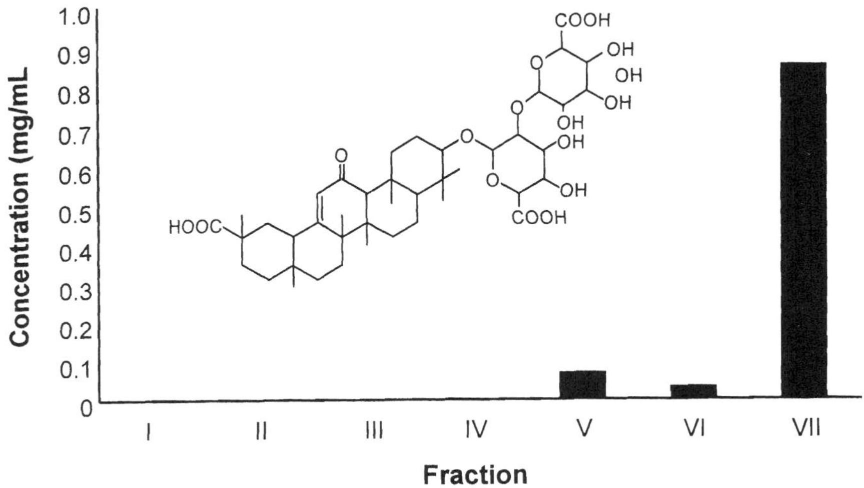

Figure 4. Amount of glycyrrhizin in fractions from an aqueous sample and its structure

The presence of glycyrrhizin in licorice root has been reported in many articles. For example, glycyrrhizin in licorice root and root extract was analyzed by HPLC with a reverse phase column (*16*). Analysis of glycyrrhizin in fresh and dried licorice root was also performed by a reverse phase HPLC (*17*). A review of licorice root constituents and their biological activities describes the role of glycyrrhizin in medicinal activity of Chinese licorice root medicine (*3*). The anti-inflammatory activity of glycyrrhizin has also been reported. Glycyrrhizin inhibited prostaglandin E2 production by activated peritoneal macrophages from rats (*18*). Glycyrrhizin reportedly possessed anti-inflammatory effect on liver cells in viral hepatitis (*19*). Antitumor-promoting and anti-inflammatory activities of licorice principles including glycyrrhizin have also been reported (*20*). Glycyrrhizin related compound, licochalcone A, exhibited anti-inflammatory activity and has been reported as a licorice constituent possessing clinical activity (*21*).

The results of the present study proved that licorice root contains medicinal substances. Aroma extract exhibited potent antioxidant activity. Aqueous extract of licorice root possessed strong anti-inflammatory activities, which may come from its main component, glycyrrhizin.

References

1. Chinese Materia Medica (Zhonghua Benchao). Shanghai Science & Technology Press; Shanghai, pp. 1998, pp. 500-514.
2. Lin, Z. J.; Qui, S.-X.; Wufuer, A.; Shum, L. *J. Chromatogr. B* **2005**, *814*, 201-207.
3. Kitagawa, I. *Pure Appl. Chem.* **2002**, *74*, 1189-1198.
4. Vaya, J.; Tamir, S.; Somjen, D. *Oxid. Stress Disease* **2004**, *14*, 615-634.
5. Roth, K. *Chem. Unser. Zeit* **2004**, *38*, 202-207.
6. Gordon, M. H; An, J. *J. Agric. Food Chem.* **1995**, *43*, 1784-1788.
7. Ujihara, H.; Ishida, K. *Bio. Indust.* **2005**, *22*, 29-38.
8. Lee, K.-G.; Shibamoto, T. *J. Agric. Food Chem.* **2002**, *50*, 4947-4952.
9. Baratta, M. T.; Dorman, H. J. D.; Deans, S. G.; Figueiredo, A. C.; Barroso, J. G.; Ruberto, G. *Flavour. Frag. J.* **1998**, *13*, 235-244.
10. Dorman, H. J. D.; Surai, P.; Deans, S. G. *J. Essent. Oil. Res.* **2000**, *12*, 241-248.
11. Hoffmann, D. In *The Herbal Handbook: A User's Guide to Medical Herbalism*; Hoffmann, D., Ed.; Healing Arts Press; Rochester, VT, 1988, pp. 93-95.
12. Macku, C.; Shibamoto, T. *J. Agric. Food Chem.* **1991**, *39*, 1990-1993.
13. Nyiligira, E.; Viljoen, A. M.; Baser, K. H.; Ozek, T.; van Vuuren, S. F. *S. African. J. Bot.* **2004**, *70*, 611-617.
14. Khatib, A.; Kim, M.-Y.; Chung, S.-K. *Food Sci. Biotechnol.* **2005**, *14*, 223-227.
15. Vaya, J.; Belinky, P. A.; Aviram, M. *Free Radical Biol. Med.* **1997**, *23*, 302-313.
16. Okada, K.; Tanaka, J.; Miyashita, A.; Imoto, K. *Jpn. Yakugaku Zasshi* **1981**, *101*, 822-828.
17. Lauren, D. R.; Jensen, D. J.; Douglas, J. A.; Follet, J. M. *Phytochem. Anal.* **2001**, *12*, 332-335.
18. Ohuchi, K.; Kamada, Y.; Levine, L.; Tsurufuji, S. *Prostag. Med.* **1981**, *7*, 457-463.
19. Fujisawa, Y.; Sakamoto, M.; Matsushita, M.; Fujita, T.; Nishioka, K. *Microbiol. Immun.* **2000**, *44*, 799-804.
20. Shibata, S. In *Food Phytochemicals for Cancer Prevention*, Ho, C.T.; Osawa, T.; Huang, M.T.; Rosen, R.T., Eds., ACS Symp. Ser. 547, American Chemical Society, Washington, D.C., **1994**, pp. 308-321.
21. Kolbe, L.; Immeyer, J.; Batzer, J.; Wensorra, U.; Dieck, K. T.; Mundt, C.; Wolber, R.; Staeb, F.; Schoenrock, U.; Ceilley, R. I.; Wenck, H. *Arch. Dermat. Res.* **2006**, *298*, 23-30.

Chapter 21

Redox Properties of Proanthocyanidins and Their Health Implications

J. L. Torres[1], L. Juliá[1], A. Carreras[1], S. Touriño[1], D. Lizárraga[2], C. Matito[2], I. Medina[3], and M. Cascante[2]

[1]IIQAB-CSIC, Jordi Girona 18–26, 08034-Barcelona, Spain
[2]University of Barcelona, Martí I Franquès 1, 08028-Barcelona, Spain
[3]IIM-CSIC, Eduardo Cabello 6, 36028-Vigo, Spain

Plant polyphenols are appreciated as dietary supplements and functional food ingredients. Their putative health promoting action may originate, at least in part, from their influence on the delicate redox balance governing cell functions. *In vivo*, flavonoids, particularly catechins, may be antioxidants or pro-oxidants depending on their structure and concentration. The accurate evaluation of the redox behavior of flavonoids should help defining their beneficial or toxic effects. Proantho-cyanidin (oligomeric catechin) fractions and bio-based products from grape, pine and witch hazel with variable proportions of catechol and pyrogallol (two and three adjacent phenolic hydroxyl groups respectively) moieties were prepared and tested on stable free radicals and for *in vitro* activity on skin and colon cells to show that there may be a relationship between the electron transfer capacity of proanthocyanidins and their influence on cell functions such as proliferation and apoptosis.

Polyphenols of plant origin present in foods and supplements (*1*) are appreciated as antioxidant chemopreventive agents against a variety of diseases in which oxidative stress plays a significant role (*2-4*) although no conclusive proof of this beneficial action has ever been provided (*5, 6*). The primary beneficial role of polyphenols is believed to be the scavenging of harmful reactive oxygen species (ROS) generated by the mitochondria. Flavanols are particularly efficient scavengers due to the presence of polyphenolic structures such as catechol, pyrogallol and gallate moieties (Figure 1). Moreover it is becoming evident that the biological effects of polyphenols involve other mechanisms in which the redox potential also appears to play a crucial role. Interestingly, the same structures may be antioxidant and pro-oxidant (see Figure 2 for a simplified model) depending on the biological microenviron-ment. For instance, pyrogallol is able to scavenge the superoxide radical and also to form it from molecular oxygen (*7*). The superoxide radical may trigger the formation harmful species such as the hydroxyl radical via the Fenton reaction (*8*) and peroxynitrites via nitric oxide (*9*), among others. In the particular case of apoptosis (programmed cell death) superoxide and oxidative stress appears to be determinant for triggering or halting the cascade of events (*10*). Whether the antioxidant/pro-oxidant effects of catechins are preventive, therapeutic or toxic will depend on where, how and how much the delicate redox balance of the aerobic organism is altered (*11*). These actions will be significant only in organs where polyphenols are bioavailable after oral administration, namely the gastrointestinal tract and particularly the colon.

Electron Transfer and Superoxide Anion

Not all phenolic groups in catechins are equally reactive in terms of superoxide formation and pro-oxidant capacity. The pyrogallol moiety (three hydroxyls) of catechins such as (–)-epigallocatechin and (–)-epigallo-catechingallate appear to be the most active pro-oxidant structure (*12, 13*) whereas the catechol moiety (two hydroxyls) of (–)-epicatechin barely participates in this so called redox cycling (*14*) and functions only as antioxidant scavenger. Because the superoxide anion is formed by transferring an electron to molecular oxygen (Figure 2) it appears that the high electron transfer capacity of pyrogallol compared to catechol is, at least in part, responsible for the pro-oxidant action of some catechins. To estimate the scavenging efficiency of catechins and, most important, their possible pro-oxidant and/or pro-apoptotic activities electron transfer capacity is an important parameter to measure.

Measurement of Electron Transfer Capacity and Apoptosis in Colon Cells

We have recently introduced a new stable radical (HNTTM, Figure 3) reactive only through electron transfer (*15, 16*). This can be used to evaluate the capacity of catechins to donate electrons to ROS and to provide an estimation of

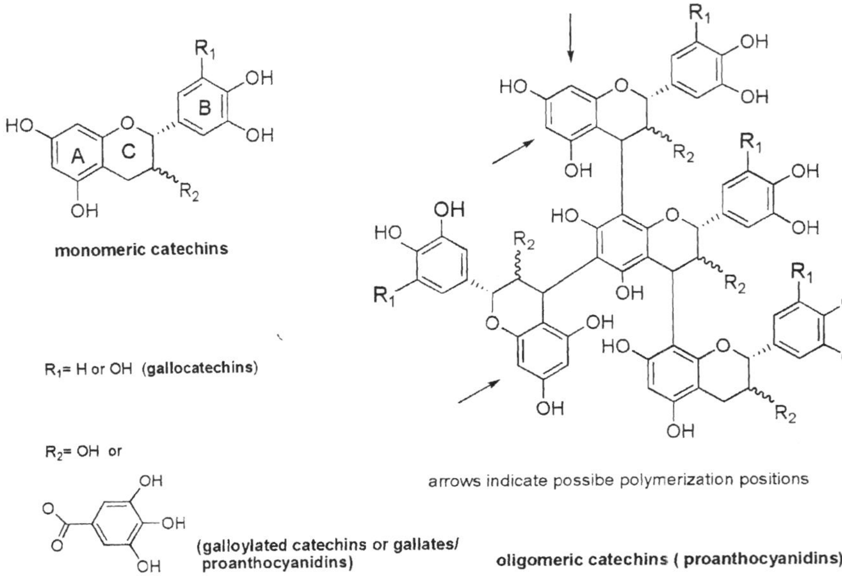

Figure 1. Structures of flavanols.

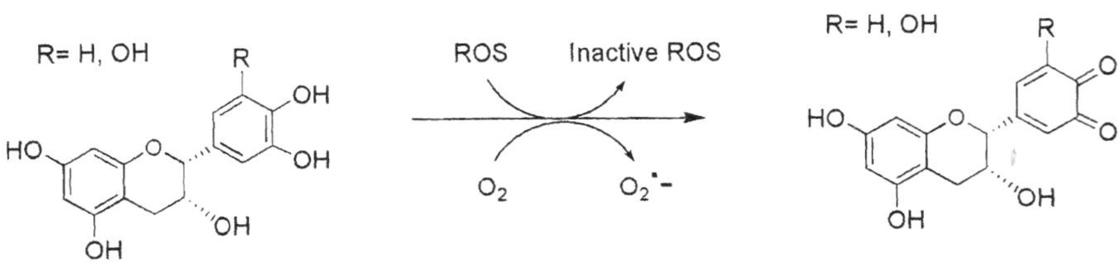

Figure 2. Scavenging of ROS and superoxide formation by catechins.

their possible pro-oxidant and pro-apoptotic activities. We have tested a variety of catechin monomers, catechin conjugates with cysteamine and cysteine (*17, 18*) and proanthocyanidin fractions for their activity against HNTTM and compared the results with the apoptosis triggered by the same polyphenols on HT29 colon carcinoma cells. First the antiproliferative potency of polyphenols was determined and then apoptosis (early and late/necrosis) was assessed by FITC-FACS after treatment of the cells for 72 h at their respective IC_{50} concentrations. To assess the possible artifactual results due to autooxidation of the catechins (*14*) control experiments with iron free medium were conducted. The catechin-gallate monomers and conjugates triggered some early apoptosis (5-10% of total cells) and showed high electron transfer capacity values (6-7 electrons per molecule) while gallocatechin-gallates also showed high electron transfer values but less early apoptosis. In agreement with this the gallocatechins were the most efficient antiproliferative compounds on colon carcinoma cells. This, together with the late apoptosis and necrosis detected for gallocatechins might be associated with the pyrogallol group on ring B.

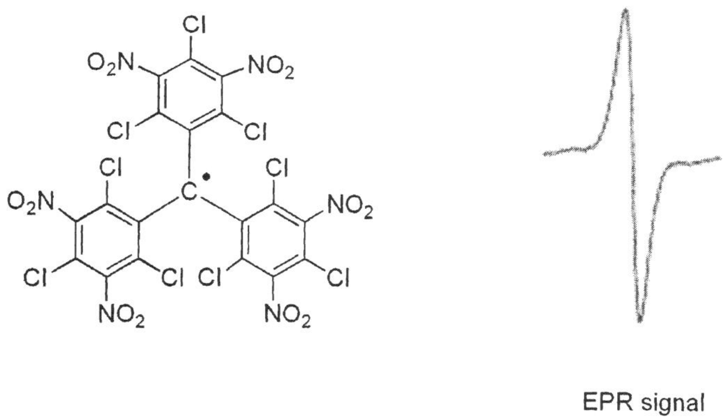

Figure 3. Tris(2,4,5-trichloro-3,5-dinitrophenyl)methyl radical (HNTTM).

We have also tested proanthocyanidin extracts and fractions with variable proportions of gallocatechins and gallates. Table I summarizes the results obtained with some significant fractions from pine (no gallocatechins, no gallates), witch hazel (high gallocatechin and gallate content) and grape (intermediate gallocatechin and gallate content). The electron transfer capacity and apoptosis were estimated as stated before and the structural analysis, including the estimation of the mean molecular weight used to calculate the mean electron transfer capacity was done by thioacidolysis with cysteamine as described (*19*). In agreement with the conclusions on monomers, the results show that apoptosis and particularly late apoptosis and necrosis may be due to the presence of pyrogallol moieties (gallocatechins) in the fractions with high electron transfer capacities.

Table I. Electron Transfer capacity and Induction of Apoptosis by Selected Fractions

Fraction	*Composition*[a]	*Electron transfer*[b]	*Apoptosis*[c]
IV Pine	0, 0	2.5	10*, 5, 2
IV Grape	<1, 25	3.5	11*, 5, 3
IV Witch Hazel	33, 21	2.9	2*, 3*, 5*
V Pine	0, 0	1.3	5*, 2, 1
V Grape	<1, <1	2.8	5, 2, 2
V Witch Hazel	2.5, 1	2.9	n.d.
VIII Pine	0, 0	~1	10*, 6, 2
VIII Grape	<1, 34	3.5	17**, 6*, 5*
VIII Witch Hazel	<4, 40	3.8	6*, 7*, 11*

a: mean percentage gallocatechins, mean percentage gallates; b: electrons transferred per molecule of catechin; c: percentage apoptotic cells (early, late, necrotic), $p<0.05$*, $p<0.001$**

Electron Transfer and Health in the Intestinal Tract

Catechins taken orally as dietary supplements are bioavailable in the intestinal tract. Monomers are absorbed in the small intestine while proanthocyanins reach the colon mainly intact and there they are depolymerized and metabolized by the intestinal microflora into smaller phenolics (*20*). Catechin gallates are more stable than the bare catechins upon metabolisation (*21*). It is then the epithelial tissue of the intestinal tract were proanthocyanidins may exert their antioxidant and/or pro-oxidant actions. The electron transfer capacity of dietary oligomeric catechins will define whether they are mainly scavengers or may participate in redox cycling through the formation of the superoxide radical. The outcome of these effects maybe protection, apoptosis or necrosis. Extracts from different plant sources contain variable amounts of putatively pro-oxidant proanthocyanidins. New chemical tools such as stable radicals sensitive only to electron transfer may help predict the effect of dietary proanthocyanidins in the tissues where they are bioavailable.

References

1. Shahidi, F.; Naczk, M. *Phenolics in Food and Nutraceuticals*; CRC Press: Boca Ratón, FL, 2004.
2. Katiyar, S. K.; Mukhtar, H. *J. Cell Biochem.* **1997**, *Suppl. 27*, 59-67.
3. Youdim, K. A.; Spencer, J. P. E.; Schroeter, H.; Rice-Evans, C. *Biol. Chem.* **2002**, *383*, 503-519.
4. Schulz, E.; Anter, E.; Keaney, J. F. *Curr. Med. Chem.* **2004**, *11*, 1093-1104.
5. Manach, C.; Mazur, A.; Scalbert, A. *Curr. Opin. Lipidol.* **2005**, *16*, 77-84.
6. Scalbert, A.; Manach, C.; Morand, C.; Remesy, C.; Jimenez, L. *Crit. Rev. Food Sci. Nutr.* **2005**, *45*, 287-306.
7. Alanko, J.; Riutta, A.; Holm, P.; Mucha, I.; Vapaatalo, H.; Metsa-Ketela, T. *Free Radical Biol. Med.* **1999**, *26*, 193-201.
8. Halliwell, B.; Gutteridge, J. M. C. *Biochem. J.* **1984**, *219*, 1-14.
9. Beckman, J. S.; Koppenol, W. H. *Am J Physiol Cell Physiol* **1996**, *271*, C1424-1437.
10. Cai, J.; Jones, D. P. *J. Biol. Chem.* **1998**, *273*, 11401-11404.
11. Halliwell, B. *Plant Physiol.* **2006**, *141*, 312-322.
12. Kondo, K.; Kurihara, M.; Miyata, N.; Suzuki, T.; Toyoda, M. *Archiv. Biochem. Biophys.* **1999**, *362*, 79-86.
13. Kondo, K.; Kurihara, M.; Miyata, N.; Suzuki, T.; Toyoda, M. *Free Radical Biol. Med.* **1999**, *27*, 855-863.
14. Long, L. H.; Clement, M. V.; Halliwell, B. *Biochem. Biophys. Res. Commun.* **2000**, *273*, 50-53.
15. Torres, J. L.; Varela, B.; Brillas, E.; Juliá, L. *Chem. Commun.* **2003**, 74-75.
16. Jiménez, A.; Selga, A.; Torres, J. L.; Juliá, L. *Org. Lett.* **2004**, *6*, 4583-4586.
17. Torres, J. L.; Bobet, R. *J. Agric. Food Chem.* **2001**, *49*, 4627-4634.
18. Torres, J. L.; Lozano, C.; Juliá, L.; Sánchez-Baeza, F. J.; Anglada, J. M.; Centelles, J. J.; Cascante, M. *Bioorg. Med. Chem.* **2002**, *10*, 2497-2509.
19. Torres, J. L.; Selga, A. *Chromatographia* **2003**, *57*, 441-445.
20. Gonthier, M. P.; Donovan, J. L.; Texier, O.; Felgines, C.; Remesy, C.; Scalbert, A. *Free Radical Biol. Med.* **2003**, *35*, 837-844.
21. Meselhy, R.; Nakamura, N.; Hattori, M. *Chem. Pharm. Bull.* **1997**, *45*, 888-893.

Chapter 22

Antioxidant Effects of Flavonoids Isolated from Young Green Barley Leaves toward Oxidative Degradation of β-Carotene

Hisao Umeda[1] and Takayuki Shibamoto[2]

[1]Japan Pharmaceutical Development Company, Ltd., Osaka, Japan
[2]Department of Environmental Toxicology, University of California, Davis, CA 95616

The stability of β-carotene was studied using a buffer or sucrose solution at pH 9. A β-carotene test solution containing an antioxidant (flavonoids, Trolox®, and vitamin C) was photo-irradiated at λ = 310 nm for various time periods at 25 °C. β-Carotene was extremely unstable and complete oxidative degradation occurred readily in 12 h. Presence of oxygen in the test solution played an important role in β-carotene degradation. If the test solution was degassed almost completely, significant β-carotene degradation did not occur and significant inhibitory effects of antioxidants were observed. Flavonoids isolated from young green barley leaves, lutonarin and saponarin inhibited β-carotene degradation by 50 - 80% in test solutions prepared with different degas methods. Saponarin inhibited β-carotene degradation for a longer time than other antioxidants in the sample degassed by both a nitrogen steam and a vacuum method, inhibiting β-carotene degradation over 50% for 60 h.

Food components that possess some biological activities, such as anti-carcinogenicity, anti-mutagenicity, anti-oxidative activity, and anti-aging activity, have recently received much attention from food and nutrition scientists as a third functional component of foods, after nutrients and flavor compounds.

It is well known that high consumption of fruits and vegetables is associated with a lowered risk of degenerative diseases, including cancer, cardiovascular disease, cataracts, atherosclerosis, diabetes, arthritis, immune deficiency diseases, and brain dysfunction (*1*). Among the functional components found in fruits and vegetables, carotenoids have been studied most intensively (*2*). β-Carotene and related compounds possess antioxidative activities (*3*), which are associated with the prevention of many diseases described above (*4, 5*).

On the other hand, the stability of β-carotene is significantly important when it is used as a nutrient supplement. β-Carotene showed some prooxidative activity with the testing system using ω-3 fatty acids, even though it had been reported to possess certain antioxidative activity (*6*).

In the present study, the stability of β-carotene with antioxidants upon photoirradiation was investigated.

Materials and Methods

Materials

β-Carotene, L-ascorbic acid, teterahydrofuran (THF), acetic acid/sodium acetate buffer, Trizma buffer, potassium phosphate dibasic/potassium phosphate tribasic buffer, sucrose, and Trolox® [2H-1-benzopyran-2-carboxylic acid, 3,4-dihydro-6-hydroxy-2,5,7,8-tetramethyl-(9Cl)] were purchased from Sigma-Aldrich (St. Louis, MO).

Saponarin and lutonarin (structures are shown in Figure 1) were isolated from young green barley leaves (*Hordium vulgare L. var. nudum Hook*) harvested two weeks after germination by using column chromatography with Amberlite XAD-2 nonionic polymeric absorbent. Young green barley powder was prepared by a previously reported method (*7*). An aqueous solution (powder/water = 1/5, w/w) of young green barley powder was boiled for 1 h and then filtered with a Buchner funnel. The filtrate was placed in a column packed with Amberlite XAD-2 resin and the solvent was eluted. A column was subsequently eluted with 20, 30, 40, and 50% methanol solution in series. Saponarin was rich in 50% methanol solution and lutonarin was rich in 20% methanol solution. After column chromatography, the saponarin and lutonarin fractions were further purified to 100% by using a preparative HPLC equipped with a 25 cm X 1 cm i.d. Develosil ODS-5 column (Phenomenex, Torrance, CA) and a UV detector at $\lambda = 280$ nm.

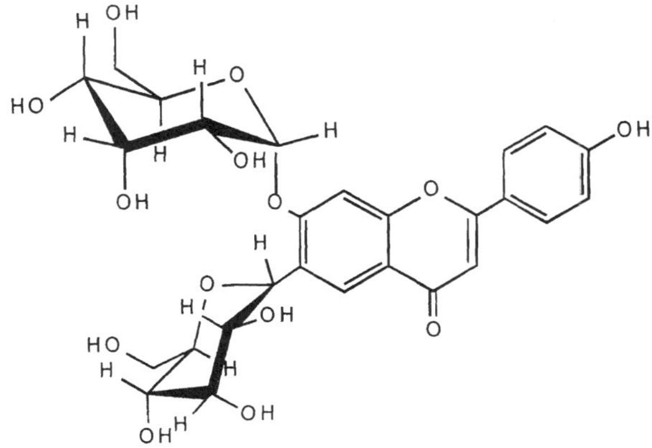

Saponarin

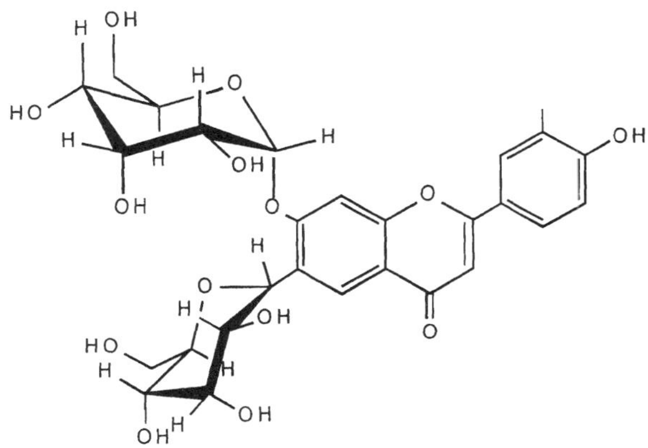

Lutonarin

Figure 1. Structure of saponarin and lutonarin.

Preparation of Testing Sample Solutions

Sample I: β-Carotene (100 mg) was dissolved into a minimal amount of dimethyl sulfoxide (DMSO) in a 100 mL volumetric flask, and then diluted to exactly 100 mL with buffer solution (pH = 9) containing 0.04 mmol/L each of saponarin, lutonarin, and ascorbic acid. Sample II: The same solution of Sample I was degassed with a purified nitrogen stream. Sample III: Trolox® (0.04 mmol/L) was added to the same solution of Sample I except the amount of β-carotene was 250 mg and the solution was degassed with a nitrogen stream and a vacuum method. Sample IV: Sucrose (25%) was added to the same solution of Sample III and the solution was degassed with a nitrogen stream and a vacuum method. Sample V: The same solution of Sample III was degassed with a nitrogen stream and with vacuum method while sonicating.

Photoirradiation of Testing Samples

The testing samples were placed in 150 mL-Pyrex bottles with Teflon-lined screw caps, then irradiated by UV using a Rayonet photochemical reactor Model RPR 100, Southern New England Ultraviolet Co. (Hamden, CT) for various period of times. The reactor was equipped with four 310-nm reactor lamps. The UVB intensity was measured near the center of the reactor with a Spectroline UVB meter (Spectronic Corp., Westbury, NY) at 0.68 ± 0.2 mW/cm^2. The value is mean ± standard deviation (n = 3).

Analysis of β-Carotene and Its Degradation Products

The sample irradiated 12 h without antioxidant was extracted with 30 mL dichloromethane using a liquid-liquid continuous extractor for 6 h. After the extract was dried over anhydrous sodium sulfate, it was condensed using a rotary flash evaporator and then further condensed under a purified nitrogen stream to 0.3 mL.

Photodegradation compounds of β-carotene were identified by comparison with the Kovats gas chromatographic retention index I (*8*) and by the MS fragmentation pattern of each component compared with those of authentic chemicals.

An Agilent 5890 series II gas chromatograph (GC) interfaced to an Agilent 5791 A mass selective detector (GC/MS) was used for mass spectral identification of the GC components at MS ionization voltage of 70 eV. GC was equipped with a 30 m x 0.25 mm i.d. (d_f = 1 μm) DB-5 bonded-phase fused-silica capillary column (J & W Scientific, Folsom, CA). The linear velocity of the helium carrier gas was 30 cm/sec. The injector and the detector temperatures were 250 °C. The oven temperature was programmed from 50 °C to 200 °C at 3 °C/min and held for 40 min.

Results and Discussion

β-Carotene has reportedly potent biological activities, including antioxidant activities, and it is often mentioned that drinking fruit and vegetable juices, such as citrus and carrot, is health beneficial because of their β-carotene content.

However, β-carotene is extremely unstable under the presence of oxygen. Therefore, it is necessary to prevent the oxidative degradation of β-carotene in order to appreciate its biological activities. When β-carotene was allowed to stand in a non-degassed solution, it degraded within 12 h and any antioxidants used did not give satisfactory inhibitory effect (Figure 2).

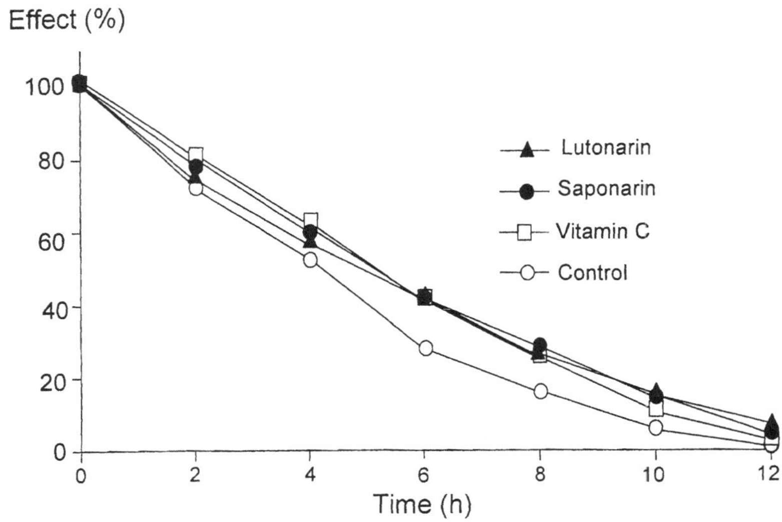

Figure 2. β-Carotene concentration in Sample I: 100 - 250 mg/100 mL of buffer solution (pH = 9).

When the testing solution was degassed with a nitrogen steam, however, the stability of β-carotene increased moderately and inhibitory activity of antioxidants was observed (Figure 3). Trolox® inhibited β-carotene degradation by over 80% for 10 h. Flavonoids, saponarin and lutonarin inhibited β-carotene degradation by over 60% for 10 h and then their effects decreased but they still inhibited over 50% after 20 h. Vitamin C exhibited slight antioxidant activity. It inhibited β-carotene degradation better than flavonoids did at stages earlier than 10 h.

If a sample was degassed thoroughly by a nitrogen steam and a vacuum method (Figure 4), the stability of β-carotene was maintained over 80 h addition of antioxidants improved stability slightly. However, β-carotene in the sample with lutonarin degraded 80% after 80 h. This may be due to the sample not having been degassed well.

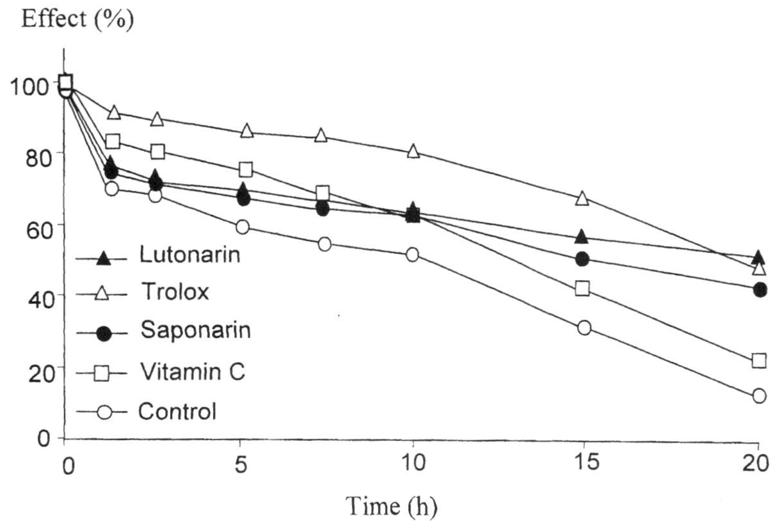

Figure 3. β-Carotene concentration in Sample II: 250 mg/100 mL of buffer solution (pH = 9). Degassed with N_2.

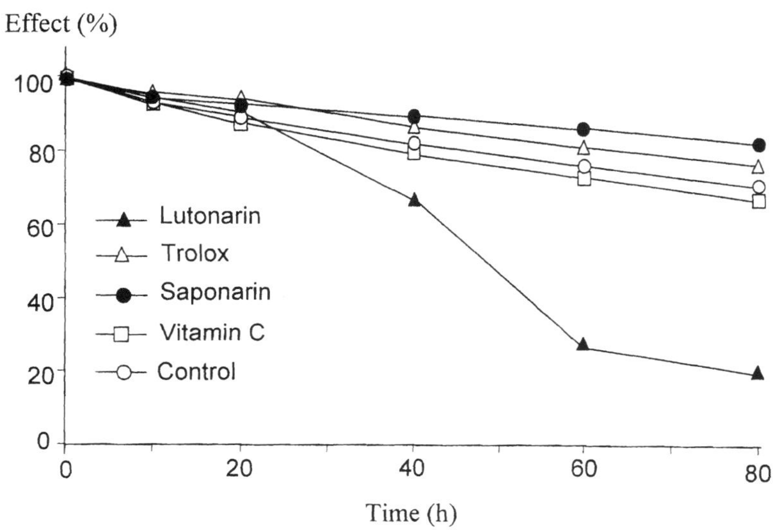

Figure 4. β-Carotene concentration in Sample III: 250 mg/100 mL of buffer solution (pH = 9). Degassed with N_2 and vacuum.

When 25% glucose was added to the testing solution (Figure 5), β-Carotene degraded completely after 60 h. The addition of antioxidants increased β-carotene stability modularly in this sample. Saponarin inhibited β-carotene degradation by over 60% for 60 h.

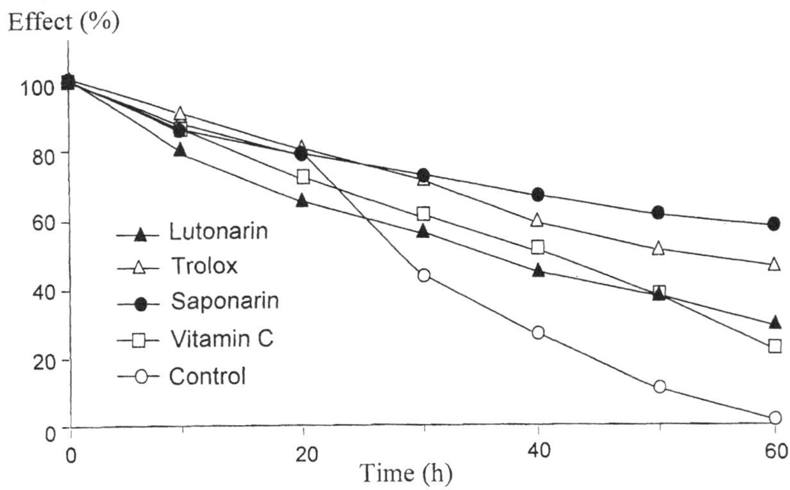

Figure 5. β-Carotene concentration in Sample IV: 250 mg/100 mL of 25% aqueous glucose solution (pH = 9). Degassed with N_2 and vacuum.

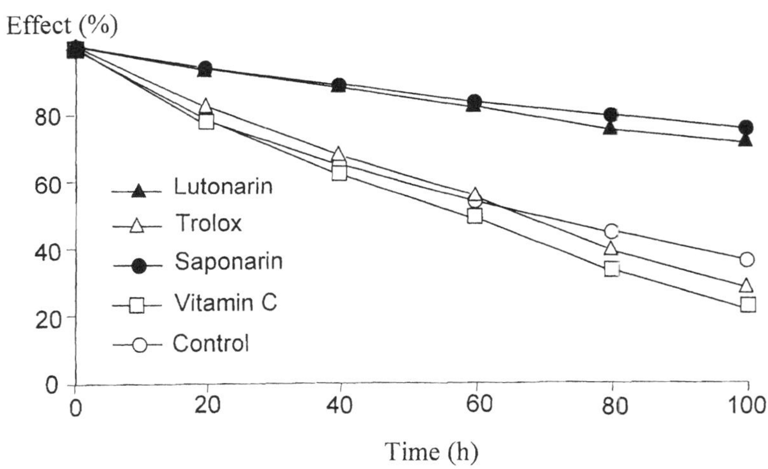

Figure 6. β-Carotene concentration in Sample V: 250 mg/100 mL of buffer solution (pH = 9). Degassed with N_2 and vacuum while sonication.

When the testing solution was degassed by a vacuum method during sonication, both flavonoids inhibited β-carotene degradation over 70% for over 100 h (Figure 6). Trolox® and vitamin C did not show any appreciable activity.

If the degassing process was not satisfactory and some level of oxygen remained in the test sample, satisfactory results were not obtained—as in the cases of lutonarin in Figure 4 and Trolox® in Figure 6. Therefore, the presence of oxygen in a test solution must play an important role in β-carotene degradation. However, the appreciable inhibitory effect of flavonoids isolated from young green barley leaves to β-carotene degradation was observed in the present study.

Table I shows the photodegradation products of β-carotene. Among over 100 peaks observed in a gas chromatogram of an extract from a photoirradiated β-carotene solution, major products identified were dihydroactnidiolide, 2,3-epoxy-β-ionone, and 2-hydroxy-2,6,6-trimethyl cyclohexanone. Dihydroactnidiolide and 2,3-epoxy-β-ionone were also reported in photodegradation products of β-ionone (*9*). 2,3-Epoxy-β-ionone is known as an aroma chemical and possesses an intensely sweet and fruity-woody aroma (*10*). β-Ionone, which has a fruity aroma, was also found in the present study. Therefore, these products may form *via* β-ionone. Some oxygen adducts of β-carotene formed in the early stage of photoirradiation, including epoxy, aldehyde, ketone, and dihydrofuran derivatives of β-carotene, as has previously been reported (*11*). Those derivatives may break down into the smaller molecular weight compounds identified in the present study. Several cyclohexanone and cyclohexenone derivatives were found in the β-carotene degradation products. These chemicals possess a powerful minty-camphoraceous odor which is generally described as an unpleasant smell (*10*). Formation of these compounds may affect the quality of foods and beverages.

Table I. Compounds Identified in Photo-irradiated β-Carotene

2,6,6-trimethyl cyclohexanone
2,6,6-trimethyl-2-cyclohexenone
1-formyl-2,6,6-trimethyl cyclohexane
2-hydroxy-2,6,6-trimethyl cyclohexanone
1-formyl-2,6,6-trimethyl-1-cyclohexene
β-ionone
4-(2,6,6-trimethyl-1-hexenyl) butane-2,4-dione
2,3-epoxy-β-ionone
dihydroactinidiolide
2,2-dimethyl heptanoic acid-6-one

References

1. Ames, B. N.; Shigenaga, M. K.; Hagen, T. M. *Proc. Natl. Acad. Sci.* **1993**, *90*, 7915-7922.
2. Khachik, F.; Cohen, L. A.; Zhao, Z. In *Functional Foods for Disease Prevention I, Fruits, Vegetables, and Teas. ACS Symp. Ser. 701;* Shibamoto, T.; Terao, J.; Osawa, T.; Eds.; American Chemical Society, Washington, D.C., 1997, pp. 71-85.
3. Palozza, P.; Krinsky, N. I. *Free Radical Biol. Med.* **1991**, *11*, 407-414.
4. Baruchel, S.; Wainberg, M. A. *J. Leukocyte Biology*. **1992**, *52*, 11-14.
5. Namiki, M.; Yamashita, K.; Osawa, T. In *Active Oxygens, Lipid Peroxides and Antioxidants;* Yagi, K.; Ed.; CRC Press, Boca Raton Ann Arbor, NY, 1993, pp. 319-332.
6. Pellett, L. J.; Andersen, H. J.; Chen, H.; Tappel, A. L. *J. Nutr. Biochem.* **1994**, *5*, 479-484.
7. Osawa, T.; Katsuzaki, H.; Hagiwara, Y.; Hagiwara, H.; Shibamoto, T. *J. Agric. Food Chem.* **1992**, *40*, 1135-1138.
8. Kovats, E. *Adv. Chromatogr*. **1965**, *1*, 229-247.
9. Etoh, H.; Ina, K. *Agric. Biol. Chem.* **1979**, *43*, 2593-2595.
10. Arctander, S. *Perfume and Flavor Chemicals*, Published by the author, Montclair, NJ, 1969.
11. Yamauchi, R.; Miyake, N.; Inoue, H.; Kato, K. *J. Agric. Food Chem.* **1993**, *41*, 708-713.

Chapter 23

The Role of Copigmentation with Phenolic Compounds on the Vitamin C Resistance and Antioxidant Activity of Anthocyanins

Pi-Jen Tsai, Chan-Chiung Liu, and Yui-Ting Huang

Department of Food Science, National Pingtung University of Science and Technology, Number 1, Shuehfu Road, Neipu, Pingtung, Taiwan 91201

Coexistence of vitamin C is known to expedite the degradation of anthocyanin and leads to serious color deterioration. The aim of this study was to elucidate whether the copigmentation occurred between anthocyanin and phenolic compounds could improve the resistance of anthocyanin against vitamin C and improve its color stability. Two model systems (pH 2 and pH 3) with five varieties of phenolic compounds (ferulic acid, gallic acid, catechin, chlorogenic acid, and caffeic acid) were investigated. The effect of copigmentation on assorted characteristics such as color qualities (Degradation Index, half-life and retention percentage of the pigment), activation energy for degradation and antioxidant capacity were studied. In addition, bathochromic shift and hyperchromic effect were also examined to quantify the degree of copigmentation. The results showed that copigmentation between anthocyanin and phenolic compounds (especially chlorogenic acid at pH 3) would significantly increase the anthocyanin stability against vitamin C, heat and sulfide, which also led to an enhanced antioxidant capacity. Further analysis revealed that FRAP increased with increased molar ratio of added polyphenols as well as the degree of bathochromic shift, and the increase of antioxidant capacity mainly came from polyphenols. HPLC analysis confirmed the existence of a novel substance of copigment complex. In conclusion, the copigmentation may significantly decrease the anthocyanin degradation caused by vitamin C, and increase the antioxidant capacity of the system.

Ascorbic acid, also known as vitamin C, is a natural anti-browning constituent in fruit products by preventing enzymatic browning or by reducing the quinone from melanin formation. However, in most anthocyanin rich foods, ascorbic acid is detrimental to color maintenance. The concurrent disappearance of ascorbic acid and anthocyanin has been explained either by colorless compounds after condensation (*1*) or the discoloration of hydrogen peroxide, a degradation product of ascorbic acid (*2, 3*). Nevertheless, very few information concerning how to maintain these two natural antioxidants at the same time has been reported.

Copigmentation for anthocyanin (ACN) after reacting with other phenolic compounds will usually lead to complex formation and enhanced color stability (*1*). The reaction in grape wine during vinification was found to be responsible for the color stability and high antioxidant capacity of red wine which contributes to the so-called "French paradox" (*4-6*). The objective of this study was to elucidate whether copigmentation could improve the resistance of anthocyanin against vitamin C and its effect on antioxidant capacity.

Factors influencing the copigmentation are quite complicated. Increase of the number of methoxy groups on the anthocyanin, the concentration of the anthocyanin or mole ratio of the copigment substance, all favor the reaction (*1, 7*). Characteristics of polyphenol, especially their pKa values, and pH levels of the system are known to affect the reaction in berry juice (*8*). A similar situation of color enhancement and correspondent change of antioxidant activity in grape juice with rosemary extract addition was also reported (*9*). Copigmentation attribute explained by different deprotonation patterns of these polyphenols was also used to explain the stabilizing effect (*10*). In this study, roselle anthocyanin at pH 2 and 3 were used to explore the effect of copigmentation with five varieties of phenolic compounds on pigment stability and antioxidant capacity. In addition, bathochromic shift and hyperchromic effect were also examined to quantify the degree of copigmentation.

Materials and Methods

Copigmentation Assay

The powder of anthocyanin extract from roselle (*Hibiscus Sabdariffa L.*) was prepared as previously reported (*11*). Aqueous solutions of 0.5 mM anthocyanin at pH 2 (buffer solution: 0.2 M KCl, 0.2 M HCl) and pH 3 (buffer solution: 0.1 M citric acid, 0.2 M Na_2HPO_4) were formulated. Assays of copigmentation were accomplished using 500 ppm of vitamin C, with or without 0.01 M, 0.02 M polyphenols (mole ratio of anthocyanin:to polyphenol 1:20 and 1:40, respectively). Samples of copigmentation were obtained by treating the above specimens at 25 °C for 1 to 72 hours. Scan mode of spectrophotometer was set to trace the maximum absorption wavelength (λ_{max}) and the absorbance at such wavelength ($A_{\lambda max}$).

Determination of Copigment Intensity

The copigment intensity was calculated as (A –Ao) / Ao according to the method of Davies and Mazza (*12*). Ao represented the maximum absorbance of anthocyanin without phenols by spectrophotometer, while A's denoted the absorbance of anthocyanin fortified by polyphenols.

Degradation Index (DI)

The degradation index of anthocyanin (DI value) was calculated as $A_{420\ nm}$ / $A_{\lambda max}$ (*13*). $A_{\lambda max}$ again was the maximum absorbance of sample detected by spectrophotometer.

Retention Percentage and Half-life of Anthocyanin

The samples were analyzed using a high performance liquid chromatograph (Hitachi L-7100 pump, Hitachi L-7420 UV-VIS detector, Lichrosorb RP-18 column 250 mm × 4.6 mm). The mobile phase was a gradient mixture of 5% acetic acid and acetonitrile (14). The retention percentage of anthocyanin was calculated with reference to the peak area of sample at time 0 h. Half-life ($t_{1/2}$) of anthocyanin was determined as the time duration when the anthocyanin absorbance ($A_{\lambda max}$) was reduced by one-half assuming first-order kinetics for anthocyanin degradation.

Heat and Sulfide Resistances

Heat resistance was conducted via heating solutions (ACN, ACN + 500 ppm vitamin C, ACN + 500 ppm vitamin C + phenolics) at various temperatures for up to 60 hours. Sulfide resistance was implemented by incorporating 5% sodium bisulfite ($NaHSO_3$) into the solutions at room temperature for 24 hours. DI (degradation index) values and ACN retention percentages were recorded to evaluate the ACN stability against heat processing and sulfide presence.

FRAP Assay

Ferric reducing ability of plasma (FRAP) assay is a method of measuring the antioxidant ability to reduce Fe^{+3} to Fe^{+2}. Freshly prepared FRAP reagent was homogeneously mixed with test sample, absorbance at 593 nm was recorded after 6 minutes of reaction at 37°C. FRAP capacity was expressed as μmole/l. The detailed procedures were previous described by Tsai *et al.* (*14*).

Statistical Analysis

As described previously (*14*), the Km and Vmax values for 5α-reductase were determined in rat liver microsomes from Lineweaver-Burk plots using different concentrations of T. The mean value for Km in our experimental data was 2.45 μM and that for Vmax 3.11 nmol of DHT/min/mg protein.

Results and Discussion

Effects of copigmentation between roselle anthocyanin and various phenolic compounds at different mole ratios on pigment retention, in the presence of 500 ppm of vitamin C at pH 2 and pH 3 are shown in Figure 1. Significantly improved pigment retention appeared in the system with higher mole ratios of copigment substances (polyphenol) and higher pH. The highest enhancements were obtained with chlorogenic acid (1:40 mole ratio) which resulted in 3.3 to 4.8 folds of increase in anthocyanin retention (compared to Control 2) at pH 2 and 3, respectively. Since anthocyanin existed as flavylium cation at pH 2 and pseudobase form at pH 3, this might be due to the decreased amount of carbinol pseudobase in the anthocyanin solution after copigmentation with polyphenols (*15*).

In order to elucidate the resistance of the copigment with different concentrations of ascorbic acid, the pigment stability was investigated with vitamin C from 0 to 2000 ppm. Table I offers the changes of anthocyanin retention percentage and DI, after copigmentation with chlorogenic acids in presence of different vitamin C contents at 25°C for 24 hours. Fortification with chlorogenic acid at 2000 ppm vitamin C improved anthocyanin retention by more than 30 times (47% vs. 1.5%), and reduced the DI value to only 1/3 of unfortified solution (0.752 vs. 2.321). There were merely 2 folds of increase in anthocyanin retention (92.0% vs. 40.3%), and 40% decline in DI value (0.560 vs. 0.899) were registered at 250 ppm vitamin C. Apparently, benefits from copigmentation were more significant at higher vitamin C concentrations. It suggested that the copigment may strongly resist to hydrogen peroxide, which is one of the degradation products from ascorbic acid, hence the anthocynin was preserved from destruction. This is consistent with the findings which stated the marked degradation of anthocyanin in sour cherry juice could be attributed to the exhausting of antioxidative substances, flavonols (*16*).

(a)

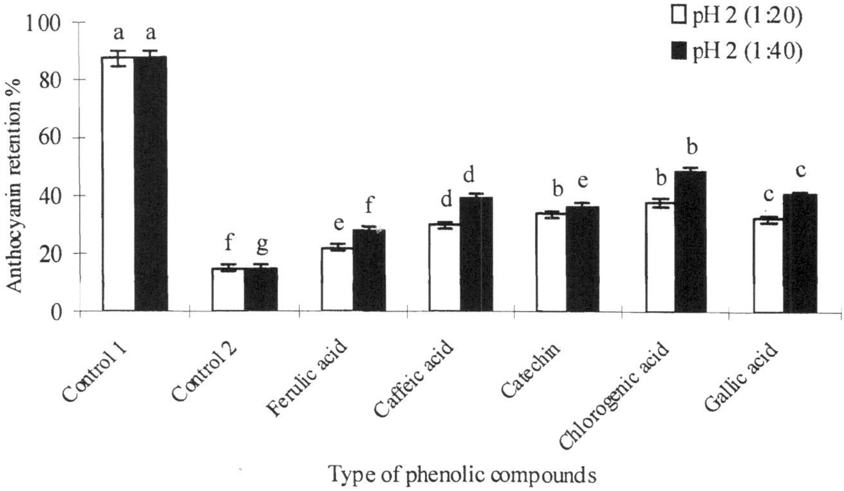

(b)

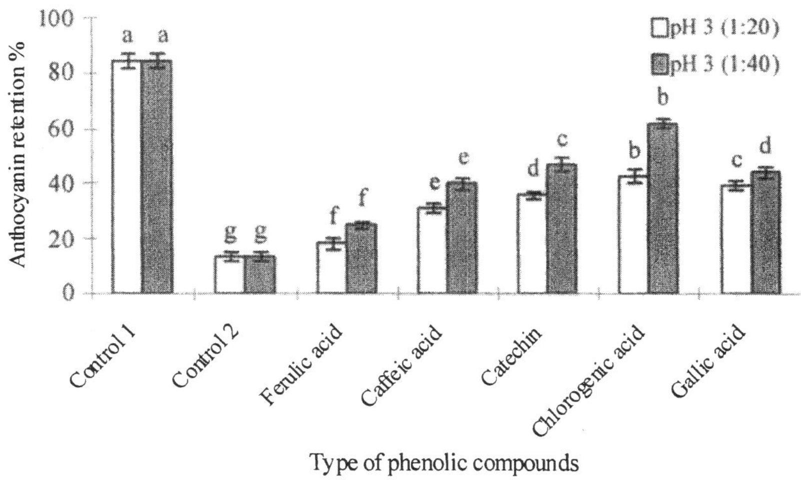

Figure 1. Effect of anthocyanins retention percentage of model system in the presence of different phenolic compounds at 500 ppm of vitamin C, at 25°C for 72 hr (a) pH 2 (b) pH 3. [a-g value marked by the different letters for each phenolic compound are significantly different by Duncan's Multiple Range Test ($p<0.05$)]. Control 1: without addition of vitamin C; Control 2: with addition of 500 ppm vitamin C.

Table I. The changes of anthocyanin retention percentage and degradation index in presence of chlorogenic acids with different vitamin C contents at 25 °C for 24 hr.

Vitamin C concentration (ppm)	*ACN Retention (%)*		*DI value*	
	ACN	*ACN + Chlorogenic Acid*	*ACN*	*ACN + Chlorogenic Acid*
0	85.0[a]	96.8[a]	0.617[e]	0.554[b]
250	40.3[b]	92.0[a]	0.899[d]	0.560[b]
500	12.8[c]	82.9[b]	1.310[c]	0.606[b]
1000	4.9[d]	73.8[c]	1.894[b]	0.629[b]
2000	1.5[d]	47.0[d]	2.321[a]	0.752[a]

*[a-e] Values values marked by the different letters within the same column are significantly different at 5% level ($p < 0.05$).

Heat tolerance is a very important feature for anthocyanin stability. Usually, color intensification with self-stacking conformation (copigment) also showed a better pigment stability due to reduced hydrolytic attacks (*17*). This is confirmed by the changes of DI value for anthocyanin copigmented with chlorogenic acid as shown in Figure 2. For instance, coexistence of anthocyanin with 500 ppm vitamin C increased the DI value from 2.5 to 4.0 after 18.5 hours of heating at 90°C, yet the chlorogenic acid helped to bring the DI value back to 2.9. Data at lower temperatures (60°C, 70°C, and 80°C) also rendered the similar outcome, not shown in Figure 2.

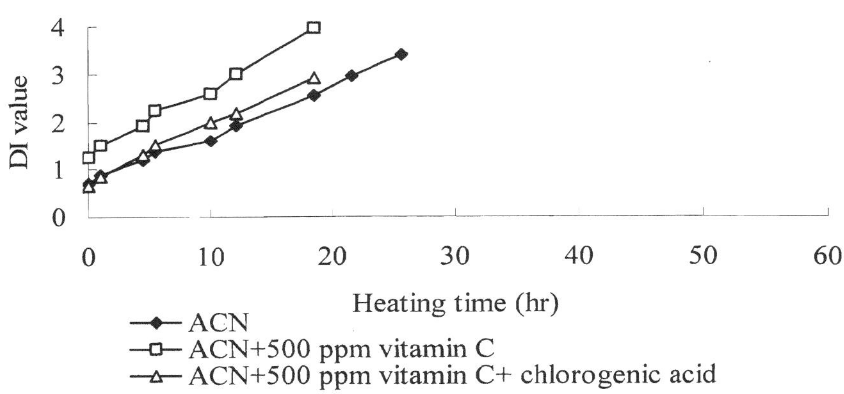

Figure 2. The changes of DI values for various anthocyanin solutions during heating at 90°C (pH 3).

Addition of ascorbic acid served to lower hyperchromic shift exhibited by copigmented anthocyanin, and caused a rapid color loss upon incorporation (*9*). This can be observed more clearly at a higher temperature or during long term storage. The changes of half-life of anthocyanin (Table II) and pigment retention during heating at 80 °C (Table III) substantiate this finding. Chlorogenic acid again proved to be most effective as the half-lives were the highest among all vitamin C concentrations at 15 and 35 °C, respectively. ACN retention percentages with chlorogenic acid were also highest among these phenolic compounds tested with 500 ppm vitamin C except 1:20 molar ratio at pH 2.

Table IV presents the sulfide resistance of the pigment after copigmentation. It indicates that copigmentation greatly improve the anthocyanin retention especially at pH 2 in the presence of vitamin C. Caffeic acid (1:40 mole ratio at pH 2) had the most prominent effect on sulfide resistance as it improved the ACN retention by almost 2 times (72.4% vs. 36.5%). These results were entirely consistent with those reported for grape wine. In which, a new pigment (pyranoanthocyanin) formed via a condensation between anthocyanin and flavanols was reported. This new compound exhibited a higher stability to pH dependent hydration and bisulfite bleaching than the original anthocyanin probably due to steric hindrance (*18*).

Table II. The Half-life ($t_{1/2}$, hr) of Roselle Anthocyanin (pH 3) in the Presence of Vitamin C and Phenolics at 15 and 35°C, Respectively

Vitamin C (ppm)	*Control*	*Ferulic acid*	*Caffeic acid*	*Catechin*	*Chlorogenic acid*	*Gallic acid*
(15 °C)						
0	91.6[e]	115.8[c]	111.7[d]	136.4[b]	140.7[a]	135.4[b]
500	21.3[e]	23.4[b]	25.6[d]	30.6[c]	35.1[a]	33.2[b]
1000	19.9[f]	21.0[b]	23.7[e]	27.9[d]	31.4[a]	29.0[c]
2000	19.0[f]	20.6[d]	20.9[e]	24.8[c]	28.6[a]	25.5[b]
(35 °C)						
0	72.0[d]	101.5[c]	101.0[c]	105.5[b]	129.5[a]	107.5[b]
500	19.6[e]	24.0[c]	23.1[d]	23.9[c]	31.9[a]	25.0[b]
1000	18.7[d]	21.7[c]	21.8[c]	21.4[c]	28.3[a]	23.5[b]
2000	18.5[d]	20.8[b]	20.6[b]	20.6[b]	23.1[a]	20.1[b]

*[a-k] Values marked by the different letters within the same row are significantly different at 5% level ($p < 0.05$).

The anthocyanin concentration was 0.5 mM, each phenolic concentration was 0.02 M (molar ratio 1:40)

Table III. The Change of Anthocyanin Retention Percentage During Heating (80°C for 1 Hour)

Anthocyanin with phenols	*Vitamin C (ppm)*	*ACN Retention %*			
		pH 2		*pH 3*	
		1:20	*1:40*	*1:20*	*1:40*
Control	0	95.64^{b}	95.64^{c}	95.13^{c}	95.13^{b}
	500	67.35^{i}	67.35^{l}	60.05^{k}	60.05^{h}
Ferulic	0	91.28^{d}	90.22^{e}	93.74^{d}	88.41^{c}
	500	79.81^{h}	77.39^{k}	72.61^{j}	57.43^{i}
Caffeic	0	91.00^{d}	86.86^{h}	90.88^{i}	85.22^{d}
	500	84.85^{f}	81.69^{j}	75.10^{i}	69.86^{g}
Catechin	0	94.73^{c}	92.41^{d}	92.72^{e}	88.65^{c}
	500	86.60^{e}	84.09^{i}	77.59^{h}	72.87^{f}
Chlorogenic acid	0	98.23^{a}	98.66^{b}	98.34^{b}	99.32^{a}
	500	84.64^{f}	87.97^{f}	85.68^{f}	88.78^{c}
Gallic acid	0	98.41^{a}	99.56^{a}	99.24^{a}	98.01^{a}
	500	83.70^{g}	86.83^{g}	84.55^{g}	81.22^{e}

* $^{a\text{-}k}$ Values marked by the different letters within the same row are significantly different at 5% level ($p < 0.05$). Control, 0: anthocyanin solution only. Control, 500 ppm: anthocyanin solution with 500 ppm vitamin C.

Table IV. Change of Anthocyanin Retention Percentage with 5% Sodium Bisulfite ($NaHSO_3$) Incorporated

Anthocyanin with phenols	*Vitamin C (ppm)*	*ACN Retention %*			
		pH 2		*pH 3*	
		1:20	*1:40*	*1:20*	*1:40*
Control	0	75.44^{d}	75.44^{d}	50.51^{h}	50.51^{h}
	500	36.50^{i}	36.50^{j}	41.32^{i}	41.32^{i}
Ferulic acid	0	80.00^{ab}	81.55^{b}	60.77^{c}	62.03^{d}
	500	63.27^{f}	69.65^{g}	57.80^{f}	54.07^{g}
Caffeic acid	0	78.52^{c}	83.36^{a}	58.25^{e}	68.94^{a}
	500	62.03^{g}	72.37^{e}	64.24^{b}	60.59^{e}
Catechin	0	80.48^{a}	80.17^{c}	59.86^{d}	67.11^{b}
	500	62.90^{f}	68.18^{h}	65.13^{a}	61.39^{d}
Chlorogenic acid	0	79.35^{b}	81.64^{b}	57.56^{f}	63.95^{c}
	500	64.23^{e}	71.24^{f}	57.47^{f}	56.50^{f}
Gallic acid	0	78.15^{c}	80.23^{c}	54.29^{g}	56.68^{f}
	500	57.97^{h}	58.90^{i}	57.39^{f}	57.09^{f}

*$^{a-j}$ Values marked by the different letters within the same row are significantly different at 5% level ($p < 0.05$). Control, 0: anthocyanin solution only. Control, 500 ppm: anthocyanin solution with 500 ppm vitamin C.

Figure 3 clearly illustrates that in the roselle anthocyanin-catechin solution system, the FRAP exhibited positive correlations with both copigment intensity (A –Ao)/Ao, and bathochromic shift ($\Delta\lambda_{max}$), with coefficients of 0.94 and 0.96, respectively (data not shown). This makes it possible to use the color enhancement as an index of antioxidant capacity.

Composition change of anthocyanin upon copigmentation with phenolic compounds was also examined using HPLC-UV/VIS analysis (A $_{\lambda max}$). The retention time for original roselle anthocyanin ranged from 13 to 15 minute, whereas new peak was identified at 21 minute upon addition of ferulic acid (chromatograms not shown). This could be rationally reasoned as a novel compound formed as anthocyanin-ferulic acid complex. There have been some reports about a new type of anthocyanin derivative, namely pyranoanthocyanin, that could be generated from anthocyanin in red wines through interactions with pyruvic acid, acetaldehyde, and caffeic acid, among others. This new derivative was believed to augment the color density of pigment, and its stability at elevated pH's, also resulted in higher tolerance of anthocyanin (*8, 18*). Isolation and structural identification of this novel complex will wait for further exploration.

In conclusion, copigmentation complex was found to decrease the damage to anthocyanin by ascorbic acid oxidation.

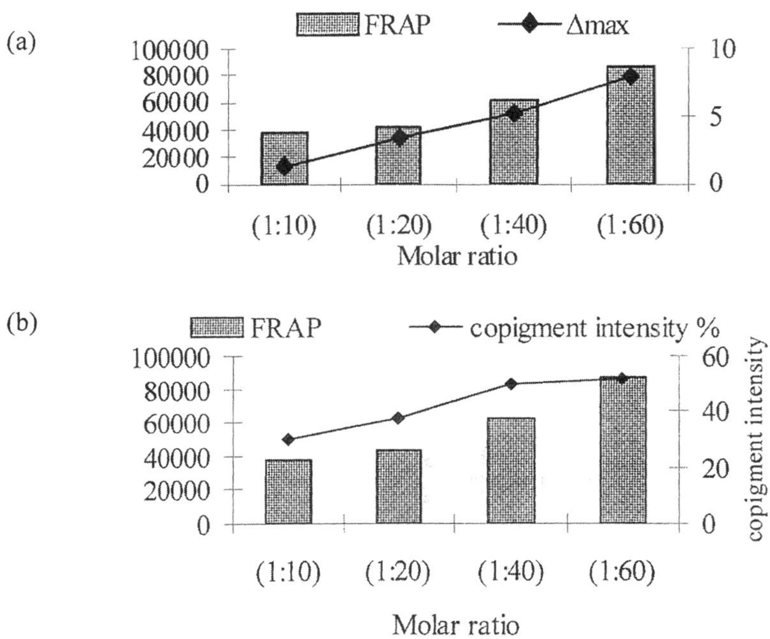

Figure 3. Effect of mole ratio of catechin added on the FRAP and (a) hyperchromic effect and (b) bathochromic shift of the roselle anthocyanin system at pH 3.

Acknowledgements

The partial financial support from National Science Foundation, Taiwan, under project number NSC-93-2214-E020-002 is greatly acknowledged.

References

1. Markakis, P. In *Anthocyanins as Food Colors*; Markakis, P. Ed.; Academic Press: New York, NY, **1982**, pp. 163-180.
2. Jackman, R. L.; Yada, R.Y. *J. Food Biochem.* **1987**, *11*, 201-247.
3. García-Viguera, C.; Bridle, P. *Food Chem.* **1999**, *64*, 21-26.
4. Frankel, E. N.; Waterhouse, A. L.; Teissedre, P. L. *J. Agric. Food Chem.* **1995**, *43*, 890-894.
5. Whitechead, T. P.; Robinson, D.; Allaway, S.; Syms, J.; Hale, A. *Clin. Chem.* **1995**, *41*, 32-35.
6. Boulton, R. *Am. J. Enol. Vitic.* **2001**, *52*, 67-86.
7. Asen, S.; Stewart, S. N.; Norris, K. H. *Phytochemistry* **1972**, *11*, 1139-1143.
8. Eiro, M. J.; Heinonen, M. *J. Agric. Food Chem.* **2002**, *50*, 7461-7466.
9. Talcott, S. T.; Brenes, C. H.; Pires, D. M.; Del Pozo-Insfran, D. *J. Agric. Food Chem.* **2003**, *51*, 957-963.
10. Dimitric-Markovic, J. M.; Petranovic, N. A.; Baranac, J. M. *J. Photochem. Photobiol. B-Biol.* **2005**, *78*, 223-228.
11. Tsai, P. J. ; Ou, S. A. *J. Food Sci.* **1996**, *23*, 629-640.
12. Davies, A. J.; Mazza, G. *J. Agri. Food. Chem.* **1993**, *41*, 716-720.
13. Sistrunk, W. A.; Gascoigne, H. L. *J. Food Sci.* **1983**, *48*, 430-440.
14. Tsai P. J.; McIntosh, J.; Pearce, P.; Camden, B.; Jordan, B. R. *Food Res. Int.* **2002**, *35*, 351-356
15. Bękowska, A.; Kucharska, A. Z.; Oszmiański, J. 2003. *Food Chem.* **2003**, *81*, 349-355.
16. Özkan, M. *Food Chem.* **2002**, *78*, 499-504.
17. Hoshino, T. *Phytochemistry* **1991**, *30*, 2049-2055.
18. Lee, D. F.; Swinny, E. E., Asenstorfer, R. E.; Rowe, S. E.; Jones, G. P. Incorporation of anthocyanins into complex wine pigments. **2005** [05-03-24]. Available from: *http://www.pi.csiro.au*.

Chapter 24

Antioxidant Activities of Polyphenol Containing Extracts from Citrus

G. K. Jayaprakasha, K. N. Chidambara Murthy, and Bhimanagouda S. Patil

Vegetable and Fruit Improvement Center, Department of Horticultural Sciences, Texas A&M University, 2119 TAMU, College Station, TX 77843

Several fruits and vegetables that posses antioxidant activity have been reported to be rich in polyphenols and are commercially promoted as functional foods. Citrus fruits contain many bioactive compounds such as phenolics, flavonoids, limonoids, carotenoids, sterols and ascorbic acid. Six different varieties of citrus fruits were extracted with five solvents. The dried extracts were screened for their radical scavenging activity and antioxidant capacity. The total phenolics have been determined by Folin-Ciocalteu method and the results have been expressed as catechin equivalents. The antioxidant capacity of the extracts is in accordance with the amount of phenolics / lycopene / vitamin C present in each fraction and may provide a good source of antioxidants.

Antioxidants can exercise their protective properties at different stages of the oxidation process and by different mechanisms. There are two main types of antioxidants, namely, "primary" (chain breaking, free radical scavengers) and secondary" or "preventive". "Secondary" antioxidant mechanisms may include deactivation of metals, inhibition of breakdown of lipid hydroperoxides to unwanted volatile products, regeneration of "primary" antioxidants, singlet oxygen quenching, etc (*1*). Compounds possessing such activity are known to possess health benefits in preventing/delaying onset of damage of various biologically significant molecules. This intern helps in prevention or delay in onset of some the diseases like, cardiovascular and carcinoma. The commonly used synthetic antioxidants for the lipid peroxidation are BHT, BHA, TBHQ, propyl gallate and trolox. Among natural compounds ascorbic acid, tocopherol, polyphenols and carotenoids are the major ones (*2*).

Phenolics are known to posses higher antioxidant activity compared to vitamins such as A, C and E, when tested against LDL and VLDL models (*3, 4*). Lipid peroxidation is a complex chain process involving a variety of radicals. The oxidation is influenced by temperature, light, air, physical and chemical properties of the substrate, and the presence of oxidation catalysts or initiators (*4, 5*). Plants are rich in phenolic compounds of different origins and functions. Most of them are biologically active as antiviral, antimicrobial, anticarcinogens and atherosclerogenic (*2, 6*). Phenolic compounds are large, heterogeneous groups of secondary plant metabolites that are widespread in the plant kingdom and they have a wide variety of structures. Flavonoids, tannins and phenolic acids are the main phenolic compounds. It is well known that diets rich in fruits and vegetables are capable of preventing or delaying the onset of certain chronic degenerative diseases of aging, including cardiovascular malfunction and common cancers (*5, 7*). The antioxidant properties of fruits and spices derived phenolic compounds have been extensively studied by using in vitro chemical systems (*7-15*). These systems have the advantage of being relatively simple and inexpensive to carry out. However, such in vitro assays are very important before going to in vivo or pre-clinical experiments. On the other hand, their relevance to in vivo health-protective activities is uncertain. It is therefore considered prudent to use more than one antioxidant assay system to measure antioxidant activities, as there may be distinct mechanisms involved, resulting in different outcomes, depending on the method of test (*16*).

Extraction Efficiency of Phenolics Using Different Solvents

Lyophilized fruit powder of blood orange, citron, pummelo, rio red, sour orange and navel orange were extracted with five different solvents. The extracts were concentrated under vacuum, lyophilized and stored at -20 °C until further use. Table 1 depicts the yield from citron, blood orange, rio red, sour orange, pummelo and navel orange using different solvents. Methanol extraction of all the citrus fruits gave highest yield. However, hexane extract gave minimum yield except citron fruit. In our recent studies, it has been reported that the yield of extractable compounds was highest in methanol extract from pomegranates in comparison with the solvents such as ethyl acetate and water (*10*).

Table I. Yield (g/100g) of Citrus Fruit Extracts Using Different Solvents*

Solvents used for extraction	Citrus varieties					
	Citron	Blood orange	Rio-red	Sour orange	Pummelo	Navel orange
Hexane	1.50	0.83	1.01	0.82	1.6	0.64
EtOAc	20.01	7.19	4.21	10.18	3.4	9.12
Acetone	1.30	2.76	5.05	2.01	6.7	2.88
MeOH	40.60	68.01	60.50	47.91	40.4	21.04
MeOH:water (80:20)	7.31	20.01	12.90	2.73	6.9	5.71

* values are mean of triplicate trials

Phenolics have been measured using Folin Cio-calteu method and results have been expressed as catechin equivalents (Table II). Ethyl acetate (EtOAc) extract of rio red, sour orange, pummelo and navel orange gave highest phenolics. Hexane extract did not show any phenolics. Water and acetone extraction gave more phenolics in case of citron and blood orange respectively. On the other hand extraction of phenolic compounds from the fruit is commonly achieved with methanol or aqueous methanol (*17, 18*). Amount of polyphenols were measured using Folin-ciocalteu method. However, the fact based on their chemical reducing capacity relative to catechin. It has been observed that the phenol antioxidant index is a combined measure of the quality and quantity of antioxidants in vegetables (*19*). Anagnostopoulou et al. (*20)* reported EtOAc extract from sweet orange peel has maximum phenolics and which confirms our results. In the present study the responses of the extracts in this assay may be ascribed to the variety and quantity of phenolics found in four extracts from two different varieties of citrus fruits.

Generally, extraction with hexane will give non-polar compounds like fatty material or some carotenoids. EtOAc will give medium polar carotenoids and phenolics. The other solvents were used for the extraction of polar compounds like aglycones and glucosides of flavonoids and limonoids depending upon their polarity. It is believed that major components for antioxidant activity in edible plants are carotenoids, vitamin C and polyphenolic compounds. Thus, it is necessary to extract these compounds effectively when antioxidant activities are measured. In this perspective, we chose five different polar solvents to extract all the antioxidants from citrus fruits for the measurement of antioxidant activity. The polarities of antioxidant components in each extract are likely to be different. Type of solvents and polarity may affect the single electron transfer and the hydrogen atom transfer, which are key aspects in the measurements of antioxidant capacity. Hence, the selection of extraction solvent is critical for the complex food samples (*21-23*).

Table II. Phenolics Present in Different Citrus Fruits Extracts in mg g^{-1}

Solvents used for extraction	Citrus varieties used					
	Citron	Blood orange	Rio-red	Sour orange	Pummelo	Navel orange
EtOAc	1.277	2.96	2.91	2.02	9.7	6.4
Acetone	2.964	4.88	0.2	1.62	4.3	2.47
MeOH	3.928	0.80	0.54	1.24	1.2	0.52
MeOH:water (80:20)	7.181	0.20	0.86	1.64	1.8	3.6

Citrus Fruits and Their Radical Scavenging Activity

Freeze dried fruit powder of six varieties of citrus fruits was successively extracted with hexane, EtOAc, acetone, MeOH and MeOH:water (80:20) for 8 h each separately. All the extracts were freeze dried and stored at -20 °C until further use. The results of free radical scavenging potentials of citrus fruit extracts from citron, blood orange, rio red, sour orange, pummelo, navel orange and ascorbic acid at 1000 µg/mL were tested by DPPH method and the results are depicted in Figure 1. Antioxidants react with DPPH, which is a nitrogen-centered radical with a characteristic absorption at 517 nm and convert to 1,1,-diphenyl-2-picryl hydrazine, due to its hydrogen donating ability at a very rapid rate (*12*). The degree of discoloration indicates the scavenging potentials of the antioxidant extracts. MeOH:water (80:20) extract of citron, sour orange and rio red; EtOAc extract of blood orange, pummelo and navel orange exhibited maximum free radical scavenging activity, respectively. On the other hand, hexane extract of citron and pummelo; MeOH extract of blood orange, sour orange, rio red and navel orange showed minimum activity (Figure 1). It is known that free radicals cause autooxidation of unsaturated lipids in food (*24*). The antioxidant activity of the fractions was attributed to their hydrogen donating ability (*25*). On the other hand, antioxidants are believed to intercept the free radical chain of oxidation and to donate hydrogen from the phenolic hydroxyl groups, thereby forming stable end product, which does not initiate or propagate further oxidation of lipid (*26*). The data obtained from this study revealed that, certain fractions isolated from citrus fruits are good free radical scavengers and primary antioxidants that react with DPPH radical, which may be attributed to its proton donating ability. Hsiu-Ling et al. (*27*) reported the antioxidants in the juice and freeze-dried flesh and peel of red pummelo and their ability to scavenge free radicals and compare them with those in white pummelo juice. The total phenolic content of red pummelo juice extracted by

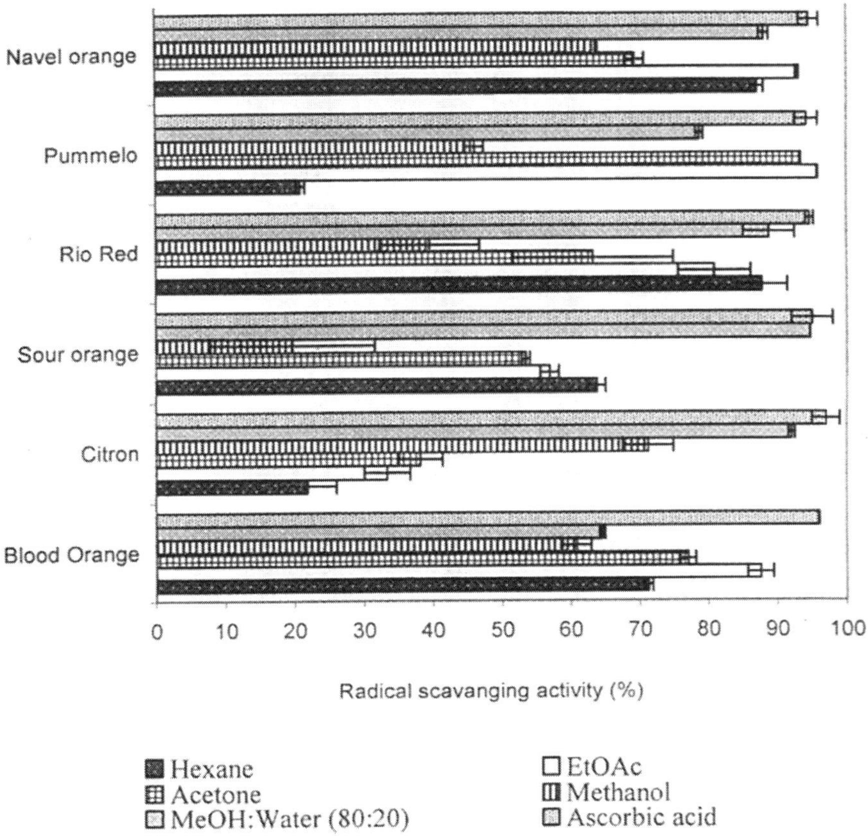

Figure 1. Radical scavenging activity of citrus fruit extracts using DPPH method at 1000 μg/mL

methanol (8.3 mg/ml) was found to be significantly higher than that of white pummelo juice (5.6 mg/ml). The carotenoid content of red pummelo juice was also significantly higher than that in white pummelo juice. The ability of methanol extracts of freeze-dried peel and flesh from red pummelo to scavenge these radicals was 20-40% that of BHA and vitamin C effects. Fresh red pummelo juice is an excellent source of antioxidant compounds and exhibited great efficiency in scavenging different forms of free radicals including DPPH, superoxide anion and hydrogen peroxide radicals.

Antioxidant Capacity of Citrus Extracts

The different citrus fruit extracts exhibited various degrees of antioxidant capacity (Figure 2). It is difficult to ascertain an order of antioxidant capacities of different extracts because of the differential responses. Maximum antioxidant capacities were observed in the MeOH:water (80:20) extract of citron, hexane extract of blood orange, MeOH extract of sour orange, EtOAc extract of pummelo, navel orange and rio red. The phosphomolybdenum method is based on the reduction of molybdenum (VI) to molybdenum (V) by the antioxidant compounds and the formation of a green molybdenum (V) complex, which has a maximal absorption at 695 nm. Variations in antioxidant capacity of different extracts may be attributed to differences in their chemical composition such as phenolics, ascorbic acid and carotenoids. Our recent results indicated that certain

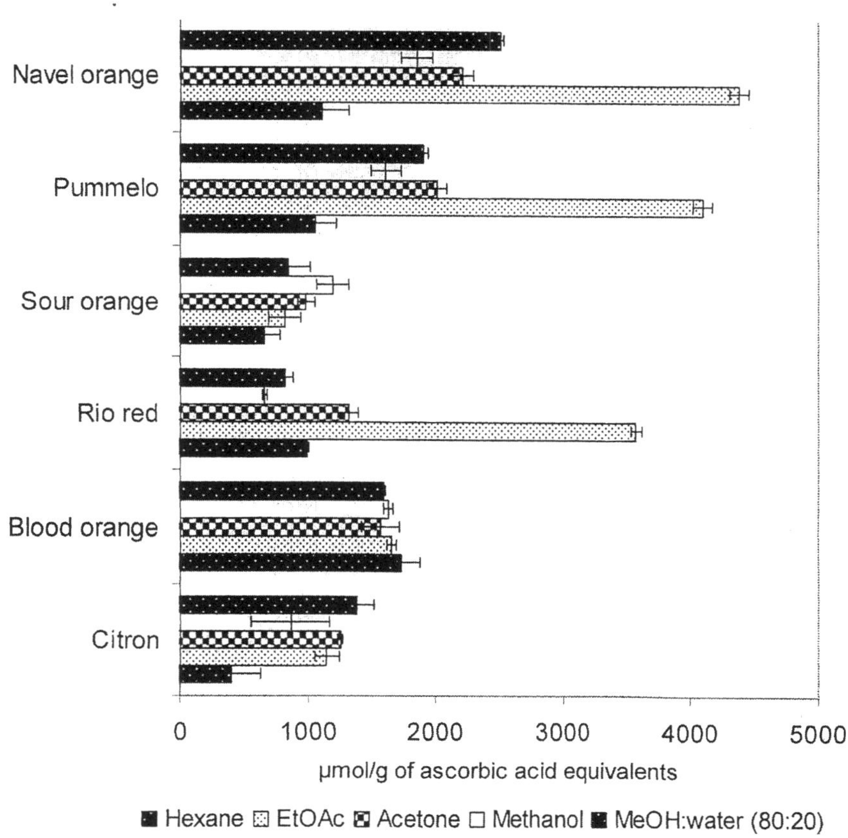

Figure 2. Antioxidant capacity of citrus fruit extracts at 400 μg/mL

citrus limonoids are found to posses moderate antioxidant activity (*28, 29*). The antioxidant activity shown by the citrus fruit extracts may be due to the presence of flavonoids, carotenoids and ascorbic acid (*21*). Rehman (*30*) reported citrus peel extracts caused significant changes free fatty acids, peroxide value, and iodine value of refined corn oil during 6 months of storage at 25 and 45 °C as compared to control. These results confirm the findings of earlier workers, who identified phenolic and flavonoid antioxidative compounds in the non-volatile fraction of methanolic extract of citrus peel (*31*). Many other researchers found that antioxidant activity in the extract of edible and non-edible plant materials due to the presence of phenolic compounds (*2, 10, 32*). The findings also revealed that the antioxidant property was observed in orange peel ultra-filtered molasses due to the presence of phenols, including numerous flavanones, flavone glycosides, polymethoxylated flavones, hydroxy cinnamates and other miscellaneous phenolic glycosides and amines (*33*).

Correlation between the content of the total phenolics and radical scavenging activity of the citrus fruit extracts has been reported by several authors (*33, 34*). Some of the studies reported that there is no correlations between the total phenolic content and the radical scavenging activity (*33*), but in the present study showed very high correlation coefficient of the total phenolics and radical scavenging activity of all the samples (r^2 = 0.99) except MeOH:water (80:20) extract of citron (*21, 22*). While there are many methods for the total antioxidant determination, most of the methods have their own limitations (*33, 34*). It was shown that some antioxidant assay methods give different antioxidant activity trends (*35*). The DPPH approach seems to be rapid and accurate method for assessing the antioxidant activity of fruit and vegetable extracts. The results are highly reproducible and comparable to other free radical scavenging methods such as ABTS (*36*). Xu et al. (*37*) correlated the fifteen varieties of citrus juice components with antioxidant capacity. Phenolic compounds and ascorbic acid were identified as possible antioxidants in orange juice. Phenolic compounds were able to scavenge radicals and to chelate metals (*38*), while ascorbic acid can play a pro-oxidant role in the presence of transition metals. These compounds can also act as antioxidants because of their ability to trap superoxide anions (*39*). Depending on the concentrations of phenolic compounds and of transition metals, a complex can be formed that facilitates the redox process (*40*). Recent literature on citrus fruits and antioxidant activity in different models have been summarized and present in Table III. In general, the phenolic compounds at low concentrations show antioxidant behavior. At higher concentrations, they show pro-oxidant behavior; upon further increasing of their concentration, they again show antioxidant behavior. This always depends on the type (position and number of hydroxyl in the molecule) and the concentration of the phenolic compound, as well on that of the transition metal (*48*). Thus, different citrus fruit extracts at different concentrations showed different degree of antioxidant activity due to the presence various compounds.

Table III. Recently Reported Research Findings on Antioxidant Activity of Citrus Fruits

Materials employed	Activity reported	Ref
Pummelo flesh and peels	DPPH, superoxide anion, and hydrogen peroxide	41
Huyou (*Citrus paradisi* Changshanhuyou) peel	$ABTS^{\cdot+}$ method, and ferric reducing antioxidant power (FRAP) assay	42
'Cara cara' navel orange (*Citrus sinensis* L. Osbeck)	Malondialdehyde , H_2O_2, Superoxide dismutase (SOD) and catalase	43
'Red Flesh' navel orange pulp	SOD, catalase, guaiacol peroxidase, Ascorbate peroxidase, and dehydroascorbate reductase , glutathione reductase	44
Pericarpium *Citri Reticulatae* of a new Citrus cultivar	DPPH scavenging, hydroxyl radical scavenging, superoxide anion radical scavenging, hydrogen peroxide scavenging and reducing power assay	45
Pomaces of *Citrus unshiu*	DPPH, Reducing power assay	46
Edibale part of citron and blood orange	DPPH, phosphomolybdenum method, NBT reduction assay	21
Edibale part of pummelo and navel orange	DPPH, phosphomolybdenum method, ORAC, ABTS, reducing power	22
Edibale part of rio red, sour orange	DPPH, phosphomolybdenum method, NBT reduction assay, reducing power	23
Citrus unshiu, Citrus reticulata, and *Citrus sinensis*	Total antioxidant capacities (TAC) by ferric reducing antioxidant power (FRAP) assay.	47
Juices from fifteen different citrus varieties	Ferric reducing antioxidant power, DPPH assay	38

The results obtained in the present study (*21-23*) demonstrated that the citrus fruit extracts can effectively scavenge various reactive oxygen species or free radicals under *in vitro* conditions. This may be due to the number of stable oxidized products that can form after oxidation or radical scavenging. The broad range of activity of the extracts suggests that multiple mechanisms are responsible for the antioxidant activity. The multiple antioxidant activity of extracts demonstrated in this study clearly indicatesd the potential application value of the citrus fruits. However, the *in vivo* safety of extracts needs to be thoroughly investigated in experimental rodent models prior to its possible application as an antioxidant ingredient, either in animal feeds or in human health foods. The above results demonstrated that, some of the extracts could exhibit antioxidant properties, which are comparable to commercial synthetic antioxidants.

Structure Activity Relation and Antioxidant Activity of Citrus Flavanones

Antioxidant activity of flavanones depends upon the arrangement of functional groups. The spatial arrangement of substitutents influences the antioxidant activity more than the flavan backbone. The configuration and the total number of hydroxyl groups influence several mechanisms of the antioxidant activity. This is important in flavanones with a substitution of the neohesperidoside group in the 7th position. This may be due to presence of the aglycone form, the 3', 4'-dihydroxy substitution does not influence much on the antioxidant activity. On the other hand, in the flavanones glycosylated with a neohesperidose of the 7th OH group, the 3', 4'-catechol structure the antioxidant power is noticeably increased (*49*).

The difference in antioxidant capacity between the polyhydroxylated and the polymethoxylated flavanones (PMF) is due to the differences either in hydrophobicity or in molecular planarity. The influence of the O-methylation in the aglyconic flavanones is negligible. On the other hand, in case of flavanones such as neohesperidoside molecule has methoxyl group at 7th position and its antioxidant activity has been decreased noticeably as compared to PMF's. Moreover, in the flavanones glycosylated with a neohesperidose of the 7th hydroxyl group, the 3′, 4′-catechol structure noticeably increased the antioxidant power and it can be observed in case of neoeriocitrin and naringenin. Oxidation of a flavonoid occurs on the B-ring when the catechol is present yielding a fairly stable ortho-semiquinone radical through facilitating electron delocalization. Flavanones glycosylated with neohesperidose lacking catechol system form relatively unstable radicals and are weak scavengers (*50*). Further, due to the presence of electron-donating groups makes the aromatic system rich in electrons. This confers a higher degree of instability to the flavanone phenoxyl radicals. Therefore, it could be hypothesized that the sugar molecule in the 7th position is able to interact with the methoxyl group in the 4th position and reduce the antioxidant power (*51*)

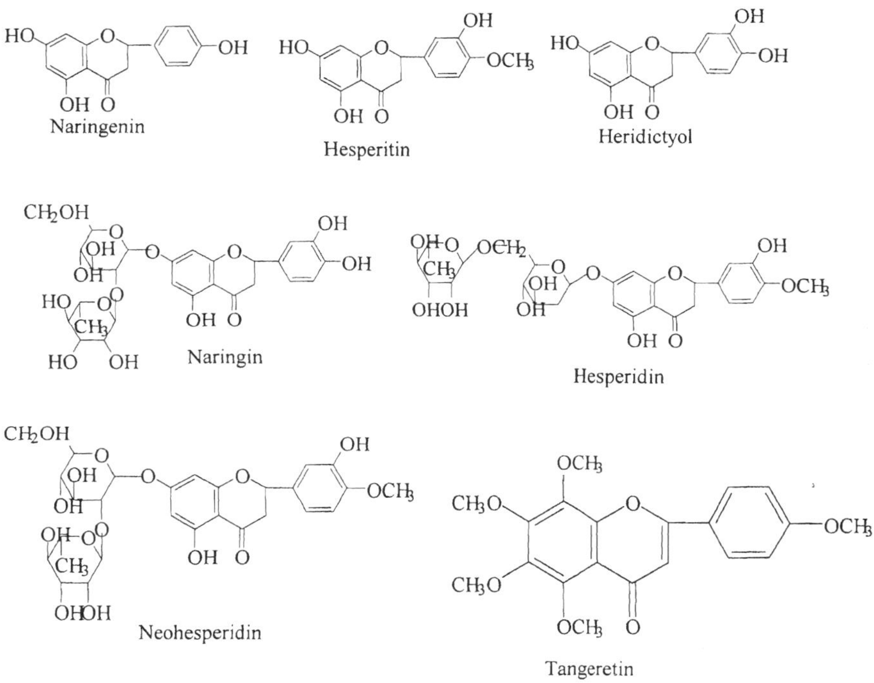

Figure. 3. Structures of citrus flavanoids

Flavanone glucosides such as neohesperidin showed an antioxidant power comparable to free flavanones. Hesperitin has shown to have an antioxidant activity higher than the neohesperidin. O-Glycosylation at hydroxyl position influences radical-scavenging activities and this could be caused by the steric effect which perturb the planarity and ability to delocalize electrons (*51*). The antioxidant activity of hesperidine with hesperitin is negligible. Therefore, it could be hypothesized that the kind of sugar in the 7th position (neohesperidose or rutinose) and the position of methoxyl group (3' or 4' position) influence the ability to delocalize electrons. Antioxidants represent a group of substances very important in diet but we cannot ignore that these compounds, when in high concentrations or in particular environmental conditions, can act like pro-oxidants inducing radicals reaction.

In conclusion, further studies are needed on the isolation and characterization of individual compounds from hexane, EtOAc, acetone, MeOH and MeOH:water extracts to elucidate their different antioxidant mechanisms and the existence of possible synergism, if any among the compounds.

Acknowledgements

This project is based upon work supported by the USDA CSREES IFAFS #2001-52102-02294 and USDA-CSREES # 2006-34402-17121 "Designing Foods for Health" through the Vegetable and Fruit Improvement Center.

References

1. Frankel, E.L.; Meyer, A.S. *J. Sci. Food Agric.* **2000**, *80*, 1925-1941.
2. Jayaprakasha, G.K.; Singh, R. P.; Sakariah, K. K. *Food Chem.,* **2001**, *73*, 285-290.
3. Vinson, J. A.; Dabbagh, Y. A.; Serry, M. M.; Jang, J. *J. Agric. Food Chem.* **1995**, *43*, 2800-2802.
4. Gordon. M.H. The mechanism of antioxidant action in vitro, In-Food Antioxidants, Hudson, B. J. F., **1990**, Elsevier, London, 1-18.
5. Steinmetz, K. A.; Potter, J. D. *Cancer Causes Control* **1996**, *2*, 325-351.
6. Moyer, R. A.; Hummer, K. E.; Finn, C. E. Frei, B.; Wrolstad, R. E. *J. Agric. Food Chem.* **2002**, *50*, 519-525
7. Cao, G.; Alessio, H. M.; Cutler, R. G. *Free Radic. Biol. Med. 14,* **1993**, 303-311.
8. Jayaprakasha, G.K.; and Jaganmohan Rao, L.; *Zeitschrift für Naturforschung,* **2000**, *55c*, 1018-1022.
9. Singh, R.P., Chidamabara Murthy, K.N., Jayaprakasha, G. K. *J. Agric. Food Chem.* **2002**, 50, 81-86.
10. Chidamabaramurthy, K. N.; Singh, R.P.; Jayaprakasha, G. K. *J. Agric. Food Chem.* **2002**, *50*, 5909-5914.
11. Jayaprakasha, G. K.; Jena, B.S.; Negi, P.S.; Sakariah, K.K. *Zeitschrift für Naturforschung,* **2002**, *57c*, 828-835.
12. Jayaprakasha, G.K.; Jaganmohan Rao, L.; Sakariah, K. K. *Bioorganic and Medicinal Chemistry,* **2004**, *12*, 5141-5146.
13. Abdille, Md. H.; Singh, R.P.; Jayaprakasha, G.K.; Jena, B.S. *Food chemistry,* **2005**, *90*, 981-896.
14. Jayaprakasha, G. K.; Mayumi Ohnishi-Kameyama.; Hiroshi Ono.; Mitsuru Yoshida.; Jaganmohan Rao, L. *J. Agric. Food Chem.* **2006**, *54*, 1672-1679.
15. Jayaprakasha, G.K.; Negi, P.S.; Jena, B.S. *Innovative Food Science and Emerging Technology,* **2006**, *7*, 246-250.
16. Wiseman, S.; Waterhouse, A. L.; Korver, O.; *Critic. Rev. Food Sci. Nutr.* **2001**, *41*, 387-412.

17. Amiot, M. J.; Fleuriet, A.; Macheix, J. J. *J. Agric. Food Chem* **1986**, *34*, 823-826.
18. Antolovich, M.; Prenzler, P.; Robards, K.; Ryan, D. *Analyst,* **2000**, *125*, 989-1009.
19. Vinson, J. A.; Hao, Y.; Zubic, S. K. *J. Agric. Food Chem* **1998**, *46*, 3630-3634.
20. Anagnostopoulou, M.A.; Kefalas, P.; Papageorgiou, V. P.; Assimopoulou, A.N.; Boskou, D. *Food chemistry,* **2006**, *94*, 19-25.
21. Jayaprakasha, G.K.; Patil, B. S. *Food chemistry,* **2007**, *101*, 410-418.
22. Jayaprakasha, G.K.; Girennavar, B.; Patil, B.S. *Food Research & Technology - LWT,* **2008**, doi:10.1016/j.lwt.2007.03.017.
23. Jayaprakasha, G.K.; Girennavar, B.; Patil, B. S. *Bio-Resource Technology,* **2008**, (In-press).
24. Kaur, H.; Perkins, J. The free radical chemistry of food additives. In O. I. Aruoma & B. Halliwell (Eds.), Free radicals and food additives (pp. 17–35). **1991**, London: Taylor and Francis.
25. Shimada, K. K.; Fujikawa, K. Y.; Nakamura, T. *J. Agric. Food Chem* **1992**, *40*, 945-950.
26. Sherwin, E. R. *Journal of American Oil Chemists Society*, 1978, *55*, 809-814.
27. Hsiu-Ling, T.; Sam, C. K. C.; Sue-Joan. C. *J. Agric. Food Chem* **2007**, 55, 2867-2872.
28. Yu, J. L.; Wang, R. W.; Walzem, R. W.; Miller, E. G.; Pike, L. M.; Patil, B. S. *J. Agric. Food. Chem.* **2005**, *53*, 2009-2014.
29. Poulose, S. M.; Harris, E. D.; Patil, B. S. *Journal of Nutrition,* **2005**, *135*, 870-877.
30. Rehman, Z. ur. *Food chemistry,* **2006**, *99*, 450-454.
31. Alexandra, B.; Marie Elisabeth, C.; Hubert, R.; Clendette, B. *J. Agric. and Food Chemistry,* **1998**, *46*, 2123-2129
32. Kaehkoenen, M. P.; Hopia, A. I.; Vuolala, H. J.; Rauh, J. R.; Pihlaja, K.; Kujala, T. S. *J. Agric. Food Chem,* **1999**, *47*, 3952-3954.
33. Manthey, J. A. *J. Agricultural and Food Chemistry,* **2004**, 52, 7586-7592.
34. Yu, L.; Haley, S.; Perret, J.; Harris, M.; Wilson, J.; Qian, M. *J. Agric. Food Chem,* **2002**, *50*, 1619-1624.
35. Gorinstein, S.; Milena, C.; Machackova, I.; Haruenkit, R.; Park, Yong-Seo.; Jung, Soon-Teck, *Food Chemistry,* **2004**, *84*, 503-510.
36. Ou, B.; Huang, D. ; Hampsch-Woodill, M.; Flanagan, J.; Deemer, E. *Journal of Agricultural Food Chemistry,* **2002**, *50*, 3122-3128.
37. Gil-Izquierdo, A.; Gil, MI.; Ferreres, F.; Tom'as-Barber'an, F. A. *Journal of Agricultural and Food Chemistry,* **2001**, *49*, 1035-1041.
38. Xu, G.; Donghong Liu.; Jianchu Chen.; Xingqian Ye.; Yaqin Ma.; John Shi. *Food Chemistry* **2008**, *106*, 545-550.

39. Halliwell, B.; Gutteridge, J.M.C. *Free Radicals in Biology and Medicine,* 3rd ed., **2002**, New York.
40. Stadler, R. H.; Turesky, R. J.; Mu¨ ller, O.; Markovic, J.; Leong-Morgenthaler, P. *Mutation Research,* **1994**, *308*, 177-190.
41. Khan, N. S.; Ahmad, A.; Hadi, S. M. *Chemico–Biological Interactions,* **2000**, *125*, 177-189.
42. Xu, G.; Liu, D.; Chen, J.; Ye, X.; Ma, Y.; Shi, J.; *Food Chemistry* **2008**, *106*, 545-551.
43. Huang, R.-H.; Liu, J.H.; Lu, J.M.; Xia, R.X. *Postharvest Biol. Technol.* **2007**, doi:10.1016/j.postharvbio.2007.06.018.
44. Huang, R.; Xia, R.; Hu, L.; Lu, Y.; Wang, M. *Scientia Horticulturae* **2007**, *113*, 166-172
45. Tsai, H. –L.; Chang, S. K. C.; Chang, S. –J. *J. Agric. Food Chem.* **2007**, *55*, 2867-2872.
46. Yi, Z.; Yu, Y.; Liang, Y.Z.; Zeng, B. *Food Science and Technology*, **2007**, doi:10.1016/j.lwt.2007.04.008.
47. Kim, J.-W.; Kim, B. C.; Lee, J. H.; Lee, K. Nam, S C Lee, *Radiat. Phys. Chem.* **2007**, doi:10.1016/j.radphyschem.2007.02.
48. Abeysinghe, D. C.; Li, X.; Sun, C.D.; Zhang, W.S.; Zhou, C.H. Chen, K. S. *Food Chemistry* **2007**, *104*, 1338-1344
49. Arena, E.; Fallico, B.; Maccarone, E. *Food Chemistry,* **2001**, *74*, 423-427.
50. Dugas, A. J.; Castaneda-Acosta, J.; Bonin, G. C.; Price, K. L.; Fischer, N. H.; Winston, G. W. *Journal of Natural Products,* **2001**, *63*, 327-331.
51. Van Acker, S. A. B. E.; Groot, M. J. de,; Berg, D. J. V.; Tromp, M. N. J. L. Kelder, G. D.; van der Vijgh, W. J. F.; Bast, A. *Chemical Research in Toxicology,* **1996**, *9*, 1305-1312.

Chapter 25

Antioxidant and Biocide Activities of Selected Mexican and Chilean Plants

Carlos L. Céspedes[1], Julio Alarcón[1], J. Guillermo Ávila[2], and Isao Kubo[3]

[1]Department of Basical Sciences, Faculty of Sciences, University of Bío-Bío, Andrés Bello Avenue, P.O. Box 447, Chillán, Chile
[2]Phytochemistry Laboratory, UIPRO, FES-Iztacala, UNAM, México
[3]ESPM Department, Division of Insect Biology, College of Natural Resources, University of California, Berkeley, CA 94720

Our phytochemical studies are biodirected in order to find botanical biocides. Some anthocyanins, diterpenes, triterpenes, sesquiterpene lactones, coumarins and flavonoids from Agavaceae, Asteraceae, Celastraceae, Elaeocarpaceae and Scrophulariaceae families and some of their chemical derivatives were isolated. The natural and some derivatives compounds possess antioxidant, antifungal, antibacterial and antifeedant activities. Very little is known about these plants and compounds on bacteria, fungal, and insect pests. In addition, these compounds also scavenge DPPH radical and crocin in TLC autographic and spectrophotometric assay, together with other *in-vitro* antioxidant bioassays. Our results indicate that these compounds have good antioxidant activity, and upset metabolism of the insect, bacteria and fungus. The natural compounds that isolated here represent a valuable resource because of their nutraceutical and biocidal activities. Progress in the biochemical and phytochemical characterization of this pathway are presented.

Antioxidants are substances that delay the oxidation process, inhibiting the polymerization chain initiated by free radicals and other subsequent oxidizing reactions (*1*). This concept is fundamental to food chemistry, where synthetic antioxidants like butylated hydroxytoluene (BHT) have long been used to preserve quality of food by protecting against oxidation-related deterioration. A growing body of literature points to the importance of natural antioxidants from many plants, that may be used to reduce oxidative damage, not only in foods, but also in the human body. This may provide protection against chronic diseases, including cancer and neurodegenerative diseases, inflammation and cardio-vascular disease. Adverse conditions within the environment, such as smog and UV-radiation, in addition to diets rich in saturated fatty acids, increase oxidative damage in the body. Given this constant exposure to oxidants, antioxidants may be necessary to counteract chronic oxidative effects, thereby improving the quality of life (*2*).

The increasing interest in the measurement of the antioxidant activity of different plant samples is derived from the overwhelming evidence of the importance of reactive oxygen species (ROS), including superoxide ($O_2^{\bullet-}$), peroxyl ($ROO^\bullet$), alkoxyl ($RO^\bullet$), hydroxyl ($HO^\bullet$), and nitric oxide ($NO^\bullet$) radicals in aging and chronic disease (*3*). Several methods have been developed to measure the antioxidant (AOX) activity in biological samples, including oxygen radical absorption capacity (ORAC), ferric reducing antioxidant power (FRAP), DPPH, crocin radical scavenging and inhibition of formation of thiobarbituric acid reactive species (TBARS) (*4*), as a more complete tool for AOX measurements (*5*).

The use of traditional medicine is widespread and plants still present a large source of novel active biological compounds with different activities, including anti-inflammatory, anti-cancer, anti-viral, anti-feedant and anti-bacterial activities. Together with enzyme inhibitions (tyrosinase, acetylcholinesterase, melanin oxidase, among others), the antioxidants may play a role in these health promoting activities as nutraceuticals (*6*). In the continuation of our general screening program of Mexican and Chilean flora with biological activities (*7, 8*), an examination of the extracts and compounds from *Araucaria araucana* (Araucariaceae), *Yucca periculosa* (Agavaceae), *Baccharis magellanica, B. salicifolius, B. linearis, Gutierrezia microcephala, G. gayana, Parastrephia quadrangularis, Podanthus ovatifolius, P. mitiqui, Roldana barba-johannis, Tagetes lucida* (Asteraceae), and two exotic species *Rhus javanica* (Anacardiaceae) and *Pimpinella anisum* (Umbelliferaceae) has been initiated.

There is a widespread effort to find new nutraceuticals and agrochemicals, and currently it has been focused on natural compounds such as flavonoids, coumarins, terpenoids and phenolics from diverse botanical families from Mexico and America (*7*).

As was mentioned, tyrosinase also known as polyphenol oxidase (PPO) enzyme (*9*), is a copper containing enzyme widely distributed in micro-organisms, animals, and plants. It catalyzes two distinct reactions of melanin synthesis (*10*), the hydroxylation of a monophenol (monophenolase activity) and the conversion of an *o*-diphenol to the corresponding *o*-quinone (diphenolase activity). Tyrosinase is responsible for browning in plants and considered to be deleterious to the color quality of plant derived foods and beverages. In addition, looking for sclerotization and moulting regulation processes in insects, tyrosinase is one of the key enzymes in that insect metamorphoses process (*11*). Acetylcholinesterase (AChE), which is the enzyme contained in nerve tissues, plays an exceeding important role in the transmission of a nerve impulse. The free acetylcholine in the inactive form bound to proteins accumulates in the ending of a nerve in vesicles. The consumed acetylcholine is constantly replenished by its synthesis (by the acetylation of choline). All these processes are occurring when an impulse is transmitted through a cholinergic synapse. Thus, the process of synaptic transmission is an involved biochemical cycle of acetylcholine exchange. AChE has a key role in this cycle because the inhibition of activity leads to the accumulation of free acetylcholine in the synaptic cleft, producing the disruption of nerve impulses, then convulsive activity of the muscles can be transformed into paralysis, and other features of self-poisoning by surplus acetylcholine appear. It occurs when AChE is inhibited by some terpenoids (*12,13*). On other hand, modifications to AChE can confer resistance to insecticides (*14*).

In addition to many chemicals (flavonoids, stilbenoids, phenylpropanoids, and other phenolics) that possess tyrosinase inhibitory activity, these compounds also show a strong antioxidant (AOX) activity against diverse reagents as model of AOX measurement (i.e. DPPH, ABTS, Trolox, TRAP, ORAC, FRAP, etc.). Mainly the activity is due to the presence of diverse moieties of the chemical structure of the molecules, for instance, orcinol or catechol groups, or hydroxyl bonded to an aromatic ring system (gallic acid and gallates in general, resveratrol, quercetin, etc), in these cases, it is possible to correlate the AOX activity with tyrosinase and AChE inhibition, and IGR activity (*15- 19*).

In addition, many of these compounds are polyphenolic secondary compounds that are ubiquitous in angiosperms and that have antifeedant effects on phytophagous insects (*20-26*). It has been assumed that phenols bind to proteins, acting as nutritional protein precipitating agents, thus reducing their digestibility (*21, 27*).

We have previously reported that diverse secondary metabolites have different sites of action and different molecular targets when they interact with enzymes and metamorphosis processes (*7, 8, 17 -19, 28-40*).

The data presented also indicate that it is possible to correlate some antioxidant activities (i.e. crocin, DPPH) with growth and development of fungus and insects; these data are important for allelopathic and chemical ecology studies (*35*). In addition to gedunin and toosendanin, compounds **1** - **11** (Figure 1) were used in tests concerned with the reduction of DPPH and other antioxidant measurements of routine.

In this chapter, a review of extracts and bioactive compounds isolated from selected endemic botanical species, their AOX and biological activities such as some enzymes inhibition and their antifeedant/IGR activities are reported. The results are classified by botanical families.

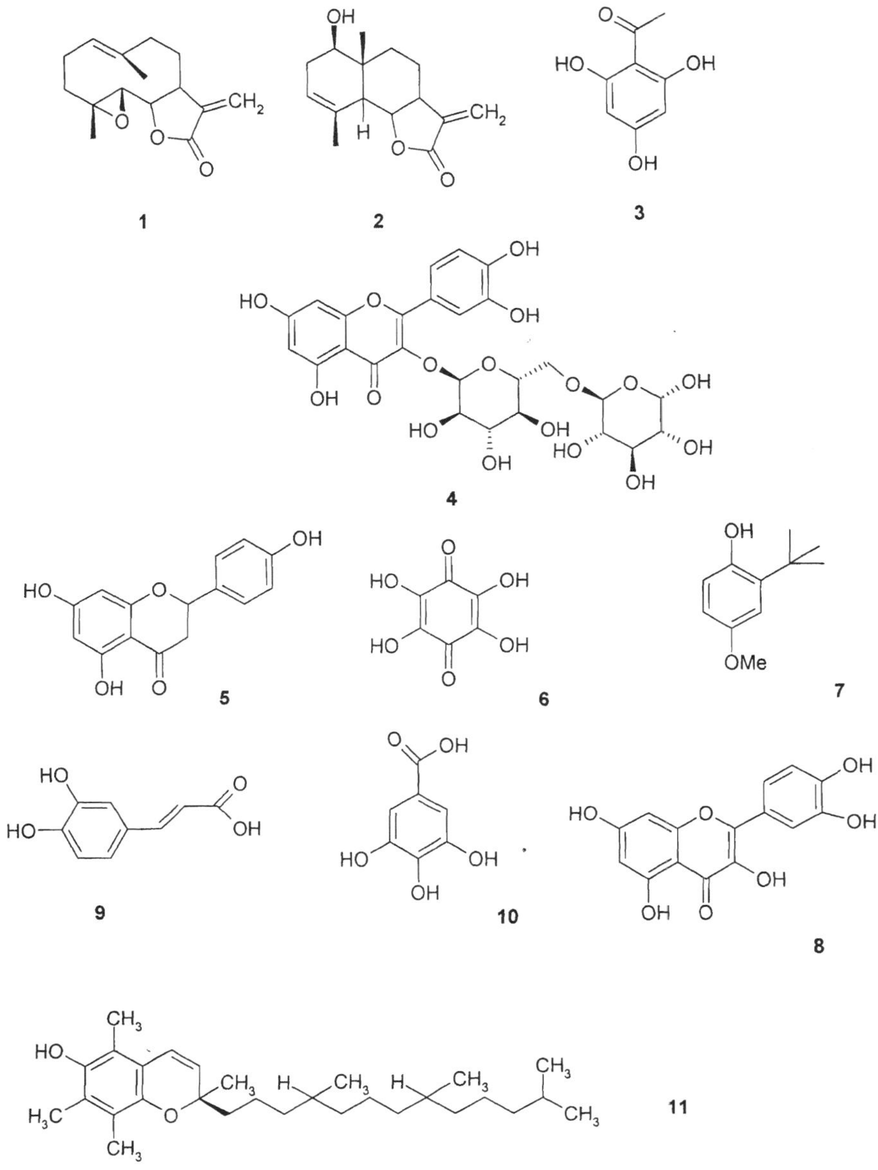

*Figure 1. Compounds used for antioxidant and bioassays. Parthenolide **1**, santamarin **2**, 2'4'6'-trihidroxy acetophenone **3**, rutin **4**, (±)-naringenin **5**, tetrahydroxyquinone (THQ) **6**, orto-butylhydroxy anisol (BHA) **7**, quercetin **8**, caffeic acid **9**, gallic acid **10** and tocopherol **11**.*

AGAVACEAE

Yucca periculosa F. Baker, known commonly as ''palmitos'' or ''izote'' is a tree, endemic to Mexico, that grows in the semi-arid regions of Tehuacan–Cuicatlan Valley, Puebla–Oaxaca States. In the wild, these plants survive under different environmental stress conditions (*42*), with a lifespan of up to 100 years or so. Our field observations indicate that this specie has a strong resistance to insect attack (*19*).

From methanol extract of bark were isolated 4,4′-dihydroxstilbene **12**, resveratrol **13** and 3,3',5,5'-tetrahydroxy-4-methoxystilbene **14** (Figure 2). In order to establish correlation between insect growth regulatory (IGR) and acute toxicity activities with the antioxidant properties of these phenolic compounds, crocin and DPPH radical scavenging tests of these stilbenes were carried out. In **14** the presence of a methoxyl group increases the strength of these compounds for scavenging of DPPH. The inhibition of DPPH activity by stilbenes is well known (*43,44*). Therefore, plant stilbenes may be considered as efficient IGR and radical scavengers (*44-46*). The antioxidant activity of these compounds was also evaluated spectrophotometrically on the bleaching of the H_2O-soluble crocin (*47*). Compounds **12**, **13**, **14**, and MeOH extract (Me-Yuc) were all active, with activities comparable to gallic acid (positive control) (*19*).

CELASTRACEAE

Celastraceae family embrace shrubs and trees of moderate height occurring in southern Chile. Plants from this family belonging to *Maytenus* genus have antifeedant, anti-cancer and anti-inflammatory effects. These species shows high allelopathic and antifeedant effects on plant and insect species into its habitat, respectively (*34, 35, 48*). *Maytenus* genus is characterized by the occurrence of different bioactive compounds, especially β-dihydroagarofurans (*49 - 51*). Until now, we focused mainly on two Southamerican species of this genus, *Maytenus disticha (Hook) Urban* and *M. boaria.* The former is commonly known as "maitencito" or "romasillo", is a small tree which grows in rainfall forests in the South Pacific slope ranging from Araucanian Region to "Tierra del Fuego" in the Patagonian Region in Chile. *M. boaria*, the sole tree among the Celastraceae family in Chile, usually grows in the arid climate in the Slope Mountains (*34, 35*).

Considering that in Celastraceae family the main biologically active compounds are β-agarofuran type sesquiterpenoids, we selected these plants due to their high resistance to insect attack and to an ample body of etno-medicinal data, until now no biological work has been carried out on this plant. Moreover, we investigated the AOX activity against DPPH and crocin of extracts and compounds, its IGR and inhibition of AchE, a key enzyme in the insect nervous system in which the colynergic system is essential.

From the aerial parts of *M. disticha* six agarofurans (*52 - 54*) and from the seeds of *M. boaria,* four β-agarofuran polyesters were isolated and chemically characterized (*32, 52*).

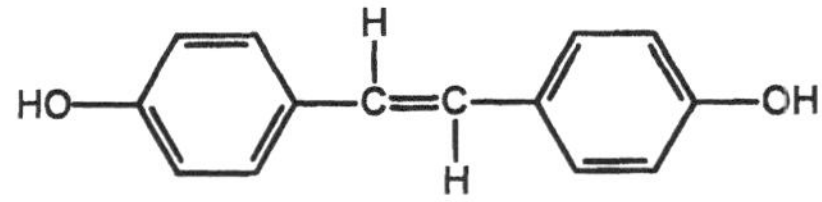

12

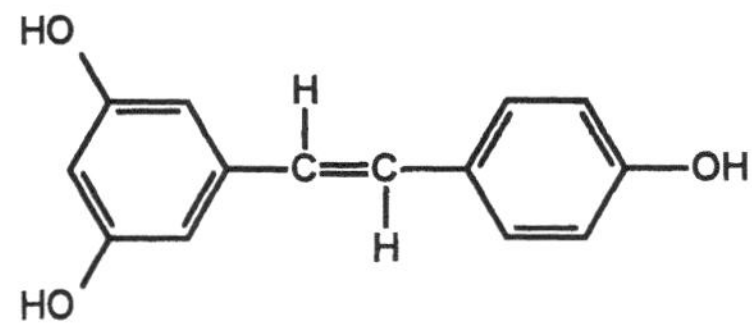

13

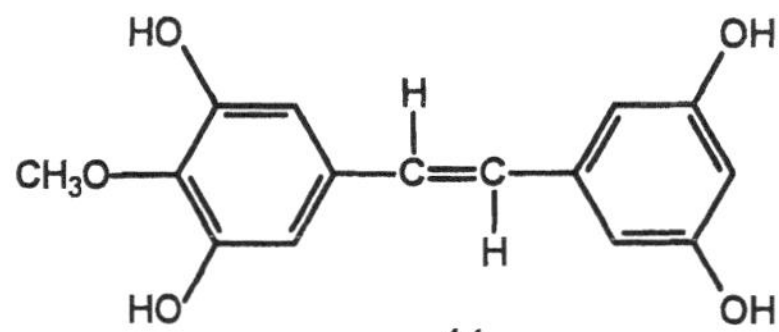

14

Figure 2 . Stilbenes from Y. periculosa.

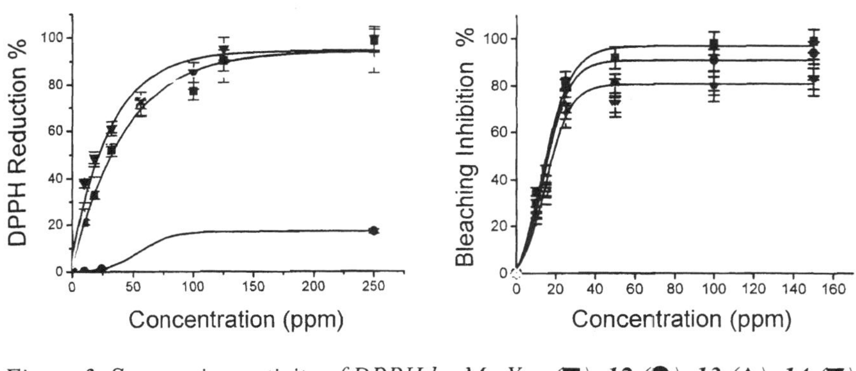

*Figure 3. Scavenging activity of DPPH by Me-Yuc (■), **12** (●), **13** (▲), **14** (▼) on radical reduction of DPPH. Measurements at 517 nm, determination after 30 min and Inhibitory activity of caffeic acid **9** (■), Me-Yuc extract (●), **14** (▲), gallic acid **10** (▼), **13** (◆) on the bleaching of crocin measurement at 440 nm of fluorimetric emission, determination after 20 min. Values of compounds and extract in ppm.*

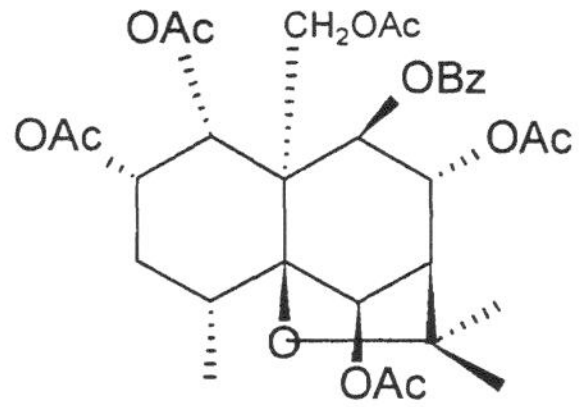

15

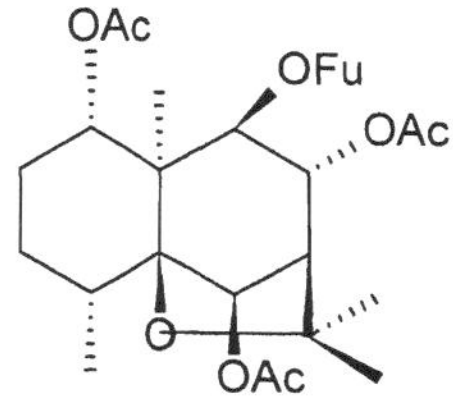

16

Figure 4. β-Agarofuran type sesquiterpenoids from Maytenus spp.

Insecticidal bioassays showed that a n-hexane/ethyl acetate (1:1, v/v) (HE) extract of aerial part of *M. disticha* produced the highest mortality (79.1%) at 12.0 ppm, similar to **15**, **16** and MeOH extract of *M. boaria* (>65%) and at 15.0 ppm **16** and HE extract showed the highest larval mortality (100%). Compound **16** was more active than the positive control (toosendanin) and compound **15**, decreasing larval growth, total length of the larvae, percentage of larvae that reached pupation and emergence. It is noteworthy that *β*-agarofuran **15**, although was not so effective as **16**, showed an appreciable reduction on larval length growth compared with the control, thus the activity of these compounds is comparable to the commercial insecticide toosendanin (*35, 55*).

In order to determine the site of action of compounds and extracts from the plants, we carried out an acetylcholinesterase (AChE) inhibitory activity assay. MeOH and n-hexane extracts, and **15** and **16** showed inhibitory AChE activity in a dose-dependant manner confirming that these compounds are active inhibitors of AChE in *Maytenus spp* (*35*).

The AOX activity of these extracts and compounds against crocin and DPPH radical scavenging test were carried out. Acetone, HE and MeOH extracts all had activities comparable to gallic acid (Figure 5) (*47*).

ASTERACEAE

Antioxidant and biocidal activities of species belonging to the Asteraceae family from Mexico and Chile, due to their excellent ethnomedicinal data, strong resistance against oxidative stress and pest attack observed in nature were of interest. Seven species of this family: *Gutierrezia microcephala*, *G. gayana*, *Podanthus ovatifolius*, *P. mitiqui*, *Parastrephia quadrangularis*, *Roldana barba-johannis* and *Tagetes lucida* were evaluated for their antioxidant, antibacterial, antifungal and IGR activities.

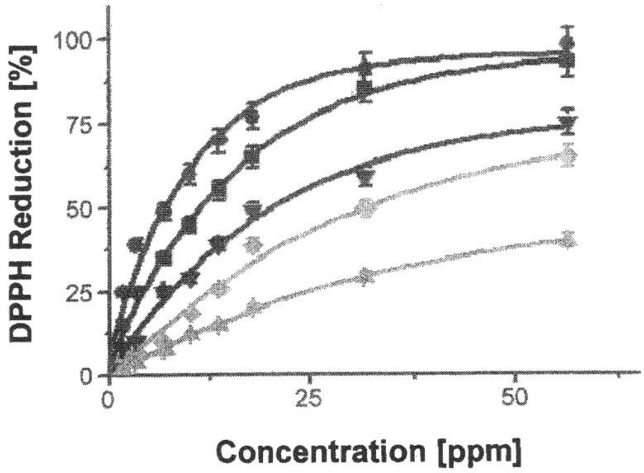

Figure 5. Antioxidant Activity (scavenging of DPPH radicals) of Extracts and Compounds from M. boaria. For explanation of the extracts and partitions, see text. Extracts n-hexane (■), acetone (●), methanol (▼), dichloromethane (◆) and agarofurane 16 (▲).

Gutierrezia spp **and** ***Parastrephia quadrangularis***

G. microcephala A. Gray is commonly known as broom-wood, grows in arid regions of the central and North of México and in the southwestern region of the United States, on areas of arbustive vegetation together with creosotebush *Larrea tridentata* (*56*). The plant collected in Saltillo, Coahuila State, is rich in flavonoids (*57*). Four flavones **17-20**, were obtained from the acetonic and methanolic extracts which were isolated previously by Fang et al. (*58, 59*). n-Hexane extract afforded bacchabolivic acid **21** (*60, 61*), a new ent-clerodane reported on this plant, which was lately submitted to esterification to yield the methyl ester **22**. Antioxidants, AChE inhibition and insecticidal effects of **17-22**, hexane and methanol extracts against larvae were evaluated (*62*).

Clerodane **21** showed the highest inhibition (100% of length and weight) at 50.0 ppm and flavonoids had lower larval inhibition at the same concentration. In many of the treatments, mean adult weight was significantly delayed in the average time to reach the adult stage relative to the control larvae. GI and RGI clearly showed a stronger effect by **21**, **22** and hexane extract. It is possible to infer that the substitution of polymethoxy flavones induce an increase on the activity of these flavones and a furan ring seems to be necessary for insecticidal activity of *ent*-clerodane type diterpenes (*62*).

The presence or absence of a methyl ester group increased or decreased, respectively, the strength of these compounds on inhibition of AChE in clerodanes. It is obvious that the nature of the ester substituent at C-8 plays an important role for the insecticidal activity of *ent*-clerodanes. The most active compound **21** contained a small and relative hydrophilic acid group at C-8, whereas **22** with a bulky and more lipophillic ester group exhibited a lower

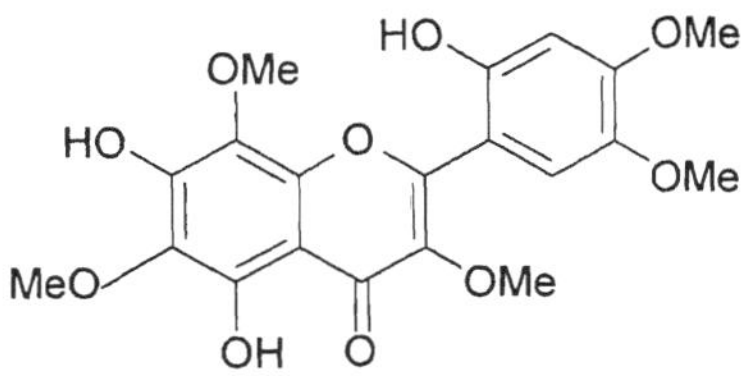

17

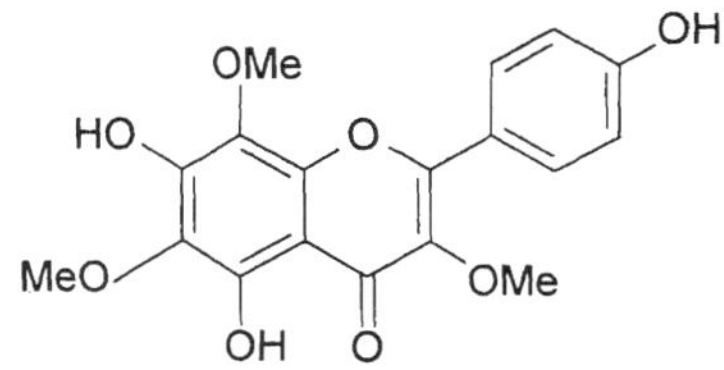

18

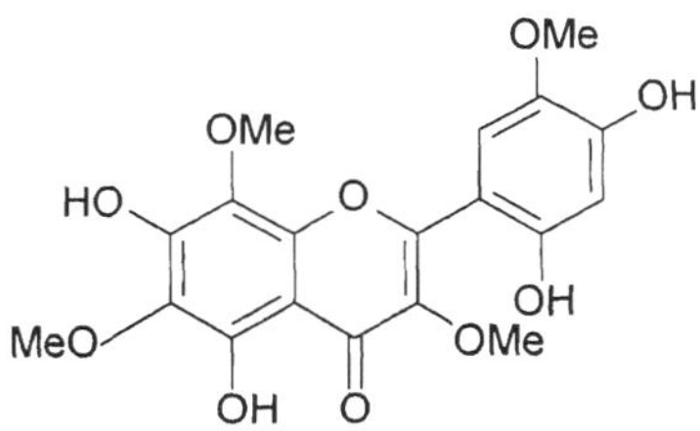

19

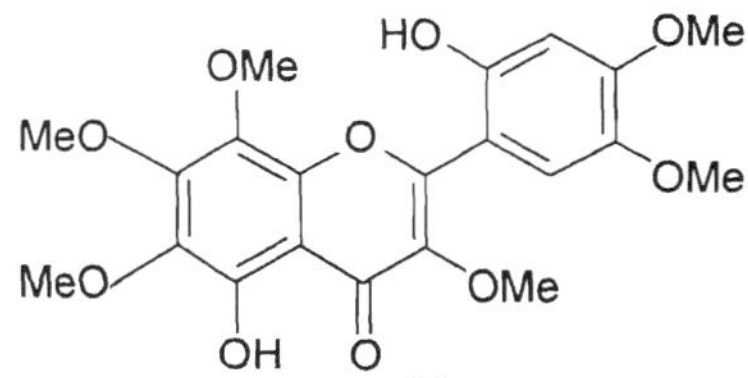

20

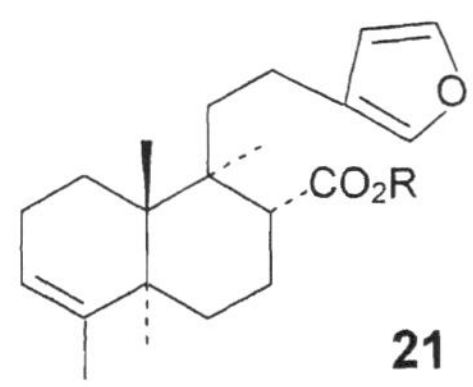

21 R = H
22 R = Me

Figure 6. Flavonoids and Clerodane from G. microcephala

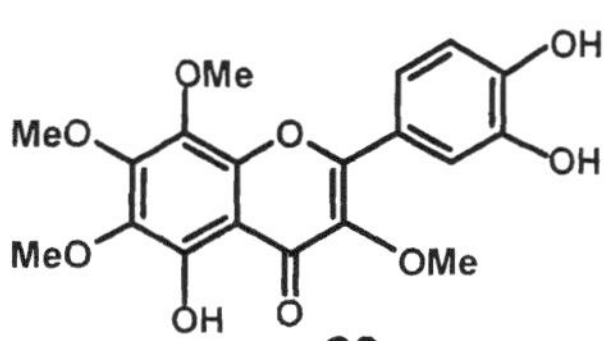

23

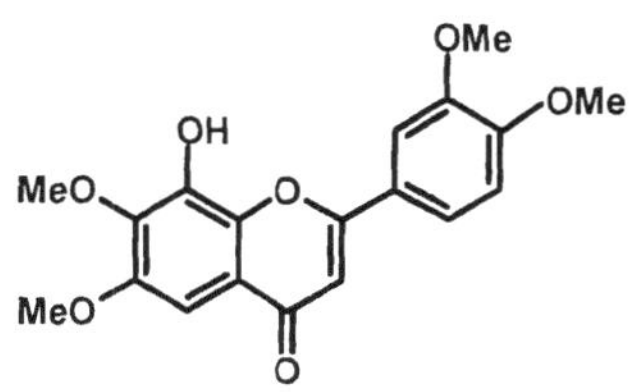

24

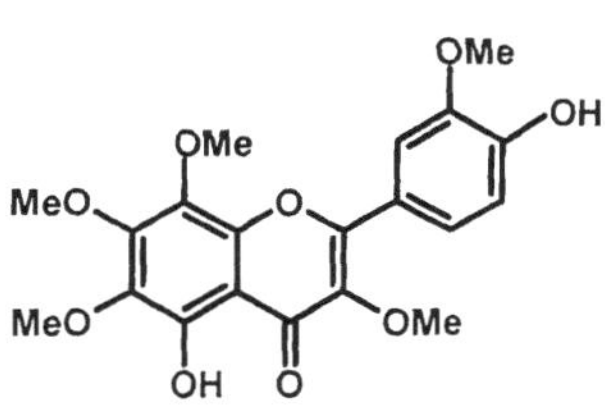

25

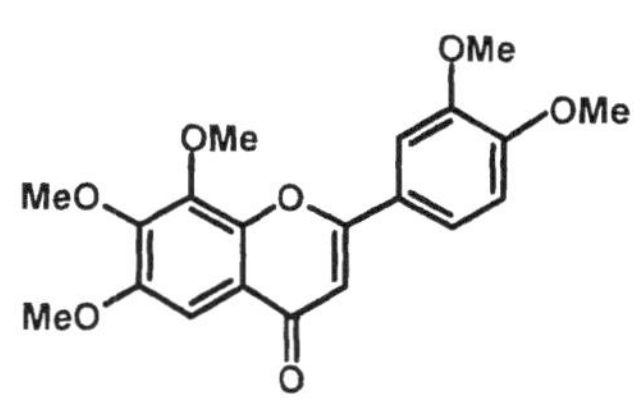

26

Figure 7. Flavonoids from G. gayana and P. quadrangularis

activity level (*63*). With respect to the flavonoids, the presence of an extra methoxyl substituent in the A ring seems to be the cause of growth inhibitory activity (length and weight) shown by **17** and **18** (*64*).

G. gayana and *Parastrephia quadrangularis*.- These species are endemic shrubs of 0.5 to 2.0 m height, growth from Atacama desert to "Aconcagua" Region in Chile, and used by the "Aymaraes" and "Araucanian" people for the treatment of multiple ailments such as emollient, balsamic and digestive injuries. From these plants the antioxidant and anti-inflammatory activities were determined. From aerial parts of *G. gayana* we obtained a MeOH extract **E1**, **E2** (a mixture of n-hexane/ethyl acetate/MeOH; 1/1/1), a fraction **F8**, 5,4',5'-trihydroxy-3,6,7,8-tetramethoxy flavone **23**, MS-983: 8-hydroxy-3',4',6,7-tetramethoxy flavone **24**, 4',5-dihydroxy-3,3',6,7,8-pentamethoxyflavone **25** and MS-982 a mixture of two flavonoids: **24** and 3',4',6,7,8-pentamethoxy flavone **26**.

Radical scavenging properties of the compounds (**23-26**), the extracts **E1**, **E2** and fraction **F8** were evaluated against the DPPH radical, using DPPH as a TLC spray reagent and a spectrophotometric measurement. Compound **25** and mixture MS-982 were less sensitive than **23**, **E2** and **F8**. It is worth mentioning that quercetin (data not show) and α-tocopherol, were used as reference compounds that exhibited strong antioxidant properties (Figure 8).

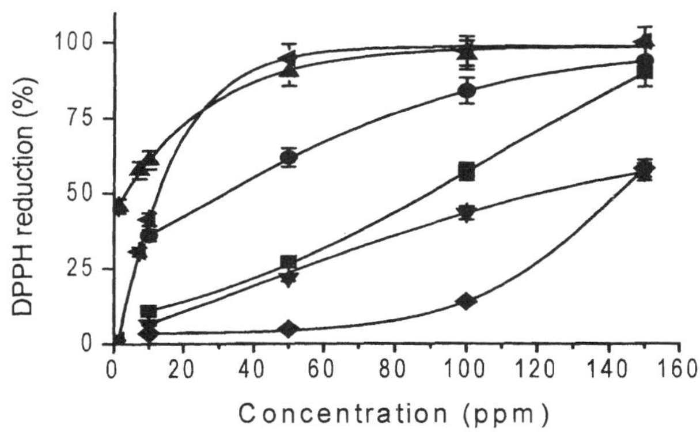

Figure 8. DPPH reduction by flavonoids from G. gayana and P. quadrangularis. E2 (■), F8 (●), MS-982 (24 + 26) (▼), 24 (◆), 23 (▲) and tocopherol (◀).

Podanthus Mitiqui* and *Podanthus ovatifolius

Several sesquiterpenoid lactones have been isolated from Chilean members of the Asteraceae (*65*, *66*). A phytochemical study of *Podanthus ovatifolius* led to the isolation of ovatifolin (*67*, *68*), whose stereochemistry was determined by Gopalakrishna's group (*69*). *P. mitiqui* and *P. ovatifolius* are small shrubs which grow on the rain forest of Pacific slopes, especially on Araucanian region of southern Chile (*70*). We have reported the plant growth inhibitory and antioxidants effects of arturin **27**, arturin acetate, ovatifolin **28**, ovatifolin acetate, deacetylovatifolin **29**, 1,10-epoxyovatifolin **30**, 11,13-dihydroovatifolin **31** and MeOH and CH_2Cl_2 extracts on germination. The seedling growth activity and the behavior of these compounds toward metabolic energetic reactions on dicotyledonous and a monocotyledon standard target species was studied (*71*).

Radical scavenging properties of compounds **27-31** (Figure 9) were evaluated against the DPPH radical, using DPPH as a TLC spray reagent (*47*). Compounds **27**, arturin acetate and ovatifolin acetate were less sensitive than **28**, **29**, **30** and **31**. It is worth mentioning that quercetin, a flavonol with three hydroxyl groups, was used as a reference compound that possesses strong antioxidant properties.

In addition to DPPH the antioxidant activity of these compounds was also evaluated spectrophotometrically on the bleaching of the H_2O-soluble crocin (*47*). Compounds **28**, **29**, **30** and **31** were all active, and showed activities comparable to gallic acid.

Roldana barba-johannis

The methanol extract from the aerial parts of *R. barba-johannis* afforded sargachromenol, sargahydroquinoic acid and sargaquinoic acid. These natural

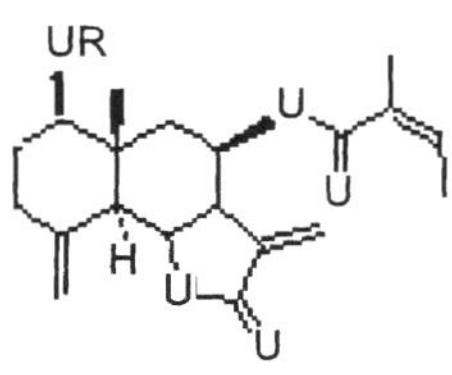

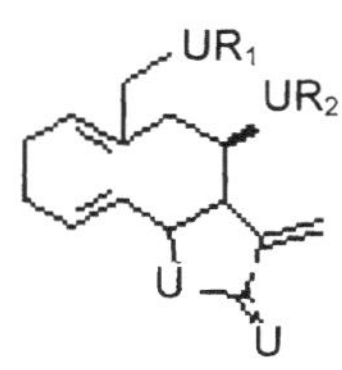

Arturin **27** (R = H) Ovatifolin **28** (R1 = Ac; R2 = H
Deacetylovatifolin **29**
(R1 and R2 = H)

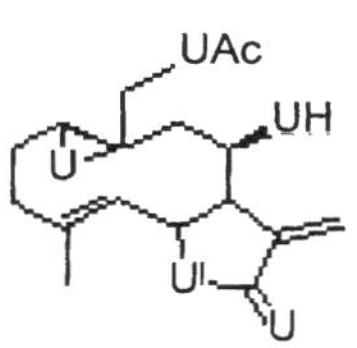

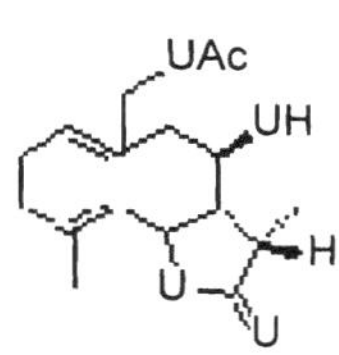

1,10-Epoxideovatifolin **30** 11,13-Dihydroovatifolin **31**

Figure 9. Sesquiterpene Lactones from P. mitiqui and P. ovatifolius

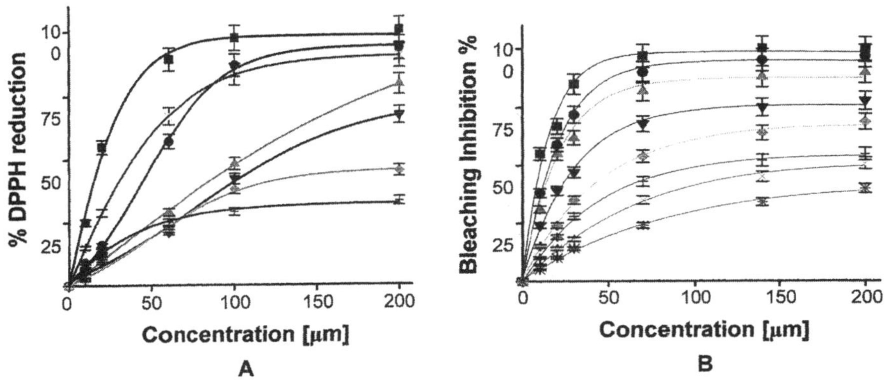

Figure 10. A: Scavenging activity of compounds (3-6). MeOH and CH_2Cl_2 extracts on DPPH radical. Measurements at 517 nm, determination after 30 min. Quercetin (●), CH_2Cl_2 extract (■), deacetylovatifolin (▲), epoxyovatifolin (▼), ovatifolin (◆), ovatifolinacetate (+), MeOH extract (×). Values of MeOH and CH_2Cl_2 extracts in ppm. B: Inhibitory activity of compound s 3-6, MeOH, and CH_2Cl_2 extracts on the bleaching of crocin measurement at 440 nm of fluorimetric emission, determination after 20 min. CH_2Cl_2 extract (■), deacetylovatifolin (●), epoxyovatifolin (▲), gallic acid (▼), ovatifolin (◆), dihydroovatifolin (+), ovatifolinacetate (×) and arturin (✳). Values of CH_2Cl_2 extract in ppm.

products and the corresponding acetylated and methylated derivatives showed insecticidal and insect growth regulatory activities against fall army worm (*Spodoptera frugiperda*). The most active compounds were sargachromenol **32,** and its acetylated derivative **34**; sargahydroquinoic acid **35** and its acetylated derivative **37**; and a mixture of sargachromenol **32**, sargahydroquinoic acid **35**, and sargaquinoic acid **38** (6:3:1) and the acetylated form of this mixture, **33** and **36** showed a minor activity. All these compounds (Figure 11) and mixtures had significant effects between 5.0 and 20.0 ppm in diets. Compounds were insecticidal to larvae, with lethal doses between 20 and 35 ppm. In addition, these substances also demonstrated scavenging properties toward DPPH in TLC autographic and spectrophotometric assays and appear to have selective effects on the pre-emergence metabolism of the insect. The results from these compounds were fully comparable in activity to known natural insect growth inhibitors such as gedunin and methanol extracts of *Cedrela salvadorensis* and *Yucca periculosa* (*38*).

Tagetes lucida

T. lucida known as "pericón", is a medicinal plant used from pre-Columbian times by Aztecs and other Mesoamericans. This specie is distributed from

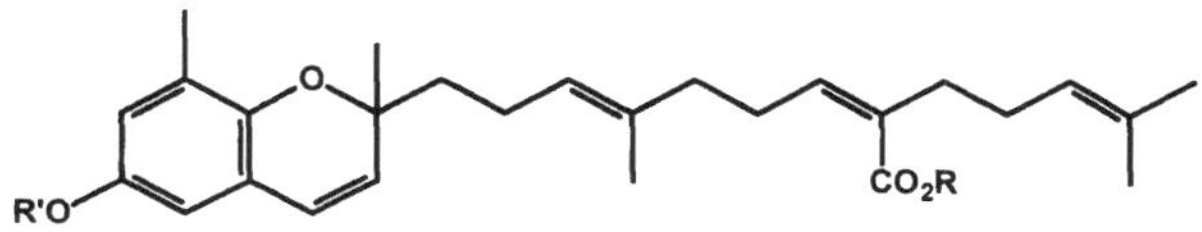

32 R, R'= H
33 R= CH_3 R' = H
34 R= H R' = Ac

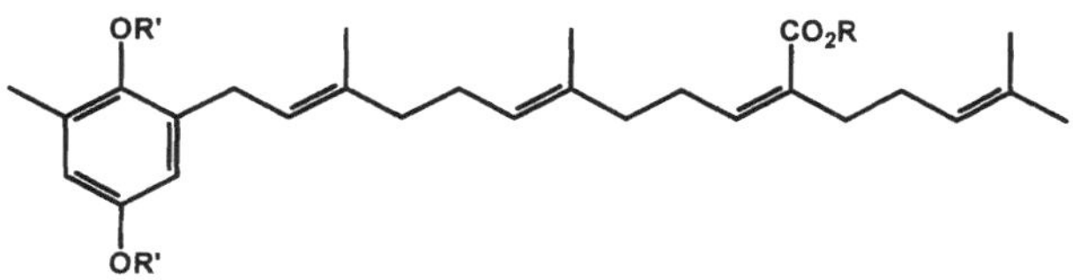

35 R, R'= H
36 R= CH_3 R' = H
37 R= H R' = Ac

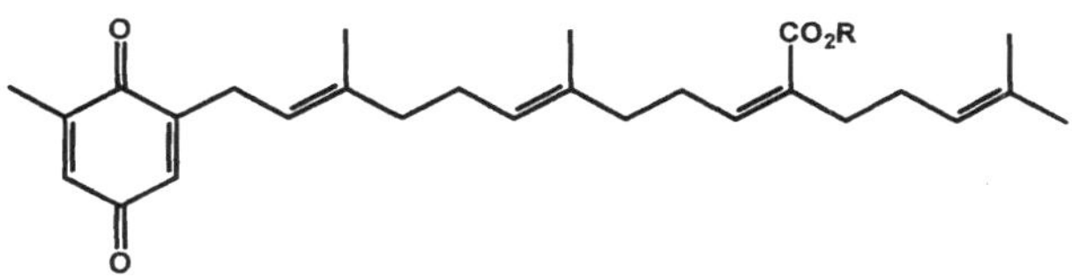

38

M = 32 + 35 + 38
Ma = M methylated

Figure 11. Plastoquinones and Tocotrienols from R. barba-johannis.

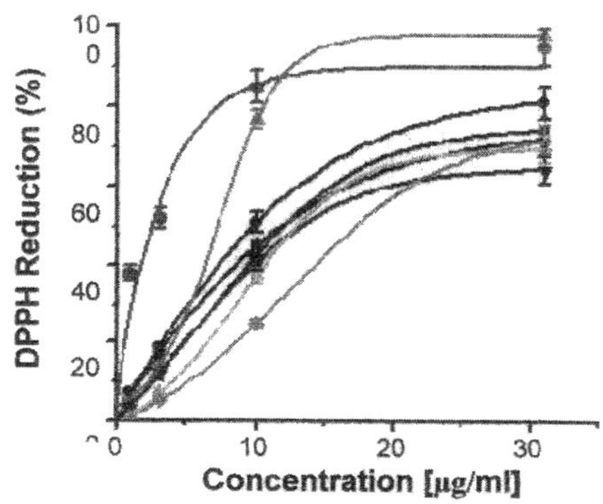

*Figure 12. Radical scavenging activity of Ma (■), M (●), **35** (▲), **34** (▼), **37** (◆), **32** (◀), tocopherol (▶) and caffeic acid (⬢) on radical reduction of DPPH. Measurements at 517 nm, determination after 30 min.*

Northern México to Northern Nicaragua (*72*). Thr leaves and inflorescence are used for stomachache and for the treatment of anti-inflammatory ailments. From a dichloromethane extract, we isolated several coumarins: scopoletin (6-methoxy-7-hydroxycoumarin) **52**, esculetin (6,7-dihydroxycoumarin) **49**, 7-methoxy-6-hydroxycoumarin **50**, scoparone (6,7-dimethoxy-coumarin) **45** and their derivatives 6,7-diacetoxy coumarin **53** and 6-methoxy-7-acetoxycoumarin **54** (Figure 8).

In the literature there is few information about insecticidal effects of coumarins. Nagaiah et al. (*73*) reported about the IGR and antifeedant effects of some coumarins from the bark of *Xeromphis uliginea* on growth of *Spodoptera litura*.

The antibacterial and antifungal activity of extracts and compounds from *T. lucida* were examined. The most active compounds against Gram positive and negative bacteria were the dihydroxylated coumarins **41** and **42**. In addition to **41** and **42**, the compounds **40**, **44**, **45** and **49** showed an interesting activity against *Vibrio cholerae* a key bacterium in the contaminated water, the most active being **40**, **41** and **42**. The most sensitive bacteria against coumarins were the Gram negative ones. The extract MeOH/CH_2Cl_2 (1:4) $\mathbf{M_2}$ at 0.34 μg/disc inhibited the growth of the enterobacteria *Escherichia coli* and *Proteus mirabilis* (40%), *Klebsiella pneumoniae* (31.1%), *Salmonella* sp. (35.5%) and *Shiguella* sp. (0%) at 72 h of culture. The dimethoxy compounds **44** and **45** showed a strong activity against fungal strains, especially *T. mentagrophytes* and *R. solani* (100% of inhibition at 125.0 and 250.0 μg/mL, respectively). These results show that this plant could be very useful as nutraceutical source (*74*).

These compounds are formed by the action of phenylalanine ammonia lyase (PAL) and then to ortho-coumaric acid by the action of cinnamic acid *o*-hydrolase (*75*). Coumarins are located within specialized structures such as secretory ducts called vittae (*76*). Different extracts from aerial parts of *T. lucida* (Asteraceae) have been used in folk medicine to treat many ailments including anti-inflammatory and migraines. We have preliminary results about the antioxidant activities of the MeOH, CH_2Cl_2 and MeOH/CH_2Cl_2 (1:4) extracts and compounds obtained from Mexican Tarragon (*T. lucida*). These samples were assayed against 2,2-diphenyl-1-picryl-hydrozyl (DPPH) radical and thiobarbituric acid reactive species (TBARS) a lipid peroxidation bioassay and were affected by the natural components present. The antioxidant capacities were strongly correlated with total polyphenols content determined by Folin-Ciocalteu method (data not shown). The highest values against the DPPH-radical reduction and TBARS formation were shown by MeOH (**A**), methanol/dichloromethane (1:4) (**B**), and CH_2Cl_2 (**C**), extracts and were comparable to BHT, quercetin and tocopherol; extract B, compounds **41**, **42**, **49**, **46** and **52** were most active against DPPH (Figure 14) and TBARS formation (data not show). The results showed that samples from this plant could also be very useful as antioxidant substances.

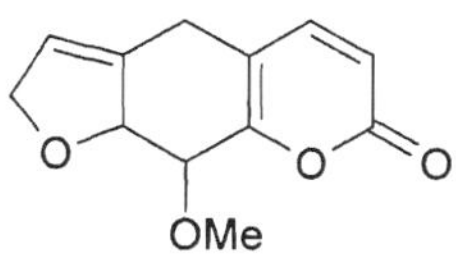

39

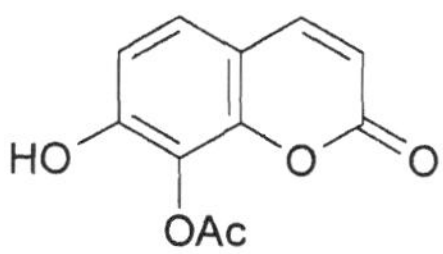

40

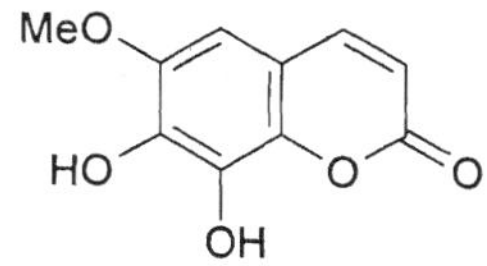

41

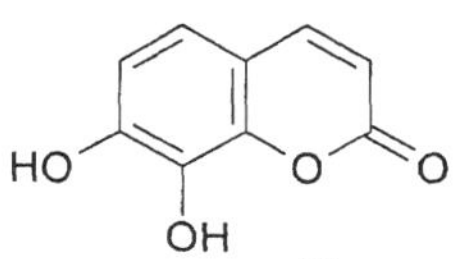

42

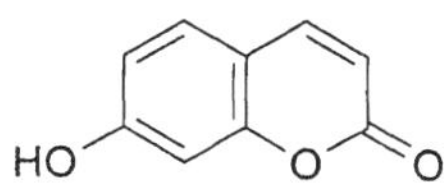

43

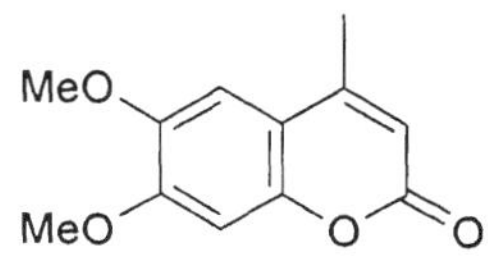

44

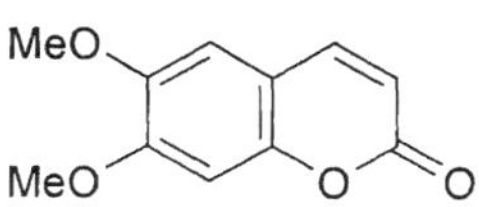

45

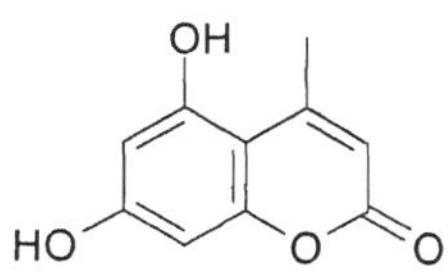

46

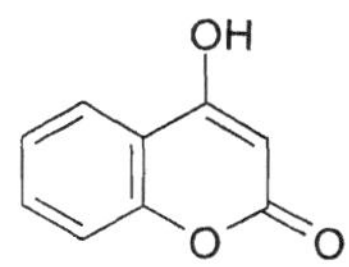

47

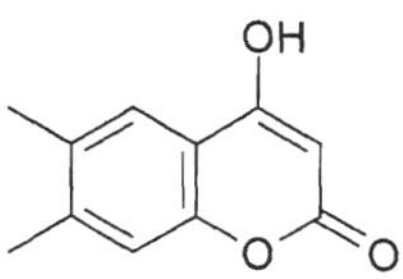

48

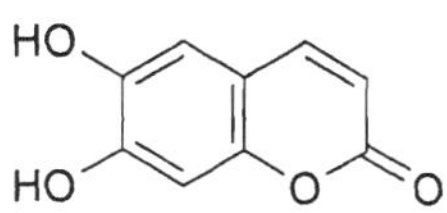

49

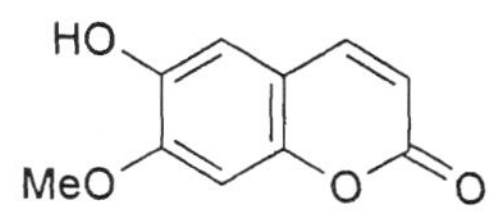

50

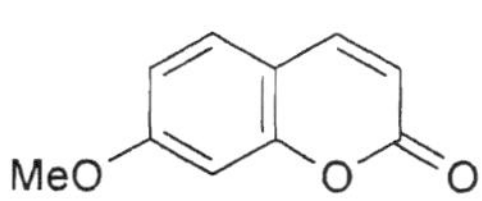

51

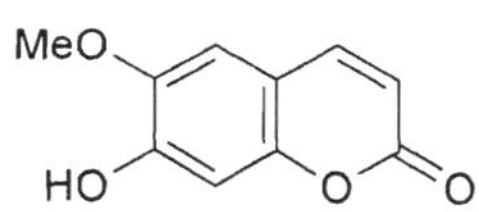

52

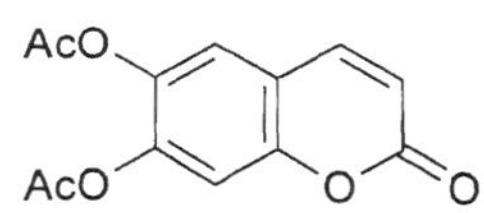

53

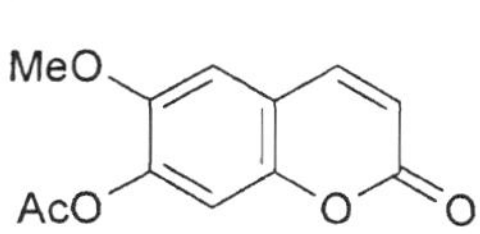

54

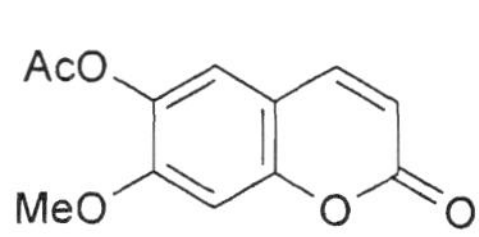

55

Figure 13. Chemical Structures of Coumarins from T. lucida

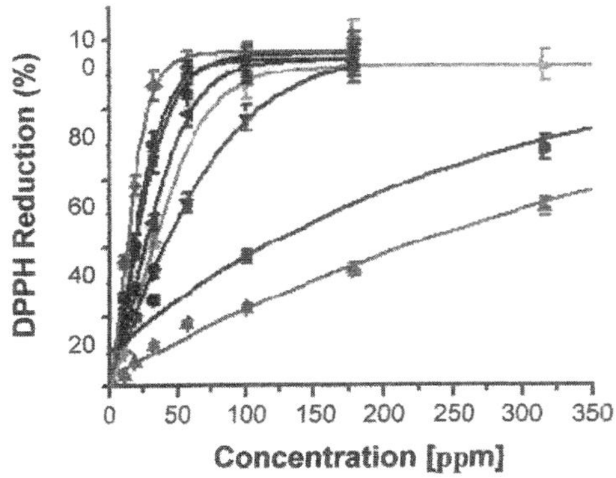

Figure 14. Scavenging activity of extracts 46 (⬢), 52 (▲), 49 (▼), extract B (◆), 41 (◀) and 42 (▶) the compounds tocopherol (■), and quercetin (●), on radical reduction of DPPH. Measurements at 517 nm, determination after 30 min.

Baccharis spp.

From the large genus *Baccharis* (Compositae-Asteraceae), strictly American genus with approximately 350 species. Only a few are important medicinal plants, and their uses seem to be rather diverse, and more than 50 species have been investigated chemically. In continuation of our investigations on South American *Asteraceae* representatives, we have studied *B. linearis*, *B. magellanica* and *B. umbelliformis* endemic species of Austral-Southern Chile and Argentina (*36*).

In the semi-arid zones from Atacama to Patagonia region of Austral Southerm Chile, grows *B. line aris, B. magellanica* and *B. umbelliformis* (Tribe: Eupatorieae, Fam: Asteraceae). *B. Linearis* is a dominant shrub which can be found growing either in communites as with associated flora as in solitary form, throughout central to south-central regions (*77*), *B. magellanica* and *B. umbelliformis* growth in semiarid regions of Austral Southern Chile (*78 – 80*). Some investigations on mechanisms of allelopathic action report that different phenolic (isoprenoid and non-isoprenoids) compounds are seedling growth and energetic inhibitors (*81, 82*) and specifically acetophenones also enhance and promote the growth root of myrtle (*83*).

We aimed to correlate structure activity relationship with the inhibitory behavior on germination-respiration, the main macroscopic parameter of the development and growth of weedy plants. Our data indicate that it is possible to correlate some antioxidant activities (i.e. crocin, DPPH) against germination, and respiration. Such data are important for allelopathic intervention by secondary metabolism studies (*71*). In addition, tetrahydroxy-1,4-quinone, quercetin and BHA, gallic acid were used as internal standard in assays concerned with the reduction of DPPH and antioxidant measurements, respectively.

As far as we know, the effects of **56 - 67** on germination and respiration of weed seeds, antioxidant, radical scavengers, oxygen uptake and photosynthesis have not been investigated. This Is the first report on the allelopathic effects of the main compounds and some of their synthetic derivatives and extracts from South American *Baccharis* species.

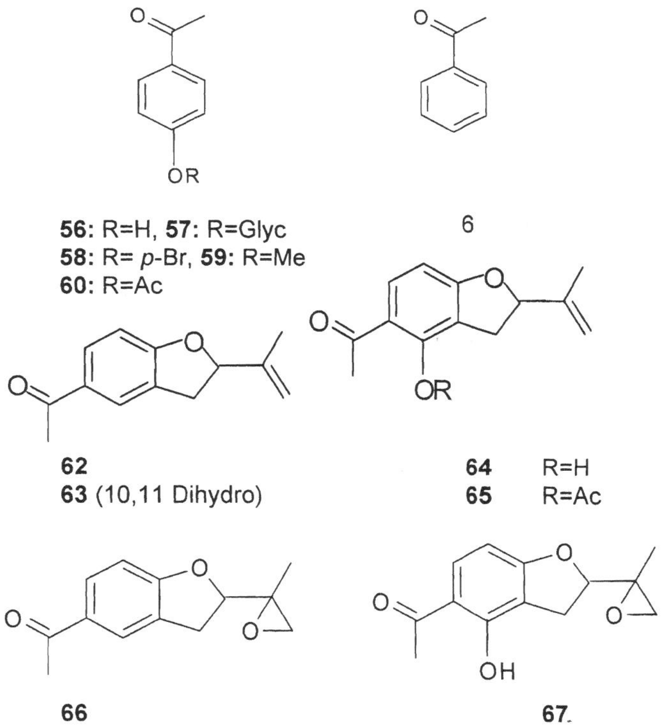

*Figure 15. Chemical structures from Baccharis spp. Structures of 4-hydroxyacetophenone **56**, 4-glucosylacetophenone **57**, 4-methylacetophenone **58**, 4-bromoacetophenone **59**, 4-acetyl acetophenone **60**, acetophenone **61**, tremetone **62**, 10,11-dehydrotremetone **63**, 4-hydroxy tremetone **64**, 4-acetyltremetone **65**, 10,11-epoxytremetone **66**, and 10,11-epoxy-4-hydroxytremetone **67**.*

ARAUCARIACEAE

In continuation of our search for natural products with possible insecticide activity, we have studied trees of the rain forest of southern Chile, due to their strong resistance against insect attack observed in nature. We have evaluated the insecticidal activity of *Araucaria araucana* and IGR activities on FAW.

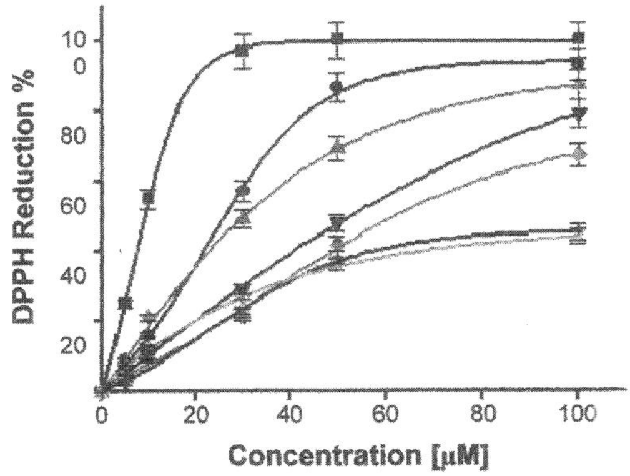

*Figure 16. DPPH Reduction of THQ (■), BHA (●), **67** (▲), **64** (▼), **66** (◆), **62** (+) and **56** (✕), compounds from Baccharis spp.*

Araucaria araucana (Mol.) K. Koch

This conifer is endemic from rain forest of southern Chile and Argentina. It has high comercial, ethnobotanical, taxonomic, and ecological value derived from its long biogeographical and remote occurrence. These characteristics offer a unique opportunity for search of secondary metabolites specially those with a defensive role against pathological and phytofagous pests. Five lignans (secoisolariciresinol **68**, lariciresinol **69**, pinoresinol **70**, eudesmin **72**, and methyl-pinoresinol **71**) were isolated from MeOH extracts of bark and wood of *A. araucana* and their structures were determined using spectroscopic methods.

In addition to antifeedant, insecticidal and IGR activities against *S. frugiperda* (FAW), the antibacterial activity of these compounds was determined against Gram (+) and Gram (-) bacteria and the antifungal activity against *Fusarium moniliforme, Aspergillus niger, Trichophyton mentagrophites, Mucor miehei, Paecilomyces variotii, Ceratocystis pirifera, Trametes versicolor* and *Penicillium notatum*. These lignans exhibited antifungal and antibacterial activity (*39*) and as antifeedant on FAW, into the range between 1.0 and 50.0 ppm.

In addition to insecticidal effects (data not show) these lignans showed a strong antioxidant activity against DPPH radical reduction, with very low concentrations, between 1 to 10 ppm, almost all lignans showed a I_{50} of inhibition around 4.5 ppm, except for methyl pinoresinol **71** and eudesmin **72** that did not show any inhibition (unpublished work).

Rhus javanica

In continuing search for alternative insect control agents, the ethanol extract of Gallae Rhois was noted to inhibit the oxidation of L-DOPA catalyzed by

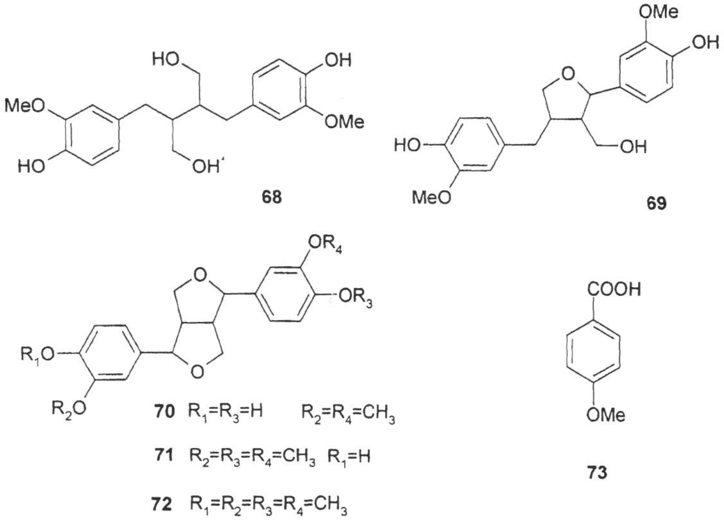

Figure 17. Lignans from A. araucana and anisol from Pimpinella anisum.

mushroom tyrosinase. The same extract was also found to exhibit growth inhibitory activity against the pink bollworm *Pectinophora gossypiella* in an artificial diet feeding assay (*84*). Our original attempt to clarify the gall formation mechanism on a molecular level could not be achieved due to lack of *M. chinensis* availability. Hence, the current study was emphasized characterization of tyrosinase inhibitors as well as insect growth inhibitors against *P. gossypiella*.

As a defense mechanism of the leaves of *Rhus javanica* (Anacardiaceae) against the aphid *Melaphis chinensis* (Aphididae) attack, tannic acid **77** is rapidly accumulated and forms galls along the midrib of the leaves resulting in a unique natural medicine Gallae Rhois. Tannic acid **77** was found to inhibit the oxidation of L-3,4-dihydroxyphenylalanine (L-DOPA) catalyzed by tyrosinase (EC 1.14.18.1) with an IC_{50} of 22 μM. The aphid would likely detoxify the ingested toxic tannic acid to relatively nontoxic gallic acid **74**, whereas the non-adapted pink bollworm *Pectinophora gossypiella* larvae are sensitive to the ingested tannic acid (*18*).

Pimpinella anisum

By bioassay-guided fractionation using mushroom tyrosinase, anisaldehyde (*p*-methoxybenzaldehyde) **73** was characterized as the principal inhibitor present in the *n*-hexane extract of the seeds of *Pimpinella anisum* L. (Umbelliferae) (*85*). In search for tyrosinase inhibitors from plants (*86*), the ethyl acetate fraction of the same plant was found to show inhibitory activity and was subjected to further characterization.

Anisic acid (*p*-methoxybenzoic acid) **73** was characterized as a tyrosinase inhibitor from aniseed, a common food spice. It inhibited the oxidation of L-3,4-dihydroxyphenylalanine (L-DOPA) catalyzed by tyrosinase with an IC50 of 0.60 mM. The inhibition of tyrosinase by anisic acid is a reversible reaction with residual enzyme activity. This phenolic acid was found to be a classical noncompetitive inhibitor and the inhibition constant, *K*I was obtained as 0.603 mM. Anisic acid also inhibited the hydroxylation of L-tyrosine catalyzed by tyrosinase. The lag phase caused by the monophenolase activity was lengthened and the steady-state activity of the enzyme was decreased by anisic acid (*17*).

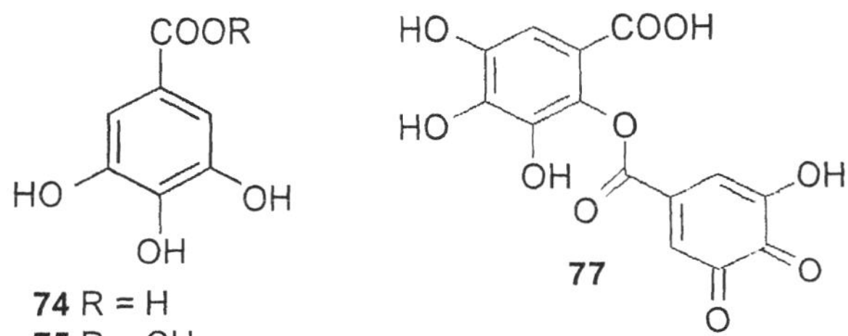

74 R = H
75 R = CH_3
76 R = CH_2CH_3

Figure 18. Chemical structure of gallic acid and its related compounds. Possible structure of the primary oxidation product (77) of gallic acid catalyzed by tyrosinase. This quinol-quinone seems to be relatively stable, but may further oxidize.

Experimental

Reduction of the 2,2-Diphenyl-1-Picrylhydrazyl Radical

Extracts and partitions were chromatographed on TLC and examined for antioxidant effects by sprying the TLC plates with DPPH reagent. Specifically, the plates were sprayed with 0.2% DPPH in methanol (*47*). Plates were examined 30 min after spraying, and active compounds appeared as yellow spots against a purple background. In addition, TLC plates were sprayed with 0.05% β-carotene solution in chloroform, then held under UV_{254} light until the background bleached. Active components appeared as pale yellow spots against a white background (*71*). Samples that showed a strong response were selected for fractionation by open column chromatography, using solvents of increasing polarity. Furthermore each fraction was analyzed with DPPH in microplates of 96 wells as follow: extracts, partitions and fractions (50 μL) were added to 150 μL of DPPH (100 μM, final concentration) in methanol (The microtiter plate was immediately placed in an Biotek™ Model ELx808, Biotek Instruments, Inc., Winooski, VT) and their absorbance read at 515 nm after 30 min (*47, 71*). Quercetin and α-tocopherol were used as standards.

Inhibition of Acetyl Cholinesterase

An enzyme extract containing Acetyl cholinesterase (AChE) was obtained according to the method of Grundy and Still (*87*). Inhibition of AChE was determined according to the Ellman's procedure (colorimetric method) (*88*) using both the control (MeOH) and test solutions (compounds and extracts). All AChE inhibition was carried out according to procedures described previously in Céspedes et al. (*35, 62*).

Tyrosinase Inhibition EnzymeAssay

The mushroom tyrosinase (EC 1.14.18.1) used for the bioassay was purchased from Sigma Chemical Co. (St. Louis, MO) and was used without further purification. Although mushroom tyrosinase differs somewhat from other sources (*89*), this fungal source was used for the entire experiment because it is readily available.

The diphenolase assay was performed as previously described (*90*) with slight modifications (*91*). First, 0.3 ml of a 5.0 mM aqueous solution of L-DOPA was mixed with 0.6 ml of 0.25 M phosphate buffer (pH 6.8) and 1.9 ml water, incubated at 30 °C for 10 min. Then, 0.1 ml of the sample solution and 0.1 ml of the aqueous solution of the mushroom tyrosinase (0.2 mg/ml in 0.1 M phosphate buffer, pH 6.8) were added in this order to the mixture. This solution was immediately monitored for the formation of dopachrome by measuring the linear increase in optical density at 475 nm. The reaction was carried out under a constant temperature of 30 °C. Absorption measurements were recorded using a Spectra MAX plus Microplate spectrophotometer.

The monophenolase assay was performed with L-tyrosine as a substrate at 30 °C. To the 3 ml assay system containing 50 mM phosphate sodium buffer (pH 6.8), 2 mM L-tyrosine, different concentrations of anisic acid and 0.1 ml of the enzyme (1.0 mg/ml in 0.1 M phosphate buffer, pH 6.8) was added, and the solution was immediately monitored at 475 nm, similar to the diphenolase assay (*17, 18*).

Measurement of oxygen consumption

The reaction mixture consisted of 0.6 ml of 0.25 M phosphate buffer (pH 6.8), 1.9 ml of water and 0.1 ml of 6 mM anisic acid or anisaldehyde DMSO solution was first incubated at 30 °C for 5 min. Then, 0.1 ml of 0.05 M phosphate buffer solution of mushroom tyrosinase (20 μg) was added and oxygen consumption was measured with an OBH 100 oxygen electrode and an oxygraph equipped with a water-jacket chamber of YSI 5300 (all from Yellow Springs Instruments Co., Yellow Springs, OH) maintained at 30 °C for 20 min. A catalytic amount (0.01 mM) of L-DOPA was added after 5 min. The oximeter was calibrated with air (100%) and sodium dithionite (0%) (*17*).

Bioassays withFall armyworm (FAW), Spodoptera frugiperda

Larvae used for the experiments were obtained from the culture at the Centro de Investigación en Biotecnología at the Universidad Autónoma del Estado de Morelos, Cuernavaca, Morelos, México, maintained under previously described conditions (*19, 33, 35, 38*).

Acute Toxicity on FAW

Acute toxicity was elucidated by topical application of larvae of last stage of *S. frugiperda*. The larvae of *S. frugiperda* were iced to stop their movement and treated on their abdomens with each one of the test compounds. The solvent used was 10.5 μl of acetone in topical form with 50 μl microsyringe, and control was only treated with 10.5 μl of acetone. After 24 hours, survival was recorded. Five larvae were used for each concentration (*19, 38*).

Relative Growth Index and Growth Index

The relative growth index (RGI) and growth index (GI) were calculated according to Zhang et al. (*92*).

Statistical Analysis

Data were analyzed by one-way ANOVA followed by Dunnett's test for comparisons against control. Values of $p \leq 0.05$ (*) and $p \leq 0.01$ (**) were considered statistically significant and the significant differences between means were identified by GLM Procedures. In addition, differences between treatment means were established with a Student-Newman-Keuls (SNK) test. The I_{50} values for each analysis were calculated by Probit analysis. Complete statistical analyses were performed using the MicroCal Origin 6.2 statistical and graphs PC program.

Discussion

The sites and mode of action as IGR of these compounds and may be due to a combination of diverse antifeedant actions as midgut esterase inhibition and postdigestive toxicity, as found for other limonoids (*25, 93, 94*) and extracts (*111*) or as enzyme inhibitors such as estearases, proteases, tyrosinase or acetylcholinesterase (*13, 17, 18, 95*).

There are mainly three types of tyrosinase inhibitors based on their inhibition mechanisms. First, the inhibitors directly interact with the binuclear copper active site of the enzyme, known as either copper chelators or substrate analogues. The inhibitors belonging to this category are mostly aromatic compounds and usually act as competitive inhibitors; the inhibitors disrupt the tertiary structure of the enzyme rather than directly interact with the active site. These inhibitors are not only aromatic but are also non-aromatic compounds, and generally act as non-competitive inhibitors. Third, the inhibitors do not interact with the enzyme directly or indirectly but act as reducing agents for dopaquinone. In addition to these three main types, there are some additional types of inhibitors. For example, inactivators of the enzyme and some tannins belong to this type of inhibitors. Hence, tyrosinase inhibitors might ultimately provide clues to control insect pests by inhibiting tyrosinase, resulting in incomplete cuticle hardening and darkening (*86, 96, 97*).

Some investigations on sites and mechanism of insecticidal or IGR action report that different phenolic compounds are enzyme and metabolism inhibitors (*103 - 105, 62, 85, 96, 97*). It is important to note that similar insect growth regulatory activity on *S. litura* (Common cutworm) was studied by Morimoto et al. (*64*). These authors claims that flavonoids from *Gnaphalium affine* (Asteraceae) have insect antifeedant properties against this insect. It is possible to infer that the substitution pattern of flavonoids induce an increase in the activity of those phenolic compounds (*62*).

In our case, the insecticidal activity of stilbene **14** could be attributed to tyrosinase inhibition, as previously reported (*104*) or to antifeedant activities of such phenolic compounds (*17 – 19, 85, 96, 97*). The presence of hydroxyl, methoxy and furan moieties, seems to be necessary for insecticidal activity as in limonoids containing several of these chemical groups (*33, 35, 37, 62, 93*) and in other phenolic compounds as alkanols, tannic, gallic, and anisic acids, respectively (*17, 18, 103, 105*).

It is worth mentioning that resveratrol **13**, the methoxylated stilbene **14**, sargahydroquinoic acid **35**, 6,7-dihydroxycoumarin **40** and pinoresinol **47** had more potent insect growth inhibitory activity than toosendanin. It is obvious that the nature of the substituents as well as hydroxyl and methoxy groups in the aromatic ring plays an important role in the insecticidal activity. These results prove previous findings on the quantitative structure activity relationship of stilbene derivatives for instance, namely that the inhibitory tyrosinase activity of the respective natural product depends on the polarity of ring A and on the size of the substituent (*106, 104*).

These facts show that the antifungal activity showed by stilbenoids (*107 - 109*), acute toxicity and growth inhibition may be due to the inhibition of tyrosinase, and this target was demonstrated also for other stilbenes and phenolics from natural sources, such as gallic acid **10** and anisic acid **50** (*17, 18, 46, 104, 110*).

In addition to enzymatic activities, IGR correlate well with antioxidant (AOX) measurements of the reduction of DPPH; in many cases mainly when the phenolic compounds show AOX activity they also show IGR activity. Our phenolics compounds could be considered as efficient IGR and radical scavengers (*33, 35, 36, 46, 44, 45, 38*). Thus, it is possible to infer that when some compounds show AOX properties they also can show IGR or tyrosinase inhibitory activity. There are some cases such as stilbenes, sargachromene and sargahydroquinoic acid, flavonoids as quercetin, naringenin and kaempferol, and coumarins with these properties. Then it can be said that some natural compounds with antioxidant, antifungal and moulting inhibitors properties, can play a role as a model for search of new botanical pesticides, nutraceuticals and bioactive compounds.

The percentage of larvae that reached pupation decreased in all tested compounds in comparison to the control. The most important effect was observed with stilbene **14**, gedunin **19**, toosendanin **21**, β-dihydroagarofurans **22**, and **23**, clerodane **28**, argentatine A **30**, sargachromenol **32**, sargahydroquinoic acid **35**, 6,7-dihydroxycoumarin **40**, and pinoresinol **47**, photogedunin epimeric mixture, photogedunin acetate mixture, which reduced successful pupation into a range between 0% and 25% and a significant delay of pupation time. The effects of these substances on reducing insect growth, decreasing the percentage of emergence, and increasing mortality of *S. frugiperda* are similar to those of other natural products (*111, 112*).

Inspection of the structure of the most active compounds isolated from these plants suggests that the presence of oxygenated function were necessary for the activity displayed by the gedunin-type and phenolic molecules against *S. frugiperda*. Almost all tested compounds except cedrelanolide and methoxylated flavones showed comparable activity to the commercial insecticide toosendanin, which suggests potential for further development of these materials. However, none of these substances have been found with the outstanding activity of azadirachtin (*113, 114*).

In the case of acetylcholinesterase (AChE), the action is directly related with the interaction between the inhibitors and acetylcholine, the inhibition of the enzyme occurs with detachment of the acid residue of the toxic compound, the nature of acid residue does not affect the structure of the inhibited enzyme, but acts strongly on the process of its inhibition. Natural compounds carrying a charge in an acid residue are very active inhibitors of cholinesterase, such as alkylamonio alcohols (*98*), alkaloids (physostigmine, phenserine, tolserine, among others) (*99*), some terpenoids (*35, 37, 62, 95*) and monoterpenoids (*13, 87, 100*). The study of AchE inhibitors as IGR substances, can help in the search of new natural compounds for the treatment of Alzheimer's disease (*101*) and Parkinson's diseases (*102*).

Conclusions

The insecticidal activity of extracts may be due to synergistic effects showed by the phenolic components of the mixtures in the test system used in this investigation. These facts are indicative of the potency of the methanol extracts. Thus, the effect of compounds on reducing insect growth, increasing development time and mortality of *S. frugiperda* is similar to that of gedunin and more potent than the MeOH extract from *Cedrela salvadorensis* (*33, 62*). The mode of action of these compounds may be due to a combination of antifeedant action as midgut phenoloxidase inhibition, moulting sclerotization toxicity and nerval system inhibition, as found for other phenolics and terpenoids (*13, 24, 85, 96, 97*) and extracts (*115*). In addition, the presence of an orcinol or catechol group seems to be important for these activities as shown for the most potent compounds in this chapter. Furthermore, a great percentage of larvae that reached pupation decreased with the application of phenolics in comparison to the control, which might be due to the inhibition by tyrosinase as well or to the accumulation of proteinase inhibitors (*116*). The activity of these plants, their metabolites and MeOH extracts is comparable to the insect growth regulators gedunin and toosendanin, which suggests potential for further development of these materials.

Acknowledgements

This work was supported in part by following grants: grants IN243802 and IN211105-3 DGAPA-UNAM, DIPRODE U. Bío Bío and Dirección de Investigación U. de Concepción, Chile.

References

1. Halliwell, B.; Aruoma, O. I. *FEBS Lett.* **1991**, *281*, 9-19.
2. Roberts, W. G.; Gordon, M. H.; Walker, A. F. *Europ. J. Clin. Nut.* **2003**, *57*, 1303- 1310.
3. Fernandes, E.; Costa, D.; Toste, S. A.; Lima, J. L. F. C.; Reis, S. *Free Radical Biol. Med.* **2004**, *37*, 1985-1905.
4. Taruscio, T. G.; Barney, D. L.; Exon, J. *J. Agric. Food Chem.* **2004**, *52*, 3169-3176.
5. Prior R. L., Wu X., Schaich K. *J. Agric. Food Chem.* **2005**, *53*, 4290-4302.
6. Schinella, G. R.; Tournier, H. A.; Prieto, J. M.; Mordugovich de Buschiazzo, P.; Rios, J. L. *Life Sciences* **2002**, *70*, 1023- 1033.

7. Céspedes, C. L.; Marín, J. C.; Domínguez, M.; Avila, J. G., Serrato, B. In *Lead molecules from natural products: Discovery and new trends*; Khan, M. T. H.; Ather, A., Eds.; Advances in Phytomedicine Series, Vol II. Elsevier: The Netherlands, 2006, pp. 385-422.
8. Céspedes, C. L.; Avila, J. G.; Marín, J. C.; Domínguez, M.; Torres, P.; Aranda, E. In *Naturally Occurring Bioactive Compounds: a new and safe alternative for management of pests and diseases*; Rai, M.; Carpinella, C. Eds.; Advances in Phytomedicine Series, Vol III. Elsevier: The Netherlands, 2006, pp. 1-27.
9. Mayer, A. M. *Phytochemistry* **1987**, *26*, 11-20.
10. Robb, D. A. In *Copper proteins and copper enzymes*, Lontie, R., Ed.; Vol. II, CRC Press: Florida, 1984, pp. 207-240.
11. Andersen, S. O. In *Molting and Metamorphosis*, Ohnishi, E.; Ishizaki, H., Eds.; Japan Sci. Soc. Press: Tokyo, 1990, pp. 133-155
12. Ryan, M. F.; Byrne, O. *J. Chem. Ecol.* **1988**, *14*, 1965-1975.
13. Miyazawa, M.; Watanabe, H.; Kameoka, H. *J. Agric. Food Chem.* **1997**, *45*, 677-679.
14. Fournier, D.; Bride, J-M.; Hoffmann, F.; Karch, F. *J. Biol. Chem.* **1992**, *267*, 14270-14274.
15. Baldwin, I. T.; Halitschke, R.; Kessler, A.; Schittko, U. *Curr. Op. Plant Biol.* **2001**, *4*, 351-358.
16. Schultz, J. C. *Nature* **2002**, *416*, 6878, 267.
17. Kubo, I.; Chen, Q-X.; Nihei, K. I.; Calderon, J. S.; Céspedes, C. L. *Z. Naturforsch* **2003**, *58*, 713-718.
18. Kubo, I.; Kinst-Hori, I.; Nihei, K. I.; Soria, F.; Takasaki, M.; Calderon, J. S.; Céspedes, C. L. *Z. Naturforsch* **2003**, *58*, 719-725.
19. Torres, P.; Avila, J. G.; Romo de Vivar, A.; García, A. M.; Marín, J. C.; Aranda, E.; Céspedes, C. L. *Phytochemistry* **2003**, *64*, 463-473.
20. Feeny, P.P. *J. Insect Physiol.* **1968**, *14*, 805-817.
21. Feeny, P. P. In *Biochemical Interactions between Plants and Insects,* Wallace, J. W.; Mansell, R. L., Eds.; Plenum Press: New York, NY, 1976, pp. 1-40.
22. Rhoades, D. F.; Cates, R. G. *Phytochemistry* **1976**, *10*, 168-213.
23. Swain, T. In *Herbivores: Their interactions with secondary plant metabolites,* Rosenthal, G. A.; Janzen, D. H., Eds.; Academic Press: New York, NY, 1979, pp. 657-682.
24. Simmonds, M. S. J. *Phytochemistry* **2003**, *64*, 21-30.
25. Champagne, D. E.; Isman, M. B.; Towers, G. H. N. In *Insecticides of Plant Origin*, Arnason, J. T.; Philogene, B. J. R.; Morand, P., Eds.; American Chemical Society: Symposium Series 387, Washington DC, 1989, pp. 95-109.
26. Champagne, D. E.; Koul, O.; Isman, M. B.; Scudder, G. G. E.; Towers, G. H. N. *Phytochemistry* **1992**, *31*, 377-394.
27. Rhoades, D. F. In *Herbivores: Their interactions with secondary plant metabolites*, Rosenthal, G. A.; Janzen, D. H., Eds.;, Academic Press: New York, NY, 1979, pp. 3-54.

28. Céspedes, C. L.; Calderón, J. S.; King-Diaz, B.; Lotina, B. *J. Agric. Food Chem.* **1998**, *46*, 2810-2816.
29. Céspedes, C. L.; Calderón, J. S.; Salazar, J. R.; King-Diaz, B.; Lotina, B. *Bol. Soc. Chil. Quim.* **1999**, *44*, 173-183.
30. Gómez-Garibay, F.; Calderón, J. S.; De La, O. A. M.; Céspedes, C. L.; Téllez, O.; Taboada, J. *Phytochemistry* **1999**, *52*, 1159-1163.
31. Céspedes, C. L.; Calderón, J. S.; Gómez-Garibay, F.; Segura, R.; King-Díaz, B.; Lotina, B. *J. Chemical* Ecology **1999**, *25*, 2665-2676.
32. Céspedes, C. L.; Alarcón, J.; Achnine, L.; Becerra, J.; Silva, M. *Rev. Latinoamer Quím.* **1999**, *27*, 41-45.
33. Céspedes, C. L.; Calderón, J. S.; Lina, L; Aranda, E. *J. Agric. Food Chem.* **2000**, *48*, 1903-1908.
34. Céspedes, C. L.; Achnine, L.; Alarcón, J.; Becerra, J.; Lotina-Hennsen, B. *Z. Naturforschung C* **2000**, *55*, 631-637.
35. Céspedes, C. L.; Alarcón, J.; Aranda, E.; Becerra, J.; Silva, M. *Z. Naturforsch* **2001**, *56*, 603-613.
36. Céspedes, C. L.; Uchoa, A.; Salazar, J. R.; Perich, F.; Pardo, F. *J. Agric. Food Chem.* **2002**, *50*, 2283-2292.
37. Céspedes, C. L.; Martínez, M.; Calderón, J. S.; Salazar, J. R.; Aranda, E. *Z. Naturforsch* **2001**, *56*, 95-105.
38. Céspedes, C. L.; Torres, P.; Marín, J. C.; Arciniegas, A.; Romo de Vivar, A.; Pérez-Castorena, A.L.; Aranda, E. *Phytochemistry* **2004**, *65*, 1963-1975.
39. Céspedes, C. L.; Avila, J. G.; García, A. M.; Becerra, J.; Flores, C.; Aqueveque, P.; Bittner, M.; Hoeneisen, M.; Martinez, M.; Silva M. *Z. Naturforsch. C.* **2006**, *61*, 35-43.
40. Céspedes, C. L.; Uchoa, A.; Calderón, J. S. *Biopestic. Internat.* **2005**, *1*, 142-164.
41. Céspedes, C. L.; Salazar, J. R.; Martínez, M.; Aranda, E. *Phytochemistry* **2005**, *66*, 2481-2493.
42. Casas, A.; Valiente-Banuet, A.; Viveros, J. L.; Caballero, J.; Cortés, L.; Dávila, P.; Lira, R.; Rodríguez, I. *Econ. Bot.* **2001**, *55*, 129-166.
43. Burns, J.; Yokota, T.; Ashihara, H.; Lean, M. E. J.; Crozier, A. *J. Agric. Food Chem.* **2002**, *50*, 3337-3340.
44. Rimando, A.; Cuendet, M.; Desmarchelier, C.; Metha, R. G.; Pezzuto, J. M.; Duke, S. O. *J. Agric. Food Chem.* **2002**, *50*, 3453-3457.
45. Stivala, L. A.; Savio, M.; Carafoli, F.; Perucca, P.; Bianchi, L; Maga, G.; Forti, L.; Pagnoni, U. M.; Albini, A.; Prosperi, E.; Vannini, V. *J. Biol. Chem.* **2001**, *276*, 22586-22594.
46. Kim, Y-M.; Yun, J.; Lee, Ch-K.; Lee, H.; Min, K. R. *J. Biol. Chem.* **2002**, *277*, 16340-16344.
47. Bors, W.; Saran, M.; Eltsner, E. F. In *Modern Methods Plant Analysis*, New Series, Acad. Press: New York, NY, 1992, 13, 277.
48. González, A. G.; Rodríguez, F. M.; Bazzocchi, I. L.; Ravelo, A. G. *J. Nat Prod.* **2000**, *63*, 48-51.

49. Itokawa, H.; Shirota, O.; Ichitsuka, K.; Morita, H.; Takeya K. *J. Nat. Prod.* **1993**, *56*, 1479-1485.
50. Backhouse, N.; Delporte, C.; Negrete, R.; Muñoz, O.; Ruiz, R. *Int. J. Pharm.* **1994**, *32*, 239-244.
51. Shirota, O.; Morita, H.; Takeya, K.; Itokawa, H.; Iitaka, Y. J. *Nat. Prod.* **1994**, *57*, 1675-1681.
52. Alarcón, J.; Becerra, J.; Silva, M.; Jakupovic, J.; Tsichritzis, F. *Rev. Latinoamer. Quim.* **1991**, *22*, 65-66.
53. Alarcón, J.; Becerra, J.; Silva, M.; Morgenstren, T.; Jakupovic, J. *Phytochemistry* **1995**, *40*, 1457-1460.
54. Alarcón, J.; Becerra, J.; Silva, M. *Bol. Soc. Chil. Quím.* **1998**, *43*, 65-71.
55. Chen, W.; Isman, M. B.; Chiu, S. F. *J. App. Ent.* **1995**, *119*, 367-370.
56. Roitman, J. N.; James, L. F. *Phytochemistry* **1985**, *24*, 835-848.
57. Wollenweber, E.; Dörr, M.; Fritz, H.; Papendieck, S.; Yatskievych, G.; Roitman, J. N. *Z. Naturforsch.* **1997**, *52*, 301-307.
58. Fang, N.; Leidig, M.; Mabry, T.J.; Munezaku, I. *Phytochemistry* **1985**, *24*, 3029-3034.
59. Fang, N.; Leidig, M.; Mabry, T. J. *Phytochemistry* **1986**, *25*, 927-934.
60. Zdero, Ch.; Bohlmann, F.; Solomon, J. C.; King, R. M.; Robinson, H. *Phytochemistry* **1989**, *28*, 531-542.
61. Zdero, Ch.; Bohlmann, F.; Niemeyer, H. *Phytochemistry* **1992**, *31*, 1723-1726.
62. Calderón, J. S.; Céspedes, C. L.; Rosas, R.; Gómez-Garibay, F.; Salazar, J. R.; Lina, L.; Aranda, E.; Kubo, I. *Z. Naturforschung* **2001**, *56*, 382-394.
63. Rodríguez, B.; Rodríguez, B.; De la Torre, M.; Simmonds, M.S.J.; Blaney, W. M. *J. Nat. Prod.* **1999**, *62*, 594-600.
64. Morimoto, M.; Kumeda, S.; Komai, K. *J. Agric. Food Chem.* **2000**, *48*, 1888-1891.
65. Hoeneisen, M.; Silva, M. *Rev. Latinoam. Quim.* **1986**, *17*, 19-24.
66. Bittner, M.; Silva, M.; Rozas, Z.; Papastergiou, F.; Jakupovic, J. *Phytochemistry* **1994**, *36*, 695-698.
67. Hoeneisen, M.; Silva, M.; Bohlmann, F. *Phytochemistry* **1980**, *19*, 2765-2766.
68. Hoeneisen, M.; Kodama, M.; Ito, S. *Phytochemistry*, **1981**, *20*, 1743.
69. Gopalakrishna, E. M.; Watson, W. H.; Hoeneisen, M.; Silva, M. *J. Cryst. Mol. Struct.* **1977**, *7*, 49-57.
70. Marticorena, C.; Quezada, H. *Gayana Botanica* **1985**, *42*, 68-75.
71. Céspedes, C. L; Hoeneisen, M.; Bittner, M.; Becerra J.; Silva, M.. *J. Agric. Food Chemistry* **2001**, *49*, 4243-4251.
72. Calderón, G.; Rzedowski, J. In *Flora Fanerogámica del Valle de México,* Instituto de Ecología; A.C., Eds.: Pátzcuaro: Michoacán, México, 2005.
73. Nagaiah, K.; Krupadanam, G. L. D.; Srimannarayana, G. *Fitoterapia* **1992**, *63*, 378-379.
74. Céspedes, C. L.; Avila, J. G.; Martínez, A.; Serrato, B.; Calderón-Mugica, J. C.; Salgado-Garciglia, R. *J. Agric. Food Chem.* **2006**, *54*, 3521-3527.

75. Seigler, D. S. In *Plant Secondary Metabolism*, Kluwer Academic Publishers: Boston, MA, 1997.
76. Berenbaum, M. R. In *Hervibores: Their interactions with secondary plant metabolites*, Vol. 1, Rosenthal, G.A.; Berenbaum, M.R. Eds.; Academic Press: San Diego, CA, 1991, pp. 221-249.
77. Labbé, C.; Rovirosa, J.; Faini, F.; San Martín, A.; Castillo. M. *J. Nat. Prod.* **1986**, *49*, 517-518.
78. Muñoz-Pizarro, C. In *Sinopsis de la Flora Chilena.* Universidad de Chile, Ed.: Santiago de Chile: Santiago, Chile, 1966.
79. Reyes, A.; Reyes, M.; Lemaire, M.; Romero, M. *Rev. Latinoamer. Quim.* **1990**, *21*, 1-2.
80. Cordano, G.; Muñoz De La-Peña, A.; Medina, J. *J. Nat. Prod.* **1982**, *45*, 653.
81. Harborne, J. B. *Nat. Prod. Rep.* **1999**, *16*, 509-523.
82. Mucciarelli, M.; Camusso, W.; Bertea, C. M.; Bossi, S.; Maffei, M. *Phytochemistry* **2001**, *57*, 91-98.
83. Klein, J.D.; Cohen, S.; Hebbe, Y. *Scientia Horticulturae.* **2000**, *83*, 71-76.
84. Kubo I. In *Phytochemical Potential of Tropical plants, Vol. 27,* Downum, K. R.; Romeo, J. T.; Stafford, H.A. Eds.; Plenum Press: New York, NY, 1993, pp. 133-151.
85. Kubo, I.; Kinst-Hori, I. *Planta Medica* **1999**, *65*, 19-22.
86. Kubo, I. In *Phytochemicals for pest control*, Hedin, P. A.; Hollingworth, R.; Masler, E.; Miyamoto, J.; Thompson, D. Eds.; American Chemical Society: ACS Symp. Ser. 658, Washington, DC, 1997, pp 310-326.
87. Grundy, D. L.; Still, C. C. *Pestic. Biochem. Physiol.* **1985**, *23*, 383-388.
88. Ellman, G. L.; Courtney, K. D.; Andres, V.; Featherstone, R. M. *Biochem. Pharmacol.* **1961**, *7*, 88-95.
89. Van Gelder, C. W. G.; Flurkey, W. H.; Wichers, H. J. *Phytochemistry* **1997**, *45*, 1309-1323.
90. Kubo, I.; Kinst-Hori, I. *J. Agric. Food Chem.* **1998**, *46*, 1268-1271.
91. Chen, Q. X.; Zhang, Z.; Zhou, X. W.; Zhuang, Z. L. *Int. J. Biochem. Cell Biol.* **2000**, *32*, 717-723.
92. Zhang, M.; Chaudhuri, S. K.; Kubo, I. *J. Chem. Ecol.* **1993**, *19*, 1109-1118.
93. Nakatani, M.; Huang, R. C.; Okamura, H.; Naoki, H.; Iwagawa, T. *Phytochemistry* **1994**, *36*, 39-41.
94. Isman, M. B.; Matsuura, H.; MacKinnon, S.; Durst, T.; Towers, G. H. N.; Arnason, J. T. In *Recent Advances in Phytochemistry., Phytochemical Diversity and Redundancy in Ecological Interactions, Vol. 30,* Romeo, J. T.; Saunders, J. A.; Barbosa P., Eds.; Plenum Press: New York, NY, 1996, pp. 155-178.
95. Ortego, F.; López-Olguín, J.; Ruíz M.; Castañera, P. *Pestic. Biochem. Physiol.* **1999**, *63*, 76-84.
96. Kubo, I. *Rev. Latinoamer. Quim.* **2000**, *28*, 7-20.
97. Kubo, I.; Xiao, P.; Nihei, K-I.; Fujita, K-I.; Yamagiwa, Y.; Kamikawa, T. *J. Agric. Food Chem.* **2002**, *50*, 3992-3998.

98. Brestkin, A. P.; Khovanskikh, A. E.; Maizal, E. B.; Moralev, S. N.; Novozhilov, K. V.; Sazonova, I. N. *Insect. Biochem. Molec. Biol.* **1992**, *22*, 769-775.
99. Yu, Q-Sh.; Zhu, X.; Holloway, H. W.; Whittaker, N. F.; Brossi, A.; Greig, N. H. *J. Med. Chem.* **2002**, *45*, 3684-3691.
100. Gracza, L. *Z. Naturforsch.* **1985**, *40*, 151-153.
101. Cummings, J. *Am. J. Psychiatry* **2000**, *157*, 4-15.
102. Weinstock, M.; Gorodetsky, E.; Poltyrev, T.; Gross, A.; Sagi, Y.; Youdim, M. *Biol. Psych.* **2003**, *27*, 555-561.
103. Hammond, D.; Kubo, I. *Bioorg. Med. Chem.* **1999**, *7*, 271-278.
104. Shimizu, K.; Kondo, R.; Sakai, K. *Planta Med.* **2000**, *66*, 11-15.
105. Panzuto, M.; Mauffette, Y.; Albert, P. J. *J. Chem. Ecol.* **2002**, *28*, 145-160.
106. Gorham, J.; Tori, M.; Asakawa, Y. In *Biochemistry of Natural Products Series, Vol. 1,* Harborne, J. B. Ed.; Chapman and Hall: London, 1995, pp. 1-110.
107. Schultz, T. P.; Hubbard, T. F.; Jin, L-H.; Fischer, T. H.; Nicholas, D. D. *Phytochemistry* **1990**, *29*, 1501-1507.
108. Schultz, T. P.; Boldin, W. D.; Fischer, T. H.; Nicholas, D. D., McMurtrey, K. D.; Pobanz, K. *Phytochemistry* **1992**, *31*, 3801-3806.
109. Pacher, T.; Seger, C.; Engelmeier, D.; Vajrodaya, S.; Hofer, O.; Greger, H. *J. Nat. Prod.* **2002**, *65*, 820-827.
110. Gilly, R.; Mara, D.; Oded, S.; Zohar, K. *J. Agric. Food Chem.* **2001**, *49*, 1479-1485.
111. Arnason, J. T.; Philogène, B. J.; Donskov, N.; Kubo, I. *Ent. Exp. Appl.* **1987**, *43*, 221-227.
112. Arnason, J. T.; Mackinnon, S.; Durst, A.; Philogène, B. J. R.; Hasbun, C.; Sánchez, P.; Poveda, L.; San Roman, L.; Isman, M. B.; Satasook, C.; Towers, G. H. N.; Wiriyachitra, P.; McLaughlin, J. L. In *Phytochemical Potential of Tropical Plants*, Downum, K. R.; Romeo, J. T.; Stafford, H. A., Eds.; Plenum Press: New York, NY, 1993, pp. 107-131.
113. Govindachari, T. R.; Narasimhan, N. S.; Suresh, G.; Partho, P. D.; Gopalakrishnan, G.; Krishna-Kumari, G. N. *J. Chem. Ecol.* **1995**, *21*, 1585-1600.
114. Kraus, W. In *The Neem Tree Azadirachta indica A. Juss and Other Meliaceous Plants.' Sources of Unique Natural Products for Integrated Pest Management, Medicine, Industry and Other purposes. XXIII,* Schmutterer, H., Ed.: VCH Verlagsgesellschaft mbH: Weinheim, Germany, 1995, pp. 35-88.
115. Feng, R. Y.; Chen, W. K.; Isman, M. B. *Pestic. Biochem. Physiol.* **1995**, *53*, 34-41.
116. Tamayo, M. C.; Rufat, M.; Bravo, J. M.; San Segundo, B. *Planta* **2000**, *211*, 62-71.

Carcinogenesis and Anticarcinogenesis

Chapter 26

Anthocyans and Chemoprevention: Evidence from Cellular Investigations

De-Xing Hou

Department of Biochemical Science and Technology, Faculty of Agriculture, Kagoshima University, Korimoto 1–21–24, Kagoshima City 890–0065, Japan

Anthocyans are polyphenolic ring-based flavonoids, and widespread in fruits and vegetables with red-blue color. In recent years, we have used culture cells to investigate their chemopreventive effects. In mouse epidermal cells (JB6), a model for screening anticarcinogenic compounds, anthocyans inhibited cell transformation by targeting mitogen activated protein kinase (MAPK)-mediated activator protein-1 (AP-1) pathway. In mouse macrophage cells (RAW264), a model for identifying the anti-inflammation compounds, anthocyans suppressed LPS-evoked inflammation through targeting nuclear factor kappa B (NF-κB), AP-1 and CCAAT/enhancer-binding protein (C/EBPδ)-mediated cyclooxygenase-2 (COX-2) expression. In human leukemia cells (HL-60), a model for testing antitumoral compounds, anthocyans induced apoptosis of cancer cells through the reactive oxygen species (ROS)/c-Jun NH_2-terminal kinase (JNK)-mediated mitochondrial dysfunction pathway. Our data provided the first molecular and cellular basis that anthocyans might have chemopreventive effects on several key steps involved in carcinogenesis.

Anthocyans (in Greek *anthos* means flower, and *kyanos* means blue) are flavanols with polyphenolic chemical structure. Anthocyan encompasses "anthocyanin" for the glycoside and "anthocyanidin" for the aglycon. The chemical structures are shown in Figure 1A (*1, 2*). Depending on their pH and the presence of chelating metal ions, they are intensely colored in blue, violet or red. Up to date, more than 400 anthocyanins have been found in nature (*1, 2*). On the other hand, the aglycon is a diphenylpropanoide-based polyphenolic ring structure, and is limited to a few structure variants. Only six kinds of anthocyanidins including delphinidin, cyanidin, petunidin, pelargonidin, peonidin and malvidin are common in fruits and vegetables (Figure 1B) (*1, 2*).

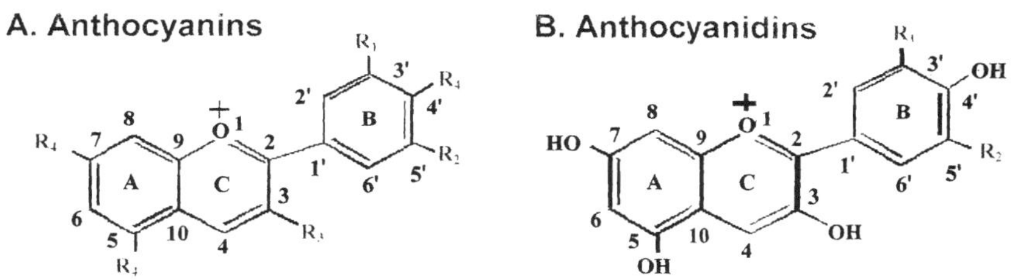

R_1 and R_2 are H, OH, or OCH_3
R_3 is a glycosyl or H
R_4 is OH or glycosyl

Malvidin, $R_1 = R_2 = OCH_3$
Pelargonidin, $R_1 = R_2 = H$
Cyanidin, $R_1 = OH$, $R_2 = H$

Peonidin, $R_1 = OCH_3$, $R_2 = H$
Petunidin, $R_1 = OCH_3$, $R_2 = OH$
Delphinidin, $R_1 = R_2 = OH$

Figure. 1. Chemical structure of anthocyanins (A) and anthocyanidins (B)

Anthocyans occur in very large amounts in some diets from colored crops, fruits and vegetables such as berries, purple sweet potatoes, red grapes and cabbages (*1, 2*). Servings of 200 g of aubergine or black grapes can provide up to 1500 mg anthocyans and servings of 100 g of berries up to 500 mg. Daily intake of anthocyans in humans has been estimated from several milligrams to grams (*3*), which depends on the nutrition customs. Mean dietary intake in Finland has been estimated to be 82 mg/day (*3*). The American diet can have much as 180-215 mg/day (*3*). Recently, daily intake of several grams can be also obtained if an individual is consuming flavonoids supplements, such as bilberry, grape or elderberry extracts (*4*).

There are some reports hinting at the potential anti-carcinogenicity of anthocyans. In a cohort of elderly individuals, who consumed large amounts of strawberries, the odds ratio for developing cancer at any site was 0.3, compared to subjects who refrained from high berry consumption (*5*). Consumption of colored fruits and vegetables has also been associated with a reduced risk of human breast cancer (*6*) and colorectal polyp recurrence (*7*). Anthocyan-containing foodstuffs have been linked with a decreased risk of coronary heart disease. They have been shown to possess beneficial effects in several parts of the organism (*8*), including the central nervous system and the eye, and have been suspected to account, at least in part, for the "French paradox", i.e. the decreased risk of cardiovascular disease despite a high-fat diet in individuals living in France.

Accumulated data indicate anthocyans have considerable bioavailability (*3*), and biological activities including anti-oxidation, anti-mutagenicity, anti-inflammatory, and anti-proliferation in some cancer cells (*9, 10*). These facts suggest that anthocyans might be potential chemoprevention agents, and that mechanisms on chemopreventive effects need to be considered at molecular level. Most of cancer chemoprevention-related compounds, such as components of tea or red wine, act to prevent tumor promotion by targeting signal transduction pathways to attenuate the expression of AP-1, and/or COX-2, or by inducing cell cycle arrest and apoptosis. Therefore, the preset study is to characterize the chemopreventive effects of anthocyans by targeting those well-accepted cellular/molecular mechanisms that can at least partially explain the effectiveness of natural compounds as chemopreventive agents. The contents include that (i) anthocyans inhibit neoplastic transformation through the inhibition of AP-1 activation; (ii) anthocyans suppress inflammation by blocking COX-2 overexpression; and (iii) anthocyans inhibit proliferation/or growth of tumor cells by inducing apoptosis.

Results and Discussion

Anti-carcinogenic Effects

A mechanism-based model is important to investigate anti-carcinogenic effects of natural bioactive components. Mouse epidermal cells (JB6) provide a cell culture-based model for studying tumor promotion (*11*). In this cell line, tumor promoters such as TPA, EGF and TNF-α induce AP-1 activity and neoplastic transformation by activating MAPK including ERK, JNK or p38 kinase (*11*). The induced AP-1 activity and neoplastic transformation can be blocked by chemopreventive agents, such as retinoids (*12*), pyrrolidine dithiocarbamate (*13*), tea polyphenols (*14*), and glycoside compounds (*15*). Many of these inhibitors have been shown to be active not only in the JB6 transformation model but also in mouse skin tumor promotion *in vivo*. Thus, JB6 cells provide a validated model to screen cancer chemopreventive agents, and to elucidate their mechanisms at the molecular level.

To investigate the anti-carcinogenic effects of anthocyans, four kinds of representative anthocyanins (Figure 2A) and six kinds of typical anthocyanidins (Figure 2B) were used to examine their effects on cell transformation in JB6

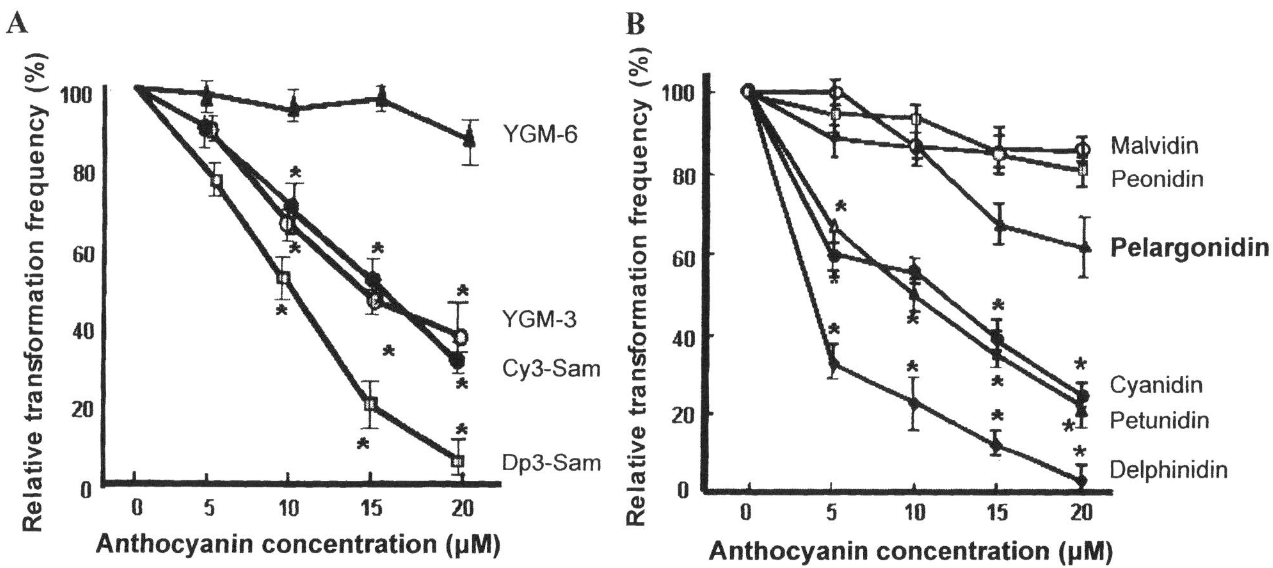

Figure 2. The inhibitory effect of anthocyanins (A) and anthocyanidins (B) on TPA-induced JB6 cell transformation.

cells. Within the concentration range of 5-20 μM, in which the cellular viability was not affect as measured by 3-(4,5-dimethylthiazol-2-yl)-2,5-diphenyl-tetrazolium bromide (MTT) assay (data not shown). Of anthocyanins and anthocyanidins tested, only those with an ortho-dihydroxyphenyl structure on the B-ring such as YGM3, Cy3Sam, Dp3Sam, delphinidin, cyanidin and petunidin (Figure 1) suppressed TPA-induced cell transformation in a dose-dependent manner (Figure 2), suggesting that the ortho-dihydroxyphenyl contributes to the inhibitory action. Molecular evidence revealed that 20 μM of delphinidin, but not peonidin, blocked the phosphorylation of protein kinases in ERK pathway at early times (2 h) and JNK signaling pathway at later times (12 h) (*16*). On the other hand, p38 kinase was not inhibited by 20 μM of delphinidin. Furthermore, 10 μM of MAPK specific inhibitors (SP600125 for JNK and UO126 for MERK1/2) could specifically block the activation of c-Jun and ERK, and cell transformation. Thus, the active anthocyans might contribute to the inhibition of cell transformation by blocking activation of ERK and JNK pathway. On the other hand, TPA treatment in JB6 cells generated reactive oxygen species (ROS) that further promotes neoplastic transformation. We identified that TPA-induced superoxide anion contribute to AP-1 activation and cell transformation. Five μM of delphinidin showed synergistic effect with 200U/ml of SOD to inhibit AP-1 activation and cell transformation (*16*). Our findings together with other reports suggest that the inhibitory effects of active anthocyans on AP-1 activation and cell transformation are due in part to their potent scavenging activity for superoxide radicals and in part to blocking ERK and JNK pathway. Both targets may be important in the cancer prevention activity of anthocyans (Figure 3).

JB6 Cl41 cells (1×10^4) were exposed to 20 ng/mLTPA with or without 5-20 μM of anthocyanins (A) or anthocyanidins (B) on soft agar medium. The cell colonies were scored with a computerized image analyzer after 14-day incubation in a 37°C, 5% CO_2 incubator. The inhibitory efficiency of cell transformation is expressed as a percentage of the transformation frequency when the cells were treated with TPA alone. Each value represents the mean ± SD of 4-5 separate experiments. *$P<0.05$, significantly different from TPA alone. Cy3-Sam: cyanidin 3-sambubioside; Dp3-Sam: delphinidin 3-sambubioside; YGM-3: cyanidin 3-(6,6′-caffeoylferuloylsophoroside)-5-glucoside; YGM-6: peonidin 3-(6,6′-caffeoyl-feruloylsophoroside)-5-glucoside.

Anthocyans might contribute to anti-carcinogenesis through ROS scanvenging and possible interaction with kinases to block downstream signal transduction pathways and gene expressions. AP-1: activator protein-1; ERK: extracellular signal-regulated kinase; JNK: c-Jun NH_2-terminal kinase; MEK: MAPK/ERK kinase; SEK:SAPK/ERK kinase; TPA: 12-*O*-tetradecanoylphorbol-13-acetate. Solid line arrow: stimulation; Dashed line arrow: possible pathway; T-bar: suppression.

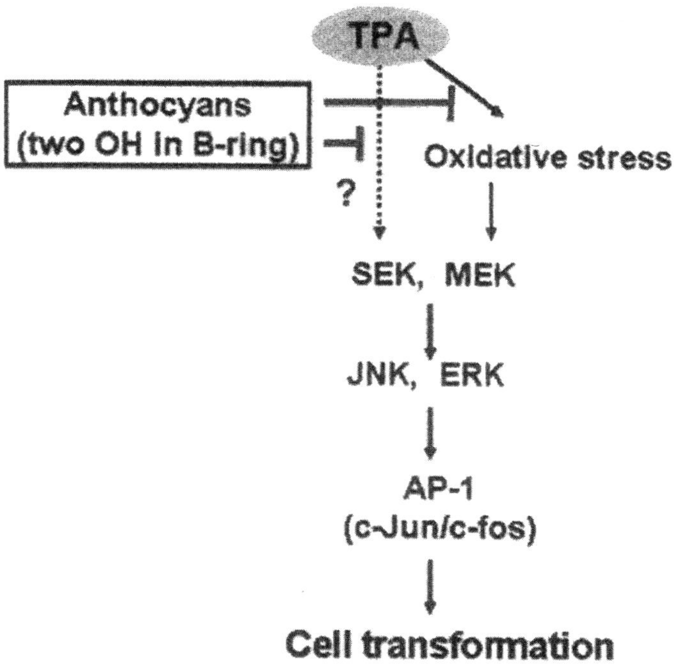

Figure 3. Diagramatic representation of the molecular mechanisms of anti-transformation in JB6 cells by anthocyans.

Anti-inflammatory Effects

COX is a rate-limiting enzyme for synthesis of dienoic eicosanoids such as prostaglandin E_2. COX exists in two isoforms (*17, 18*). COX-1 is expressed constitutively in many types of cells and is responsible for the production of prostaglandins under physiological conditions. COX-2 is induced by pro-inflammatory stimuli, including mitogens, cytokines and bacterial LPS in macrophages (18) and epithelial cells (*19*). Accumulated data indicate that COX-2 is involved in many inflammatory processes and induced in various carcinomas, suggesting that COX-2 plays a key role in inflammation and tumorigenesis (*20*). Interestingly, some antioxidants with chemopreventive effects inhibit COX-2 expression by interfering with the signaling mechanisms that regulate the COX-2 gene (*21*). Thus, COX-2 gene has been used as a biomarker to investigate the cancer chemopreventive effects of phytochemicals.

The effects of anthocyan extracts and anthocyanidins on the expression of COX-2 were investigated in LPS-activated murine macrophage RAW264 cells

(*22*). Anthocyan extracts from bilberry showed a dose-dependent suppression of LPS-induced COX-2 expression in the concentration range of 0.25-1 mg/mL (Figure 4A). Of five anthocyanidins, delphinidin and cyanidin inhibited LPS-induced COX-2 expression at 75 μM. However, pelargonidin, peonidin and malvidin did not at this concentration (Figure 4B). The structure-activity relationship suggests that the ortho-dihydroxyphenyl structure of anthocyanidins on the B-ring appears to be related with the inhibitory actions. Delphinidin, the most potent inhibitor, caused a dose-dependent inhibition of COX-2 expression at both mRNA and protein levels in the concentration range of 25-100 μM *(22)*. Western blotting analysis indicated that 50-100 μM of delphinidin inhibited the degradation of Iκ-B, nuclear translocation of p65 and C/EBPδ, and phosphorylation of c-Jun, but not CREB. Moreover, 50-100 μM of delphinidin suppressed the activations of MAPK including JNK, ERK and p38 kinase. MAPK inhibitors (U0126 for MEK1/2, SB203580 for p38 kinase and SP600125 for JNK) specifically blocked LPS-induced COX-2 expression at 20 μM *(22)*. Thus, our results demonstrated that LPS induced COX-2 expression by activating MAPK pathways, and delphinidin suppressed COX-2 by blocking MAPK-mediated pathways with the attendant inhibition of NF-κB, AP-1 and

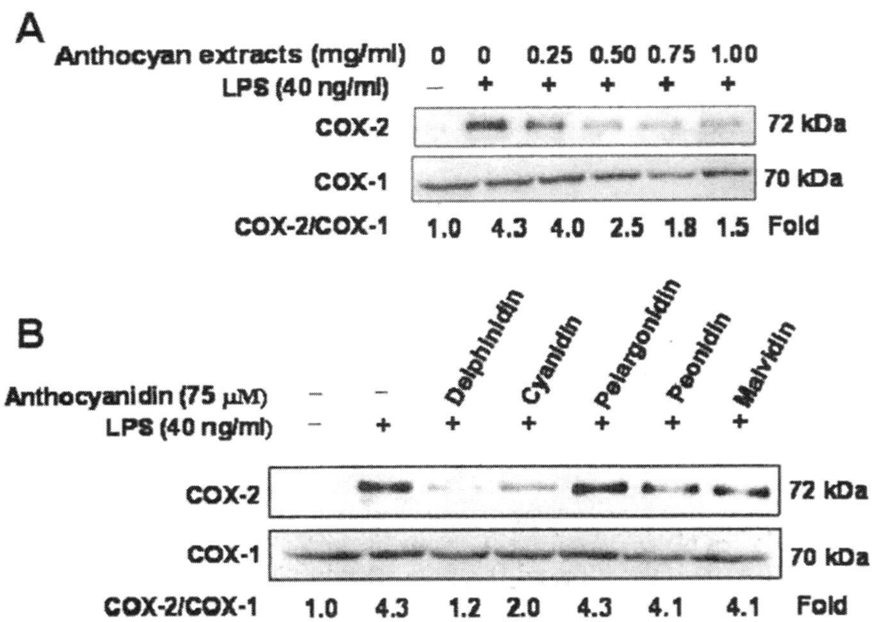

Figure 4. The inhibitory effect of anthocyan extracts (A) and anthocyanidins (B) on COX-2 expression in LPS-activated RAW264 cells.

C/EBPδ (Figure 5). These findings provide the first molecular basis that anthocyans with ortho-dihydroxyphenyl structure may have anti-inflammatory properties through the inhibition of MAPK-mediated COX-2 expression.

After RAW264 cells (1×10^6 cells) were starved in serum-free medium for 2.5 h, the cells were treated with 0.25-1.0 mg/mL of anthocyans extracts from bilberry (A) or 75 μM of the indicated anthocyanidins and 0.1% DMSO vehicle (B) for 30 min, and then exposed to 40 ng/mL LPS for 12 h. Cellular lysate was applied on 10% SDS-PAGE. The proteins of COX-2 and COX-1 were detected with corresponding specific antibodies, and visualized by chemiluminescence's ECL kit. The relative amounts of the proteins were quantified using Imager Gauge Software (Fuji Photo Film).

Anthocyans might suppress COX-2 expression by targeting MAPK-mediated C/EBPδ and AP-1 pathway, and NF-κB-mediated pathway. AP-1: activator protein-1; C/EBP: CCAAT/enhancer-binding protein; COX-2: cyclooxygenase-2; ERK: extracellular signal-regulated kinase; JNK: c-Jun NH_2-terminal kinase; LPS: lipopolysaccharide; MAPK: mitogen-activated protein kinase; NF-κB: nuclear factor-kappa B. Solid line arrow: stimulation; T-bar: suppression.

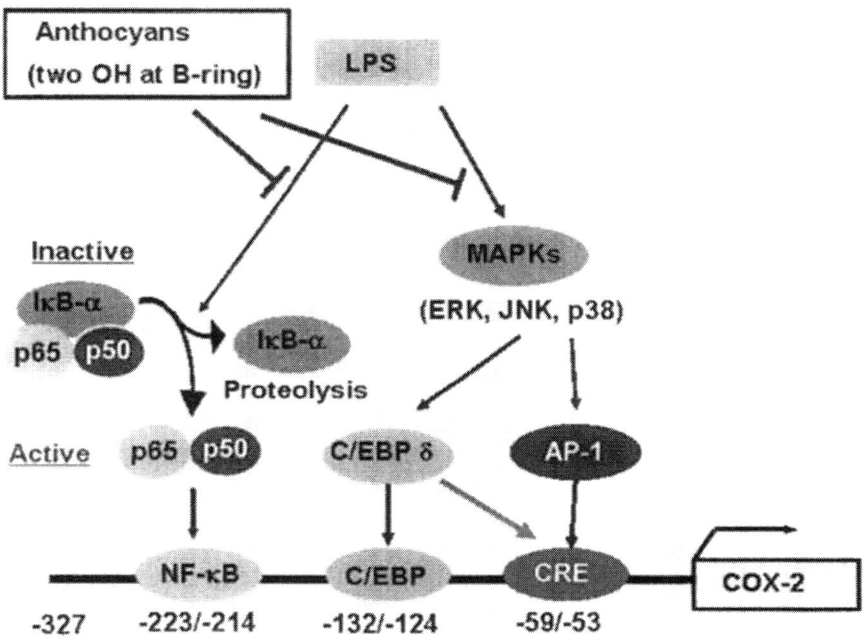

Figure 5. Diagramatic representation of the molecular mechanisms of inhibitory effects on COX-2 expression by anthocyans.

Apoptosis Induction

Apoptosis has been reported to pay an important rule in elimination of seriously damaged cells or tumor cells by chemopreventive agents (*23, 24*). The cells that have undergone apoptosis are typically shown in chromatin condensation, and DNA fragmentation (*25*). They are rapidly recognized by macrophages before cell lysis, and then can be removed without inducing inflammation (*23*). Therefore, apoptosis-inducing agents are expected to be ideal chemopreventive or chemotherapeutical agents. Many therapeutic agents for cancer, such as cisplatin (*26*), paclitaxel (*27*), isothiocyanates (*28*) and adriamycin (*29*), have been reported to eliminate tumor cells by inducing apoptotic cell death.

To investigate their anti-cancer effects of anthocyans, induction of apoptosis was examined in human leukemia cells (HL-60), which is a valid model for testing antileukemic or general antitumoral compounds. Anthocyans purified from bilberry induced apoptosis in HL-60 cells at concentration range of 0.5-1.5 mg/mL for 6-12h (Figure 6A). To identify the structure-activity relationship, HL-60 cells were treated with 100 μM of six kinds of anthocyanidins (Figure 6B).

HL-60 cells ($2x10^6$) were plated on 60-mm dishes and were cultured for 24 h. The cells were then exposed to 0.5-1.5 mg/mL of anthocyan extracts from bilberry (A) or 100 μM of the indicated anthocyanidins and 0.1% DMSO vehicle (B) for 6 h. Cells were harvested by centrifugation, and DNA was extracted. The DNA was separated on 2% agarose gel, and digitally imaged after staining with ethidium bromide. DNA markers are øX174/*Hae*III digests.

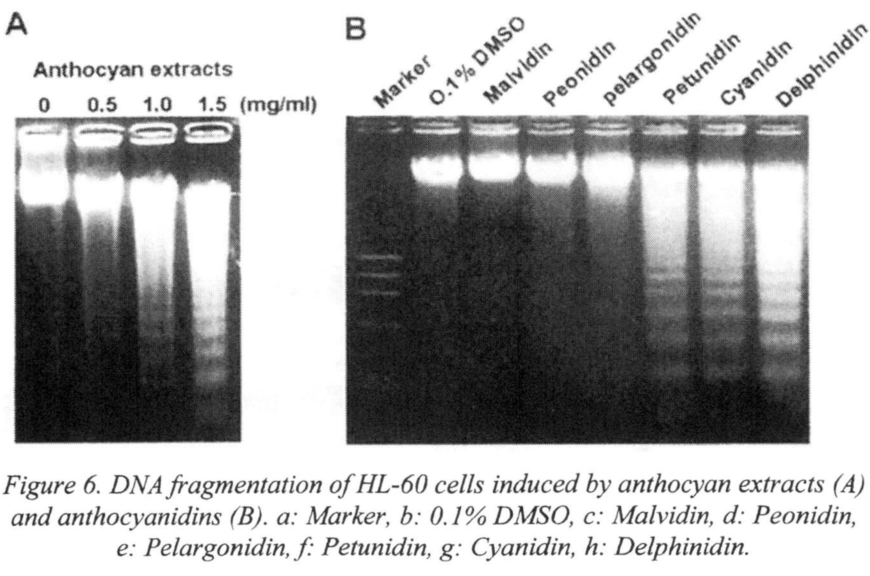

Figure 6. DNA fragmentation of HL-60 cells induced by anthocyan extracts (A) and anthocyanidins (B). a: Marker, b: 0.1% DMSO, c: Malvidin, d: Peonidin, e: Pelargonidin, f: Petunidin, g: Cyanidin, h: Delphinidin.

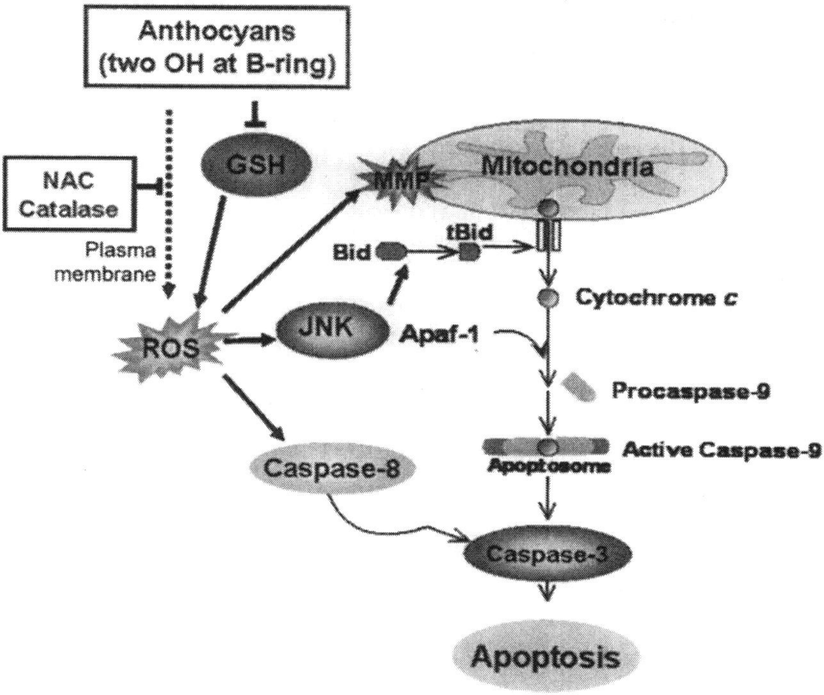

Figure 7. Diagramatic representation of the molecular mechanisms of apoptotic induction in HL-60 cells by anthocyans.

Anthocyans might induce apoptosis in HL-60 cells via ROS/JNK-mediated mitochondria death pathway. GSH: glutathione; JNK: c-Jun NH_2-terminal kinase; MMPs: matrix metalloproteinases; NAC: *N*-acetyl-L-cysteine; ROS: reactive oxygen species. Solid line arrow: stimulation; T-bar: suppression.

The results revealed that only those with an ortho-dihydroxyphenyl structure on the B-ring induced apoptosis at 100 μM for 6-12h, suggesting that the ortho-dihydroxyphenyl structure of anthocyanidins contribute to the induction of apoptosis. Delphinidin and delphinidin 3-sambubioside (Dp3-Sam), the most potent inducer, were used to investigate the molecular mechanism underlie apoptosis. Concomitant with the apoptosis, 100 μM of delphinidin or Dp3-Sam stimulated ROS generation, JNK phosphorylation, *c-jun* gene expression, Bid truncation, mitochondrial membrane potential ($\Delta\Psi m$) loss, cytochrome *c* release and activation of caspases. Antioxidants including *N*-acetyl-L-cysteine (NAC) and catalase effectively blocked a series of events such as ROS generation, JNK phosphorylation, caspase-3 activation, and DNA fragmentation (*30, 31*). Thus, anthocyans trigger an apoptotic death program in HL-60 cells through ROS/JNK-mediated mitochondria death pathway (Figure 7).

Conclusions

The effects of cancer chemoprevention of anthocyans at cellular and molecular levels through targeting the key steps involved in carcinogenesis were characterized. The conclusions include that:

1. Anthocyans inhibited cell transformation induced by TPA in mouse epidermal cells through targeting ERK- and JNK-mediated AP-1 pathway.
2. Anthocyans suppressed COX-2 overexpression in LPS-evoked mouse macrophage RAW264 cells through blocking ERK-, JNK-, and p38 kinase-mediated activation of transcriptional factors such as NF-κB, AP-1 and C/EBPδ.
3. Anthocyans induced apoptosis in human leukemia cell via ROS-dependent mitochondrial death pathway.
4. Structure-activity study indicated that the ortho-dihydroxyphenyl structure on the B-ring of anthocyans appears essential for these actions.

These findings provided the first molecular basis for the inhibitory effects of anthocyans on carcinogenesis including promotion, inflammation and progression, suggesting that anthocyans have potency for cancer chemoprevention.

Acknowledgements

This work was supported by grant-in-aid for scientific research of the Ministry of Education, Culture, Sports, Science and Technology (MEXT) of Japan (13660130 & 18580125), and by Kagoshima University Fund for Education and Research Activities to D.-X. Hou.

References

1. Harborne, J.B.; Williams, C. A. *Nat. Prod. Rep.* **2001**, *18*, 310-333.
2. Mazza, G. *Crit. Rev. Food Sci. Nutr.* **1995**, *35*, 341-371.
3. Manach, C.; Williamson, G.; Morand, C.; Scalbert, A.; Rémésy, C. *Am. J. Clin. Nutr.* **2005**, *81*, 230S-242S.
4 Skibola, C.; Smith, M. *Free Radic. Biol. Med.* **2000**, *29*, 375-383.
5. Colditz, G. A.; Branch, L. G.; Lipnick, R. J. *Am. J. Clin. Nutr.* **1985**, *41*, 32-36.
6. Adlercreutz, H. *Baillieres Clin. Endocrinol. Metab.* **1998**, *12*, 605-623.
7. Almendingen, K.; Hofstad, B.; Vatn, M. H. *Nutr. Cancer* **2004**, *49*, 131-138.
8. Detre Z.; Jellinek H.; Miskulin M. *Clin. Physiol. Biochem.* **1986**, *4*, 143-149.

9. Hou, D-X. *Curr Mol Med.* **2003**, *3*, 149-159.
10. Cooke, D.; Steward, W. P.; Gescher, A. J.; Marczylo, T. *Eur. J. Cancer* **2005**, *41*, 1931-1940.
11. Young, M. R.; Yang, H. S.; Colburn N. H. *Trends Mol. Med.* **2003**, *9*, 36-41.
12. Dong, Z.; Birrer, M. J.; Watts, R. G.; Matrisian, L. M.; Colburn, N. H. *Proc. Natl. Acad. Sci.* **1994**, *91*, 609-613.
13. Li, J. J.; Westergaard, C.; Ghosh, P.; Colburn, N. H. *Cancer Res.* **1997**, *57*, 3569-3576.
14. Dong, Z.; Ma, W.; Huang, C.; Yang, C. S. *Cancer Res.* **1997**, *57*, 4414-4419.
15. Liu, G.; Bode A.; Ma, W. Y.; Sang, S.; Ho, C. T.; Dong, Z. *Cancer Res.* **2001**, *61*, 5749-5756.
16. Hou, D-X.; Kai, K.; Li, J. J.; Lin, S.; Terahara, N.; Wakamatsu, M.; Fujii, M.; Young, M. R.; Colburn, N. *Carcinogenesis* **2000**, *25*, 29-36
17. Funk, C. D.; Funk, L. B.; Kennedy, M. E.; Pong, A. S.; Fitzgerald, G. A. *FASEB J.* **1991**, *5*, 2304-2312.
18. Hempel, S. L.; Monick, M. M.; Hunninghake, G, W. *J. Clin. Invest.* **1994**, *93*, 391-396.
19. Kelley, D. J.; Mestre, J. R.; Subbaramaiah, K.; Sacks, P. G.; Schantz, S. P.; Tanabe T.; Inoue H.; Ramonetti, J. T.; Dannenberg, A. J. *Carcinogenesis* **1997**, *18*, 795-799.
20. Hla, T.; Ristimaki, A.; Appleby, S.; Barriocanal, J. G. *Ann. N. Y. Acad. Sci.* **1993**, *696*, 197-204.
21. Chinery, R.; Beauchamp, R. D.; Shyr. Y.; Kirkland, S. C.; Coffey, R. J.; Morrow, J, D. *Cancer Res.* **1998**, *58*, 2323-2327.
22. Hou, D. X.; Yanagita, T.; Uto, T.; Masuzaki, S.; Fujii, M. *Biochem. Pharmacol.* **2005**, *70*, 417-425.
23. Thompson, C. B. *Science* **1995**, *267*, 1456-1462.
24. Galati, G.; Teng S.; Moridani, M. Y.; Chan, T. S.; O'Brien, P. J. *Drug Metabol. Drug Interact* **2000**, *17*, 311-349.
25. Kerr, J. F. R.; Wylie, A. H.; Currie, A. R. *Br. J. Cancer* **1972**, *26*, 239-257.
26. Sanchez-Perez, I.; Murguia, J. R.; Perona, R. *Oncogene* **1998**, *16*, 533-540.
27. Lee, L. F.; Li, G.; Templeton, D. J.; Ting, J. P. *J. Biol. Chem.* **1998**, *273*, 28253-28260.
28. Chen, Y. R.; Wang, W.; Kong, A. N.; Tan, T. H. *J. Biol. Chem.* **1998**, *273*, 1769-1775.
29. Yu, R.; Shtil, A. A.; Tan, T. H.; Roninson, I. B.; Kong, A. N. *Cancer Lett.* **1996**, *107*, 73-81.
30. Hou, D. X.; Ose, T.; Lin, S.; Harazoro, K.; Imamura, I.; Kubo, M.; Uto, T.; Terahara, N.; Yoshimoto, M.; Fujii, M. *Int. J. Oncol.* **2003**, *23*, 705-712.
31. Hou, D, X.; Tong, X.; Terahara, N.; Luo, D.; Fujii, M. *Arch. Biochem. Biophys.* **2005**, *440*, 101-109.

Chapter 27

DNA Intercalation, Topoisomerase I Inhibition, and Oxidative Reactions of Polyphenols

Michael R. Webb, Kyungmi Min, and Susan E. Ebeler

Department of Viticulture and Enology, University of California, One Shields Avenue, Davis, CA 95616

Polyphenols occur naturally in many foods and have a range of biological effects. Although the mechanisms for these effects are not clear, inhibition of metabolic enzymes may be one possible mode of action in the cell. Using a gel electrophoresis method validated in our laboratory we evaluated the effects of several polyphenols on the DNA-maintenance enzyme topoisomerase IB (topo I). This enzyme (along with topoisomerase II) regulates supercoiling of chromosomal DNA and plays a pivotal role in chromosome replication, transcription, recombination, segregation, condensation and repair. In our studies, the flavones and flavonols had the greatest intercalating and poisoning activity. On the other hand, no evidence for DNA intercalation or topo I inhibition was observed with the anthocyan(id)ins we evaluated. Experimental conditions, including pH, ionic strength and the presence of reductants, free radical scavengers, and trace metals may all have an effect on the solution properties and reactivities of the polyphenols. These experimental conditions must be carefully considered when relating results of *in vitro* studies to potential biological effects *in vivo*.

Polyphenols are widely distributed in fruits and vegetables and contribute bitterness, astringency, and color to foods and beverages. Numerous studies have shown that diets rich in fruits and vegetables are associated with a decreased risk of many degenerative diseases, including cardiovascular disease and cancer (*1-6*). The compounds responsible for the health protective effects are not fully known, however, dietary polyphenols have been widely studied for their antiviral, anti-inflammatory, antimutagenic and antiproliferative activity (*7-12*). In addition, the antioxidant activity of polyphenols, both *in vitro* and *in vivo* is thought to play at least a partial role in the observed biological effects (*13-17*). However, in general, the mechanisms by which polyphenols confer their health protective effects remain poorly understood. This is due to a number of factors including the dynamic and unstable nature of polyphenols in aqueous, aerobic solutions and difficulties in relating *in vitro* conditions to those that exist *in vivo*.

In our laboratory we have previously observed that dietary polyphenols in red wine, including the flavonoid, catechin, can delay tumor onset in a transgenic mouse model of neurofibromatosis (*9, 18*). In our most recent efforts we have focused on understanding how dietary polyphenols can interact with DNA and DNA-maintenance enzymes to influence cellular processes (*19-21*). The enzymes topoisomerase IB (topo I) and topoisomerase II (topo II) regulate chromosomal DNA supercoiling and are important in chromosome replication, transcription, recombination, segregation, condensation and repair (*22*). DNA supercoiling or relaxation is regulated by controlling the breakage of the phosphodiester bond for one strand (topo I) or both strands (topo II) of duplex DNA.

In our studies we have focused on the effects of polyphenols on the activity of the topo I enzyme. Topo I is thought to regulate relaxation/unwinding of DNA strands through a multi-step process. In the first step, topo I binds at the DNA reaction site. Next, strand scission occurs and a tyrosine residue of the enzyme is covalently attached to the 3' end of the broken strand. Finally, the DNA strand unwinds, the broken strand is re-ligated and the enzyme released. The topo I enzyme can be inhibited at various steps in the process. Catalytic (or relaxation) inhibitors bind to the enzyme active site and prevent reaction with the DNA. Another type of inhibition occurs when the inhibitor prevents religation and/or detachment of the enzyme from the DNA strand. This type of inhibition is referred to as "poisoning" or cleavable complex stabilization and results in DNA strand breakage (also termed "nicking"). There is much interest in the identification of various types of topo I inhibitors due to their potential to have antitumor effects.

Finally, topoisomerases are useful for studying the intercalative binding of compounds to DNA. Because intercalators can prevent full relaxation by the topo enzyme, assays designed to study relaxation and supercoiling of DNA in the presence and absence of intercalating compounds can provide valuable information on how exogenous compounds interact with DNA to influence cellular processes. However, in many assays it can be difficult to fully distinguish the effects of inhibition, poisoning, and intercalation unless careful controls and reference standards are introduced. Therefore, our work in this area has been focused on developing a gel electrophoresis assay that can be used for simultaneous determination of inhibition, poisoning, and intercalation by flavonoids. We subsequently used the assay to compare activity of several structurally related flavonoids. In this chapter we present a brief summary and overview of our studies in this area (*19-21*).

Development of a Screening Assay for Determining DNA Intercalation and Topoisomerase I Inhibition and Poisoning

The full screening assay has been previously described (*19*). The assay uses closed, circular plasmid DNA which can supercoil to different degrees, ranging from fully relaxed (Rel), to fully supercoiled (SC) forms (Figure 1). The degree of supercoiling defines distinct forms of the plasmid referred to as topoisomers (R_n). The topoisomers are separated using gel electrophoresis and result in distinct bands as the various supercoiled forms migrate through the gel (more supercoiled forms travel further through the gel since their tightly wound and compact shape results in less resistance to migration). Intercalators insert into the DNA strand, alter the extent of supercoiling and change the topoisomer distribution (Figure 1). For example, addition of ethidium bromide (EB), a known intercalator, to fully relaxed plasmid results in increasing degrees of supercoiling as the concentration of EB is increased (Figure 2, Lanes 1-5). Similarly, intercalation by the flavonoid, luteolin is observed in Lanes 14-18 (Figure 2). Reference/controls showing fully relaxed plasmid without (Lane 8) and with (Lanes 9 and 10) addition of topo I are included in the assay for comparison. Additional references/controls using fully supercoiled plasmid without (Lane 11) and with topo I (Lane 12) are also shown. In the absence of topo I, no unwinding of the supercoiled plasmid, which forms as a result of luteolin intercalation, is observed (Lane 13).

To observe poisoning activity (which results in formation of 'nicked' DNA strands), the covalently closed, relaxed plasmid must be separated from the nicked form (Nik). This is done by performing a second electrophoresis step. Following the first electrophoresis (to distinguish intercalation and catalytic inhibition activity), the gel is incubated with ethidium bromide (EB), which intercalates into the DNA in the gel causing the relaxed form to supercoil but leaving the nicked form relatively unchanged since, being nicked, it cannot supercoil. In this way the shape of the two forms becomes radically different and the nicked and relaxed forms will then separate from each other rapidly if the electrophoresis is continued. The poisoning activity of the naturally occurring alkaloid, camptothecin, can be observed in Figure 2 (Lanes 6 and 7) as an increase in the intensity of the nicked plasmid bands (Nik). Fully linear-ized plasmid (breakage of both DNA strands) is also observed as a distinct band

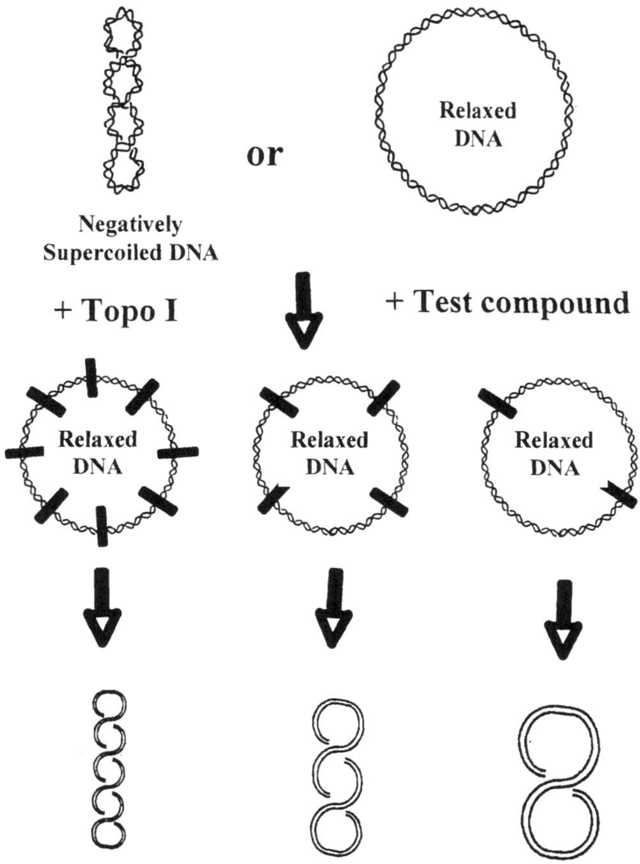

Figure 1. DNA topoisomerase assay showing intercalation of test compounds into plasmid DNA. Topoisomerase relaxes the supercoiled DNA and the test compound (portrayed here as a short, solid black line) inserts into the relaxed DNA). The topoisomerase is inactivated and the test comound is then either extracted and/or removed during gel electrophoresis. The plasmid rewinds to a degree determined by the amount of compound which was intercalated at equilibrium. The degree of supercoiling determines how fast the plasmid migrates in the gel. If no intercalation occurs, the plasmid remains relaxed.

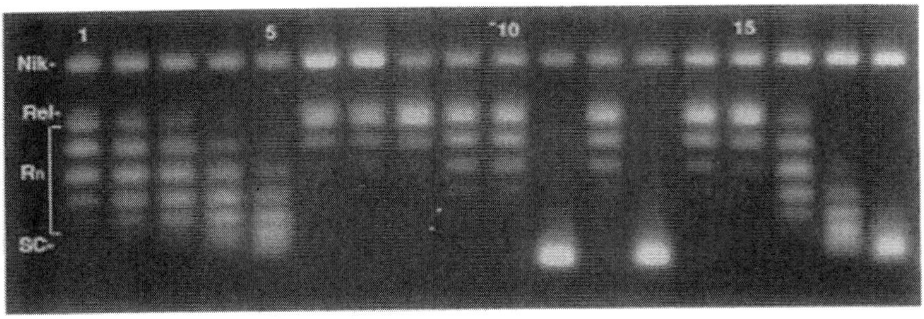

Figure 2. Intercalation activity by ethidium bromide and luteolin using relaxed plasmid DNA. ***Lanes 1-5****: Ethidium bromide (0.1, 0.13, 0.15, 0.17, 0.2 μg/mL) + relaxed DNA;* ***Lanes 6,7 (duplicates)****: Camptothecin + relaxed DNA + 10X topo;* ***Lane 8****: Control-relaxed DNA, no topo;* ***Lanes 9,10****: Control-relaxed DNA + 10X topo;* ***Lane 11****: Control-supercoiled DNA;* ***Lane 12****: Control-supercoiled DNA + 10X topo;* ***Lane 13****: Luteolin + supercoiled DNA, no topo;* ***Lanes 14-18****: Luteolin (10, 25, 75, 150, 200 μM) + relaxed DNA + topo. Reproduced with permission from reference 19. Copyright 2003 Elsevier Inc.*

between the nicked and relaxed forms, but no linear plasmid is observed in these samples.

Finally, the ability of topo I to relax supercoiled DNA is the basis for the inhibition portion of the assay (see Figure 3, Lane 1 vs Lane 2 containing supercoiled plasmid without and with topo I, respectively). By comparing the amounts of relaxed and supercoiled forms, in the presence of test compounds, catalytic inhibition relaxation activity can be readily observed. If topo I is inhibited it will not be able to relax the supercoiled form and inhibition is indicated by the progressive increase in the amount of supercoiled plasmid still present as inhibitor concentration is increased (Lanes 9-13 show inhibition by luteolin; Lanes 14-17 show inhibition by morin). By comparing these results to those using fully relaxed plasmid (Lanes 4-8), the catalytic inhibition activity can be distinguished from intercalative activity. An additional comparison is also included showing that luteolin alone, when added to fully supercoiled DNA, does not influence the extent of relaxation if no topo I is added (Lane 3).

Effects of Flavonoids on Topo I Poisoning and Inhibition and DNA Intercalation

Using the screening assay described above, 34 polyphenols, from a range of flavonoid classes (Figure 4; Table I) were surveyed for their ability to act as topo I poisons, inhibitors, and DNA intercalators (*20*). Of the compounds studied, the

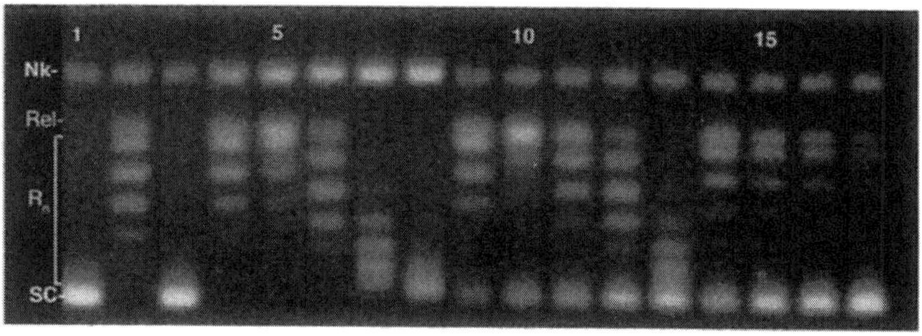

Figure 3. Catalytic relaxation inhibition by luteolin and morin. ***Lane 1****: Control. supercoiled DNA;* ***Lane 2****: Control, supercoiled DNA + 10X topo;* ***Lane 3****: Luteolin (150 μM) + supercoiled DNA;* ***Lanes 4-8****: Luteolin (10, 25, 75, 150, 200 μM) + relaxed DNA + 10X topo;* ***Lanes 9-13****: Luteolin (0, 25, 50, 75, 150 μM) + supercoiled DNA + 2X topo;* ***Lanes 14-17****: Morin (50, 75, 100, 150 μM) + supercoiled DNA + 2X topo. Reproduced with permission from reference 19. Copyright 2003 Elsevier Inc.*

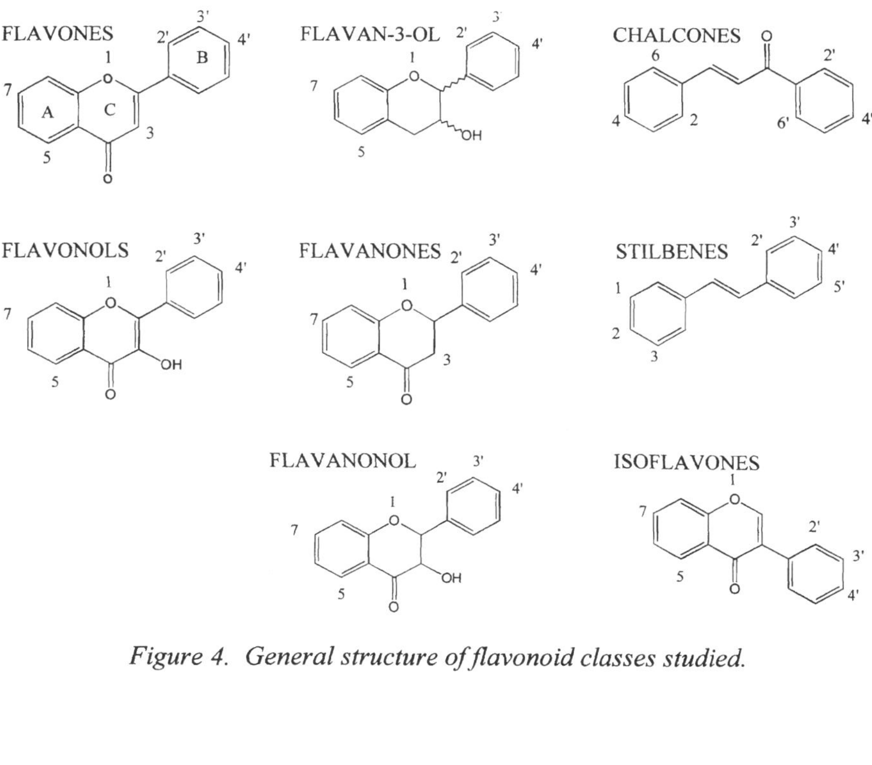

Figure 4. General structure of flavonoid classes studied.

flavones and flavonols were the most potent poisons (Figure 4; Table I). These compounds generally showed a dose-response relationship with higher concentrations resulting in increased poisoning activity. The flavones and flavonols were also strong intercalators (Table I), however, there was no clear relationship between poisoning activity and intercalating activity (R^2 = 0.213 when degree of intercalation was plotted against degree of poisoning for compounds showing both activities). There was no or only weak poisoning and intercalation activity for the flavanonols, flavan-3-ols, flavanones, isoflavones, chalcones, and stilbenes studied (Table I). These results indicate that DNA intercalation is not required for stabilization of the topo-I-DNA cleavable complex (*i.e.,* poisoning), although intercalation may be associated with poisoning in some cases.

Structure activity determinates of the intercalating activity indicated that the presence of a double bond between the 2 and 3 carbon of the C-ring (Figure 4) results in a sufficiently planar structure to allow intercalaction between adjacent bases of the DNA duplex. For example, no intercalating activity was observed for the flavanonol, taxifolin, which, lacks this double bond, while quercetin, the structural analog containing this double bond, is a strong intercalator (see also Table I). The presence of a free hydroxy at the 4′ position of B-ring of the flavones and flavonols also appears necessary for intercalation (Figure 4; Table I). For example, chrysin, which lacks this hydroxyl is unable to intercalate.

In this study, inhibition activity could not be accurately assessed due to flavonoid aggregation in solution. Based on these observations, we determined that solution conditions are critical for understanding chemical reactions of the flavonoids.

Flavonoids (FH) will ionize (F^-), oxidize, and aggregate (FH_n) in solution according to the following relationships:

$$FH_{(precip)} \longleftarrow (FH)_n \longleftrightarrow FH \longleftrightarrow F^- + H^+$$

Therefore conditions of pH, ionic strength, redox potential, and the presence of other binding agents (*e.g.*, proteins) must be carefully controlled and may make it difficult to compare results from one assay to another. When considering *in vivo* effects of the polyphenols, those polyphenols with a pK_a near physiological pH (*e.g.*, naringen, pK_a ~7) will be partially ionized and the reactivities and interactions of the protonated and unprotonated forms may vary considerably.

Anthocyan(id)in Interactions with DNA

Anthocyanins (and the aglycon anthocyanidins) are responsible for the red and blue color of many fruits and vegetables, and dietary consumption of these compounds can be significant (12.5 mg/day/person in the United States; *23*). Therefore we extended our study of flavonoid/DNA interactions to include an evaluation of a series of anthocyan(id)ins of varying structures (Figure 5) (*21*).

Table I. Topo I Poisoning and Intercalation Activity of Tested Polyphenols

Flavonoid Class	*Compound*	*Poisoning Activity at 100 μM (± SD)**	*Intercalation Activity***
Flavones	Chrysin (5,7-dihydroxyflavone)	2 ± 2.1	0
	3′,4′-dihydroxyflavone	21 ± 12.7	158 (31.8)
	Apigenin (5,7,4′-trihydroxyflavone)	44 ± 8.9	NQ
	Luteolin (5,7,3′,4′-tetrahydroxyflavone)	55 ± 8.2	97 (5.6)
	Luteolin-4'-glucoside (5,7,3′-trihydroxy-4′-*O*-glucoside)	19 ± 4.3	310 (10.4)
	Luteolin-7-glucoside (5, 3′,4′-trihydroxy-7-*O*-glucoside)	51 ± 14	71 (16)
	Orientin (5,7,3′,4′-tetrahydroxy-8-*C*-glucoside	69 ± 5.6	170 (14.1)
	Diosmin (5,3′-dihydroxy-7,4′-dimethoxyflavone)	13 ± 4.9	0
	5,7,3′,4′-tetramethoxyflavone	< 20	0
	5,6,7,3′,4′,5′-hexamethoxyflavone	< 20	0
Flavonols	Fisetin (7,3′,4′-trihydroxyflavonol)	48 ± 1	198 (23.6)
	Kaempferol (5,7,4′-trihydroxyflavonol)	56 ± 6.2	140 (6.3)
	Quercetin (5,7,3′,4′-tetrahydroxyflavonol)	57 ± 15.6	91 (18.2)
	Myricetin (5,7,2′,3′,4′-tetrahydroxyflavonol)	62 ± 14.9	225 (30.5)
	Gossypetin (5,7,8,3′,4′-pentahydroxyflavonol)	9 ± 4.9	0
	Morin (5,7,2′,4′-tetrahydroxyflavonol)	16 ± 5.5	0

Continued on next page.

Table I. *Continued.*

Flavonoid Class	*Compound*	*Poisoning Activity at 100 μM (± SD)**	*Intercalation Activity***
Flavonols (con't)	Rhamnetin (7-methoxy-5,3′,4′-trihydroxyflavonol)	< 25	NQ
	Tamarixetin (4'-methoxy-5,7,3′-trihydroxyflavonol)	6 ± 2.5	0
	Rutin (5,7,3′,4′-tetrahydroxy-3-*O*-rutinoside)	5 ± 1.8	0
	Quercitrin (5,7,3′,4′-tetrahydroxy-3-*O*-rhamnoside)	15 ± 12	0
Flavanonols	(+)-Taxifolin (5,7,3′,4′-tetrahydroxyflavanonol (2R,3R))	14 ± 9.9	0
	Silibinin	5 ± 4	0
Flavan-3-ols	(+)-Catechin (2R,3S) (5,7,3′,4′-tetrahydroxyflavan-3-ol)	1 ± 1	0
	(-)-Epicatechin (2R, 3S) (5,7,3′,4′-tetrahydroxyflavan-3-ol)	9 ± 5.4	0
	(-)-Epicatechingallate (2R, 3S) (5,7,3′,4′-tetrahydroxyflavan(-3-*O*-gallate (2R, 3R))	3	0
	(-)-Epigallocatechingallate (5,7,3′,4′,5′-tetrahydroxyflavan(3-*O*-)gallate (2R, 3R))	5	0
Flavanones	Naringenin (5,7,4′-trihydroxyflavonone)	0	0
	Hesperitin (5,7,3′-trihydroxy-4′-methoxyflavonone)	4 ± 2.5	0
Isoflavones	Genistein (5,7,4′-trihydroxyisoflavone)	22 ± 9.5	NQ
	Daidzein 4′,7-dihydroxyisoflavone)	14 ± 2.1	0

Table I. *Continued.*

Flavonoid Class	*Compound*	*Poisoning Activity at 100 μM (± SD)**	*Intercalation Activity***
Chalcones	Phloretin (4,2′,4′,6′-tetrahydroxychalcone)	19 ± 2.8	0
	Phloridzin 4,4′,6′-trihydroxy-2′-glucoside)	19 ± 0.7	0
Stilbenes	*trans*-Resveratrol 1,3,4′-trihydroxystilbene)	26 ± 2.1	0
	trans-Piceatannol (1,3,3′,4′-tetrahydroxystilbene)	23 ± 6.4	0

*Relative to nicking induced by 10 μM Camptothecin (%) (larger values reflect higher potency)

** μM necessary to equal unwinding produced by 0.15 μg/mL ethidium bromide (larger values represent decreasing degree of intercalation)

NQ non-quantifiable due to poor response and/or non-reproducibility

Table adapted with permission from reference *20*. Copyright 2004, the Biochemical Society.

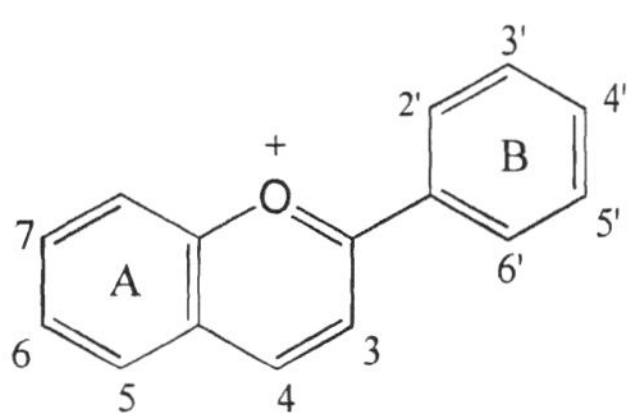

	Substitution Pattern					
Compound	3	5	7	3Ő	4Ő	5Ő
Luteolinidin	H	OH	OH	OH	OH	H
Cyanidin	OH	OH	OH	OH	OH	H
Cyanidin 3-O-glucoside	Glucose	OH	OH	OH	OH	H
Cyanidin 3,5-O-diglucoside	Glucose	Glucose	OH	OH	OH	H
Malvidin 3-O-glucoside	Glucose	OH	OH	OCH_3	OH	OCH_3

Figure 5. Structures of anthocyan(id)ins studied.

We observed that none of the anthocyanidins were able to intercalate plasmid DNA in our gel electrophoresis assay in the concentration range studied (6 – 300 μM) (Figure 6, data for malvidin 3-O-glucoside). In our experiments, the pH was 7.5, a pH where the anthocyanins would be uncharged and nonplanar (carbinol/pseudobase predominates) and therefore would not be expected to insert between the base pairs as readily as the planar flavylium form which predominates at lower pHs.

Inhibition of catalytic relaxation activity was observed only at relatively high concentrations (> 50 μM for cyanidin and > 125 μM for the other compounds (Figure 7; shown for cyanidin and cyanidin-3,5-O-diglucoside only). Cleavable complex stabilization (poisoning) was also not observed. However, we did observe that all of the anthocyanins induced significant strand breakage in the absence of topoisomerase enzyme (see strong nicked band intensity in Lanes 4 and 5 of Figure 7). When antioxidants such as DMSO (which can act as a free radical scavenger, blocking hydroxyl radical attack on the DNA) were added to the assay, the extent of nicking decreased. The presence of transition metals (e.g., Fe^{+3}) also resulted in a nearly two-fold increase in anthocyanin induced nicking of the plasmid.

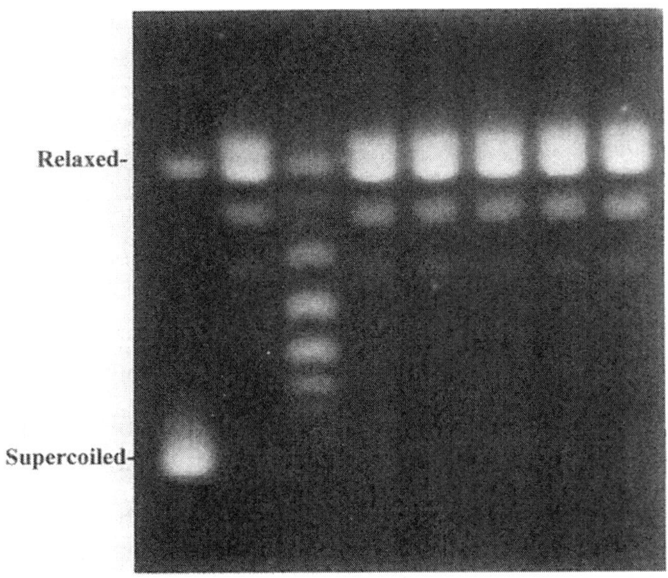

Figure 6. Intercalation activity of Malvidin-3-O-glucoside. ***Lane 1**: Control, supercoiled DNA;* ***Lane 2**: Control, supercoiled DNA + topo;* ***Lane 3**: Control, relaxed + topo + Ethidium Bromide (0.13 μg/mL);* ***Lanes 4-8**: Malvidin-3-Glu (6, 12, 30, 60, 120 μM, respectively) + relaxed DNA + topo. Reproduced with permission from reference 21. Copyright 2007, Blackwell Publishing.*

These experiments suggest that using assay conditions which mimic physiological pH (7.5) and redox conditions, the anthocyan(id)ins may induce DNA strand breakage, possibly through hydroxyl radical production in the presence of trace amounts of transition metals. These results are consistent with previous studies showing that anthocyanins can induce DNA and chromosomal damage in cell cultures (*24-26*), but contrast with numerous *in vitro* studies that indicate that anthocyaains can act as potent antioxidants (*13-17, 27*). Our results again point to the critical need for understanding the effects of experimental conditions on observed chemical reactions *in vitro* and *in vivo*.

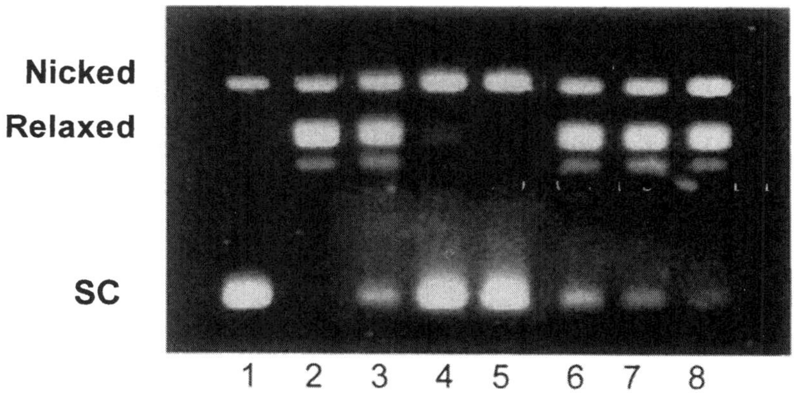

Figure 7. Topoisomerase inhibition and nicking by cyanidin chloride and cyanidin 3,5-O-diglucoside. ***Lane 1****: Control, supercoiled DNA;* ***Lane 2****: Control, relaxed DNA;* ***Lanes 3,6****: Control, supercoiled DNA + topo;* ***Lanes 4-5****: Relaxed DNA + topo + cyanidin chloride (63 μM, 125 μM);* ***Lanes 7-8****: Relaxed DNA + topo + cyanidin 3, 5-O-diglucoside (63 μM, 125 μM). Reproduced with permission from reference 21. Copyright 2007, Blackwell Publishing.*

Summary

Using a gel electrophoresis assay validated in our laboratory, large numbers of compounds can be screened for their ability to intercalate DNA and inhibit topoisomerase IB activity (both catalytic relaxation inhibition and posoning/cleavable complex stabilization can be estimated). Flavonoid structure is important for determining activity, and flavones and flavonols were found to be the most potent intercalators and poisons under conditions which mimic physiologic pH. Anthocyanins had no intercalation or poisoning activity in the conditions of our assay, but they did induce oxidative DNA strand breakage.

Most importantly, our studies point to the critical need to consider the solution properties of the flavonoids when interpreting *in vitro* and *in vivo* experimental results. In solution, the polyphenols exist in a complex equilibrium between ionized, unionized, and aggregated forms. Therefore the relative amounts of each species present will be influenced by subtle changes in pH, ionic strength, redox conditions, etc., and the resulting effects on the ability of the various polyphenol forms to interact with cellular macromolecules (*e.g.*, DNA and proteins) and to undergo oxidative and precipitation reactions may significantly impact *in vivo* biologic activity.

Acknowledgements

This work was partially supported by an NIEHS Fellowship (NIH#5T32E507059) for MRW and Wine Spectator and Jastro Shields Scholarships for MRW and KM, the American Vineyard Foundation and California Competitive Grant Program for Research in Viticulture and Enology.

References

1. Liu, S.; Manson, J. E.; Lee, I.-M.; Cole, S. R.; Hennekens, C. H.; Willett, W. C.; Buring, J. E. Fruit and vegetable intake and risk of cardiovascular disease: the Women's Health Study. *Am. J. Clin. Nutr.* **2000**, *72*, 922-928.
2. Greenwald, P.; Clifford, C.K.; Milner, J. A. Diet and cancer prevention. *Eur. J. Cancer* **2001**, *37*, 948-965.
3. Ferguson, L. R. Role of plant polyphenols in genomic stability. *Mutat. Res.* **2001**, *475*, 89-111.
4. Le Marchand, L. Cancer preventitive effects of flavonoids-a review. *Biomed Pharmacother.* **2002**, *56,* 296-301.
5. Peterson, J.; Lagiou, P.; Samoli, E.; Lagiou, A.; Katsouyanni, K.; LaVecchia, C.; Dwyer, J.; Trichopoulos, D. Flavonoid intake and breast cancer risk: a case-control study in Greece. *Br. J. Cancer* **2003**, *89*, 1255-1259.
6. Halliwell, B. Dietary polyphenols: Good, bad, or indifferent for your health? *Cardiovascular Res.* **2007**, *73*, 341-347.
7. Kuntz, S.; Wenzel, U.; Daniel, H. Comparative analysis of the effects of flavonoids on proliferation, cytotoxicity, and apoptosis in human colon cancer cell lines. *Eur. J. Nutr.* **1999**, *38*, 133-142.
8. Wang, I. K.; Lin-Shiau, S. Y.; Lin, J. K. Induction of apoptosis by apigenin and related flavonoids through cytochrome c release and activation of caspase-9 and caspase-3 in leukemia HL-60 cells. *Eur. J. Cancer* **1999**, *35*, 1517-1525.

9. Ebeler, S. E.; Brenneman, C.A.; Kim, G. S.; Jewell, W. T.; Webb, M. R.; Chacon-Rodriguez, L.; MacDonald, E. A.; Cramer, A. C.; Levi, A.; Ebeler, J. D.; Islas-Trejo, A.; Kraus, A.; Hinrichs, S. H.; Clifford, A. J. Dietary catechin delays tumor onset in a transgenic mouse model. *Am. J. Clin. Nutr.* **2002**, *76*, 865-872.
10. Edenharder, R.; Krieg, H.; Kottgen, V.; Platt, K. L. Inhibition of clastogenicity of benzo[a]pyrene and of its *trans*-7,8-dihydrodiol in mice *in vivo* by fruits, vegetables, and flavonoids. *Mutat. Res.* **2003**, *537*, 169-181.
11. Uhl, M.; Ecker, S.; Kassie, F.; Lhoste, E.; Chakraborty, A.; Mohn, G.; Knasmuller, S. Effect of chrysin, a flavonoid compound, on the mutagenic activity of 2-amino-1-methyl-6-phenylimidazo[4,5-b]pyridine(PhIP) and benzo(a)pyrene (B(a)P) in bacterial and human hepatoma (HepG2) cells. *Arch. Toxicol.* **2003**, *77*, 477-484.
12. Williams, R. J.; Spencer, J. P.; Rice-Evans, C. Flavonoids: antioxidants or signaling molecules? *Free Rad. Biol. Med.* **2004**, *36*, 838-849.
13. Satué-Gracia, M. T.; Heinonen, M.; Frankel, E. N. Anthocyanins as antioxidants on human low density lipoprotein and lecithin-liposome systems. *J. Agric. Food Chem.* **1997**, *45*, 3362—3367.
14. Wang, H.; Cao, G.; Prior, R. L. The oxygen radical absorbing capacity of anthocyanins. *J. Agric. Food Chem.* **1997**, *45*, 304-309.
15. Russo, A.; Acquaviva, R.; Campisi, A.; Sorrenti, V.; Di Giacomo, C.; Virgata, G.; Barceellona, M. L.; Vanella, A. Bioflavonoids as antiradicals, antioxidants and DNA cleavage protectors. *Cell. Biol. Toxicol.* **2000**, *16*, 91-98.
16. Ramirez-Turtosa, C.; Andersen, Ø.M.; Gardner, P. T.; Morrice, P. C.; Wood, S. G.; Duthie, S. J.; Collins, A. R.; Duthie, G. G. Anthocyanin-rich extract decreases indices of lipid peroxidation and DNA damage in vitamin E-depleted rats. *Free Rad. Biol. Med.* **2001**, *31*, 1033-1037.
17. Acquaviva. R.; Russo, A.; Galvano, F.; Galvano, G.; Barcellona, M. L.; Li Volti, G.; Vanella, A. Cyanidin and cyanidin 3-*O*-β-D-glucoside as DNA cleavage protectors and antioxidants. *Cell. Biol. Toxicol.* **2003**, *19*, 243-252.
18. Clifford, A. J.; Ebeler, S. E.; Ebeler, J. D.; Bills, N. D.; Hinrichs, S. H; Teissedre, P. L.; Waterhouse, A. L. Delayed tumor onset in transgenic mice fed an amino acid-based diet supplemented with red wine solids. *Am. J. Clin. Nutr.* **1996**, *64*, 748-756.
19. Webb, M. R.; Ebeler, S. E. A gel electrophoresis assay for the simultaneous determination of topoisomerase I inhibition and DNA intercalation. *Anal. Biochem.* **2003**, *321*, 22-30.
20. Webb, M. R.; Ebeler, S. E. Comparative analysis of topoisomerase IB inhibition and DNA intercalation by flavonoids and similar compounds: structural determinates of activity. *Biochem. J.* **2004**, *384*, 527-541.
21. Webb, M. R.; Min, K.; Ebeler, S. E. Anthocyanin interactions with DNA: Intercalation, topoisomerase I inhibition and oxidative reactions. *J. Food Biochem.* **2007**, *In press.*

22. Wang, J. C. Cellular roles of DNA topoisomerases: a molecular perspective. *Nat. Rev. Mol. Cell Biol.* **2002**, *3*, 430-440.
23. Wu, X.; Beecher, G. R.; Holden, J. M.; Hayatowitz, D. B.; Gebhardt, S. E.; Prior, R. L. Concentrations of anthocyanins in common foods in the United States and estimation of normal consumption. *J. Agric. Food Chem.* **2006**, *54*, 4069-4075.
24. Pool-Zobel, B. L.; Bub, A.; Schröder, N.; Rechkemmer, G. Anthocyanins are potent antioxidants in model systems but do not reduce endogenous oxidative DNA damage in human colon cells. *Eur. J. Nutr.* **1999**, *38*, 227-234.
25. Popp, R.; Schimmer, O. Induction of sister-chromatid exchanges (SCE), polyploidy, and micronuclei by plant flavonoids in human lymphocyte cultures. A comparative study of 19 flavonoids. *Mutat. Res.* **1991**, *246*, 205-213.
26. Habermeyer, M.; Fritz, J.; Barthelmes, H. U.; Christensen, M. O.; Larsen, M. K.; Boege, F.; Marko, D. Anthocyanidins modulate the activity of human DNA topoisomerases I and II and affect cellular DNA integrity. *Chem. Res. Toxicol.* **2005**, *18*, 1395-1404.
27. Suda, I.; Oki, T.; Masuda, M.; Nishiba, Y.; Furuta, S.; Matsugano, K.; Sugita, K.; Terahara, N. Direct absorption of acylated anthocyanins in purple-fleshed sweet potato in rats. *J. Agric. Food Chem.* **2002**, *50*, 1672-1676.

Chapter 28

The Cancer Preventive Potential of Tea Polyphenol EGCG in HER2-Positive Breast Cancer

Min-Hsiung Pan[1] and Wei-Jen Chen[2]

[1]**Department of Seafood Science, National Kaohsiung Marine University, Kaohsiung, Taiwan**
[2]**Department of Biomedical Sciences, Chung Shan Medical University, Taichung 402, Taiwan**

Breast cancer is recognized as one of the leading causes of death among women in many countries and its incidence is closely linked to *HER2 (human epidermal growth factor receptor 2) gene* amplification. HER2, a second member of receptor tyrosine kinase of erbB family, is overexpressed in about 25% of human breast cancers with poor prognosis and chemoresistance, and considered as a target for breast cancer therapy. (-)-Epigallocatechin 3-gallate (EGCG), the most bioactive and abundant green tea catechin, has been thought as a chemopreventive agent by inhibition of growth and induction of apoptosis in various cancers including breast cancer. However, the molecular mechanisms responsible for cancer preventive effects of EGCG on breast cancer development are not fully elucidated. EGCG was found to suppress heregulin-β1-stimulated HER2/HER3 hetero-dimerization in breast cancer cells which initiates mitogenic signal transduction required for cancer progression. Also, the aggressive downstream phenotypes controlled by HER2 were inhibited by treatment of EGCG in breast cancer cells. These findings suggest that blockade of HER2/HER3 co-receptor formation by EGCG may be one of the possible cancer preventive mechanisms of EGCG in HER2-positive breast cancer.

Breast cancer is the most common and frequent cause of death among women in the industrialized world and its incidence closely correlates with genetic abnormalities. Several genes (such as *p53*, *BRCA1* and *BRCA2*), frequently identified in primary human breast tumor, have been characterized in pathogenesis of breast cancer (*1, 2*). One of them is *HER2* (human epidermal growth factor receptor 2) *oncogene*, which is amplified in about 25 % of human breast cancers with poor prognosis and chemoresistance (*3, 4*). HER2-positive breast cancer is highly proliferative and invasive with metastatic potential, as demonstrated by ectopic overexpression of HER2 in mouse embryo fibroblast 3T3 cells (*5*). Although amplification of HER2 is originally identified in breast cancer, aberrant HER2 expression has been found in a variety of other human cancers such as ovarian, gastric, and salivary cancers (*6*), implicating a critical role for HER2 in the development of human cancers.

The *HER2* gene encodes a transmembrane receptor tyrosine kinase (RTK) and belongs to the second member of erbB receptor family. ErbB receptor family, also known as epidermal growth factor (EGF) receptor family, comprises four homologous transmembrane receptor tyrosine kinases (erbB1-4 or Her1-4) which organizes a complex growth factor-mediated cellular signaling and appears to be important regulators of cell proliferation and differentiation (*7*). Ligand binding initiates signaling via erbB receptor homo- or heterodimerization, which in turn recruits specialized adaptors and kinases, thereby triggers and activates a network of signaling pathways (*8*). Increased expression and activation of erbB are tightly associated with development and progression in various types of cancer (*9*), suggesting erbB family receptors as targets for cancer therapy. Despite many intrinsic erbB family ligands have been recognized (*10*), HER2 remains an orphan receptor different from the other members and none of erbB family ligands can bind to HER2 with high affinity. In fact, activation of HER2 bases on its own overexpression to form the active homodimer or trans-activation by heregulins (HRGs) (*11*). HRGs are natural EGF-like ligands for HER3 or HER4 and highly expressed in breast cancer biopsies (*12*). Under HRG stimulation, HER2 prefers to heterodimerize to HER3 and proceeds signaling through activating intracellular signal pathways including phosphatidylinositol 3-kinase (PI3K)/Akt and mitogen-activated protein kinase (MAPK) pathways (*13*). Increasing evidences have indicated that suppression of HRGs expression down-regulates HER2 activation and results in the inhibition of downstream mitogenic responses, suggesting a pivotal role of HRGs in the development and the pathogenesis of breast cancer (*14*).

Tea (*Camellia sinensis*) is the most worldwide consumed beverage and its constituents have been extensively investigated. Epidemiological studies suggest that tea consumption may have a protective effect against human cancer development (*15*). Among a number of bioactive compounds of tea, (-)-epigallocatechin gallate (EGCG) (Figure 1) is thought to be the most powerful cancer chemopreventive agent. Studies in animal models have demonstrated that EGCG can block all stages of carcinogenesis (*16*) and inhibit tumor angiogenesis, metastasis and invasion (*17, 18*). One possible mechanism responsible for the cancer preventive effects of EGCG is known to inhibit growth factor-related proliferation (*19*). EGCG has been shown to block the activation of EGF receptor tyrosine kinase and lead to the inhibition of cell growth in A431 epidermoid carcinoma cells (*20*). EGFR-dependent kinases, including ERK1/2 and Akt, have been demonstrated as molecular targets for EGCG (*21*). In human head and neck squamous cell carcinoma (HNSCC) cells, EGCG inhibits TGF-α-mediated *cyclin D1* and *c-fos* promoter activities responsible for cell cycle progression, accompanied by cell cycle arrest and apoptosis (*22*). In addition, several investigations have demonstrated that EGCG suppresses activation of HER2, HER3 and the corresponding downstream signaling pathways in HNSCC, breast and human colon cancer cells (*23-25*). These findings implicate that EGCG may inhibits the downstream phenotype controlled by growth factor receptor, including ErbB receptor family.

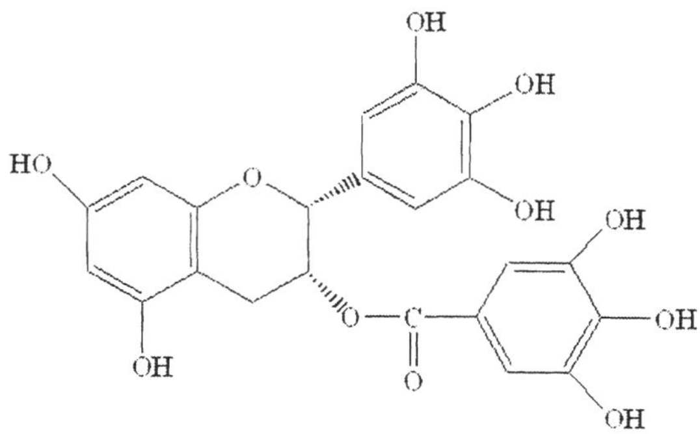

Figure 1. Chemical structure of (-)-epigallocatechin-3 catechin (EGCG)

In spite of many reports regarding the anti-cancer properties of EGCG, the molecular mechanisms responsible for cancer preventive effects of EGCG on breast cancer carcinogenesis is not fully known. In this brief contribution, we selected several human breast cancer cell lines expressing basal or over level of HER2 protein to evaluate the effects of EGCG on the interaction and activation of HER2 and HER3 under heregulin-β1 (HRG-β1, a specific HER3 ligand) stimulation. The inhibition of HER2/HER3 heterodimer function by EGCG may be an effective strategy for suppression of HER2-mediated carcinogenesis in breast cancer cells.

Materials and Methods

Materials

Recombinant human heregulin-β1 was purchased from R&D Systems (Minneapolis, MN). EGCG was obtained from Sigma (St. Louis, MO). Antibody against FAS was obtained from BD Biosciences (Los Angeles, CA). The antibodies to HER2, HER3, MMP-2, VEGF-A and β-actin were obtained from Santa Cruz Biotechnology (Santa Cruz, CA).

Cell Culture

Monolayer cultures of MCF-7 and AU565 cells were grown in Dulbecco's minimal essential medium (DMEM), and MDA-MB-453 cells were maintained in DMEM/F12 (Invitrogen). All cells were supplemented with 10% fetal calf serum (Gibco BRL, Grand Island, NY), 100 units/mL of penicillin, 100 μg/mL of streptomycin, and kept at 37°C in a humidified atmosphere of 5% CO_2 in air.

Immunoprecipitation

Five hundred micrograms of total cellular proteins in MCF-7 cell lysates were first pre-cleared by incubating with protein A-agarose (10 μL, 50% slurry; Santa Cruz Biotechnology, Santa Cruz, CA) for 15 min. The clarified supernatants were collected by microfugation, and then incubated with HER2 antibody for 2 h at 4°C. The reaction mixtures were added with 20 μL of protein A-agarose to absorb the immunocomplexes at 4°C overnight. Immunoprecipitated proteins were subjected to 8% SDS-PAGE, and then transferred onto PVDF membrane (Millipore). The HER3 proteins were visualized by Western blotting.

Western Blotting

Cell extracts were prepared in a lysis buffer (50 mM Tris-HCl, pH 8.0; 5 mM EDTA; 150 mM NaCl; 0.5% NP-40; 0.5 mM phenylmethylsulfonyl fluoride; and 0.5 mM dithiothreitol) for 30 min at 4°C. Equal amounts of total cellular proteins (50 μg) were resolved by SDS-polyacrylamide gel electrophoresis (SDS-PAGE) (6% for FAS; 10% for MMP-2 and β-actin; 12% for VEGF-A), transferred onto polyvinylidene difluoride (PVDF) membranes (Immobilon P, Millipore, Bedford, MA), and then probed with primary antibody followed by secondary antibody conjugated with horseradish peroxidase. The immunocomplexes were visualized with enhanced chemiluminescence kits (Amersham, UK).

Results and Discussion

Tyrosine phosphorylation of HER3 can be observed under HRG-β1 stimulation; however, HER3 defects the intrinsic tyrosine kinase activity (*26*) and it means that receptor heterodimerization is needed for HER3 activation. In fact, HER2 is the preferred partner of co-receptor for HER3 and HER2/HER3 heterodimer appears to be the most transforming and mitogenic receptor complex (*27*). To determine the effect of EGCG on the interaction between HER2 and HER3, MCF-7 breast cancer cells expressing basal level of HER2 were treated with different concentrations of EGCG and then stimulated by HRG-β1. HER2/HER3 interaction was detected by co-immunoprecipitation using a HER2 antibody. After HRG-β1 stimulation, HER3 protein was apparently observed compared with that in unstimulated cells as demonstrated by Western blotting with a HER3 antibody, and the co-immunoprecipitated HER3 with HER2 by HRG-β1 stimulation were gradually disappeared under EGCG pre-treatment (Figure 2).

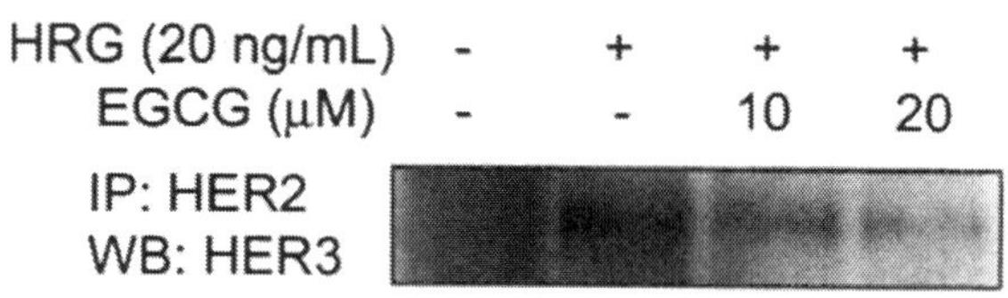

Figure 2. EGCG down-regulates HER2/HER3 heterodimerization due to HRG-β1 stimulation. Serum-starved MCF-7 cells were pre-incubated with various doses of EGCG for 30 min, and then stimulated with 20 ng/mL of HRG-β1 for 10 min. At the end of incubation, cell lysates were harvested and immunoprecipitated with HER2 antibody, and then the co-precipitated HER3 was determined by Western blotting using a HER3 antibody.

Moreover, the tyrosine phosphorylation of HER2 and HER3 was positively correlated with the HER2/HER3 heterodimer formation (unpublished data). These findings indicate that EGCG could inhibit HER2/HER3 signaling by preventing the coordination of HER2 and HER3 by HRG-β1 in MCF-7 breast cancer cells.

To further investigate the effects of EGCG on the downstream phenotype controlled by HER2/HER3 signaling, the protein levels of fatty acid synthase (FAS), matrix metalloproteinase-2 (MMP-2) and vascular endothelial growth factor A (VEGF-A) were determined by Western blotting in several breast cancer cell lines. All of these proteins have been considered as the downstream effectors up-regulated by HER2 or HER3 (*28-30*). As shown in Figure 3, EGCG markedly inhibited HRG-β1-induced FAS expression in breast cancer cell lines, MDA-MB-453 and AU565. The increased expression of MMP-2 and VEGF-A by HRG-β1 were also inhibited by EGCG treatment in MDA-MD-453 and MCF-7 breast cancer cells. These data suggest that EGCG down-regulates the induction of FAS, MMP-2 and VEGF-A by HRG-β1 via the suppression of HER2/HER3 coordination. Interestingly, serum-starved MDA-MB-453 breast cancer cells expressed high level of FAS protein; even there was no HRG-β1 stimulus. The reason may be that MDA-MB-453 cells are more malignant than AU565 cells, and HER2 in MDA-MD-453 cells are overexpressed and highly active through HER2 homodimer formation despite serum deprivation. Therefore, the expression of fully level of FAS due to HER2 overexpression is predictable in serum-starved MDA-MD-453 cells.

Based on these preliminary data, we propose a molecular model for cancer preventive action of EGCG in HER2- or/and HER3-overexpresing breast cancer (Figure 4). EGCG inhibits the formation of HER2/HER3 co-receptor by HRG-β1 and finally leads to down-regulation of FAS, MMP-9 and VEGF-A levels in breast cancer cells. Inhibition of these HER2- or HER3-related proteins by treatment of EGCG (Figure 3) may contribute to cancer prevention before breast cancer onset, because FAS, MMP-9 and VEGF-A have been shown as important mediators for malignant progression of tumor cells. FAS has been implicated in tumorigenesis through its role in cell proliferation and membrane lipid incorporation of neoplastic cells, and suppression of FAS function in cancer cells leads to growth inhibition and the induction of apoptosis (*31*). Also, MMP-2 activity is involved in tumor invasion and metastasis by its capacity for degradation of extracellular matrix (ECM) of basement membrane. Statistics indicates that MMP-2 is associated with a poor prognosis in breast carcinoma patients (*32*) and thought to be a target of developed MMP inhibitors. Additionally, VEGF-A secreted by tumor cells has been demonstrated as a key regulator of angiogenesis (*33*). Accumulating studies indicate that either HER2 gene amplification or HRG-β1 stimulation regulates elevated VEGF-A expression in breast cancer cells (*30*), implicating that VEGF-A may be a crucial mediator for the aggressive phenotype of HER2 or/and HER3-overexpressing breast cancer. Based on these observations, we suggest that anti-proliferation, metastasis and angiogenesis will be accompanied by suppression of FAS, MMP-2 and VEGF-A, respectively, by EGCG via inhibition of HER2/HER3 signaling. This hypothesis may explain why drinking green tea can lower a recurrence rate among breast cancer patients.

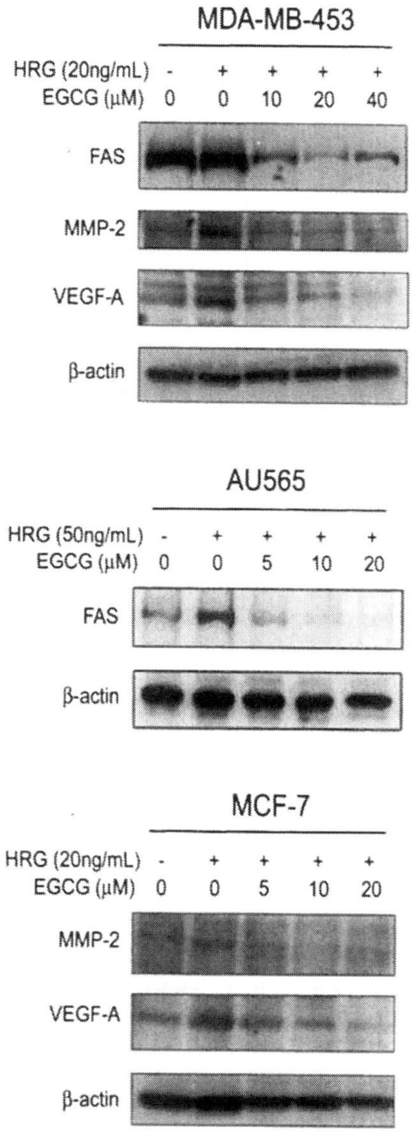

Figure 3. EGCG inhibits the expression of FAS, MMP-2 and VEGF-A proteins by HRG-β1 in breast cancer cell lines. Serum-starved breast cancer cells as indicated were pre-incubated with various doses of EGCG for 30 min, and then stimulated with HRG-β1 (20 or 50 ng/mL) for 9 h. Western blotting analysis was performed using specific antibodies to FAS, MMP-2, VEGF-A or β-actin. β-actin represented as a loading control.

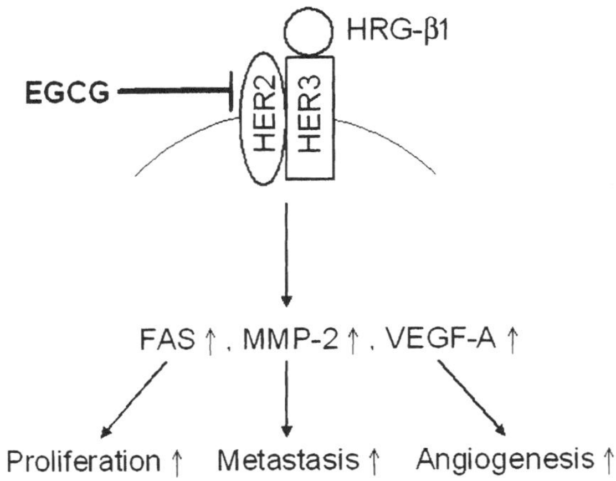

Figure 4. Proposed model for cancer preventive action of EGCG in HER2 or/and HER3-overexpressing breast cancer. EGF-like growth factor HRG-β1, a HER3 ligand, binds to HER3, initiates signaling via HER2/HER3 heterodimerization, and consequently increases the expression of FAS, MMP-2 and VEGF-A which are responsible for the aggressive phenotype such as proliferation, metastasis and angiogenesis, respectively, in breast cancer cells. EGCG may achieve its cancer preventive effects on breast cancer by inhibiting the formation of HER2/HER3 co-receptor by HRG-β1 and finally down-regulating the downstream effectors that contribute to malignancy of breast cancer.

Although our current study suggests that EGCG may be a useful chemopreventive agent for breast cancer carcinogenesis, how EGCG interrupts HER2/HER3 interaction essential for tumor progression still remains elusive. Recent evidence showed that EGCG can block EGFR-mediated proliferation in A431 epidermoid carcinoma by competition with EGF binding to EGFR (*20*). It raises the possibility that EGCG may prevent HRG-β1 binding to HER3 receptor by similar pathway; however, this issue requires further study. To sum up, our findings point out a new insight on the molecular mechanism by which EGCG blocks breast cancer development or progression and suggest that EGCG may by an effective agent in the prevention of cases of breast carcinoma where HER2 or/and HER3 overexpression.

Acknowledgement

This study was supported by the National Science Council NSC 94-2312-B-040-001 and NSC 95-2320-B-040-037.

References

1. Thor, A. D.; Moore, D. H., II; Edgerton, S. M.; Kawasaki, E. S.; Reihsaus, E.; Lynch, H. T.; Marcus, J. N.; Schwartz, L.; Chen, L. C.; Mayall, B. H. *J. Natl. Cancer. Inst.* **1992**, *84*, 845-855.
2. Radice, P. *J. Exp. Clin. Cancer Res.* **2002**, *21*, (3 Suppl), 9-12.
3. Slamon, D. J.; Clark, G. M.; Wong, S. G.; Levin, W. J.; Ullrich, A.; McGuire, W. L. *Science* **1987**, *235*, (4785), 177-82.
4. Menard, S.; Tagliabue, E.; Campiglio, M.; Pupa, S. M. *J. Cell. Physiol.* **2000**, *182*, 150-162.
5. Yu, D. H.; Hung, M.C. *Oncogene* **1991**, *6*, 1991-1996.
6. Holbro, T.; Hynes, N. E. *Annu. Rev. Pharmacol. Toxicol.* **2004**, *44*, 195-217.
7. Oltvai, Z, N.; Milliman, C. L.; Korsmeyer, S. J. *Cell.* **1993**, *74*, 609-619.
8. Yarden, Y.; Sliwkowski, M. X. *Nat. Rev. Mol. Cell. Biol.* **2001**, *2*, 127-137.
9. Hynes, N. E.; Lane, H. A. *Nat. Rev. Cancer* **2005**, *5*, 341-354.
10. Harris, R. C.; Chung, E.; Coffey, R. J. *Exp. Cell. Res.* **2003**, *284*, 2-13.
11. Peles, E.; Yarden, Y. *Bioessays* **1993**, *15 (12)*, 815-824.
12. Dunn, M.; Sinha, P.; Campbell, R.; Blackburn, E.; Levinson, N.; Rampaul, R.; Bates, T.; Humphreys, S.; Gullick, W. J. *J. Pathol.* **2004**, *203*, 672-680.
13. Carraway, K. L., 3rd; Cantley, L. C. *Cell* **1994**, *78*, 5-8.
14. Tsai, M. S.; Shamon-Taylor, L. A.; Mehmi, I.; Tang, C. K.; Lupu, R. *Oncogene* **2003**, *22*, 761-768.
15. Ahmad, N.; Mukhtar, H. *Nutr. Rev.* **1999**, *57*, 78-83.
16. Yang, Y. A.; Han, W. F.; Morin, P. J.; Chrest, F. J.; Pizer, E. S. *Exp. Cell. Res.* **2002**, *279*, 80-90.
17. Fassina, G.; Vene, R.; Morini, M.; Minghelli, S.; Benelli, R.; Noonan, D. M.; Albini, A. *Clin. Cancer Res.* **2004**, *10*, 4865-4873.
18. Garbisa, S.; Sartor, L.; Biggin, S.; Salvato, B.; Benelli, R.; Albini, A. *Cancer* **2001**, *91*, 822-832.
19. Lambert, J. D.; Yang, C. S. *J. Nutr.* **2003**, *133*, 3262S-3267S.
20. Liang, Y. C.; Lin-shiau, S. Y.; Chen, C. F.; Lin, J. K. *J. Cell Biochem.* **1997**, *67*, 55-65.
21. Sah, J. F.; Balasubramanian, S.; Eckert, R. L.; Rorke, E. A. *J. Biol. Chem.* **2004**, *279*, 12755-12762.
22. Masuda, M.; Suzui, M.; Weinstein, I. B. *Clin. Cancer Res.* **2001**, *7*, 4220-4229.
23. Masuda, M.; Suzui, M.; Lim, J. T.; Weinstein, I. B. *Clin. Cancer Res.* **2003**, *9*, 3486-3491.
24. Pianetti, S.; Guo, S.; Kavanagh, K. T.; Sonenshein, G. E. *Cancer Res.* **2002**, *62*, 652-655.
25. Shimizu, M.; Deguchi, A.; Lim, J. T.; Moriwaki, H.; Kopelovich, L.; Weinstein, I. B. *Clin. Cancer Res.* **2005**, *11*, 2735-2746.

26. Citri, A.; Skaria, K. B.; Yarden, Y. *Exp. Cell. Res.* **2003**, *284*, 54-65.
27. Pinkas-Kramarski, R.; Soussan, L.; Waterman, H.; Levkowitz, G.; Alroy, I.; Klapper, L.; Lavi, S.; Seger, R.; Ratzkin, B. J.; Sela, M.; Yarden, Y. *Embo. J.* **1996**, *15*, 2452-2467.
28. Kumar-Sinha, C.; Ignatoski, K. W.; Lippman, M. E.; Ethier, S. P.; Chinnaiyan, A. M. *Cancer Res.* **2003**, *63*, 132-139.
29. Pellikainen, J. M.; Ropponen, K. M.; Kataja, V. V.; Kellokoski, J. K.; Eskelinen, M. J.; Kosma, V. M. *Clin. Cancer Res.* **2004**, *10*, 7621-8.
30. Yen, L.; You, X. L.; Al Moustafa, A. E.; Batist, G.; Hynes, N. E.; Mader, S.; Meloche, S.; Alaoui-Jamali, M. A. *Oncogene* **2000**, *19*, 3460-3469.
31. Kuhajda, F. P. *Nutrition* **2000**, *16*, 202-208.
32. Talvensaari-Mattila, A.; Paakko, P.; Hoyhtya, M.; Blanco-Sequeiros, G.; Turpeenniemi-Hujanen, T. *Cancer* **1998**, *83*, 1153-1162.
33. Dvorak, H. F.; Brown, L. F.; Detmar, M.; Dvorak, A. M. *Am. J. Pathol.* **1995**, *146*, 1029-1039.

Chapter 29

Induction of Apoptosis by Acetylated Black Tea Polyphenol through Reactive Oxygen Species Production, Cytochrome *c* Release, and Caspases Activation in Human Leukemia HL-60 Cells

Min-Hsiung Pan[1], Hui-Jun Kang[1], Wei-Jen Chen[2], Chih-Yu Lo[3], Shiming Li[3], Shengmin Sang[4], and Chi-Tang Ho[3]

[1]Department of Seafood Science, National Kaohsiung Marine University, Kaohsiung, Taiwan
[2]Department of Biomedical Sciences, Chung Shan Medical University, No, 110, Section 1, Chien-Kuo N. Road, Taichung 402, Taiwan
[3]Department of Food Science, Rutgers, The State University of New Jersey, New Brunswick, NJ 08901
[4]Department of Chemistry Biology, Ernest Mario School of Pharmacy, Rutgers, The State University of New Jersey, Piscataway, NJ 08854

The growth inhibitory effects of black tea polyphenol, theaflavin (TF-1) and its peracetylated derivative (ATF-1) in human leukemia cancer cells were examined. TF-1 and ATF-1 displayed strong growth inhibitory effects against human leukemia HL-60 cells. ATF-1 was more potent against the growth of HL-60 cells and induction of apoptosis through modulation of mitochondrial functions regulated by reactive oxygen species (ROS). ROS generation occurs in the early stages of ATF-1-induced apoptosis, preceding cytochrome c release, caspase activation, and DNA fragmentation. The molecular mechanism of ATF-1-induced apoptosis was also investigated.

Tea (*Camellia sinensis*) is the most popular beverage, after water, consumed worldwide; the major tea beverage is black tea, especially in the western nations. Black tea leaves are produced through extensive enzymatic oxidation of polyphenols to polymerized products, such as theaflavines and thearubigins. The major theaflavins in black tea are theaflavin, theaflavin-3-gallate, theaflavin-3′-gallate, and theaflavin-3,3′-digallate (*1*). Green tea polyphenols such as (-) epigallocatechin-3-gallate (EGCG) have been demonstrated to have several inhibitory properties on the growth of tumor cell lines (*2*). The molecular mechanism of antitumor growth might operate through blocking the signal transduction pathway (*3*). However, other tea constituents such as theaflavins from black tea also have anti-proliferative or anti-carcinogenic activities (*4*). Recently, Lu *et al.* (*5*) reported that black tea significantly inhibited proliferation and enhances apoptosis in mouse skin tumor models. Halder and Bhaduri (*6*) reported that theaflavins and thearubigins have antioxidative properties on human red blood cells. Among black tea components, theaflavins are generally considered to be the more effective components for the inhibition of carcinogenesis, but it is unclear which of these theaflavins is the most effective one.

Chemoprevention, a promising strategy to prevent cancer is the use of either natural or synthetic substances or their combination to block, reverse or retard the process of carcinogenesis (*7*). Apoptosis plays a fundamental role in the maintenance of tissues and organ systems by providing a controlled cell deletion to balanced cell proliferation. A common feature of cancer cells is their ability to evade apoptosis as a result of alterations that block cell death signaling pathways (*8*). It is now apparent that many dietary chemopreventive agents with promise for human consumption can also preferentially inhibit the growth of tumor cells by targeting one or more signaling intermediates leading to induction of apoptosis (*9*).

In the current study, was first examined the antiproliferative effects of TF-1 and ATF-1 on human leukemia cells. The results clearly demonstrate that ATF-1 can induce apoptosis in a dose-dependent manner in HL-60 cells. The molecular mechanisms of the apoptotic effects induced by ATF-1 were further evaluated. It is suggested that ATF-1 modulates the production of ROS, Bcl-2 family proteins, the release of cytochrome c, and the activation of caspases in ATF-1 induced apoptosis. The results of the present study will provide molecular basis for designing compounds and understanding the healthy effect of black tea.

Materials and Methods

Cell Culture

Human promyelocytic leukemia (HL-60) cells obtained from American Type Culture Collection (Rockville, MD) were grown in 90% RPMI 1640 and

10% fetal bovine serum (GIBCO BRL, Grand Island, NY), supplemented with 2 mM glutamine (GIBCO BRL), 1% penicillin/streptomycin (10000 units of penicillin/mL and 10 mg streptomycin/mL). Medium was normally changed to phenol red-free RPMI 1640 before polyphenol treatment. Propidium iodide was obtained from Sigma Chemical Co. (St. Louis, MO).

Materials

Anhydrous pyridine, acetic anhydride, 4-(dimethylamino)-pyridine, (4-DMAP), 99+% Sephadex LH-20 gel, and HPLC sorbent (C-18) were purchased from Sigma-Aldrich Chemical Co (St. Louis, MO). Ethyl acetate (EtOAc), methanol and water (HPLC grade) were purchased from Fisher Scientific (Springfield, NJ). Black tea extract was obtained from WellGen, Inc (New Brunswick, NJ).

Ethyl acetate fraction of theaflavin extract was subjected to a Sephadex LH-20 column and eluted with acetone solution (40%, v/v). During the elution process, their unique benzotropolone structure of theaflavin with orange-red color can be easily monitored. Crude theflavin (TF-1) was collected. Crude TF-1 was further purified by a self pack C-18 column in a 20% methanol isocratic method. Fractions were checked by TLC and pure TF-1 fractions were collected. TF-1 was ready for further acetylation reaction.

Procedure for the Synthesis of TF-1 and Its Peracetylated Derivative

TF-1 (0.21 mmol) was dissolved in 4 mL EtOAc and the 4-DMAP (1.05 mmol) was dissolved in anhydrous pyridine (0.86 mL). The 4-DMAP/pyridine and acetic anhydride (10.63 mmol) was subsequently added dropwise to TF-1/EtOAc which had been stirred and placed in ice bath. The mixture was refluxed for 24 hours.

The mixture was refluxed for 6 h. Then, the reaction mixture was cooled to room temperature and quenched with water. The mixture was then extracted with EtOAc, washed with brine and dried over sodium sulfate. After removing the solvent, the mixture was dissolved in 1.00 mL acetone and applied to LH-20 column which was eluted with 50% acetone solution. Their purities as determined by HPLC were at least greater than 95% and 93% for ATF-1 (HPLC assay). AFT-1 showed positive ESI-MS *m/z* 943 $[M+H]^+$.

Determination of Cell Viability

Cell viability was determined for indicated compounds based on the trypan blue exclusion method. Briefly, HL-60 cells were plated at a density of 1×10^5

cells/mL into 24-well plates. After overnight growth, cells were pretreated with a series of concentrations of TF-1 and ATF-1 for 24 h. The final concentrations of dimethyl sulfoxide (DMSO) in the culture medium were <0.05%. The viability percentage was calculated based on the percentage of unstained cells as described previously (*10*).

Acridine Orange Staining Assay

Cells (5×10^5) were seeded into 60-mm Petri dishes and incubated at 37 °C for 24 h. The cells were harvested after treatment for 24 h, and 5 μL of cell suspension was mixed on a slide with an equal volume of acridine orange solution (10 μg/mL in PBS). Green fluorescence was detected between 500 and 525 nm by using an Olympus microscope (Olympus America, Inc., Lake Success, NY). Bright-staining condensed chromatin was detected in apoptotic cells.

Flow Cytometry

HL-60 cells (2×10^5) were cultured in 60-mm Petri dishes and incubated for 24 h. The cells were treated with various concentrations (0, 10, 15, 30, 60, and 100 μg/mL) of compounds for 24 h or 20 μg/mL ATF-1 for the indicated time (3, 6, 9, and 12 h). The cells were then harvested, washed with PBS, resuspended in 200 μL of PBS, and fixed in 800 μL of 100% ethanol at -20 °C. After being left to stand overnight, the cell pellets were collected by centrifugation, resuspended in 1 mL of hypotonic buffer (0.5% Triton X-100 in PBS and 0.5 μg/mL RNase), and incubated at 37 °C for 30 min. Next, 1 mL of propidium iodide solution (50 μg/mL) was added, and the mixture was allowed to stand on ice for 30 min. Fluorescence emitted from the propidium iodide-DNA complex was quantitated after excitation of the fluorescent dye by FACScan cytometry (Becton Dickinson, San Jose, CA).

DNA Extraction and Electrophoretic Analysis

The HL-60 cells were harvested, washed with phosphate-buffered saline (PBS), and then lysed overnight at 56 °C with a digestion buffer containing 0.5% sarkosyl, 0.5 mg/mL proteinase K, 50 mM tris(hydroxymethyl) aminomethane (pH 8.0), and 10 mM EDTA. Following lyses, the cells were then treated with RNase A (0.5 μg/mL) for 3 h at 56 °C. The DNA was then extracted using phenol/chloroform/isoamyl (25:24:1, v/v/v) prior to loading and was analyzed by 2% agarose gel electrophoresis. The agarose gels were run at 50 V for 120 minutes in Tris-borate/EDTA electrophoresis buffer (TBE). Approximately 20 μg of DNA were loaded in each well with 6x loading buffer containing 0.25% bromophenol blue, 0.25% xylene cyanol FF, and 40% sucrose, DNA was stained with ethidium bromide and visualized under UV light at 260 nm, and the plates photographed.

Production of Reactive Oxygen Species (ROS)

ROS production was monitored by flow cytometry using 2′,7′-dichlorodihydrofluorescein-diacetate (DCFH-DA). This dye is readily diffuses into cells and is hydrolyzed by intracellular esterase to yield DCFH, which is trapped within the cells. Hydrogen peroxide or low molecular weight peroxides produced by the cells oxidize DCFH to the highly fluorescent compound 2′,7′-dichlorofluorescein (DCF). Thus, the fluorescence intensity is proportional to the amount of peroxide produced by the cells. Cells were treated with ATF-1 (20 μg/mL) for different time periods, and DCFH-DA (30 μM) was added into the medium for a further 30 min at 37 °C.

Analysis of Mitochondrial Transmembrane Potential

The change of the mitochondrial transmembrane potential was monitored by flow cytometry. Briefly, HL-60 cells were exposed to ATF-1 (20 μg/mL) for different time periods and the mitochondrial transmembrane potential was measured directly using 40 nM 3,3′-dihexyloxacarbocyanine [DiOC6(3)] (Molecular Probes, Eugene, OR). Fluorescence was measured after staining of the cells for 30 min at 37 °C. Histograms were analyzed using Cell Quest software and were compared with histograms of control untreated cells.

Western Blotting

The nuclear and cytosolic proteins were isolated from HL-60 cells after treatment with 20 μg/mL ATF-1 for 0, 3, 6, 9, and 12 h. The total proteins were extracted via the addition of 200 μL of gold lysis buffer (50 mM Tris-HCl, pH 7.4; 1 mM NaF; 150 mM NaCl; 1 mM ethylene glycol bis (β-aminoethyl ether) N,N,N',N'-tetraaceteic acid (EGTA); 1 mM phenylmethanesulfonyl fluoride; 1% NP-40; and 10 μg/mL leupeptin) to the cell pellets on ice for 30 min, followed by centrifugation at 10,000 g for 30 min at 4 °C. The cytosolic fraction (supernatant) proteins were measured by Bio-Rad protein assay (Bio-Rad Laboratories, Munich, Germany). The samples (50 μg of protein) were mixed with 5× sample buffer containing 0.3 M Tris-HCl (pH 6.8), 25% 2-mercaptoethanol, 12% sodium dodecyl sulfate (SDS), 25 mM EDTA, 20% glycerol, and 0.1% bromophenol blue. The mixtures were boiled at 100 °C for 5 min and then subjected to 12% SDS-polyacrylamide minigels at a constant current of 20 mA. Subsequently, electrophoresis was ordinarily carried out on SDS-polyacrylamide gels. After electrophoresis, proteins on the gel were electrotransferred onto an immobile membrane (PVDF; Millipore Corp., Bedford, MA) with transfer buffer composed of 25 mM Tris-HCl (pH 8.9), 192 mM glycine, and 20% methanol. The membranes were blocked with blocking solution containing 20 mM Tris-HCl and then immunoblotted with primary antibodies including anti-Bcl-2, anti-Bcl-XL, anti-Bad, anti-Bax, anti-β-actin (Santa Cruz Biotech.), anti-PARP (UBI, Inc., Lake Placid, NY), anti-Bid, anti-caspase-8, anti-caspase-3, anti-Fas and Fas-L (Transduction Laboratory, Lexington, KY), and anti-DFF45/ inhibitor of caspase-activated DNase (ICAD) antibody (MBL, Naka-Ku, Nagoya, Japan) at room temperature for 1 h. Detection was achieved by measuring the chemiluminescence of blotting agent (ECL, Amersham Corp., Arlington Heights, IL), after exposure of the filters to Kodak X-Omat films. The cytochrome c protein was detected by using anti-cytochrome c antibody (Research Diagnostic Inc., Flanders, NJ).

Activity of Caspase

HL-60 cells (2 $\times 10^5$) were cultured in 100-mm Petri dishes. After 24 h of incubation, cells were treated with ATF-1 (20 μg/mL) for various times. Cells were collected and washed with PBS and suspended in 25 mM HEPES (pH 7.5), 5 mM $MgCl_2$, 5 mM EDTA, 5 mM dithiothione, 2 mM phenylmethanesulfonyl fluoride, 10 μg/mL pepstatin A, and 10 μg/mL leupeptin after treatment. Cell lysates were clarified by centrifugation at 12000 *g* for 20 min at 4 °C. Caspase activity in the supernatant was determined by a fluorogenic assay (Promeaga's CaspACE Assay System Corp., Madison, WI). Briefly, 50 μg of total protein, as determined by Bio-Rad protein assay (Bio-Rad Laboratories, Munich, Germany), were incubated with 50 μM substrate Ac-Try-Val-Ala-Asp-AMC (Ac-YVAD) (caspase-1-specific substrate), Ac-Asp-Glu-Val-Asp-AMC (Ac-DEVD-AMC) (caspase-3-specific substrate), Ac-Ile-Glu-Thr-Asp-AMC (Ac-IETD-AMC) (caspase-8-specific substrate), or Ac-Leu-Glu-His-Asp-AMC (Ac-LEHD-AMC) (caspase-9-specific substrate) at 30 °C for 1 h. The release of methylcoumaryl-7-amine (AMC) was measured by excitation at 360 nm and emission at 460 nm using a fluorescence spectrophotometer (Hitachi, F2000, Tokyo, Japan).

Results

Treatment with TF-1 and ATF-1 Causes Dose-dependent Reduction in Cell Growth

Previous studies have shown that tea polyphenols are potent antiproliferation and anticancer agents (*11*). Here, the cytotoxicity of TF-1 and ATF-1 was investigated. The structures of compounds of interest are illustrated in Figure 1.

Human leukemia HL-60 cells were treated with different concentrations of selected compounds. After 24 h of treatment, the number of live cells was determined by means of trypan blue exclusion method. As shown in Figure 2, ATF-1 is a more potent inhibitor of cell viability than other compounds. The inhibition of cell viability of ATF-1 was found to be dose-dependent with an IC_{50} of 5 μg/mL.

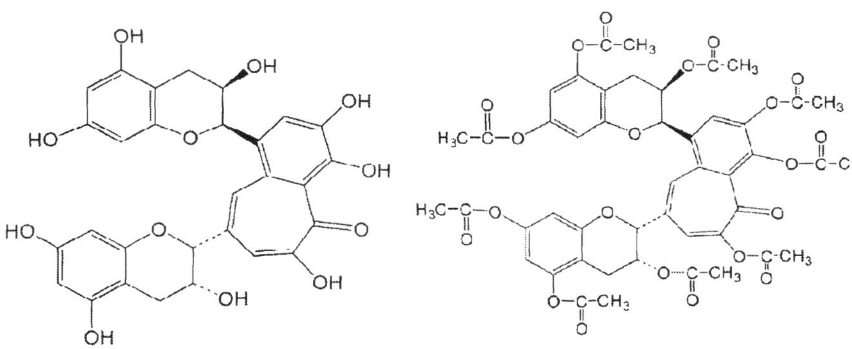

Figure 1. Chemical structures of (Left) TF-1 and (Right) Acetylated TF-1 (ATF-1).

Induction of Apoptosis by TF-1 and ATF-1 in Human Leukemia Cells

To investigate the induction of a sub-G1 cell population, the DNA content of HL-60 cells treated with selected compounds for various concentrations was analyzed by flow cytometry. ATF-1 obviously induced apoptosis after incubation with 10 μg/mL, the percentages of apoptotic cells were 80% (Figure 3A). Among them, ATF-1 was the most potent inducer of apoptosis in HL-60 cells. As shown in Figure 3B, ATF-1-induced apoptosis in HL-60 cells in a time dependent manner.

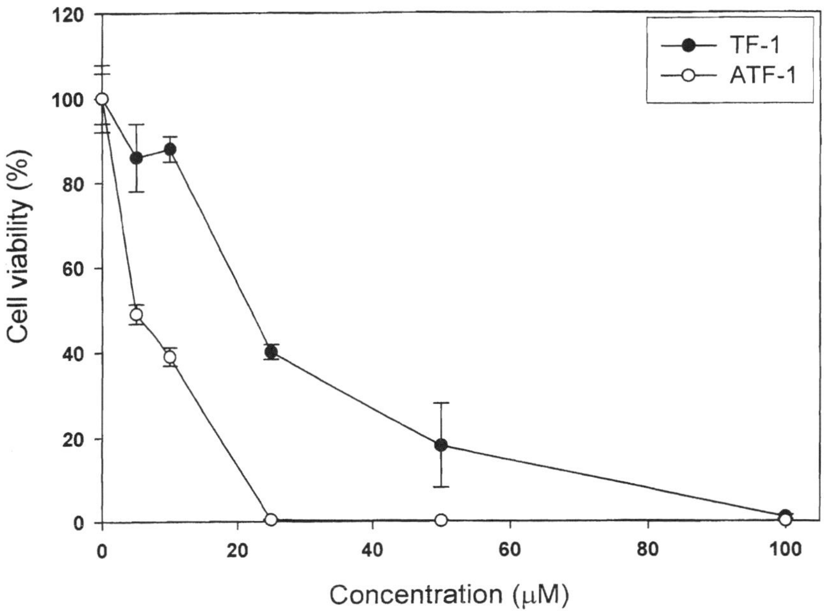

Figure 2. Effect of T-1 and ATF-1 on the cell survival of HL-60 cells. HL-60 cells were treated with different concentration of TF-1 and ATF-1 for 24 h. HL-60 cells were either treated with 0.05% DMSO as vehicle control. Cell viability then was determined by trypan blue assay. As described under Materials and Methods. Data were represented as means ± SD for three determinations.

ATF-1-induced Apoptosis in HL-60 Cells

Physiological cell death is characterized by apoptotic morphology, including chromatin condensation, membrane blebbing, inter-nucleosome degradation of DNA, and apoptotic body formation. To characterize the cell death induced by ATF-1, we examined the DNA ladder and nuclear morphology of dying cells with a fluorescent DNA-binding agent, acridine orange. HL-60 cells were treated with ATF-1 (5–20 μg/mL) for 12 h and DNA fragmentation analyses were performed. As shown in Figure 4A, significant DNA ladders were observed in HL-60 cells after 10 μg/mL of ATF-1 treatment for 12 h. After treatment with 20 μg/mL ATF-1, digested genomic DNA was evident at 6 h (Figure 4B).Treatment with 20 μg/mL ATF-1, cells clearly exhibited significant chromosomal condensation and morphological changes (Figure 4C), which is indicative of apoptotic cell death.

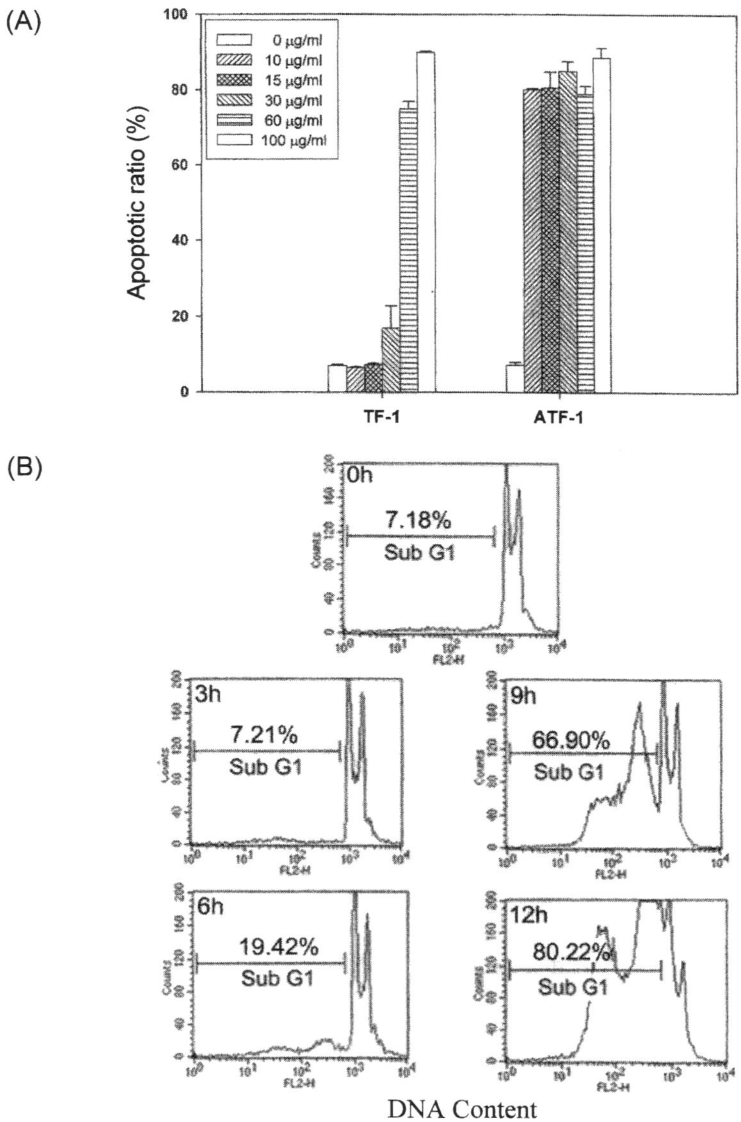

Figure 3. Induction of apoptosis in HL-60 cells by different concentration of TF-1 and ATF-1. HL-60 cells were treated with (A) 0, 10, 15, 30, 60, and 100 μg/mL of TF-1 and ATF-1. (B) Treated with 20 μg/mL ATF-1 for indicated time and Sub-G1 cells in compounds-treated HL-60 cells were determined by flow cytometry. The method of flow cytometry used is described under Materials and Methods. Sub-G1 represents apoptotic cells with a lower DNA content. The data presented are representative of three independent experiments.

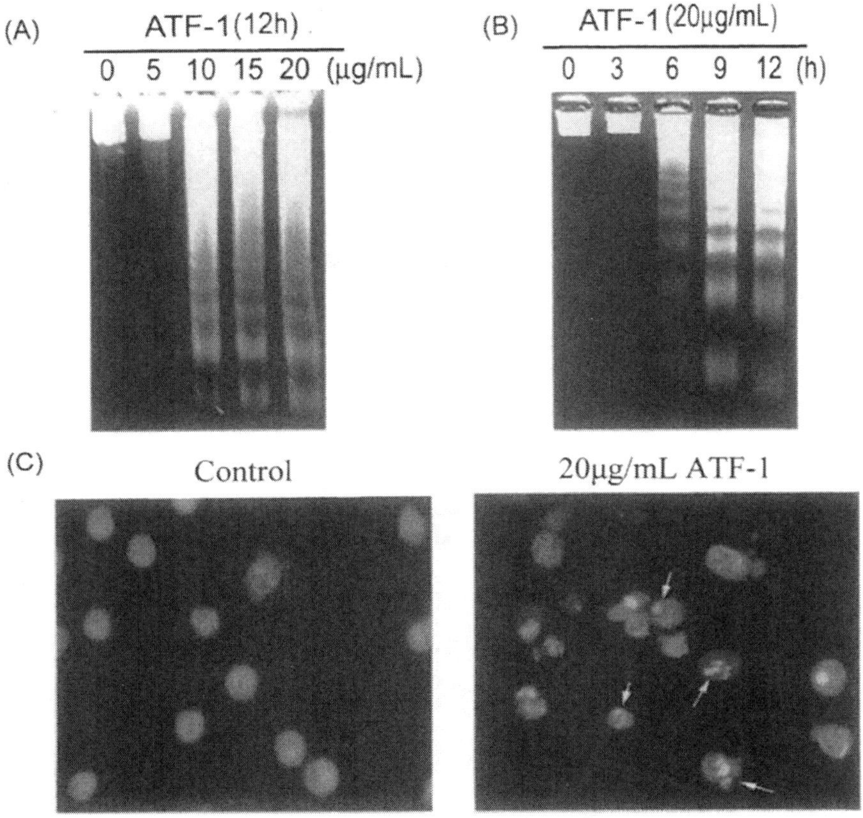

Figure 4. Induction of DNA fragmentation by ATF-1 in HL-60 cells. (A) Cells were treated with different concentration of ATF-1 for 12 h. (B) Cells were treated with 20 μg/mL ATF-1 for indicated time periods. Cellular DNA was extracted and analyzed by agarose electrophoresis. M, marker. Data shown are representative of three independent experiments. (C) HL-60 cells were treated with 0.05% DMSO as vehicle control or treated with 20 μg/mL ATF-1 for 12 h, and cells were harvested and washed with PBS following by staining with acridine orange. The nuclear staining was examined by fluorescence microscopy.

Activation of Caspase-9, Caspase-3, and Caspase-8 but not Caspase-1, Are Involved in ATF-1-induced Apoptosis.

In order to see whether caspases were involved in the cell death response induced by ATF-1, caspases were activated in a sequential cascade of cleavages from their inactive forms *(12)*. As shown in Figure 5A, caspase-9, caspase-3, and caspase-8 were time-dependently activated by ATF-1. In contrast to the increase in caspase-3 activity, negligible caspase-1 activity was observed. As was already described, ICAD is a mouse homologue of human DFF-45. Caspase-3 cleaves DFF-45, and, once caspase-activated deoxyribonuclease (CAD) is released, it can enter the nucleus, where it degrades chromosomal DNA to produce interchromosomal DNA fragmentation *(13, 14)*. Activation of caspase-3 leads to the cleavage of a number of proteins, one of which is poly(ADP-ribose) polymerase (PARP). The cleavage of PARP is the hallmark of apoptosis. PARP (116-kDa) is cleaved to produce a 85-kDa fragmentation during apoptosis *(15)*. Figure 5B shows that exposure of HL-60 cells to ATF-1 causes time- and dose-dependent cleavage of DFF-45 and degradation of 116 kDa PARP to 85 kDa fragments.

Involvement of ROS Production, Mitochondrial Dysfunction, and Release of Cytochrme *c* from Mitochondria to Cytosol in ATF-1-induced Apoptosis

Growing evidence has indicated that ROS play an important role in the induction of apoptosis. The generation of ROS using the fluorescent probe DCFH-DA and monitoring by flow cytometry was studied. HL-60 cells were treated with 20 μg/mL ATF-1 for 15 min. As shown in Figure 6A, the fluorescence intensity shifted to the right from 135.95 to 655.77 in ATF-1-induced apoptotic HL-60 cells. These data indicated that the increment of ROS might play a role as an early mediator in ATF-1-induced apoptosis. The effects of ATF-1 on the mitochondrial transmembrane potential ($\Delta\Psi_m$) and the release of mitochondrial cytochrome *c* into cytosol were then evaluated. We measured $\Delta\Psi_m$ using the fluorescent probe DiOC6(3) fluorescence and monitored it via flow cytometry. As shown in Figure 6B, which compares HL-60 cells exposed to ATF-1 and control cells, the DiOC6(3) fluorescence intensity shifted to the left from 156.55 to 69.21 in ATF-1-induced apoptotic HL-60 cells at 30 min. These results demonstrate that ATF-1 caused a decrease in mitochondrial transmembrane potential in HL-60 cells. As shown in Figure 6C, the release of mitochondrial cytochrome *c* into the cytosol was detected at 30 min in ATF-1-treated HL-60 cells.

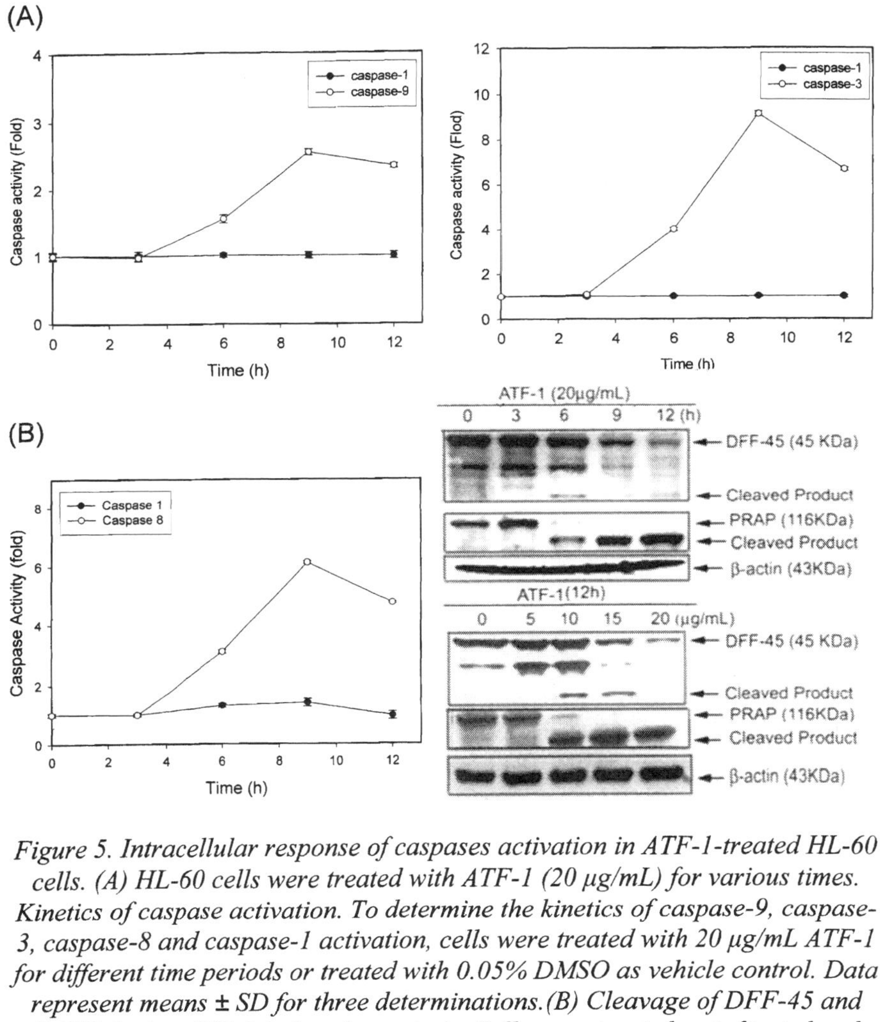

Figure 5. Intracellular response of caspases activation in ATF-1-treated HL-60 cells. (A) HL-60 cells were treated with ATF-1 (20 μg/mL) for various times. Kinetics of caspase activation. To determine the kinetics of caspase-9, caspase-3, caspase-8 and caspase-1 activation, cells were treated with 20 μg/mL ATF-1 for different time periods or treated with 0.05% DMSO as vehicle control. Data represent means ± SD for three determinations.(B) Cleavage of DFF-45 and PARP during ATF-1-induced apoptosis. Cells were treated as indicated and examined by western blotting as described in Materials and methods. This experiment was repeated three times with similar results.

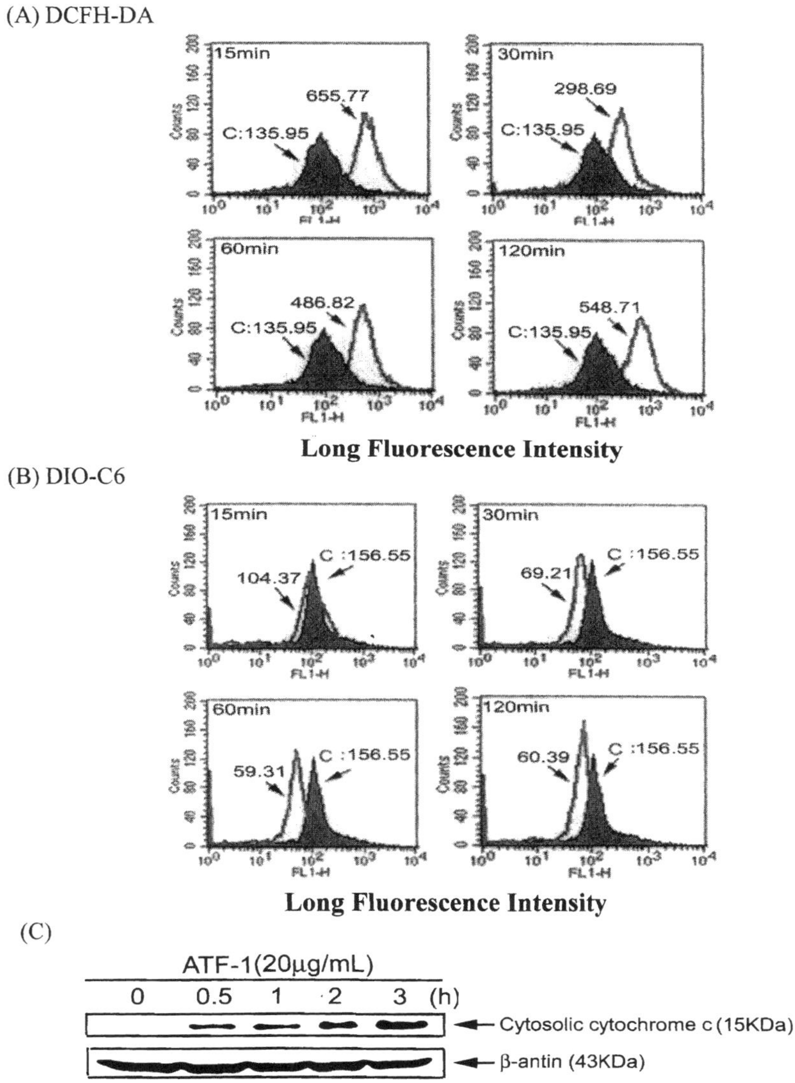

Figure 6. Induction of ROS generation, mitochondrial dysfunction, and cytochrome c release in ATF-1-induced apoptosis. (A) HL-60 cells were treated with 20 μg/mL ATF-1 for indicated time and were then incubation with 20 μM DCFH-DA and 40 nM 3,3'-dihexyloxacarbocyanine (DiOC6), respectively, and analyzed by flow cytometry. Data are presented as log fluorescence intensity. C, control. (B) HL-60 cells were treated with 20 μg/mL ATF-1 at indicated periods. Subcellular fractions were prepared as described under Materials and Methods, and cytosolic cytochrome c was detected by cytochrome c antibody.

Effect of ATF-1 on the Expression of Bcl-2 Family, Fas, Fas Ligand (FasL), and Bid Protein in HL-60 Cells

The imbalance of expression of anti- and pro-apoptotic members of the Bcl-2 family members regulates the mitochondrial pathway. It is one of the major mechanisms that determine the ultimate fate of cells in the apoptotic process. We examined the expression of the Bcl-X_L, Bax, Bad, and Bcl-2 at different time points in ATF-1-treated cells. As shown in Figure 7A, there are a marked increase of Bax and Bad protein in a time dependent manner and slight change of Bcl-X_L expression, and Bcl-2 and MCl-1 were cleaved to produce small size fragments. To assess whether ATF-1 promoted apoptosis via receptor-mediated pathway, the Fas and FasL protein were determined by Western blotting. The result showed that ATF-1 could stimulate the expression of Fas and FasL in a time dependent manner (Figure 7B). To verify whether the activation of caspase-8 was associated with Fas and FasL production in response to ATF-1 treatment, the degradation of caspase-8 and the cleavage of caspase-3 were detected after treatment of HL-60 cells with ATF-1 at the indicated time points. We next examined the pro-apoptotic protein, Bid, which upon cleavage by caspase-8, produced the truncated Bid fragment. These data suggested that the cleavage of Bid by active caspase-8 may be one of the mechanisms that contributed to the activation of mitochondrial pathway during ATF-1-induced apoptosis.

Discussion

Dietary chemopreventive compounds offer potential in the fight against cancer by blocking the carcinogenesis process through modulation of cell defensive and apoptotic machineries. Apoptosis is a complex process comprising of extrinsic and intrinsic pathways with numerous specific targets. Accumulating evidence clearly indicates that apoptosis is a critical molecular target for dietary bioactive agents for chemoprevention of cancer. In the further, a careful and well designed plan of clinical development ensures that apoptosis-targeted strategies are probably integrated into the anti-cancer armamentarium *(9)*.

Our previous studies have demonstrated that the TF-1 , isolated from black tea induced apoptosis in human lymphoma U937 and leukemia Jurkat cells *(11)*. The present study demonstrated that ATF-1, acetyl group to replace the hydroxyl group of TF-1, is more cytotoxic in HL-60 cells than TF-1 (Figure 1). The result suggests that acylation of the hydroxyl group is important to its bioactivity. Therefore, it is possible that such useful agents can be used as an adjuvant to enhance the efficacy of other known chemotherapeutic regimens to human leukemia.

The results presented herein account for ATF-1-triggered apoptosis might be increased the expression of Bax, Bad, Fas, and FasL protein that affects

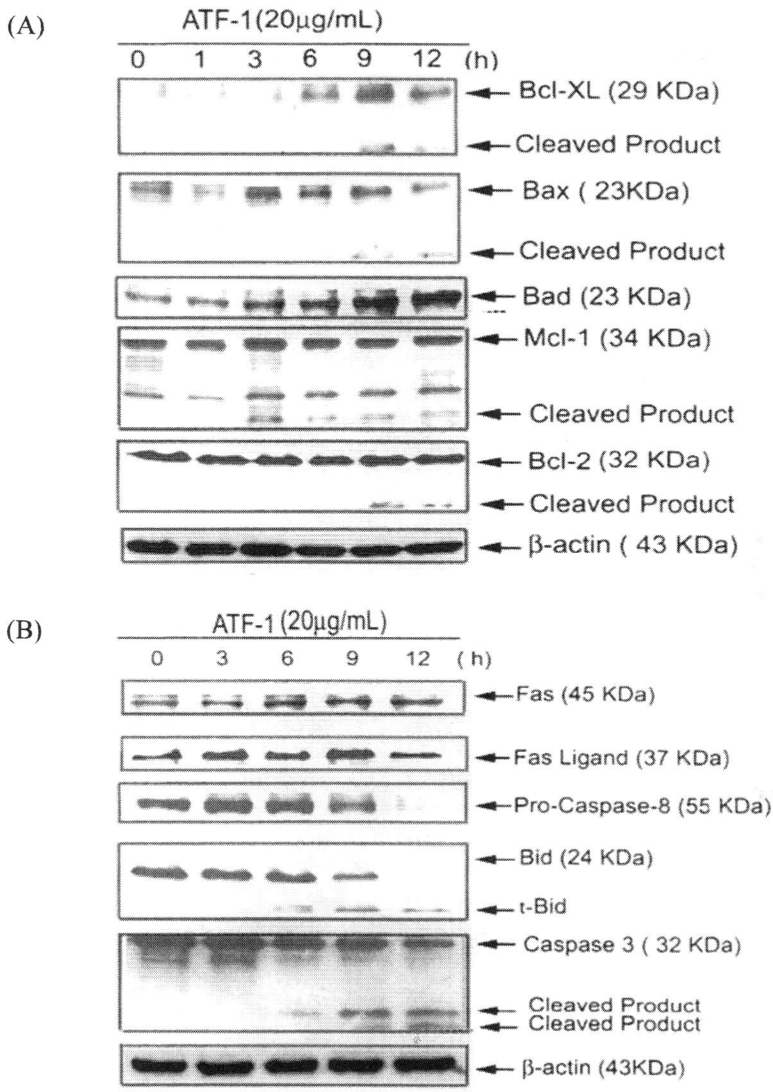

Figure 7. Effect of ATF-1 on Bcl-2 protein family, and Fas and FasL expression in HL-60 cells. (A) HL-60 cells were treated with20 μg/mL ATF-1 for the indicated time point. Expression of Bcl-xL, Bax, Bad, Mcl-1, and Bcl-2 was detected by Western blotting analysis using specific antibodies and then detected using the ECL system. (B) Expression of Fas, FasL, caspase-8, Bid, and caspase-3 was detected by Western blotting as described under Materials and Methods. Each blot is representative results of three similar experiments.

mitochondrial function, raising the possibility that the expression of proapoptotic proteins could be transcriptionally regulated in response to ATF-1 treatment, but this issue should be elucidated. ATF-1 is readily diffused into cells and is hydrolyzed by intracellular esterase to release TF-1. This is a possible mechanism that ATF-1 is more potent apoptosis-inducing agent than TF-1. Although ATF-1 could change the integrity of the mitochondrial membrane by regulating the expression of Bcl-2 family proteins, we did not rule out the possibility that ATF-1 could penetrate into cells and directly target mitochondria to increase membrane permeability and decrease $\Delta\Psi_m$ accompanied by ROS production. On the basis of these data, we propose a possible apoptotic mechanism induced by acacetin (Figure 8). In summary, our results clearly demonstrate that the cancer chemopreventive agent ATF-1 triggered apoptosis in a dose- and time- dependent manner in HL-60 cells. Analyses of expression of the Fas, FasL, and Bcl-2 family proteins, generation of ROS, subcellular location of cytochrome *c*, and the status of various caspases activities suggested that

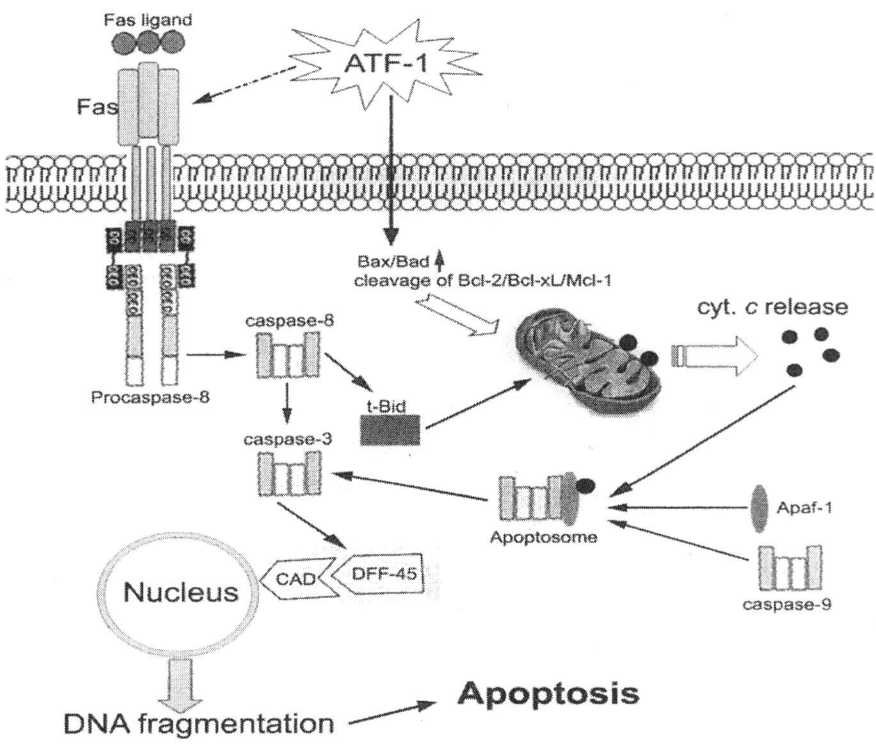

Figure 8. Tentative model for ATF-1-induced apoptosis in HL-60 cells proposed in the present study. See the text for details.

apoptosis induced by ATF-1 in HL-60 cells was mainly associated with ROS production, mitochondrial dysfunction, and Fas activation. Investigations to further elucidate the mechanisms associated with ATF-1-induced apoptosis should provide increased opportunities to develop novel for delaying cancer growth. Therefore, we speculate that the induction of apoptosis observed in this study may provide a distinct mechanism for the cancer therapeutic and chemopreventive functions of ATF-1.

Acknowledgement

This study was supported by the National Science Council NSC 95-2321-B-022-001 and NSC 95-2313-B-022-003-MY3

References

1. Pan, M. H.; Lin-Shiau, S. Y.; Ho, C. T.; Lin, J. H.; Lin, J. K. *Biochem. Pharmacol.* **2000**, *59*, 357-367.
2. Liang, Y. C.; Tsai, D. C.; Lin-Shiau, S. Y.; Chen, C. F.; Ho, C. T.; Lin, J. K. *Nutr. Cancer* **2002**, *42*, 217-223.
3. Lin, J. K.; Liang, Y. C.; Lin-Shiau, S. Y. *Biochem. Pharmacol.* **1999**, *58*, 911-915.
4. Kundu, T.; Dey, S.; Roy, M.; Siddiqi, M.; Bhattacharya, R. K. *Cancer Lett.* **2005**, *230*, 111-121.
5. Lu, Y. P.; Lou, Y. R.; Xie, J. G.; Yen, P.; Huang, M. T.; Conney, A. H. *Carcinogenesis* **1997**, *18*, 2163-2169.
6. Halder, J.; Bhaduri, A. N. *Biochem. Biophys. Res. Commun.* **1998**, *244*, 903-907.
7. Surh, Y. J. *Nat. Rev. Cancer* **2003**, *3*, 768-780.
8. Meng, X. W.; Lee, S. H.; Kaufmann, S. H. *Curr. Opin. Cell Biol.* **2006**, *18*, 668-676.
9. Khan, N.; Afaq, F.; Mukhtar, H. *Carcinogenesis* **2007**, *28*, 233-239.
10. Pan, M. H.; Chen, W. J.; Lin-Shiau, S. Y.; Ho, C. T.; Lin, J. K. *Carcinogenesis* **2002**, *23*, 1677-1684.
11. Pan, M. H.; Liang, Y. C.; Lin-Shiau, S. Y.; Zhu, N. Q.; Ho, C. T.; Lin, J. K. *J. Agric. Food Chem.* **2000**, *48*, 6337-6346.
12. Enari, M.; Talanian, R. V.; Wong, W. W.; Nagata, S. *Nature* **1996**, *380*, 723-726.
13. Liu, X.; Zou, H.; Slaughter, C.; Wang, X. *Cell* **1997**, *89*, 175-184.
14. Sakahira, H.; Enari, M.; Nagata, S. *Nature* **1998**, *391*, 96-99.
15. Tewari, M.; Quan, L. T.; O'Rourke, K.; Desnoyers, S.; Zeng, Z.; Beidler, D. R.; Poirier, G. G.; Salvesen, G. S.; Dixit, V. M. *Cell* **1995**, *81*, 801-809.

Chapter 30

Liver Carcinogenesis and 8-Hydroxy-deoxyguanosin Formation by Oxidized Lard and Dietary Oils

Kaori Sadakane[1], Takamichi Ichinose[1], Kazutoshi Fujioka[2], and Takayuki Shibamoto[2]

[1]Department of Health Sciences, Oita University of Nursing and Health Sciences, 2944–9, Notsuharu, Oita 870–1201, Japan
[2]Department of Environmental Toxicology, University of California, Davis, CA 95616

Oxidized lard and dietary oils (soybean oil and sardine oil) were orally administered to C3H/HeN mice in order to examine the role of oxidized oils in cancer formation. After 12 months, oxidized sardine oil caused the highest tumor incidence among the oxidized dietary oils tested. The amounts of 8-hydroxy-deoxyguanosine (8-OH-dG) in the mice treated with oxidized sardine oil correlated with the rates of tumor incidence. The lard and sardine oil-fed groups exhibited significant increases of 8-oxo-dGTPase mRNA levels. The lard-fed group alone showed a slight increase of the expression levels of OGG1 mRNA, but generally the expression level was not changed by addition of oxidized fat or oils. Oxidized lard and dietary oils might be harmful for our health because of the increased risk of liver carcinogenesis related to 8-OH-dG formation. It is preferable to avoid re-used oils.

Certain fats and dietary oils are used in large quantities in food preparation and consequently become components of foods and are ingested by people. Therefore, the toxicity of deteriorated or oxidized fats and oils has received much attention among not only food chemists but also consumers.

There have been numerous reports on the toxicity of oxidized fats and oils since the 1950s (*1*). In particular, heat-treated oils, such as oils used in cooking, have been known to cause various adverse effects, including organotoxicity of the internal organs (*2*) and reproductive toxicity (*3*). Dietary oxidized frying oil reportedly up-regulates the expression of PPARα in downstream genes and alters lipid metabolism in rats (*4*). High intakes of a mixture of lard and oxidized cod liver oil caused biological implications in rats (*5*). The toxicity of oxidized fats and oils is caused by oxidative secondary products because they are readily absorbed by the intestines (*6*). Among the many products formed in oxidized fats and oils, formaldehyde, acetaldehyde, acrolein, malonaldehyde (MA), glyoxal, and methylglyoxal have received much attention as chemicals implicated in various diseases (*7*).

In a previous study (*8*), oxidized dietary oils of lard, soybean, and particularly sardine were shown to increase spontaneous liver tumor development and the formation of 8-OH-dG in the liver DNA of C3H/HeN male mice. Therefore, it was suggested that the progression of liver tumors promoted by oxidized dietary oils was due to 8-OH-dG accumulated in the liver DNA and that the accumulation of 8-hydroxy-deoxyguanosine (8-OH-dG) is associated with DNA mutation and cancer promotion (*9*).

In the present study, oxidized lard and dietary oils (soybean and sardine) were orally administered to C3H/HeN mice. The formation of 8-OH-dG and mRNA of DNA repair enzymes (8-oxo guanine DNA glycosylase 1, OGG1, and 8-oxo dGTPase) in the livers was subsequently investigated.

Materials and Methods

Experimental Animals

C3H/HeN male mice (5 weeks of age) were purchased from Charles River Japan, Inc. (Kanagawa, Japan). Body weight and absence of infection were checked for 1 week. The C3H/HeN mice were divided into four groups (n = 8). The mice were fed four different diets (F-2, lard, soybean, and sardine) for 6 or 12 months by a previously reported method (*8*). Mice were housed in plastic cages lined with soft wood chips. The cages were placed in a conventional room, which was air conditioned at 23 °C and 55-70% humidity with a light/dark (12 h/12 h) cycle. The study adhered to the U.S. National Institutes of Health guidelines for the use of experimental animals. The animal care method was approved by the Animal Care and Use Committee at Oita University of Nursing and Health Science in Oita, Japan.

Diets

The four diets used were commercial F-2 diet containing 4% general fats (Funabashi Farm Co., Ltd., Chiba, Japan); F-2 diet with 4% oxidized lard added; F-2 diet with 4% oxidized soybean oil added; and F-2 diet with 4% oxidized sardine oil added. All oxidized dietary oils tested contained various levels of secondary lipid peroxidation products as reported previously (*8*). Water was accessible *ad libitum*. The F-2 diet (34% of calories derived from fat) totaled 359 kcal/100 g. These pellet-diets were prepared by the Funabashi Farm Co., Ltd. (Chiba, Japan) and were vacuum packed and sealed in vinyl bags. They were stored at 4 °C until use. The special diet was changed once every two days.

Preparation of Oxidized Lard, Soybean Oil, and Sardine Oil

The dietary oils were oxidized by a previously reported method (*10*). Oils were placed in a round-bottom flask. The flask was rotated using a rotary evaporator at 37 °C for 15 days to allow the oils complete contact with the air. The oxidized oils were stored at 5 °C until used for animal studies and analysis for toxic dicarbonyl compounds.

Measurement of 8-OH-dG in Liver DNA, RNA in 8-Oxoquanine-DNA Glycosylase (OGGI), and 8-Oxo-dGTPase

Levels of 8-OH-dG in Liver DNA, RNA in 8-Oxoquanine-DNA glycosylase (OGGI), and 8-Oxo-dGTPase were measured using previously reported methods (*11*). Eight mice from each group were used for the measurement of 8-OH-dG levels. The total RNA was isolated from the livers of the 8 mice used for 8-OH-dG analysis.

Statistical Analysis

The statistical analysis of body weight, expression of repair enzymes mRNA, and amount of 8-OH-dG in each strain were performed using a one-way ANOVA with post-hoc Bonferroni/Dunn-test. The unpaired t-test was used for the statistical comparison with the results between C3H/HeN and C57BL/6.

Differences among the groups were considered statistically significant at a level of $p < 0.05$. Data are shown as mean ± SD.

Results and Discussion

Table I shows the spontaneous mortality rate of the experimental animals. After 6 months, the mortality rate in the soybean and sardine groups (3.3 and 6.6%, respectively) were higher than those in the F-2 (0%) and in lard (0%) groups. After 12 months, the spontaneous mortality rates increased significantly. Almost half of the mice in the lard group (45%) and the soybean group (43%) died after 12 months, whereas the mortality rates of the F-2 group (10%) and the sardine group (18%) were much lower.

Table I. Rate of Tumor Incidence in the Livers of Experimental Mice after 12 Months

	Number of mice with tumor and incidence rate (%)		
Diet	Total	Benign	Malignant
F-2	7 (17.5)	4 (10)	3 (7.5)
Lard	13 (32.5)	6 (15)	7 (17.5)
Soybean	10 (25.0)	3 (7.5)	7 (17.5)
Sardine	14 (35.0)	3 (7.5)	11 (27.5)[a]

[a]$p < 0.05$ vs F-2 group.

Table II shows the rate of tumor incidence in the livers of experimental mice after 12 months. The highest total tumor incidence was in the sardine (35%) group, followed by the lard (32.5%) group. However, neither of them was statistically significant relative to the F-2 control group. Malignant tumor incidence (7 – 11%) was higher than benign tumor incidence (3 – 6%) except in the case of the F-2 group. The results of the 6 months experiment are not shown, but benign tumors were found in all groups. The tumor incidence for the 6 months experiment was lower that that of the12 months experiment and the tumor sizes were smaller than those from 12 months.

The 8-OH-dG content in the liver DNA after 6 and 12 months is shown in Figure 1. The levels of 8-OH-dG increased significantly in the sardine-fed group ($p < 0.001$ vs F-2 group), which had the highest rate of malignant tumor

Table II. Mortality Rate of Experimental Mice

	Mortality rate (%)	
Diet	6 Months	12 Months
F-2	0	10
Lard	0	45[a]
Soybean	3.3	43[a]
Sardine	6.6	18

[a]$p < 0.001$ vs F-2 group.

incidence (Table II). The results indicate that there is a correlation between the level of 8-OH-dG formation and rates of tumor incidence. The soybean oil-fed group exhibited moderate increase ($p < 0.05$ vs F-2 group) after 12 months but did not show appreciable increase after 6 months. The lard-fed group showed over 30% increase both after 6 and 12 months ($p < 0.01$ vs F-2 group). Also, there has been clear evidence that the accumulation of 8-OH-dG in DNA is closely correlated with carcinogenesis (*12-14*).

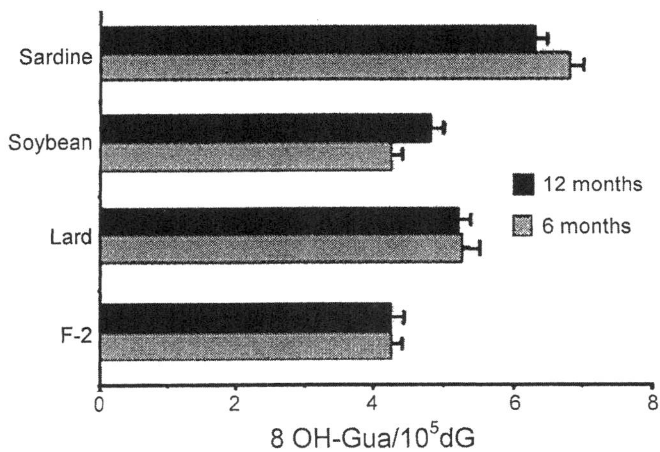

Figure 1. The level of 8-OH-dG in liver DNA from C3H/HeN mice fed with F-2 diet and three different oxidized oils. The values were mean ± SD (n = 8). The content of 8-OH-Gua in DNA was expressed as the ratio of 8-OHdG × 10^5 to total dG.

The expression levels of 8-oxo-dGTPase mRNA and OGG1 mRNA in the livers are shown in Figure 2.

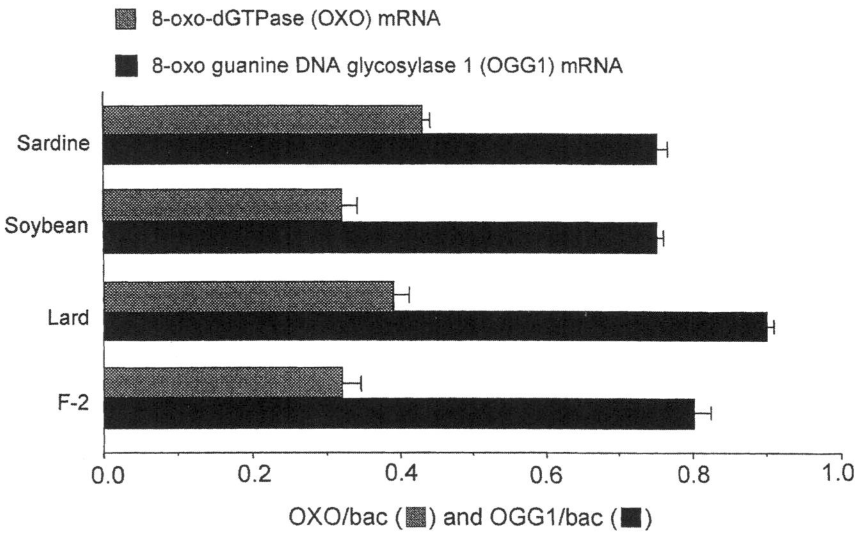

Figure 2. The expression of 8-oxo-guanine-DNA glycosylase 1 (OGG1) mRNA and 8-oxo-dGTPase mRNA in the livers from C3H/HeN mice fed with F-2 diet and three different oxidized oils. The values were mean ± SD (n = 8).

The expression levels of 8-oxo-dGTPase mRNA relative to the control (F-2-fed group) were 123% in the lard-fed group, 100% in the soybean oil-fed group, and 134% in the sardine oil-fed group. The lard and sardine oil-fed groups exhibited significant increases of 8-oxo-dGTPase mRNA levels. The expression levels of OGG1 mRNA relative to the control (F-2-fed group) were 113% in the lard-fed group, 94% in the soybean oil-fed group, and 94% in the sardine oil-fed group. Only the lard-fed group showed slight increase of the expression levels of OGG1 mRNA but generally the expression level was not changed by the addition of oxidized fat or oils. Decreasing mRNA in OGG1 and 8-oxo-dGTPase by oxidized lard and sardine oil caused the accumulation of 8-OH-dG in liver DNA, which might promote spontaneous liver tumorigenesis after 12 months (*8*). In the present study, the low response of the DNA repair enzyme system in C3H/HeN against oxidized dietary oils might promote accumulation of 8-OH-dG in DNA, and consequently cause a high susceptibility toward liver tumorigenesis in this strain.

References

1. Benes, V.; Sedlacek, B. A. J. *J. Hygi. Epid. Microbiol. Immun.* **1959**, *3*, 96-105.
2. Ebong, P. E.; Owu, D. U.; Isong, E. U. *Plant Foods Human Nutr.* **1999**, *53*, 209-222.
3. Isong, E. U.; Ebong, P. E.; Ifon, E. T.; Umoh, I. B.; Eka, O. U. *Plant Foods Human Nutr.* **1997**, *51*, 159-166.
4. Chao, P.-M.; Chao, C.-Y.; Lin, F.-J.; Huang, C.-J. *J. Nutr.* **2001**, *131*, 3166-3174.
5. Zidkova, J.; Sajdok, J.; Kontrava, K.; Kotrbova-Kozak, A.; Hanis, T.; Zidek, V.; Fucikova, A. *Czech. J. Food Sci.* **2004**, *22*, 108-120.
6. Frankel, E. N.; Neff, W.E. *Biochim. Biophys. Acta* **1983**, *754*, 264-270.
7. Okado-Matsumoto, A. I.; Fridovich, J. *Biol. Chem.* **2000**, *275*, 34853-34857.
8. Ichinose, T.; Nobuyuki, S.; Takano, H.; Abe, M.; Sadakane, K.; Yanagisawa, R.; Ochi, H.; Fujioka, K.; Lee, K.-G.; Shibamoto, T. *Food Chem. Toxicol.* **2004**, *42*, 1795-1803.
9. Wei, L.; Wei, H.; Frenkel, K. *Carcinogenesis* **1993**, *14*, 841-847.
10. Kaneda, T.; Sakai, H.; Ishii, S. *J. Biochem.* **1955**, *42*, 561-565.
11. Sadakane, K.; Ichinose, T.; Takano, H.; Abe, M.; Sera, N.; Yanagisawa, R.; Ochi, H.; Fujioka, K.; Lee, K.-G.; Shibamoto, T. *Food Chem. Toxicol.* **2006**, *44*, 1372-1376.
12. Takeuchi, T.; Morimoto, K. *Carcinogenesis* **1994**, *15*, 645-639.
13. Salim, E. I.; Wanibuchi, H.; Morimura, K.; Murai, T.; Makino, S.; Nomura, T.; Fukushima, S. *Carcinogenesis* **2002**, *23*, 1519-1529.
14. Schurkes, C.; Brock, W.; Abel, J.; Unfried, K. *Mutat. Res.* **2004**, *553*, 59-65.

Chapter 31

Suppressive Effects of Flavonoids on Activation of the Aryl Hydrocarbon Receptor Induced by Dioxins

Itsuko Fukuda and Hitoshi Ashida

Graduate School of Agricultural Science, Kobe University, 1–1 Rokkodai-cho, Nada-ku, Kobe, Hyogo 657–8501, Japan

Dioxins develop various adverse effects through transformation of the aryl hydrocarbon receptor (AhR). Since dioxins invade the body mainly with diet, it is important to search for AhR antagonists from a dietary source. Our findings have demonstrated that the suppressive effects of flavonoids on AhR transformation was classified into their subclasses with the following order; flavones = flavonols = theaflavins > flavanones > catechins > isoflavones and anthocyans. Regarding flavonols, the competitive ligand binding analysis revealed that kaempferol, quercetin, and galangin inhibited the specific binding between the AhR and its agonist, 3-methylcholanthrene (MC) in a dose-dependent manner, indicating that at least flavonols acts as competitive antagonists of the AhR. These results suggest that the mechanisms by which flavonoids suppress AhR transformation differs in their subclasses.

Dioxins are the toxic members of halogenated aromatic hydrocarbons (HAHs) and polycyclic aromatic hydrocarbons (PAHs), and develop various adverse effects such as body weight loss, immunosuppression, cancer promotion, and teratogenesis (*1*, *2*). These toxicological effects of dioxins are mainly mediated by the aryl hydrocarbon receptor (AhR) (*3*, *4*). The unliganded AhR exists as a complex with 2 molecules of heat shock protein 90 (Hsp90), the X-associated protein 2 (XAP2), and p23. Following binding of agonists including dioxins to the AhR complex, the AhR translocates into the nucleus, dissociates the complex, heterodimerizes with AhR nuclear translocator (ARNT), and binds to a specific DNA sequence called dioxin responsive element, 5'-GCGTG-3', resulting in transcriptional activation of adjacent genes (*5-8*). These sequential actions of the AhR are called as 'transformation (i. e., activation)' and lead the expression of various proteins including drug metabolizing enzymes such as cytochrome P4501A (CYP1A) subfamily, glutathione *S*-transferase (GST), and NAD(P)H:quinine-oxidoreductase (NQO1) (*9-11*). Transformed AhR also leads protein phosphorylation and disrupts intracellular signal transduction (*12*).

Because dioxins invade the body mainly with diet, it is difficult to protect us from the dioxin toxicity by certain drugs. Others and we have demonstrated that natural flavonoids (*13-17*), resveratrol (*18*), curcumin (*19*), and certain vegetable constituents (*20-22*) act as antagonists of the AhR. Flavonoids have a favorite structure to the AhR binding pocket, whose van der Waals dimensions are reported to the size of 14 x 12 x 5 Å (*5*), and galangin (*17*) inhibit binding between 2,3,7,8-tetrachlorodibenzo-*p*-dioxin (TCDD) and the AhR. These results indicate that flavonoids have a possibility to suppress the dioxin toxicity by inhibiting agonist binging to the AhR and following AhR transformation. In this study, we examined the suppressive effects of flavonoids on AhR transformation induced by TCDD, and the results from compepetitive ligand binding assay proposed putative mechanisms of flavonols.

Materials and Methods

Materials

TCDD was purchased from AccuStandard (New Haven, CT) and dissolved in dimethylsulfoxide (DMSO). For electrophoretic mobility shift assay (EMSA), dioxin responsive element (DRE; 5'GAT CTG GCT CTT CTC ACG CAA CTC CG-3'(non-coding) and 5'GAT CCG GAG TTG CGT GAG AAG AGC CA-3' (coding)) probe were synthesized. [^{3}H]3-Methylcholanthrene (MC) (1.9 Ci/mmol) and 2,3,7,8-tetrachlorodibenzofuran (TCDF) were obtained from Moravek Biochemicals, Inc. (Brea, CA) and Cambridge Isotope Laboratories (Andover, MA), respectively. All other reagents used in this study were of the highest purity grade available from commercial source.

Preparation of Rat Hepatic Cytosolic Fraction

Animal treatments in the present study was approved by the Institutional Animal Care and Use Committee and carried out according to the Guidelines of Animal Experimentation of Kobe University. Livers from male Sprague-Dawley rats (six weeks old, 140-170 g; Japan SLC, Shizuoka, Japan) were subjected to the preparation of cytosol fraction as previously described (13) and used as the AhR source.

Electrophoretic Mobility Shift Assay (EMSA)

To induce AhR transformation, rat hepatic cytosol fraction (4 mg/ml) was incubated with 1 nM TCDD at 20°C for 2 h, and transformed AhR was detected by EMSA using the ^{32}P-labeled DRE oligonucleotide probe as previously described (14). To estimate the suppressive effects of flavonoids on AhR transformation, all compound was dissolved in DMSO, and added to the cytosolic fraction 10 min prior to the addition of TCDD. After electrophoresis, the gels were dried and exposed to the X-ray films. The density of the specific AhR/DRE complex bands was analyzed and evaluated the ratio of the transformation.

Competitive Ligand Binding Analysis

Specific binding of [^{3}H]MC to the AhR was determined by ligand-binding assay using hydroxyapatite (HAP) as previously described (23). Briefly, the cytosol fraction (2 mg protein/ml) was incubated with 0.25 nM [^{3}H]MC at 20°C for 2 h in 2 ml of HEDG buffer. Non-specific binding was defined by incubation of the cytosol with [^{3}H]MC in the presence of 200-fold molar excess of TCDF. Aliquots of 250 μl of the reaction mixture were transferred to the scintillation vial to measure total [^{3}H]MC. Remained aliquots of 500 μl of the mixture were incubated with 300 μl of HAP, which was suspended in double volume of HEDG buffer, at 4°C for 30 min. The incubation was stopped by the washing process with 1 ml of HEDG buffer containing 0.5 % Tween 80 for 5 times, and the final precipitation of HAP mixture was transferred to the scintillation vials with 2 ml of ethanol to measure [^{3}H]MC-bound AhR. To determine the inhibitory effects of flavonols on specific binding between the AhR and [^{3}H]MC, each compound at indicated concentrations was pretreated to the cytosol fraction 10 min prior to the addition of [^{3}H]MC.

Results and Discussion

We determined the 50% inhibitory concentration (IC_{50}) values of natural flavonoids against TCDD-induced AhR transformation by EMSA, and the suppressive effects of flavonoids on the transformation were summarized (Table 1).

As we have reported previously, flavones and flavonols suppressed AhR transformation effectively at their dietary levels (*13*). Anthocyans, one of the flavonoid groups distributed widely in plant kingdom, did not suppress the transformation (*15*). Theaflavins, which are contained in black tea, strongly suppressed the transformation as flavones and flavonols (*16*). The order of the suppressive effects of flavonoids were classified into their subclasses as follows; flavones = flavonols = theaflavins > flavanones > catechins > isoflavones and anthocyans, indicating that the suppressive mechanisms might differ among subclasses.

Table I. Suppressive Effects of Flavonoids on TCDD-induced AhR Transformation

+++	++	+	−
Flavone	Luteolin	Luteolin-7,3'-*O*-bis-glucoside	Daidzein
Chrysin	Tangeretin	Rutin	Genistein
Baicalein	Fisetin	Eriodictyol	Puerarin
Apigenin	Morin	Hesperetin	(+)-Catechin
Flavonol	Myricetin	Naringin	(—)-Gallocatechin
Galangin	Isorhamnetin	(—)-Gallocatechin gallate	(—)-Catechin gallate
Kaempferol	Naringenin	(—)-Epicatechin gallate	(—)-Epicatechin
Quercetin		(—)-Epigallocatechin gallate	(—)-Epigallocatechin
Tamarixetin			*Anthocyans (20 kinds)*
Quercitrin			
Flavanone			
Theaflavin			
Theaflavin-3-gallate			
Theaflavin-3'-gallate			
Theaflavin-3,3'-digallate			

Suppressive effects were estimated by the IC_{50} value against 0.1 or 1 nM TCDD-induced AhR transformation. +++, IC_{50} value < 5 μM; ++, 5 μM < IC_{50} value < 10 μM; +, 10 μM < IC_{50} value < 100 μM; −, 100 μM < IC_{50} value.

The AhR binding pocket is report to favor the hydrophobic compounds whose van der Waals dimensions of 14 x 12 x 5 Å (*5*). Flavones and flavonols have a favorite structure to the AhR binding pocket, and galangin was reported to inhibit TCDD binding to the AhR (*17*). We have reported that flavonoids suppress AhR transformation by treatment before TCDD (*13*). These results suggest that the flavonoids have an ability to suppress the transformation by binding to the AhR binding pocket before binding of agonists. To examine whether flavones and flavonols inhibit ligand binding to the AhR, the competitive ligand binding analysis was performed. As shown in Figure 1, all flavonoid tested here at 1 μM inhibited [^{3}H]MC binding to the AhR. Moreover, kaempferol, quercetin, and galangin inhibited the specific binding of [^{3}H]MC to the AhR in a dose-dependent manner, and the IC_{50} values were determined as 295, 25.7, and 20.9 nM, respectively (Figure 2). These results suggest that at least flavonols act as competitive antagonists of the AhR, and suppress following sequential steps of AhR transformation.

Regarding theaflavins and catechins, they do not have a coplanar structure, and they are too big for suitable size of the AhR binding pocket. The number of a gallate moiety affected the suppressive effects. However, pyrogallol-type chemical, gallic acid did not show the suppressive effect (data not shown), suggesting that the gallate moiety with part of theaflavin and catechin structure would be adjusted to the AhR binding pocket and play a critical role in the suppression of AhR transformation. Recently, it was demonstrated that (−)-epigallocatechin gallate (EGCg) suppresses AhR transformation and its

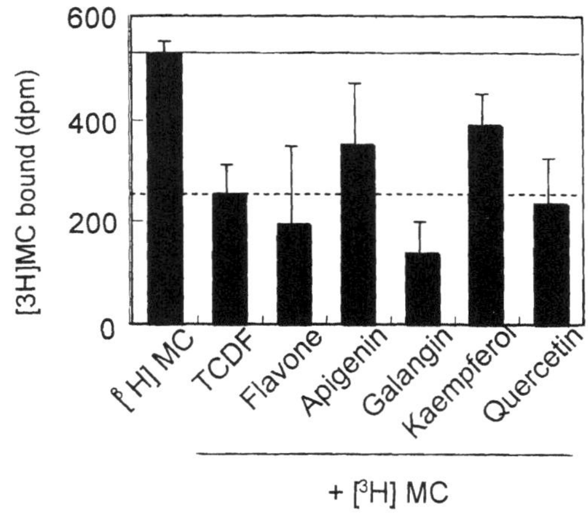

Figure 1. Flavonoids inhibit [^{3}H]MC binding to the AhR. Rat hepatic cytosol fraction was pretreated with 1 μM flavonoids 10 min prior to the addition of 0.25 nM [^{3}H]MC. Solid and broken lines indicate total and non-specific binding, respectively. Data are shown as mean ± SD (n=3).

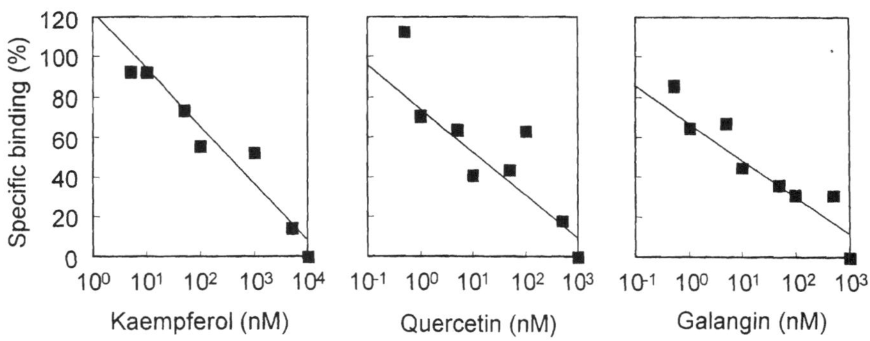

Figure 2. Flavonols inhibit specific binding of [^{3}H]MC to the AhR in a dose-dependent manner. Rat hepatic cytosol fraction was pretreated with each compound at indicated concentrations 10 min prior to the addition of [^{3}H]MC. The inhibitory effects of each compound on [^{3}H]MC binding to the AhR were indicated as % of specific binding as described in the Material and Methods.

downstream events by binding to Hsp90 proteins and not affecting the AhR (*24*). Taken together, it is suggested that theaflavins and catechins suppress the transformation by not directly affecting the receptor protein but the AhR complex.

In conclusion, flavonoids suppressed AhR transformation by the mechanisms of which at least flavonols bind to the AhR prior to agonists binding, and others do not directly affect the receptor protein. Although the further investigations are needed to elucidate fully the interaction between natural flavonoids and the receptor protein, flavonoids have the possibility to reduce the dioxin toxicity through the AhR transformation pathway.

References

1. Safe, S. H. *Crit. Rev. Toxicol.* **1990**, *21*, 51-88.
2. De Vito M. J.; Birnbaum L. S. In *Dioxins and Health,* Schecter, A., Ed.; Plenum; New York, NY, 1994, pp. 139-162.
3. Whitlock, J. P. Jr. *Annu. Rev. Pharmacol. Toxicol.* **1990**, *30*, 251-277.
4. Denison, M. S.; Phelan D.; Elferink C. J. In *Toxicant-Receptor Interactions,* Denison, M. S.; Helferich W. G., Eds.; Taylor and Francis: Philadelphia, PA, 1998, pp.3-33.
5. Denison, M. S.; Pandini A.; Nagy S. R.; Baldwin E. P.; Bonati, L. *Chem. Biol. Interact.* **2002**, *141*, 3-24.
6. Whitlock, J. P. Jr. *Chem. Res. Toxicol.* **1993**, *6*, 745-753.
7. Hankinson, O. *Annu. Rev. Pharmacol. Toxicol.* **1995**, *35*, 307-340.
8. Denison, M. S.; Fisher J. M.; Whitlock J. P. Jr. *Proc. Natl. Acad. Sci.* **1998**, *85*, 2528-2532.
9. Schrenk, D. *Biochem. Pharmacol.* **1998**, *55*, 1155-1162.
10. Xu, L.; Li, A. P.; Kaminski, D. L.; Ruh, M. F. *Chem. Biol. Interact.* **2000**, *124*, 173-189.
11. Rowlands, J. C.; Gustafsson, J.-Å. *Crit. Rev. Toxicol.* **1997**, *27*, 109-134.
12. Ashida, H; Nagy, S.; Matsumura, F. *Biochem. Pharmacol.* **2000**, *59*, 741-751.
13. Ashida, H.; Fukuda, I.; Yamashita, T.; Kanazawa, K. *FEBS Lett.* **2000**, *476*, 213-217.
14. Fukuda I.; Sakane I.; Yabushita Y.; Kodoi R.; Nishiumi S.; Kakuda T.; Sawamura S.; Kanazawa K.; Ashida H. *J. Agric. Food Chem.* **2004**, *52*, 2499-2506.
15. Mukai, R.; Fukuda, I.; Hosokawa, K.; Nishiumi, S.; Kaneko, A.; Ashida, H. *Biosci. Biotechnol. Biochem.* **2005**, *69*, 896-903.
16. Fukuda, I.; Sakane, I.; Yabushita, Y.; Sawamura, S.; Kanazawa, K.; Ashida, H. *Biosci. Biotechnol. Biochem.* **2005**, *69*, 883-890.
17. Quadri, S. A.; Qadri, A. N. ; Hahn, M. E. ; Mann, K. K. ; Sherr, D. H. *Mol. Pharmacol.* **2000**, *58*, 515-525.
18. Ciolino, H. P.; Yeh, G. C. *Mol. Pharmacol.* **1999**, *56*, 760-767.
19. Ciolino, H. P.; Daschner, P. J.; Wang, T. T. Y.; Yeh, G. C. *Biochem. Pharmacol.* **1998**, *56*, 197-206.
20. Amakura, Y.; Tsutsumi, T.; Sasaki, K.; Yoshida, T.; Maitani, T. *Biol. Pharm. Bull.* **2003**, *26*, 1754-1760.
21. Park, Y. K.; Fukuda, I.; Ashida, H.; Nishiumi, S.; Guzman, J. P.; Sato, H. H.; Pastore, G. M. *Biosci. Biotechnol. Biochem.* **2004**, *68*, 935-938.
22. Nishiumi, S.; Yabushita, Y.; Fukuda, I.; Mukai, R.; Yoshida, K.; Ashida, H. *Food Chem. Toxicol.* **2006**, *44*, 250-260.
23. Gasiewicz, T. A.; Neal, R. A. *Anal. Biochem.* **1982**, *124*, 1-11.
24. Palermo C. M.; Westlake C. A.; Gasiewicz T. A. *Biochemistry* **2005**, *44*, 5041-5052.

Other Health Effects

Chapter 32

Antiobesity Effect of Fucoxanthin from Edible Seaweeds and Its Multibiological Functions

Hayato Maeda, Masashi Hosokawa, Tokutake Sashima, and Kazuo Miyashita

Laboratory of Biofunctional Material Chemistry, Faculty of Fisheries Sciences, Hokkaido University, Hakodate, Japan

Fucoxanthin has a unique structure including an unusual allenic bond and 5,6-monoepoxide in its molecule. It induced a remarkable reduction in the growth of leukemia and prostate cancer cells. Furthermore, we have found that fucoxanthin showed anti-obesity effect with a new molecular mechanism. Uncoupling protein (UCP) is inner-membrane mitochondrial protein that has the ability to dissipate energy through uncoupling of oxidative phosphorylation which, instead of ATP, produces heat. A great deal of interest has focused on adaptive thermogenesis by UCP families (UCP1, 2 and 3) in several tissues and organs as a physiological defense against obesity, hyperlipidemia, and diabetes. In fact, UCP1 expression in brown adipose tissue (BAT) is known as a significant component of whole body energy expenditure, at least small rodents, and its dysfunction contributes to the development of obesity. However, adult humans have very little BAT, making it unlikely to be a major contributor to human weight regulation. In humans, most of fat is stored in white adipose tissue (WAT). Considered as breakthrough discoveries for an ideal therapy of obesity, regulation of UCP expression in white adipose tissue (WAT) by food constituent should be studied. Here, we show significant reduction of WAT in wistar rats and obese KK-Ay mice by feeding fucoxanthin (0.05 or 0.2 wt%/feed).

Fucoxanthin, concentrated from edible seaweed, *Undaria Pinnatifida* (Japanese name is Wakame). The daily intake of fucoxanthin to KK-Ay mice also caused a significant reduction of body weight and a significant increase in BAT weight. Clear signals of UCP1 protein and mRNA were detected by Western and Northern blot analyses in WAT in mice fed fucoxanthin, although there is little expression of UCP1 in WAT in mice fed control diet. UCP1 expression in WAT by fucoxanthin intake leads to oxidation of fatty acids and heat production in WAT mitochondria. The substrate oxidation can directly reduce WAT in animals.

Fucoxanthin has a unique structure including an allenic bond and a 5,6-monoepoxide in the molecule (Figure 1).

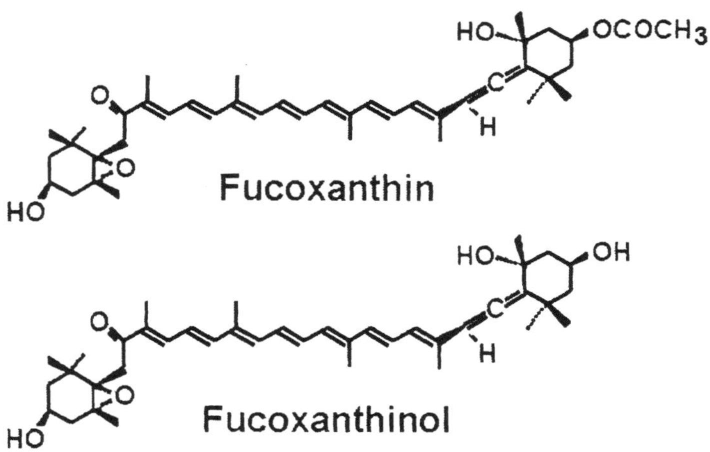

Figure 1. Structure of fucoxanthin and fucoxanthinol.

It is the most abundant of all carotenoids accounting for >10% of estimated total natural production of carotenoids. In Southeast Asian countries, some seaweeds containing fucoxanthin are often used as a food source. Among them *Undaria* (Japanese name is Wakame) and *Laminaria* (Japanese name is Konbu) are most popular edible seaweeds in Japan. Fucoxanthin is easily converted fucoxanthinol in human intestinal cells and in mice (*1*), suggesting that the active form of fucoxanthin in biological system would be fucoxanthinol (Figure 1).

Studies involving quantification of fucoxanthin in different brown seaweeds, both wild and cultured, are limited. In a study involving quantitative and qualitative analysis of six different brown seaweeds, Haugan and Liaeen-Jensen (*2*) reported that fucoxanthin is the major contributing carotenoid to the total carotenoids in those seaweeds with the fucoxanthin content varying between 43 to 83% of total carotenoids. Fucoxanthin when present in the thallus of seaweeds was found to be quite stable in the presence of organic ingredients apart from surviving the drying process and storage at ambient temperature; although, fucoxanthin in pure form is susceptible to oxidation (*3*). Further, fucoxanthin content in seaweeds exhibits seasonal variation (*4,5*) and also varies depending on the life cycle of the seaweeds (*3*) indicating the possible biological significance of this pigment in seaweeds.

Cancer chemoprevention is one of the promising methods for cancer control. Among the chemopreventive agents, carotenoids, especially β-carotene, have been investigated extensively (*6*). Since Muto *et al.* (*7*) discovered that β-carotene induces apoptosis in cervical dysplastic cells *via* down-regulation of an epidermal growth factor receptor, several carotenoids such as lycopene, β-cryptoxanthin, lutein, and canthaxanthin have been reported to induce apoptosis in certain cancer cells. Hosokawa *et al.* (*8*) found that HL-60 human promycelocytic leukemia cells underwent apoptosis by fucoxanthin. Its activity was higher than that of β-carotene. The strong inhibitory effect of fucoxanthin has also been confirmed using human prostate cancer cells (*9,10*). In their study, the effect of 15 kinds of carotenoids (phytoene, phytofluene, ξ-carotene, lycopene, α-carotene, β-carotene, β-cryptoxanthin, canthaxanthin, astaxanthin, capsanthin, lutein, zeaxanthin, vioaxanthin, neoxanthin, and fucoxanthin) present in foodstuffs was evaluated on the growth of human prostate cancer cell lines (PC-3, DU 145 and LNCap). Among the carotenoids evaluated, fucoxanthin and neoxanthin, which has a structure similar to fucoxanthin, most remarkably reduced viability by inducing apoptosis as compared with other carotenoids. Furthermore, these two carotenoids have also been shown to suppress chemically induced carcinogenesis in experimental animals.

On the other hand, fucoxanthin shows anti-obesity effect, anti-diabetic effect, and promotion effect of docosahexaenoic acid (DHA) synthesis in the liver. These activities are specific for fucoxanthin and have not been found in common carotenoids such as β-carotene and astaxanthin. Among the specific functionalities of fucoxanthin, the mechanisms for the anti-obesity effect have been well studied. Therefore, this chapter describes the anti-obesity effect of fucoxanthin.

Obesity and Anti-Obesity Natural Components

Obesity is now recognized as a worldwide problem, with ominous implications for public health and health-related costs. It may be a second-most important preventable cause of death, exceeded only by cigarette smoking. Obesity is defined as accumulation of body fat. Especially, the accumulation of fat around the internal organs is a major risk factor causing many kinds of diseases. Because when the fat cell differentiates and accumulates the excess fat into the cell, the cell secretes various bioactive components, adipo-cytokines. Some of these adipo-cytokines induce various health problems such as type-2 diabetes, hypertension, and dyslipidemia, co-morbidities that markedly increase the risk of cardiovascular disease. These problems have been regarded as metabolic syndrome. Therefore, safe and effective anti-obesity component has been keenly expected to find from food materials.

Diacylglycerol (DG) with a 1,3-configuration (*11,12*) and medium-chain triacylglycerol (MCT) (*13*) have been used for the prevention of obesity. Several studies have demonstrated that conjugated linoleic acids (CLA) reduce body fat accumulation in growing animals but not all CLA isomers contributed to this effect equally (*14-16*). The reported mechanism of these lipids action includes characteristic digestion and absorption pathways, stimulation of lipolysis, reduction of lipid synthesis, and direct action on adipocytes (*17,18*). Caffeine is naturally consumed substance that is widely contained in beverages. It has thermogenic properties and increases the metabolic rate in humans (*19-23*). This effect can be explained by the stimulation of the secretion of catecholamine such as noradrenarine from the nerve endings. Noradrenaline stimulates β_3-adrenergic receptor (β_3-AR) and then induces promotion of energy expenditure through uncoupling protein 1 (UCP1) expression in brown adipose tissue (BAT) (*24-27*) (Figure 2). Capsaicin, the major pungent principle of red pepper, also up-regulates UCP1 in BAT by release of catecholamine (*28-31*). Green tea extract is reported to increase energy expenditure and fat oxidation in humans (*32*). The tea extract contains caffeine and catechin. Epigallocatechin gallate, a main tea catechin, promotes fat oxidation and decreases fat synthesis, but does not activate β_3-adrenergic receptor (*33*). Anti-obesity activity of green tea extract is attributed to both effects of UCP1 up-regulation by caffeine and of lipid metabolism control by catechin.

Adaptive Thermogenesis through UCP1 Expression

Although anti-obesity compounds render their activity by different molecular mechanisms; adaptive thermogenesis through UCP1 expression is most important. UCP1 is a member of UCP families (UCP1, UCP2 and UCP3) (*34,35*). UCPs are found in BAT (UCP1, UCP2 and UCP3), white adipose tissue (WAT) (UCP2), skeletal muscle (UCP2 and UCP3), and brain (UCP4 and UCP5) (*35,36*). UCP2 and UCP3 are expressed in BAT, muscle and other organs, thus, are candidates to influence energy efficiency and expenditure (*35*). Therefore, researchers have produced mice lacking UCP2 (*36*) and UCP3 (*37,38*). However, despite lack of UCP2 or UCP3, no consistent phenotypic

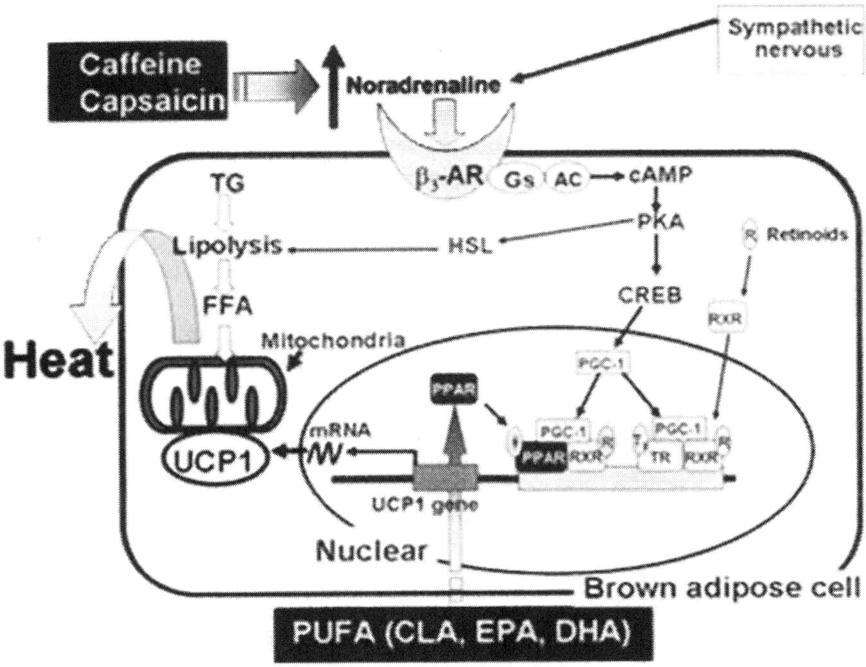

Figure 2. Possible Mechanism for Up-regulation of UCP1 in BAT

abnormality was observed in the knockout mice. They were not obese and had normal thermogenesis. These results suggest that UCP2 and UCP3 are not a major determinant of metabolic rate in normal condition, but rather, have other functions (*35,36,39-44*). Although more studies are required on the relationship between anti-obesity and the role of UCP2 and UCP3, it is certain that UCP1 can potentially reduce excess abdominal fat (*45*).

UCP1 is a dimeric protein present in the inner mitochondrial membrane of BAT, and it dissipates the pH-gradient generated by oxidative phosphoryration, releasing chemical energy as heat. UCP1 is exclusively expressed in BAT, where the gene expression is increased by cold, adrenergic stimulation, β_3-agonists, retinoids and thyroid hormone (*46*) (Figure 2). Thermogenic activity of BAT is dependent on UCP1 expression level controlled by the sympathetic nervous system via noradrenaline (*45,47-49*) As a consequence of noradrenaline binding to the adipocyte plasma membrane, protein kinase (PKA) is expressed, and then, cyclic AMP response element binding protein (CREB) and hormone-sensitive lipase (HSL) are expressed. HSL stimulates lipolysis and free fatty acids liberated serve as substrate in BAT thermognesis (*49*). They also act as cytosolic second messengers which activated UCP1 as PPARγ ligand. The same activity is expected in dietary polyunsaturated fatty acids including CLA, EPA

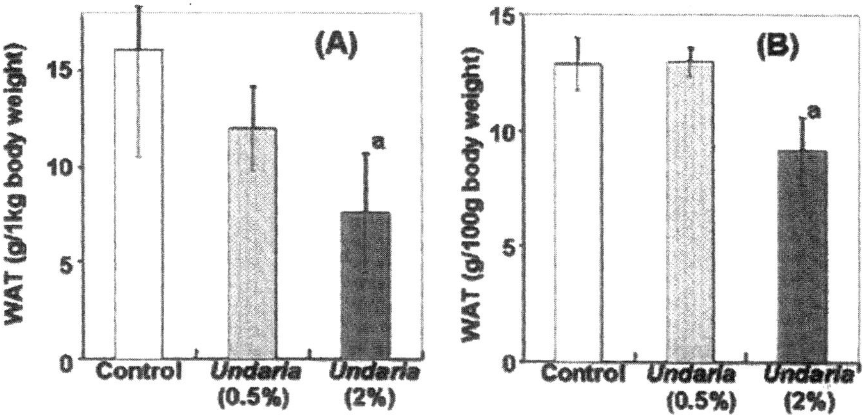

Figure 3. Weight of WAT of Rats (A) and Mice (B) Fed Undaria Lipids and Control Diet. [a]Significant different from control (P<0.01). A diet was prepared according to the recommendation of American Institute of Nutrition (AIN-93G). The dietary fats for rats were 7% soybean oil (control), 6.5% soybean oil + 0.5% Undaria lipids, and 5% soybean oil + 2% Undaria lipids. Those for mice were 13% soybean oil (control), 12.5% soybean oil + 0.5% Undaria lipids, 11% soybean oil + 2% Undaria lipids. (Adapted from Ref. 52)

and DHA (*50*). As shown in Figure 3, UCP1 expression in BAT controlled by sympathetic nervous system via noradrenaline and expressions of PPARγ, RXR, and TR. Anti-obesity effect is shown by regulation of these molecular actions.

UCPI Expression in WAT Induced by Fucoxanthin

As revealed above, UCP1 usually express only in BAT not in other biological systems. Although adult humans have a little BAT, it is still very important for anti-obesity to up-regulate UCP1 in BAT induced by food components such as caffeine, capsaicin, and polyunsaturated fatty acids. However, considered as breakthrough discoveries for an ideal therapy of obesity, expression of UCP1 expression in tissues other than BAT by food constituents would be important. UCP1, usually expressed only in BAT, has also been found in white adipose tissue (WAT) of mice overexpressing *Foxc2*, a winged helix gene, with a change in steady-state levels of several WAT and BAT derives mRNAs (*51*). This result suggests the possibility of UCP1 expression in WAT, which would be an increasingly attractive target for the development of antiobesity therapies. From this viewpoint, the antiobesity effect of edible seaweed carotenoid, fucoxanthin, is very interesting, as its activity depends on the protein and gene expressions of UCP1 in WAT (*52*).

In the study on anti-obesity effect of seaweed carotenoid, fucoxanthin, lipids were separated from edible seaweed, Wakame (*Undaria pinnatifida*), one of the most popular edible seaweed in Japan and Korea. *Undaria* lipids, containing 10% fucoxanthin, reduced significantly the weight of WAT (comprising perirenal and epididymal abdominal adipose tissues) of both rats and mice (Figure 4). Furthermore, body weight of mice fed 2% *Undaria* lipid was significantly ($P<0.05$) lower than that of the control, although there was no significant difference in the mean daily intake of diet between the two groups. In order to confirm the active component of *Undaria* lipids, fucoxanthin and *Undaria* glycolipids, the main fraction of the lipids, was administered to obese KK-Ay mice. The WAT weight of fucoxanthin-fed mice was significantly lower than that of control mice. However, there was no difference in WAT weight of mice fed *Undaria* glycolipids and control diet. This result indicates that fucoxanthin is the active component in the *Undaria* lipids resulting in the anti-obesity effect.

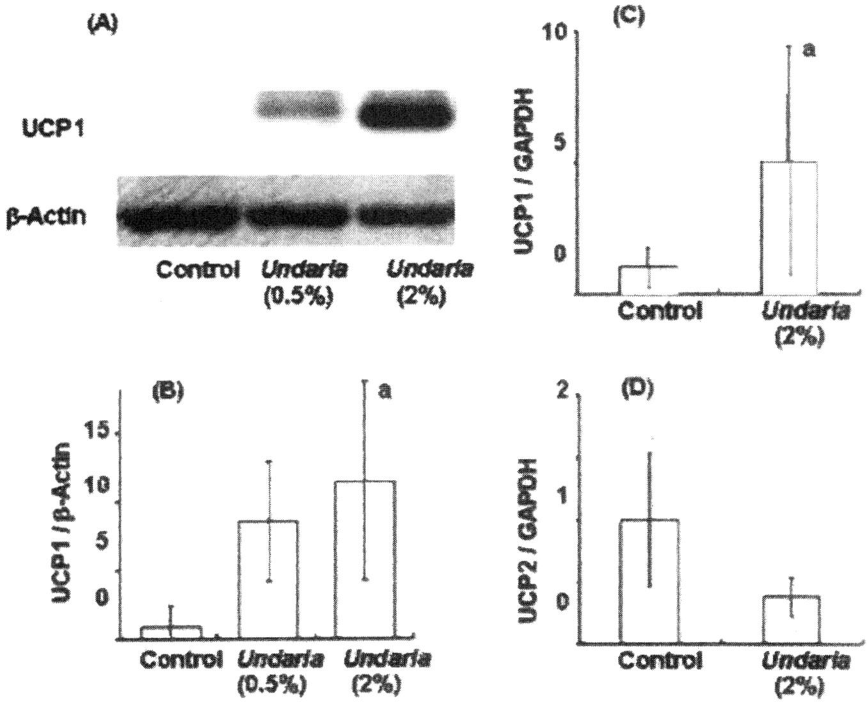

Figure 4. UCP1 and UCP2 Expressions in WAT of Mice Fed Undaria Lipids and Control Diet. (A), Western blot analysis of UCP1. (B), UCP1 protein expression. (C), UCP1 mRNA expression. (D), UCP2 mRNA expression. [a]Significant different from control (P<0.05). (Adapted from Ref. 52)

Furthermore, as shown in Figure 4, UCP1 expression was found in WAT of *Undaria* lipids-fed mice, although there was little expression in that of control mice. Expression of UCP1 mRNA was also found in WAT of *Undaria* lipids-fed mice, but little expression in that of the control (Figure 6). UCP1 is known as a specific protein inducing fat oxidation and conversion of the energy to heat. Therefore, the decrease in WAT weight of *Undaria* lipids-fed mice would be due to the adaptive thermogenesis through UCP1 expression in WAT. To confirm the active component for the UCP1 expression in WAT, purified fucoxanthin (purity>98% by HPLC) was given to animals at concentration 0.05-0.2 % (wt/diet). Significant decrease in abdominal WAT was observed with clear UCP1 expression in the WAT.

The finding that fucoxanthin induces both protein and mRNA expressions of UCP1 in WAT provides a clue for new dietary antiobesity therapy. An enormous amount of data has been collected on thermogenesis in BAT through UCP1 expression. However, there had been no information on UCP1 expression in WAT induced by a diet component until the above report had appeared. An excessive accumulation of fat in WAT induces some diseases such as Type II diabetes. Direct heat production by fat oxidation in WAT, therefore, may reduce risk of these diseases in humans. When purified fucoxanthin (purity>98% by HPLC) was given to diabetes model mouse at concentration 0.05-0.2 % (wt/diet), significant decrease in blood glucose was observed.

Reducing Effect of Fucoxanthin and Its Main Metabolite, Fucoxanthinol, on Adipocyte Differentiation

Fucoxanthin is easily converted to fucoxanthinol in the human intestinal cells and in mice (*53*).

Both fucoxanthin and fucoxanthinol inhibited intercellular lipid accumulation during adipocyte differentiation of 3T3-L1 cells (Figure 5). Fucoxanthin and fucoxanthinol also decreased glycerol-3-phosphate dehydrogenase activity, an indicator of adipocyte differentiation (*54*). The effects of fucoxanthinol were stronger than those of fucoxanthin. When 3T3-L1 cells treated with fucoxanthin and fucoxanthinol, PPARγ, a regulater of adipogenic gene expression, was down-regulated by both carotenoids in a dose-dependent manner (*54*) (Figure 6).

These results suggest that fucoxanthin and fucoxanthinol inhibit the adipocyte differentiation of 3T3-L1 cells through down-regulation of PPARγ and fucoxanthinol would be an active compound for the anti-obesity effect of fucoxanthin.

PPARγ plays an important role in the early stages of differentiation of 3T3-L1 cell, because it is a nuclear transcription factor that regulates adipogenic gene expression. Regulation of PPARγ would be one of the expected mechanisms

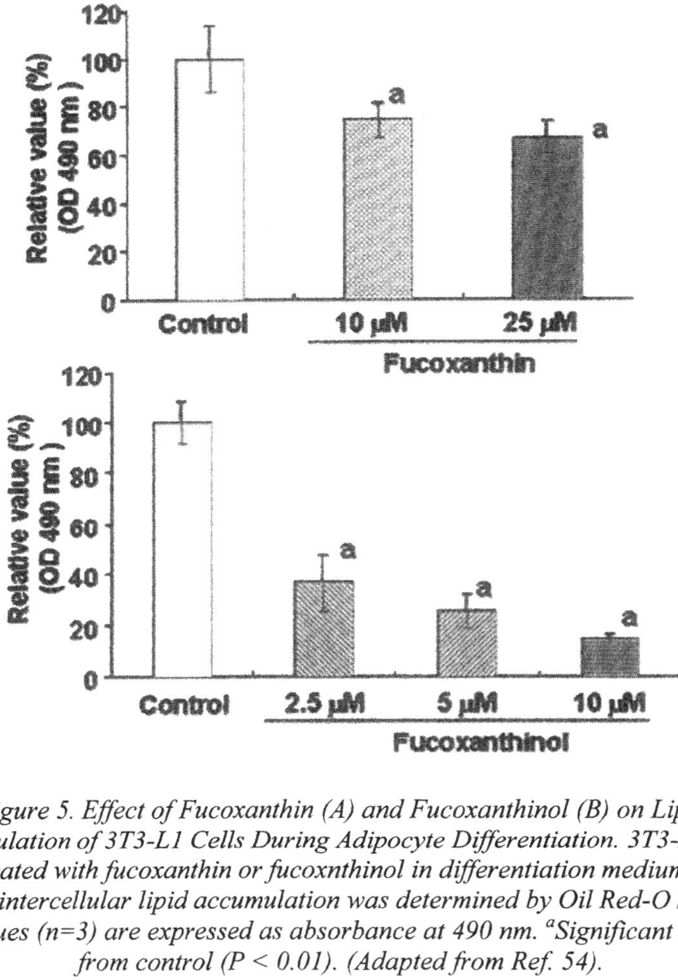

Figure 5. Effect of Fucoxanthin (A) and Fucoxanthinol (B) on Lipid Accumulation of 3T3-L1 Cells During Adipocyte Differentiation. 3T3-L1 cells were treated with fucoxanthin or fucoxnthinol in differentiation medium for 120 hr. The intercellular lipid accumulation was determined by Oil Red-O staining. The values (n=3) are expressed as absorbance at 490 nm. [a]Significant different from control ($P < 0.01$). (Adapted from Ref. 54).

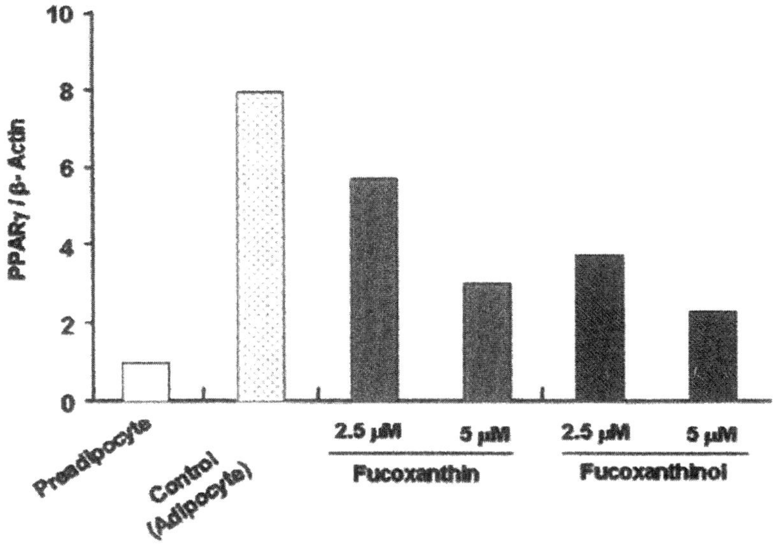

Figure 6. Expression of PPARγ in 3T3-L1 Cells Treated with Fucoxanthin and Fucoxanthinol The PPARγ protein expression level was normalized to the β-Actin level and expressed as the value relative to preadipocyte PPA levels. (Adapted from Ref. 54)

underlying the anti-obesity effect of dietary fucoxanthin. Catechin (*55*), sterols (*56*), tannic acid (*57*), phenolic lipids (*58*), and red yeast rice extracts (*59*) also inhibit 3T3-L1 differentiation. Retinoids inhibit the early stage of differentiation of 3T3-L1 cells (*60*). However, absorption rate of some of these compounds are very low and have not been fully investigated for their anti-obesity effects *in vivo*. Fucoxathin is absorbed into the animal body as fucoxanthinol and mainly accumulated in WAT and liver. On the other hand, when other carotenoids such as β-carotene and astaxanthin were given to animals, little accumulation was observed in WAT. The accumulation of fucoxanthin in WAT and regulation of fucoxanthin against PPARγ in adipose cell will be a clue for the elucidation of the anti-obesity mechanism of this interesting marine carotenoid, fucoxanthin.

References

1. Sugawara, T.; Baskaran, V.; Tsuzuki, W.; Nagao, A. *J. Nutr.* **2002,** *132*, 946-951.
2. Haugan, J; Liaeen-Jensen, S. *Biochem. Syst. Ecol.* **1994**, *22*, 31-41.

3. Mori, K.; Ooi, T.; Hiraoka, M.; Oka, N.; Hamada, H.; Tamura, M.; Kusumi, T. *Marine Drugs* **2004**, *2*, 63-72.
4. Jensen, A. In *Norwegian Institute of Seaweed Research*, Report No. 31, Tapir, Trodenheim, 1966.
5. Liaeen-Jensen, S. In *Marine natural products – Chemical and biological perspectives*. Scheuer, P. Ed.; Academic Press: New York, NY, 1978, pp. 1-73.
6. Yan, X.; Chuda, Y.; Suzuki, M; Nagata, T. *Biosci. Biotechnl. Biochem.* **1999**, *63*, 605-607.
7. Shiratori, K.; Ohgami, K.; Ilieva, I.; Jin, X.-H.; Koyama, Y.; Miyashita, K.; Yoshida, K.; Kase, S; Ohno, S. *Exp. Eye Res.* **2005**, *81*, 271-277.
8. Peto, R.; Doll, R.; Buckley, J. D.; Sporn, M. B. *Nature* **1981**, *290*, 201-208.
9. Hosokawa, M.; Wanezaki, S.; Miyauchi, K.; Kurihara, H.; Kohno, H.; Kawabata, J.; Odashima, S.; Takahashi, K. *Food Sci. Technol. Res.*, **1999**, *5*, 243-246.
10. Kotake-Nara, E.; Kushiro, M.; Zhang, H.; Sugawara, T.; Miyashita, K.; Nagao, A. *J. Nutr.* **2001**, *131*, 3303-3306.
11. Flickinger, B. D.; Matsuo, N. *Lipids* **2003**, *38*, 129-132.
12. Flickinger, B. D. In *Nutraceuticals and Specialty Lipids and Their Co-Products*, Shahidi, F. Ed.; CRC Press: Boca Raton, FL, 2006, pp. 181-186.
13. Che Man, Y. B.; Manaf, A. A. In *Nutraceuticals and Specialty Lipids and Their Co-Products*, Shahidi, F. Ed.; CRC Press: Boca Raton, FL, 2006, pp. 27-56.
14. Atkinson, R. A. In *Advances in Conjugated Linoleic Acid Research*, Vol. 1, Yurawecz, M. P.; Mossoba, M. M.; Kramer, J. K. G.; Pariza, M. W.; Nelson, G. J. Eds.; AOCS Press: Champaign, IL, 1999, pp. 348-353.
15. Keim, N. L. In *Advances in Conjugated Linoleic Acid Research*, Vol. 2, Sébédio, J.-L.; Christie, W. W.; Adlof, R. Eds.; AOCS Press: Champaign, IL, 2003, pp. 316-324.
16. Watkins, B. A.; Li, Y. In *Nutraceuticals and Specialty Lipids and Their Co-Products*, Shahidi, F. Ed.; CRC Press: Boca Raton, FL, 2006, pp. 187-200.
17. Miner, J. L.; Cederberg, C. A.; Nielson, M. K.; Chen, X.; Baile, C. A. *Obes. Res.* **2001**, *9*, 129-134.
18. Tsuboyama-Kasaoka, N.; Takahashi, M.; Tanemura, K.; Kim, H.J.; Tange, T.; Okuyama, H.; Kasai, M.; Ikemoto, S.; Ezaki, O. *Diabetes* **2000**, *49*, 1534-1542.
19. Miller, D. S.; Stock, M. J.; Stuart, J. A. *Proc. Nutr. Soc.* **1974**, *33*, A28-29.
20. Acheson, K. J.; Zahorska-Markiewicz, B.; Pittet, P.; Anantharman, K.; Jéquier, E. *Am. J. Clin. Nutr.* **1980**, *33*, 989-997.
21. Yoshida, T.; Sakane, N.; Umekawa, T.; Kondo, M. *Int. Obes. Relat. Metab. Disord.* **1994**, *18*, 345-350.
22. Astrup, A. *Endocrine* **2000**, *13*, 207-212.
23. Molnar, D.; Torok, K.; Erhardt, E.; Jeges, S. *Int. Obes. Relat. Metab. Disord.* **2000**, *24*, 1573-1578.

24. Bellet, S.; Roman, L.; Decastro, O.; Kim, K. E.; Kershbaum, A. *Metabolism* **1969**, *18*, 288-291.
25. Berkowitz, B. A.; Spector, S. *Eur. J. Pharmacol.*, **1971**, *13*, 193-197.
26. Yoshida, K.; Yoshida, T.; Kamanaru, K.; Hiraoka, N.; Kondo, M. *J. Nutr. Sci. Vitaminol.* **1990**, *36*, 173-178.
27. Kogure, A.; Sakane, N.; Takakura, Y.; Umekawa, T.; Yoshida, K.; Nishino, H.; Yamamoto, T.; Kawada, T.; Yoshikawa, T.; Yoshida, T. *Clin. Exper. Pharm. Phys.* **2002**, *29*, 391-394.
28. Kawada, T.; Hagihara, K.; Iwai, K. *J. Nutr.* **1986**, *116*, 1272-1278.
29. Kawada, T.; Sakabe, S.; Aoki, N.; Watanabe, T.; Higeta, K.; Iwai, K. *J. Agric. Food Chem.* **1991**, *39*, 651-654.
30. Watanabe, T.; Kawada, T.; Kato, T.; Harada, T.; Iwai, K. *Life Sci.* **1994**, *54*, 369-374.
31. Masuda, Y.; Haramizu, S.; Oki, K.; Ohnuki, K.; Watanabe, T.; Yazawa, S.; Kawada, T.; Hashizume, S.; Fushiki, T. *J. Appl. Phys.* **2003**, *95*, 2408-2415.
32. Duloo, A.G.; Duret, C.; Rohrer, D.; Girardier, L.; Mensi, N.; Fathi, M.; Chantre, P.; Vandermander, *J. Am. J. Clin. Nutr.* **1999**, *70*, 1040-1045.
33. Klaus, S.; Pültz, S.; Thöne-reineke, C.; Wolfram, S. *Int. J. Obesity* **2005,** *19*, 615-623.
34. Dulloo, A. G.; Samec, S. *British J. Nutr.* **2001**, *86*, 123-139.
35. Jeek, P. *Inter. J. Biochem. Cell Biol.* **2002**, *34*, 1190-1206.
36. Arsenijevic, D.; Onuma, H.; Pecqueur, C.; Raimbault, S.; Manning, B.S.; Miroux, B.; Couplan, E.; Alves-Guerra, M.-C.; Goubern, M.; Surwit, R.; Bouillaud, F.; Richard, D.; Collins, S.; Ricquier, D. *Nat. Genetics* **2000**, *26*, 435-439.
37. Gong, D.-W.; Monemdjou, S.; Garvilova, O.; Leon, L.R.; Marcus-Samuels, B.; Chou, C. J.; Everett, C.; Kozak, L. P.; Li, C.; Deng, C.; Harper, M.-E.; Reitman, M. L. *J. Biol. Chem.* **2000**, *275*, 16251-16257.
38. Vidal-Puig, A. J.; Grujic, D.; Zhang, C.-Y.; Hagen, T.; Boss, O.; Ido, Y.; Szczepanik, A.; Wade, J.; Mootha, V.; Cortright, R.; Muoio, D. M.; Lowell, B. B. *J. Biol. Chem.* **2000**, *275*, 16258-16266.
39. Lowell, B. B.; Spiegelman, B. M. *Nature* **2000**, *404*, 652-660.
40. Harper, M.-E.; Dent, R. M.; Bezaire, V.; Antoniou, A.; Gauthier, A.; Monemdjou, S.; McPherson, R. *Biochem. Soc. Trans.* **2001**, *29*, 768-773.
41. Rodríquez, A. M.; Roca, P.; Palou, A. *Euro. J. Phys.* **2002**, *444*, 484-490.
42. Schrauwen, P.; Hesselink, M. *J. Exper. Biol.* **2002**, *205*, 2275-2285.
43. Ricquier, D. *Pro. Nutri. Soc.* **2005**, *64*, 47-52.
44. Mills, E. M.; Banks, M. L.; Sprague, J. E.; Finkel, T. *Nature* **2003**, *426*, 403-404.
45. Nedergaard, J.; Golozoubova, V.; Matthias, A.; Asadi, A.; Jacobsson, A.; Cannon, B. *Biochim. Biophys. Acta.* **2001**, *1504*, 82-106.
46. Silva, J.E.; Rabelo, R. *Eur. J. Endocrinol.* **1997**, *136*, 251-264.
47. Del Mar Gonzalez-Barroso, M.; Ricquier, D.; Cassard-Doulcier, A.-M. *Obesity Rev.* **2000**, *1*, 61-72.

48. Argyropoulos, G.; Harper, M.-L. *J. Appl. Physiol.*, **2002**, *92*, 2187-2198.
49. Mozo, J.; Emre, Y.; Bouillaud, F.; Ricquier, D.; Criscuolo, F. *Biosci. Reports* **2005**, *25*, 227-249.
50. Clarke, S.D. *British J. Nutr.* **2000**, *83*, 59-66.
51. Cederberg, A.; Grønning, L. M.; Ahrén, B.; Taskén, K.; Carlsson, P.; Enerbäck, S. *Cell* **2001**, *106*, 563-573.
52. Maeda, H.; Hosokawa, M.; Sashima, T.; Funayama, K.; Miyashita, K. *Biochim. Biophys. Res. Commu.*, **2005**, *332*, 392-397.
53. Sugawara, T.; Baskaran, V.; Tsuzuki, W.; Nagao, A. *J. Nutr.* **2002**, *132*, 946-951.
54. Maeda, H.; Hosokawa, M.; Sashima, T.; Takahashi, N.; Kawada, T.; Miyashita, K. *Int. J. Mole. Med.* **2006**, *18*, 147-152.
55. Paul, A.G. *Prog. Lipid Res.*, **2001**, *40*, 269-281.
56. Furuyashiki, T.; Nagayasu, H.; Aoki, Y.; Bessho, H.; Hashimoto, T.; Kanazawa, K.; Ashida, H. *Biosci. Biotechnol. Biochem.* **2004**, *68*, 2353-2359.
57. Awad, A. B.; Begdache, L. A.; Fink, C. S. *J. Nutr. Biochem.* **2000**, *11*, 153-158.
58. Liu, X.; Kim, J. K.; Li, Y.; Liu, F.; Chen, X. *J. Nutr.* **2005**, *135*, 165-171.
59. Rejman, J.; Kożubek, A. *J. Agric. Food Chem.* **2004**, *52*, 246-250.
60. Jeon, T.; Hwang, S. G.; Hirai, S.; Matsui, T.; Yano, H.; Kawada, T.; Lim, B. O.; Park, D. K. *Life Sci.* **2004**, *75*, 3195-203.

Chapter 33

Synergistic Antibacterial Effect of 5-Methylthiopentyl Isothiocyanate and (–)-Limonene on Periodontal Pathogen

Hideki Masuda[1] and Saori Hirooka[2]

[1]Ogawa and Company, Ltd., 4–1–11, Nihonbashi Honcho Chuo-ku, Tokyo 103–0023, Japan
[2]Ogawa and Company, Ltd., 15–7, Chidori Urayasushi, Chiba 279–0032, Japan

The combination of 5-methylthiopentyl isothiocyanate, the minor component in volatiles obtained by the hydrolysis of wasabi, and (-)-limonene, the minor component of peppermint or spearmint essential oil, was found to show a synergistic antibacterial effect on *Porphyromonas gingivalis*, the primary bacterium responsible for causing periodontal disease using an isobolographic analysis, a time-kill method and a bio-luminescence method. The synergistic effect was shown to be significantly bacteriocidal within a few minutes. The different antibacterial mechanism leading to the synergistic effect was clarified from the different concentration of extracellular-adenosine 5'-triphosphate leaked out by the addition of 5-methylthiopentyl isothiocyanate and that of (-)-limonene.

Periodontal disease, which is composed of light and severe symptoms, that is, gingivitis and periodontitis, respectively, is an oral infection caused by a variety of pathogens (*1-3*). Among them, *Porphyromonas gingivalis* (*P. gingivalis*), a kind of obligate anaerobe, is reported as the most frequently found periodontal bacteria in adult periodontitis patients. *P. gingivalis* grows, produces colonization in the deep pocket between dentin and gingival, and makes an inflammatory and destructive lesion in the periodontal tissue.

To prevent periodontal disease, the habit of tooth-brushing is thought to be the easiest method. Recently, the use some foods and crude drugs have gradually increased in addition to the synthetic antibacterial materials (*4-8*).

As a popular and traditional food, we have been focusing on wasabi (*Wasabia japonica matum*). In some of the volatiles obtained by the hydrolysis of wasabi, allyl isothiocyanate is well-known because of its high quantity and high antibacterial activity (*9*, *10*). However, its flavor is strongly pungent and stimulative, so it is difficult to practically use. On the other hand, the ω-methylthioalkyl isothiocyanates, the minor components in the wasabi hydrolysate, were reported to have a radish-like and sweet flavor, and higher thresholds compaired to the threshold of allyl isothiocyanate (*11*). Therefore, from the viewpoint of flavor characteristics, the practical use of ω-methylthioalkyl isothiocyanates is considered to be more interesting. However, their antibacterial effects have hardly been studied (*12*, *13*). In addition, as for their antibacterial effect on periodontal pathogens, no report has been found. Therefore, in this study, we focused on the ω-methylthioalkyl isothiocyanates, especially 5-methylthiopentyl isothiocyanate (5-MITC).

In order to enhance the pharmacological action, a combination of chemicals may be used. The synergistic effect obtained by combination of different components provides not only a pharmacological merit, but also a good flavor. As for the peppermint or spearmint essential oil and its flavor components, such as (-)-menthol, (-)-carvone, or thymol, some studies of their effect on the periodontal pathogens have already been reported (*14-16*). However, the antibacterial effect of (-)-limonene, the minor component in peppermint or spearmint essential oil, on periodontal pathogens has never been studied. In this combination study on *P. gingivalis*, we used (-)-limonene in combination with 5-MITC (*17*).

Experimental

Chemicals

5-Methylthiopentyl isothiocyanate (5-MITC) was prepared from 4-pentenyl isothiocyanate according to a reported method (*18*, *19*). (-)-Limonene and chlorhexidine were purchased from Sigma-Aldrich, Inc. (St. Louis, MO) and used as received.

5-Methylthiopentyl isothiocyanate

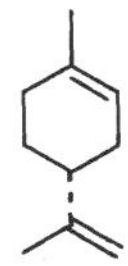

(-)-Limonene

Microorganism and Media

P. gingivalis ATCC33277 was obtained from the American Type Culture Collection (Manassas, VA, USA). A bacterial suspension for inoculation was prepared from an overnight culture on modified GAM broth (Nissui Pharmaceutical Co., Ltd., Tokyo, Japan) to produce a concentration of $1\text{-}8\times10^7$ colony forming units (CFU)/mL. The composition of the sterile diluent was as follows: KH_2PO_4 (4.5 g), Na_2HPO_4 (6.0 g), L-cystein·HCl·H_2O (0.5 g), Tween 80 (0.5 g), agar (1.0 g), and distilled water (1000 mL). Each compound mentioned above was mixed and sterilized at 115 °C for 20 min in an autoclave (*20*).

MIC_{90} Determination

The minimum inhibitory concentration (MIC) determinations were performed using a microdilution technique (serial 2-fold dilution) in 96-well microplates (*21*, *22*). The test compound solutions (26 μL), which were diluted with 100% N,N'-dimethylformamide (DMF), were added to 96-well microplates with 234 μL of modified GAM broth inoculated with $1\text{-}8\times10^7$ CFU/mL. All the microplates were covered with silicon plate seals (Fisher Scientific Japan, Ltd., Tokyo, Japan) to avoid any cross-contamination. The microplates were incubated at 37 °C for 48 h in an anaerobic growth chamber in 10% H_2, 10% CO_2, and 80% N_2. The MIC against *P. gingivalis* was defined as the minimum concentration of the test compound limiting the turbidity to < 0.05 absorbance at 655 nm using a (BIO-RAD, Model 550, Hercules, CA, USA) microplate reader. MIC_{90} was the MIC endpoint determined as the minimum concentration having greater than a 90% antibacterial activity compared to that of the control using a probit analysis. All tests were run in triplicate. The averages and standard errors were calculated from these triplicate measurements.

Isobolographic Analysis

Combination studies were performed using the broth checkerboard method (*21*, *22*). A series of 2-fold dilutions of 5-MITC was tested in combination with a

2-fold dilution of (-)-limonene. The MICs were determined using the same method as described above. All tests were run in triplicate. The averages and standard errors were calculated from these triplicate measurements.

Time-kill Method

Combination studies were also studied by a time-kill method in modified GAM broth containing 5-MITC (10 ppm) and (-)-limonene (25 ppm) (*21*, *22*). The initial inoculum was $1\text{-}8\times10^{7}$ CFU/mL. Samples were taken at specific times. The number of CFU/mL was then counted by serial 10-fold dilutions with a sterile diluent and plated onto modified GAM agar plates. The plates were incubated at 37 °C for 48 h in an anaerobic growth chamber in 10% H_2, 10% CO_2, and 80% N_2 before counting. All tests were run in triplicate.The averages and standard errors were calculated from these triplicate measurements.

Bioluminescence Method

The suspension of *P. gingivalis* was prepared by the method described in "Microorganism and Media".

The control was prepared as follows: DMF (250 µL) and the modified GAM broth (750 µL). The control solution (100 µL) was then added to the suspension of *P. gingivalis* (4.9 mL).

The 5-MITC sample was prepared as follows: 2.6% 5-MITC/DMF solution (250 µL) was diluted with the modified GAM broth (750 µL). The obtained 0.65% 5-MITC solution (50 µL) was added to the suspension of *P. gingivalis* (4.9 mL), followed by the addition of 50 µL of control solution.

The sample of (-)-limonene was prepared as follows: a 14% (-)-limonene/DMF solution (250 µL) was diluted with the modified GAM broth (750 µL). The obtained 3.5% (-)-limonene solution (50 µL) was then added to the suspension of *P. gingivalis* (4.9 mL), followed by the addition of 50 µL of the above-mentioned control solution.

The combination sample of 5-MITC and (-)-limonene was prepared as follows: 0.65% 5-MITC solution (50 µL) and 3.5% (-)-limonene solution (50 µL) were added to the *P. gingivalis* suspension (4.9 mL).

The control, the 5-MITC sample, the sample of (-)-limonene, and the sample of 5-MITC and (-)-limonene were allowed to stand for 6 min at room temperature, subsequently filtered using the filter (0.45 µm, Advantec Toyo Kaisha, Ltd., Tokyo, Japan) and pipetted into the microplates. The extracellular-adenosine 5'-triphosphate (ATP) concentrations were determined by the addition of luciferin-luciferase (100 µL, Luciferol 250 plus 60312, Kikkoman Corporation, Tokyo, Japan) to the filtrate (100 µL), followed by the measurement with a luminometer (ATTO Luminescencer-JNRII, Atto

Corporation, Tokyo, Japan) (*23*). To calculate the extracellular-ATP concentrations, a standard ATP curve ranging from 10^{-1} to 10^{-3} μM portions was used to obtain a linear relationship between the ATP concentration (μM) and relative light unit (RLU) which resulted in an r^2 (coefficient of determination) value of 0.9998. The following regression equation was obtained: ATP concentration = (0.000004 × RLU) – 0.0009. All tests were run in triplicate. The averages and standard errors were calculated from these triplicate measurements.

Results and Discussion

An isobolographic analysis is well-known to clarify the pharmacological activity resulting from the drug combination (*24*, *25*). It provides an effect as to whether the pharmacologically active response caused by the combined use of agents is higher, equal, or lower than would have been expected from each pharmacological activity by their single use and the concept of dose additivity. In Figure 1, the solid and dashed line connecting the MIC_{90} of 5-MITC and that of (-)-limonene is the average additive line and additive line showing the standard errors, respectively. Therefore, the area surrounded by the dashed lines

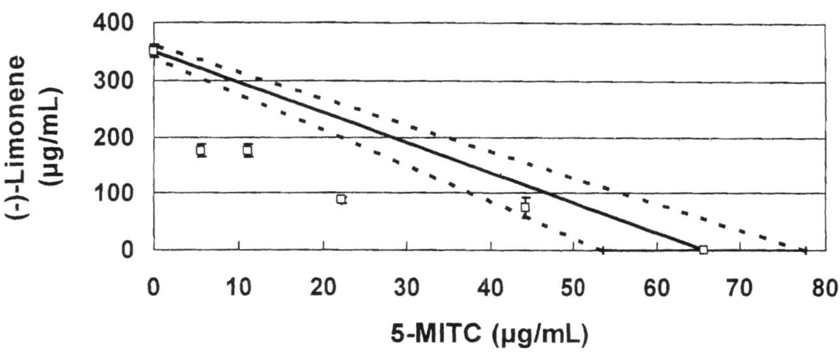

Figure 1. MIC_{90} isobologram of P. gingivalis obtained by combination of 5-methylthiopentyl isothiocyanate (5-MITC) and (-)-limonene after incubation for 48 h . Values are means ± SE (n = 3); error bars represent ± 1 SE.

shows an additive antibacterial effect. From the data shown in Figure 1, the combination of 5-MITC and (-)-limonene was found to provide a synergistic antibacterial effect.

The isobolographic analysis shown in Figure 1 provided only the static data obtained after a 48 h incubation. Therefore, in order to know whether the synergistic antibacterial effect obtained by the isobolographic analysis was bacteriostatic or bacteriocidal, we studied the variation in the viable microbe cell number. Figure 2 shows that the single use of 5-MITC and (-)-limonene provided a bacteriostatic and slightly bacteriocidal effect, respectively. However, interestingly, their combined use provided a significant bacteriocidal effect.

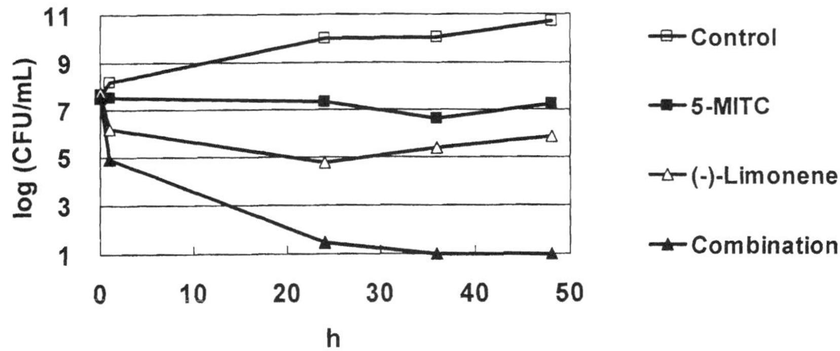

Figure 2. Time-kill curves of P. gingivalis showing synergistic effect between 5-methylthiopentyl isothiocyanate (5-MITC) and (-)-limonene. The compounds were added to the P. gingivalis cultures at the following concentrations: 5-MITC (10 ppm), (-)-limonene (25 ppm), and combination of 5-MITC (10 ppm) and of (-)-limonene (25 ppm). Values are means (n = 3).

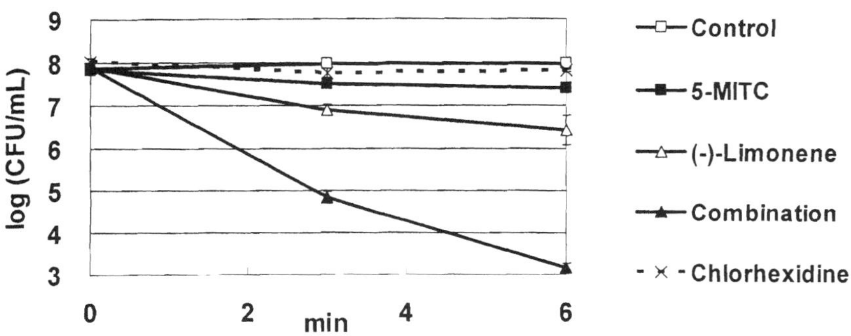

Figure 3. Time-kill curves of P. gingivalis showing synergistic effect between 5-methylthiopentyl isothiocyanate (5-MITC) and (-)-limonene. The compounds were added to the P. gingivalis cultures at the following concentrations: 5-MITC (65 ppm), (-)-limonene (350 ppm), combination of 5-MITC (65 ppm) and (-)-limonene (350 ppm), and chlorhexidine (3.2 ppm: MIC_{90} on P. gingivalis). Values are means ± SE (n = 3); error bars represent ± 1 SE.

A high effect within a short time is a very important factor for practical use in toothpastes or functional foods. Figure 3 shows that only a few minutes after the addition, the combination of 5-MITC and (-)-limonene was found to provide a marked bacteriocidal effect. In addition, its antibacterial effect was found to be significant when compared to that obtained by chlorohexidine, a well-known antibacterial agent in a dental treatment.

For the purpose of studying the action mechanism, we carried out the bioluminescence method. The disruption of the bacterium surface-structure, such as the cell wall or cell membrane followed by the leakage of the internal substrate, is known as one cause for the bacterium damage (*26*). After the surface-structure disruption caused by the addition of some agent, the internal ATP leaks out from the cell inside (Figure 4). The leaked ATP, that is, the extracellular-ATP, then reacts with luciferine in the presence of luciferase and magnesium ion to provide the luminescence.

As shown in Figure 5, the different antibacterial mechanism leading to the synergistic effect was revealed by the different extracellular-ATP concentration

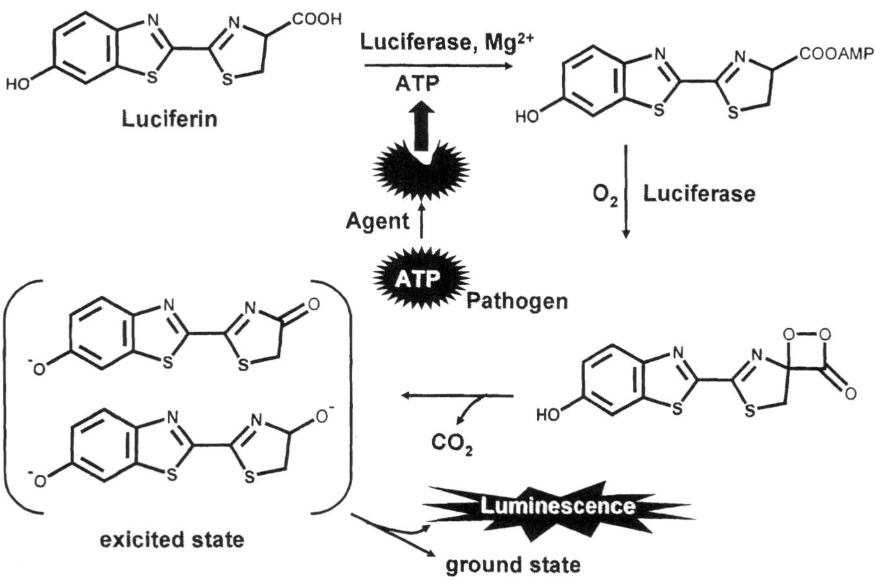

Figure 4. Measurement of extracellular-adenosine 5'-triphosphate (ATP) concentration using bioluminescence method.

between 5-MITC and (-)-limonene. As for the high ATP-leakage by (-)-limonene, the high hydrophobicity of the *P. gingivalis* surface is considered to induce the easier attachment of (-)-limonene followed by the surface-structure disruption of the bacterium and ATP-leakage (*27-29*). In addition, Figure 5 shows that the extracellular-ATP concentration obtained by the combined use of 5-MITC and (-)-limonene was much higher than the sum of the extracellular-ATP concentration obtained by their single use. Therefore, based on this result, the possibility that the change in the internal protein caused by 5-MITC is related to the promotion of the ATP-leakage from the cell inside is considered (*30-32*).

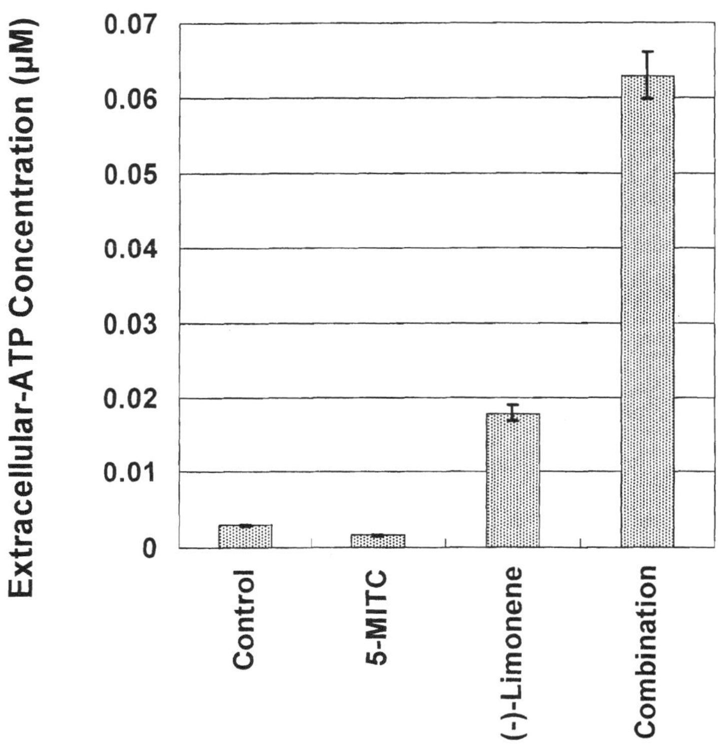

Figure 5. Extracellular-adenosine 5'-triphosphate (ATP) concentration after incubation for 6 min. The compounds were added to the P. gingivalis cultures at the following concentrations: 5-MITC (65 ppm), (-)-limonene (350 ppm), and combination of 5-MITC (65 ppm) and (-)-limonene (350 ppm). Values are means ± SE (n = 3); error bars represent ± 1 SE.

Conclusions

The isobolographic analysis clarified the synergistic effects of 5-methylthiopentyl isothiocyanate and (-)-limonene on *P. gingivalis*. The time-kill method clarified that the single use of 5-methylthiopentyl isothiocyanate and (-)-limonene produced the bacteriostatic and slightly bacteriocidal effects, respectively. However, their combined use provided a significant bacteriocidal effect in a few minutes. The extracellular-ATP concentration obtained by the bioluminescence method clarified that a different action mechanism leading to the synergistic effect. Considering their low use and preferable flavor, its practical value should be in toothpaste and functional foods for preventing the periodontal disease.

References

1. Slots, J.; Bragd, L.; Wikstroem, M.; Dahlen, G. *J. Clin. Periodontol.* **1986**, *13*, 570-577.
2. Socransky, S. S.; Haffajee, A. D.; Ximenez-Fyvie, L. A.; Feres M.; Mager, D. *Periodontology 2000.* **1999**, *20*, 341-362.
3. Haffajee, A. D.; Socransky, S. S. *Periodontology 2000.* **1994**, *5*, 78-111.
4. Kakuda, T.; Takahara, T.; Sakane, I.; Mortelmans, K. *Nippon Nogeikagaku Kaishi* **1994**, *68*, 241-243.
5. Ikematsu, T. *Nihon Univ. J. Oral Sci.* **1998**, *24*, 21-28.
6. Yamamoto, H.; Ogawa, T. *Biosci. Biotechnol.Biochem.* **2002**, *66*, 921-924.
7. Motomura, Y.; Araki, K.; Shin, K.; Miyata, T. *Nippon Shishubyo Gakkai Kaishi.* **1997**, *39*, 72-76.
8. Mizrahi, B.; Shapira, L.; Domb, A. J.; Houri-Haddad, Y. *J. Periodontol.* **2006**, *77*, 963-968.
9. Isshiki, K.; Tokuoka, K.; Mori, R.; Chiba, S. *Biosci. Biotech. Biochem.* **1992**, *56*, 1476-1477.
10. Delaquis, P. J.; Mazza, G. *Food Technol.* **1995**, *November*, 73-84.
11. Masuda, H.; Harada, Y.; Tanaka, K.; Nakajima, M.; Tateba, H. In *Biotechnology for Improved Foods and Flavors*; Takeoka, G. R.;. Teranishi, R.; Williams, P. J.; Kobayashi, A., Eds.; ACS Symposium Series 637; American Chemical Society: Washington, DC, 1996; pp 67-78.
12. Matsuoka, H.; Takita, J.; Kawakami, S.; Takahashi, A.; Ozawa, Y.; Uda, Y. *J. Antibact. Antifung. Agents* **1999**, *27*, 81-87.
13. Masuda, H.; Harada, Y.; Kishimoto, N.; Tano, T. In *Aroma Active Compounds in Foods: Chemistry and Sensory Properties*; Takeoka, G. R.; Guentert, M.; Engel, K.-H., Eds.; ACS Symposium Series 794; American Chemical Society: Washington, DC, 2001; pp 229-250.

14. Saeki, Y.; Ito, Y.; Shibata, M.; Sato, Y.; Okuda, K.; Takazoe, I. *Bull. Tokyo dent. Coll.* **1989**, *30*, 129-135.
15. Shapiro, S.; Meier, A.; Guggenheim, B. *Oral Microbiol. Immunol.* **1994**, *9*, 202-208.
16. Shapiro, S.; Guggenheim, B. *Oral Microbiol. Immunol.* **1995**, *10*, 241-246.
17. Mosandl, A.; Fischer, K.; Hener, U.; Kreis, P.; Rettinger, K.; Schubert, V.; Schmarr, H.-G. *J. Agric Food Chem.* **1991**, *39*, 1131-1134.
18. Masuda, H.; Tsuda, T.; Tateba, H.; Mihara, S. Japan Patent 02221255, 1990.
19. Harada, Y.; Masuda, H.; Kameda, W. Japanese Patent 07215931, 1995.
20. Ueno. K. In *Separation and Identification of Anaerobic Bacteria*; The Board of Education of Japan Bacterial Institute Ed.; Tokyo, Japan, 1982; pp 94.
21. Ueda, Y.; Shimizu, K. *Chemotherapy Handbook*; Nagai Publishing Co., Ltd.: Osaka, Japan, 1982; pp 24-34.
22. Eliopoulos, G. M.; Moellering, R. C. Jr. In *Antibiotics in Laboratory Medicine*; Lorian, V., Ed.; Williams & Wilkins: Maryland, USA, 1996; pp 102-103, 104-105, 333-348.
23. Lee, J.-Y.; Kim, Y.-S.; Shin, D.-H. *J. Agric. Food Chem.* **2002**, *50*, 2193-2199.
24. Tallarida, R. J.; Porreca, F.; Cowan, A. *Life Science.* **1989**, *45*, 947-961.
25. Gessner, P. K. *Toxicology* **1995**, *105*, 161-179.
26. Lin, C.-M.; Preston, III J. F.; Wei, C.-I. *J. Food Protec.* **2000**, *63*, 727-734.
27. Haapasalo, M.; Kerosuo, E.; Lounatmaa, K. *Scand. J. Dent. Res.* **1990**, *98*, 472-481.
28. Sundqvist, G.; Figdor, D.; Hänström, L.; Sörlin, S.; Sandström, G. *Scand. J. Dent. Res.* **1991**, *99*, 117-129.
29. Naito, Y. *Shika Gakuho* **1992**, *92*, 1289-1300.
30. Kawasaki, S.; Kaneko, T. *Phytochemistry* **1985**, *24*, 715-718.
31. Kawasaki, S.; Kaneko, T. *J. Agric. Food Chem.* **1987**, *35*, 85-88.
32. Ahn, E.-S.; Kim, Y.-S.; Shin, D.-H. *Food Sci. Biotechnol.* **2001**, *10*, 31-35.

Chapter 34

Immunomodulatory Activities of β-Glucan in Mushroom

Masashi Mizuno

Graduate School of Agricultural Science, Kobe University, Nada-ku, Kobe 657–8501, Japan

Hot water extracts from *Agaricus brasiliensis* S. Wasser *et al.* (ABSW) were dissolved in distilled water and fed to 8-weeks male ICR mice as drinking water for 14 days. The cumulative number of scratching behavior monitored for 20 min in saline-treated mice (control) was approximately 2,300 after intradermal injection of compound 48/80 which is a pruritogenic agent, whereas those in ABSW decreased to be approximately 960. To determine the effects of ABSW on degranulation in mast cell activation, histamine contents in blood were measured. ABSW treatment suppressed histamine release to 36% compared to control. When splenocytes were incubated with concanavalin A (Con A; T cell mitogen), contents of IFN-γ and IL-4 significantly increased compared with non-treatments of Con A, but IL-12 did not show any difference with/without Con A. Moreover, ABSW significantly induced IFN-γ production from splenocytes incubated with Con A. These results indicate that ABSW possess immunomodulating property that might be involved in the development of Th1 cells, culminating in an inhibition of immediate type allergy caused by compound 48/80.

Allergic diseases, such as asthma, allergic rhinitis, atopic dermatitis and food allergies, are steadily increasing especially in the industrialized countries. Key factors driving these rising trends are increased exposure to sensitizing allergens and reduced stimulation of the immune system during critical periods of development. Allergic responses involving IgE-dependent mast cell degranulation and eosinophil accumulation in the sites of inflammation are considered to be due to the development and activation of Th2 cells (*1*).

The production of two distinct cytokine patterns recognized in subsets of helper T cells (Th) is especially important. The set designated as Th1 is characterized by interleukin (IL)-12, interferon (IFN)-γ and IL-2 production and activates macrophages. The Th1 cytokines augment cell-mediated immunity. The other set, designated as Th2, is characterized by IL-4, -5, -6, -10, and -13 syntheses and Th2 cytokines promote humoral immunity (*2*). The representative cytokines of Th2 cells are IL-4 and IL-5. IL-4 is the major inducer of class-switching to IgE biosynthesis in B lymphocytes. IL-5 is the principal eosinophil-activating factor. On the other hand, IFN- γ, which is a representative cytokine of Th1 cells, is known to suppress the development of Th2 cells (*3*). Since it is suggested that Th1 and Th2 types of reactions are reciprocally regulated *in vivo* (*3*), the modulation of Th1/Th2 balance, namely shifting the balance from Th2 to Th1 dominance, should be a strategy for the therapy of allergic diseases involving Th2 cells.

Allergic reactions, especially immediate type allergy, are genetically determined disorders characterized by an increased ability of B-lymphocytes to synthesize IgE antibodies towards ubiquitous antigens (allergens), able to activate the immune system after inhalation or ingestion and after penetration through the skin. IgE antibodies are able to bind to high affinity Fcε receptors (FcεRI) present on the surface of mast cells/basophils (*2*). The mast cells, which are constituents of virtually all organs and tissue, are thought to play a major role in the development of many physiologic changes during allergic responses. Among the preformed and newly synthesized inflammatory substances released on degranulation of mast cells, histamine remains the best-characterized and most potent vasoactive mediator implicated in the acute phase of immediate type allergy (*4*). Compound 48/80 has been used as a direct and convenient reagent to study the mechanism of the anaphylactic reaction (*5*).

Interest in the medicinal characters of natural products has increased due to their popular use in traditional medicine. Many available substances have been found in the foods, particularly in mushrooms. *Agaricus brasiliensis* S. Wasser *et al.* (ABSW) is well-known among these mushrooms. It occurs naturally in a mountain region near Sao Paulo in Brazil and has become popularly known as "Himematsutake" in Japan and it has been traditionally used in food and folk medicine. It is reported that ABSW extract might be an effective stimulator for T cell and macrophage (*6*). Moreover, the expression of interleukin (IL)-12 (a key role in Th1 differentiation) and IL-18 (a proinflammatory cytokine in enhancing Th1 immune response) mRNA in macrophage-like cell line, RAW264.7, stimulated with an polysaccharide purified in ABSW were investigated by RT-PCR in a previous study (*7*). It is reported that within 12 h there was no drastic changes mRNA expression in IL-18 during stimulation with polysaccharide purified in ABSW. However, mRNA expression in IL-18 levels rapidly increased to be 5.6-fold 24 h post treatment. The level of IL-12 p40 mRNA expression was different from those of IL-18, which started to show increase only 12 h and through 24 h post treatment. Moreover, polysaccharides from ABSW changed the percentage of splenic Thy 1.2- and L3T4 (CD4)-positive cells in the T cell subsets in ABSW-treated mice (*6*). These results led to the presumption that this mushroom possessed the ability stimulate differentiation of naive T cells into T-helper type1 (Th1), resulting in antiallergic activity. The objectives of the present study were to examine the immunomodulating effect on immunocompetent cell and mast cell in 8-weeks male ICR mice orally feed hot water extract from ABSW.

Materials and Methods

Mice

Seven week-old male ICR mice were purchased from Japan SLC (Sizuoka, Japan). The mice were housed in cages under specific pathogen-free conditions (air temperature at 25 °C, 12h/12h light-dark cycle). They were given a laboratory chow (Nihon Nosan, Yokohama, Japan) and water *ad libitum*. Care and handling of the animals were in accordance with Kobe university animal experiment guidelines.

Preparation ABSW

ABSW was provided by Iwade Mushroom Institute (Mie, Japan). Mycelia were extracted with distilled water. The fruiting bodies were extracted with hot-water. Moreover, the residue after hot-water extraction was extracted with ammonium oxalate and NaOH according to the method of Kawagishi et al. (*8*). After alkali extraction, the solution was neutralized and desalted. Three fractions were mixed and lyophilized. For administration to ICR mice, ABSW were suspended in sterilized ion exchanged water at concentrations of 3.6 mg/ml.

Compound 48/80-Induced Systemic Anaphylactic Reaction

Compound 48/80 (Sigma, St Louis, MO, U.S.A.) was dissolved in saline at a concentration of 10 mg/ml. Compound 48/80 (10 μl) was intradermally injected on the back between the scapulae (*9*).

Observation of Scratching Behavior

Scratching behavior was observed according to the method described by Inagaki *et al.* (*5*). After elicitation of the cutaneous reactions, mice were placed in an observation chamber. The behavior was recorded in the absence of an observer using a video camera for 20 min between 30 and 50 min after elicitation. Mice generally scratched several times for about one second and a series of scratchings is indicated cumulatively.

Measurement of Histamine Contents in Plasma

The blood was collected from the heart to obtain plasma for histamine contents in 1 h after compound 48/80 inductions. The blood was centrifuged at 12,000 rpm for 1min at 4 °C with EDTA (20 μl in 1.5ml tube concentration of 50 mg/ml). Supernatant fluid is collected and plasma samples were stored at –80 °C until assay. The histamine levels in plasma were determined using an immunoassay kit (IBL, Hamburg, Germany) following the manufacturer's specifications.

Histopathological Examinations

Biopsy specimens were taken from the nape skin, in which the most prominent lesions usually appear, processed by the conventional method and stained with toluidine blue.

Cytokine Assays of Splenocytes

To investigate the immunoregulatory effect of ABSW on ICR mice, cytokine production from splenocyte stimulated with or without ConA *in vitro* was examined. Splenocytes were prepared for flow cytometry by crushing freshly dissected tissues between flat forceps in PBS (137 mM NaCl, 8.1 mM Na_2HPO_4, 2.68 mM KCl, and 1.47 mM KH_2PO_4). Cells were passed through 75

μmesh to get single cell suspension. Red blood cells were removed by adding hemolytic agent [0.16M NH_4Cl and 0.17M Tris-HCl (pH 7.65) are mixed by 9:1] for 3 min at room temperature. After washing with PBS for three times, cells (1×10^7 cells/ml) were cultured in a 24-well culture plate with RPMI 1640 (GIBCO BRL, Grand Island, NY) medium, supplemented with 10% heat-inactivated fetal bovine serum (GIBCO BRL, Grand Island, NY) and 1.0 mg/ml $NaHCO_3$. To determine antigen-nonspecific T-cell responses, spleen cells were stimulated with/without concanavalinA (ConA: 5 μg/ml, Wako Pure Chemical Industries, Osaka, Japan) at 37 °C for 3 days. The levels of IL-4, IL-12 (p70) and IFN-γ in the culture medium were measured with an OptEIA mouse cytokine ELISA set (BD Pharmingen, San Diego, CA) according to the manufacture's protocol using by EPICS XL/XL-MC (Beckman Coulter Inc., Fillerton, CA).

Results

Effects of Oral Administration of ABSW on Scratching Behavior

Oral administration of ABSW did result in no deference in body weight compared with the control, suggesting that mice grew normally. The cumulative number of scratching behavior for 20 min in saline-treated mice (control) was approximately 2,300 after intradermal injection of compound 48/80, whereas those in ABSW decreased to be approximately 960 (Figure 1).

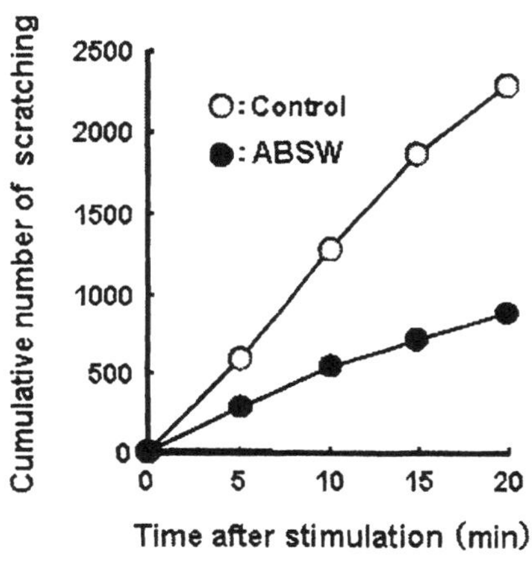

Figure 1. Scratching behavior caused by compound 48/80 in ICR mice. Each value represented the mean ± S.E. for 6 mice. Each error bar hides behind symbol. The asterisk indicated significant different from the control group ($p<0.01$).

Histamine Levels in Plasma and Histochemical Analysis

ABSW-treatment suppressed histamine release to 36% compared to control (Figure 2).

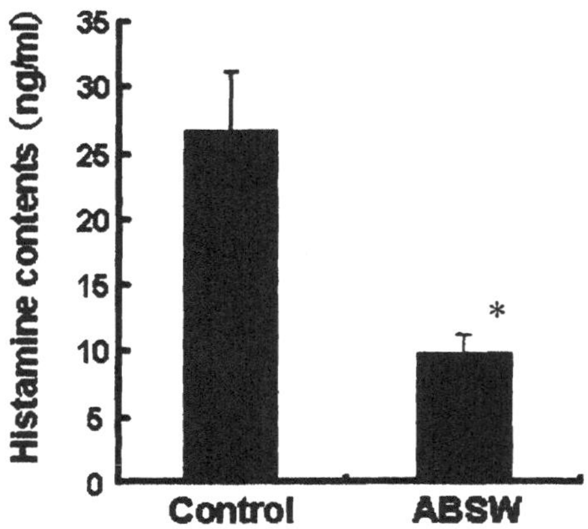

Figure. 2. Effects of ABSW on compound 48/80-induced histamine release in plasma. The asterisk indicated significant different from the control group ($p<0.01$).

Similarly, ABSW treatment suppressed mast cell degranulation (Figure 3). These results indicate that ABSW inhibited the degranulation of mast cells facilitated with compound 48/80 and histamine release.

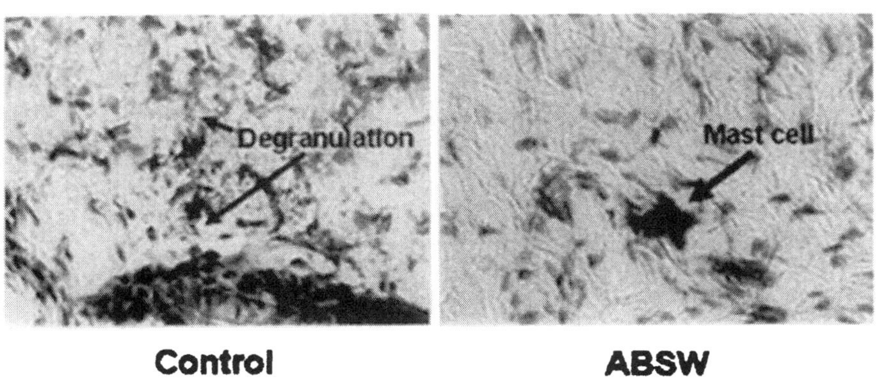

Figure 3. Toluidine blue stained sections of the nape skin. Mice were intradermally injected with compound 48/80 as described in Materials and Methods.

Cytokine Production from Splenocytes

The splenocytes stimulated with ConA released comparable levels of IFN-γ and IL-4 which are secreted by helper T cells. However, those without ConA had levels scarcely detected by ELISA. On the other hand, the production of IL-12, usually secreted by macrophages, did not show drastic changes with or without ConA (Figure 4). ABSW significantly induced IFN-γ production from splenocytes incubated with Con A compared with control in mice injected compound 48/80.

Discussion

The results indicate that oral administration of ABSW significantly inhibited the compound 48/80 induced systemic anaphylaxis-like reaction. The scratching behavior induced with compound 48/80 decreased to approximately 42% compared with saline-feeding (Figure 1). Moreover, the degranulation and histamine release of mast cells were inhibited by ABSW-treatment. The release of histamine from mast cells is a prominent feature of acute inflammatory processes such as the immediate-type anaphylaxis (*10*). These results could be explained to mean that ABSW may inhibit the anaphylaxis reaction by blocking histamine release from mast cells.

It has been reported that ABSW contained the immunomodulatory polysaccharide which could elicit the production of IL-12 and 18 from macrophages (*7*). As Il-12 and 18 are a key role in T-helper type1 (Th1) differentiation and a proinflammatory cytokine in enhancing Th1 immune response in macrophage respectively, ABSW might possess the ability to stimulate differentiation of naive T cells into Th1. Moreover, polysaccharides from ABSW changed the percentage of splenic CD4-positive cells in the T cell subsets (*6*) and activated macrophage to produce tumor necrosis factor-α through toll-like receptor (TLR) 4 (*7*). TLR 4 is one of the important receptors for innate immunity (*11*). These evidences might indicate that ABSW affected Th1 and Th2 balance.

The splenocytes stimulated with ConA enhanced the production of IFN-γ and IL-4 compared with non-treatment of ConA (Figure 4). Because ConA is T-cell mitogen, it was reasonable to increase two cytokines production from splenocytes. On the contrary, IL-12 did not increase even if splenocytes were treated with ConA, because IL-12 was produced from macrophages. Interestingly, IFN-γ production significantly enhanced in the splenocytes only when mice administered with ABSW were stimulated by compound 48/80 (Figure 4). However, in the case of control, IFN-γ production did not show any difference

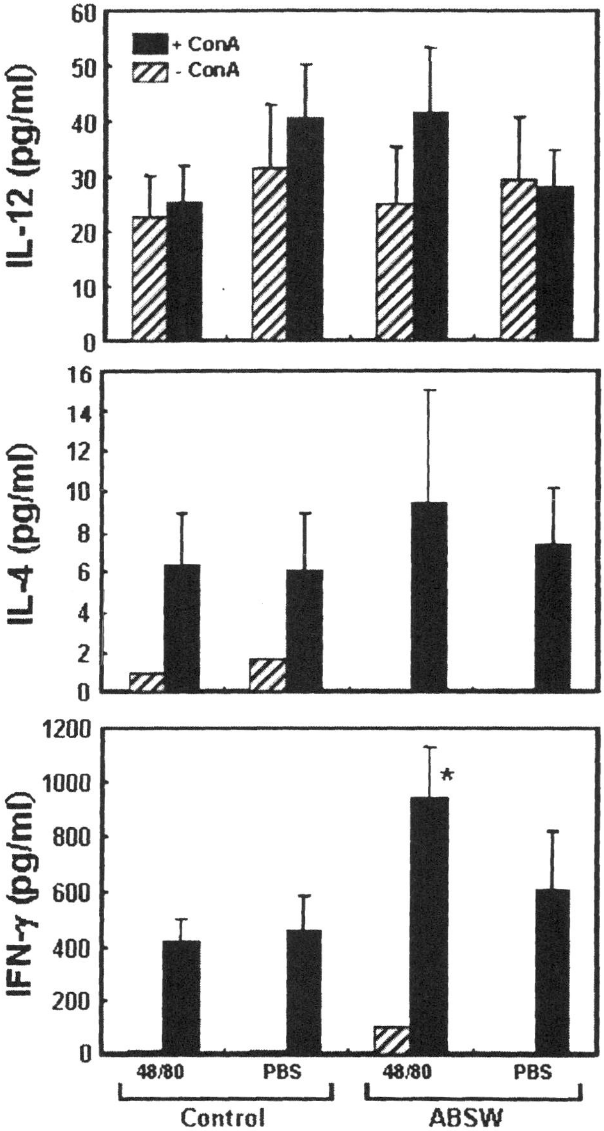

Figure 4. Effects of ABSW on the production of IL-12, IFN-γ and IL-4 from splenocytes.The asterisk indicated significant different from the control group ($p<0.05$).

between treatment and non-treatment of compound 48/80. IFN-γ is the characteristic Th1-type cytokine and inhibits Th2 response (*3*). The cumulative study on the functional properties of helper T cell subset is known as the Th2 hypothesis for the pathogenesis of allergic reactions (*12*).

In conclusion, it was ascertained that ABSW inhibited the degranulation and histamine release from mast cells *in vivo*, and enhanced IFN-γ production from splenocytes *in vitro*. These findings suggested that ABSW contained potent Th1 cytokine producing properties.

Acknowledgement

This work was supported by Iwade Mushroom Institute, Japan.

References

1. Holgate, S. T. *Nature* **1999**, *402*, B2-B4.
2. Maggi, E. *Immunotechnology* **1998**, *3*, 233-244.
3. Abbas, A. K.; Murphy, K. M.; Sher, A. Nature **1996**, *383*, 787-793.
4. Petersen, L. J.; Mosbech, H.; Skov, P. S *J. Allergy Clin. Immunol.* **1996**, *97*, 672-679.
5. Inagaki, N.; Nakamura, N.; Nagao, M.; Musoh, K.; Kawasaki, H.; Nagai, H. *Eur. J. Pharmacol.* 1999, *367*, 361-371.
6. Mizuno, M.; Morimoto, M.; Minato, K.; Tsuchida, H. *Biosci. Biotech. Biochem.* **1998**, *62*, 434-437
7. Mizuno M.; Kawakami, S.; Sakamoto Y.; Fujitake N. *Int. J. Medic. Mushrooms* **2003**, *5*, 397-403.
8. Kawagishi, H.; Inagaki, R.; Kanao, T.; Mizuno, T.; Shimura, K.; Ito, H.; Hagiwara, T.; Nakamura, T. *Carbohydr. Res.* **1989**, *186*, 267-273.
9. Orito, K.; Chida, Y.; Fujisawa, C.; Arkwright, P. D.; Matsuda. H. *Bri. J. Dermatol.* **2004**, *150*, 33-38.
10. Li, G. Z.; Chai, O. H.; Lee, M. S.; Han, E. H.; Kim, H. T.; Song, C. H. *Biol. Pharm. Bull.* **2005**, *28,* 1864-1868.
11. Takeda, K.; Akira, S. *Semin. Immunol.* **2004**, *16*, 3-9.
12. Romagnani, S. *Curr. Opin. Immunol.* **1994**, *6*, 838-846

Chapter 35

Osteoporosis Prevention by β-Cryptoxanthin

Masayoshi Yamaguchi and Satoshi Uchiyama

Laboratory of Endocrinology and Molecular Metabolism, Graduate School of Nutritional Sciences, University of Shizuoka, 52–1 Yada, Suruga-ku, Shizuoka 422–8526, Japan

Pharmacologic and nutritional factors may play a role in the prevention of osteoporosis with aging. β-Cryptoxanthin, a carotenoid, is abundant in Satsuma mandarin orange. Among various carotenoid including β-cryptoxanthin, lutein, lycopene, β-carotene, astaxanthin, and rutin, β-cryptoxanthin has been found to have a unique anabolic effect on bone calcification. β-Cryptoxanthin has stimulatory effects on osteoblastic bone formation and inhibitory effects on osteoclastic bone resorption *in vitro*. β-Cryptoxanthin has an effect on the gene expression of various proteins which are related to osteoblastic bone formation and osteoclastic bone resorption. Oral administration of β-cryptoxanthin has been shown to have the anabolic effect on bone components in young and aged rats, the preventive effect on bone loss in streptozotocin-diabetic rats and ovariectomy-induced bone loss. Moreover, the intake of β-cryptoxanthin-reinforced juice for longer periods has been shown to have stimulatory effects on bone formation and inhibitory effects on bone resorption in healthy human and postmenopausal women as estimated based on serum biochemical markers of bone metabolism *in vivo*. The intake of dietary β-cryptoxanthin may have a preventive effect on osteoporosis.

Aging induces a decrease in bone mass. Osteoporosis with its accompanying decrease in bone mass is widely recognized as a major public health problem. The most dramatic expression of the disease is represented by fractures of the proximal femur (*1*). Nutritional and pharmacologic factors may be important in preventing bone loss with increasing age (*2*). Isoflavones, which are contained in soybeans, menaquinone-7 (vitamin K_2), which is abundant in fermented soybeans, or other food and plant factors have been demonstrated to stimulate osteoblastic bone formation and to inhibit osteoclastic bone resorption *in vitro* (*3,4*). The supplementation of these food factors has preventive effects on bone loss induced in animal model of osteoporosis and in humans. Food chemical factors thus play a role in bone health and may be important in the prevention of bone loss with aging. Meanwhile, retinol (vitamin A) is known to have a detrimental effect on bone at high doses. In laboratory animals, high levels of vitamin A lead to accelerated bone resorption, bone fractures, and osteoporotic bone lesions (*5*). More recent studies have shown the anabolic effect of β-cryptoxanthin, a kind of carotenoid which is abundant in Satsuma mandarin (*Citrus unshiu* MARC.) on bone metabolism. In this chapter, the role of β-cryptoxanthin in preventing osteoporosis is introduced.

β-Cryptoxanthin Stimulates Bone Formation and Inhibits Bone Resorption *In Vitro*

The chemical structure of β-cryptoxanthin is shown in Figure 1.

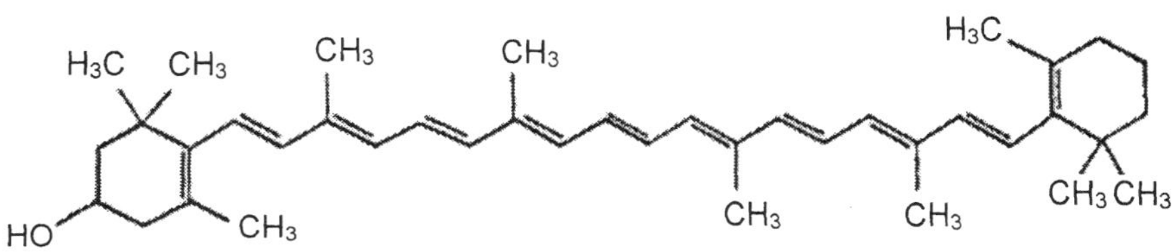

Figure 1. Chemical structure of β-cryptoxanthin.

The effects of various carotenoids and rutin on calcium content and alkaline phosphatase activity in the femoral-diaphyseal (cortical bone) and metaphyseal (trabecular bone) tissues of young rats *in vitro* were investigated *in vitro* (*6*). Culture with β-cryptoxanthin (10^{-7} or 10^{-6} M) caused a significant increase in calcium content and alkaline phosphatase activity in the femoral-diaphyseal and -metaphyseal tissues. Lutein, lycopene, and rutin at 10^{-8} to 10^{-6} M did not have anabolic effects on calcium contents and alkaline phosphatase activity in rat femoral-diaphyseal and -metaphyseal tissues. Astaxanthin and β-carotene (10^{-6} or 10^{-5} M) did not have an effect on the femoral calcium contents. Alkaline phosphatase participates in mineralization in bone tissues. β-Cryptoxanthin had a unique anabolic effect on bone calcification *in vitro*. The effects of β-cryptoxanthin increasing bone components was completely prevented with cycloheximide, an inhibitor of protein synthesis, suggesting that the effect is needed newly protein synthesis (*7*).

The effects of β-cryptoxanthin on bone resorption were investigated using bone tissues *in vitro*. Culture with the bone-resorbing factor parathyroid hormone (PTH) or prostaglandin E_2 (PGE_2) caused a significant decrease in calcium content in the diaphyseal and metaphyseal tissues (*7*). This decrease was completely inhibited by β-cryptoxanthin (10^{-8} - 10^{-6} M) (*7*). In addition, β-cryptoxanthin completely inhibited the PTH- or PGE_2-induced increase in medium glucose consumption and lactic acid production by bone tissues (*7*). β-Cryptoxanthin had inhibitory effects on bone resorption in tissue culture *in vitro*. Thus β-cryptoxanthin was found to have stimulatory effects on bone formation and inhibitory effects on bone resorption in bone tissue culture *in vitro,* as shown in Figure 2.

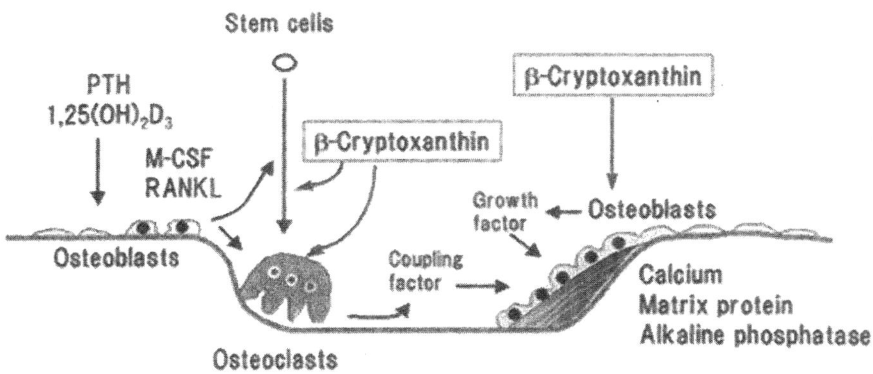

Figure 2. β-Cryptoxanthin stimulates bone formation and inhibits bone resorption, thereby increasing bone mass.

It has been reported that the serum concentration of β-cryptoxanthin due to consumption of vegetable juice in women is in the range of 1.3×10^{-7} to 5.3×10^{-7} M (*8*). β-Cryptoxanthin in the range of 10^{-8} to 10^{-6} M caused a significant anabolic effect on biochemical components in rat femoral tissues *in vitro*, suggesting a physiologic role in the regulation of bone metabolism.

Action of β-Cryptoxanthin in Osteoblastic Cells

To clarify the cellular mechanism by which β-cryptoxanthin stimulates bone formation in bone tissues, the effect of carotenoid was examined using osteoblastic cells. β-Cryptoxanthin was found to stimulate the proliferation of osteoblastic cells in subconfluent monolayers in a medium containing 10% FBS (*9*). Culture with β-cryptoxanthin also caused a significant increase in biochemical components of osteoblastic cells (*9*). The effects of β-cryptoxanthin in increasing protein content, alkaline phosphatase activity, and DNA content in osteoblastic cells are completely inhibited in the presence of DRB (5,6-dichloro-1-beta-D-ribofuranosyl benzimidazole), an inhibitor of RNA polymerase II, suggesting that β-cryptoxanthin has a stimulatory effect on transcriptional activity in osteoblastic cells. Prolonged culture with β-cryptoxanthin was found to stimulate mineralization in osteoblastic cells (*12*). β-Cryptoxanthin was found to have stimulatory effects on cell differentiation and mineralization in osteoblastic cells.

Culture with β-cryptoxanthin stimulates the expression of insulin-like growth factor (IGF-I) or transforming growth factor (TGF)-β-1 mRNA in osteoblastic cells using RT-PCR analysis (*9*). This finding may support the view that β-cryptoxanthin has a stimulatory effect on transcriptional activity in osteoblastic cells. IGF-I or TGF-β-1, which is a bone growth factor, is produced from osteoblasts (*10, 11*). The anabolic effect of β-cryptoxanthin in osteoblastic cells may be partly mediated through the action of IGF-I or TGF-β1 produced from the cells.

β-Cryptoxanthin (10^{-7} or 10^{-6} M) has also been found to increase the expression of Runx2, α1(I) collagen, and alkaline phosphatase mRNAs in osteoblastic MC3T3-E1 cells (*12*). Runx2 (Cbfa 1) is a member of the runt domain family of transcription factors, and it is involved in bone development (*13*). α1(I) Collagen is a matrix protein that is related to bone formation and mineralization in osteoblast lineage cells. Alkaline phosphatase participates in the mineralization process in osteoblastic cells. β-Cryptoxanthin has a stimulatory effect on the expression of genes for proteins involved in osteoblasic bone formation.

The effects of β-cryptoxanthin in stimulating Runx2, α1(I) collagen, and alkaline phosphatase mRNA expression in osteoblastic MC3T3-E1 cells was found to be prevented completely in the presence of DRB (*12*), supporting the view that the carotenoid stimulates transcriptional activity in osteoblastic MC3T3-E1 cells.

Vitamin A (retinol) may be able to bind to nuclear receptors in cells. β-Cryptoxanthin (10^{-7} or 10^{-6} M) caused a significant increase in alkaline phosphatase activity and protein content in osteoblastic cells. This effect was also seen in the presence of vitamin A (10^{-6} M) (*12*). Moreover, the stimulatory effect of β-cryptoxanthin on the expression of Runx2 type 1 and α1 (I) collagen mRNA was also observed in the presence of vitamin A (*12*). Vitamin A did not have a significant effect on Runx2 type 1 mRNA expression in osteoblastic MC3T3-E1 cells. Thus the mode of action of β-cryptoxanthin on gene expression in osteoblastic cells may differ from that of vitamin A, which is mediated through the RXR receptor in the nucleus of the cells (*12*). It is speculated that β-cryptoxanthin may be able to bind other receptors (including orphan receptors), and that the carotenoid may stimulate transcriptional activity in osteoblastic cells.

The mechanism of β-cryptoxanthin action in stimulating mineralization in osteoblastic cells is summarized in Figure 3.

Action of β-Cryptoxanthin in Osteoclastogenesis and Mature Osteoclasts

The receptor activator of NF-κB ligand (RANKL) plays a pivotal role in osteoclastogenesis from bone marrow cells. RANKL expression is induced in osteoblastic cells and bone marrow stromal cells in response to osteoporotic factors, such as PTH, PGE_2, and 1,25-dihydroxy vitamin D_3 (VD_3), and combined treatment of hematopoietic cells with macrophage colony-stimulating factor (M-CSF), and the soluble form of RANKL (sRANKL) induces osteoclast differentiation *in vitro* (*13*). The receptor protein RANK is expressed on the surface of osteoclast progenitors.

β-Cryptoxanthin (10^{-8} - 10^{-6} M) was shown to have a potent inhibitory effect on osteoclast-like cell formation in mouse marrow culture *in vitro* (*14*). The inhibitory effect of β-cryptoxanthin on osteoclast-like cell formation was seen at the later stage of osteoclast differentiation in bone marrow cultures. Culture with β-cryptoxanthin caused a marked inhibition of osteoblast-like cell formation induced in the presence of PTH, PGE_2, VD_3, LPS, or tumor necrosis factor-α (TNF-α). β-Cryptoxanthin had a significant inhibitory effect on osteoclast-like cell formation induced by RANKL (*14*). The inhibitory effect of β-cryptoxanthin was equal to that of 17 β-estradiol, calcitonin, genistein, and zinc sulfate, which can inhibit osteoclast-like cell formation induced by bone-resorbing factors.

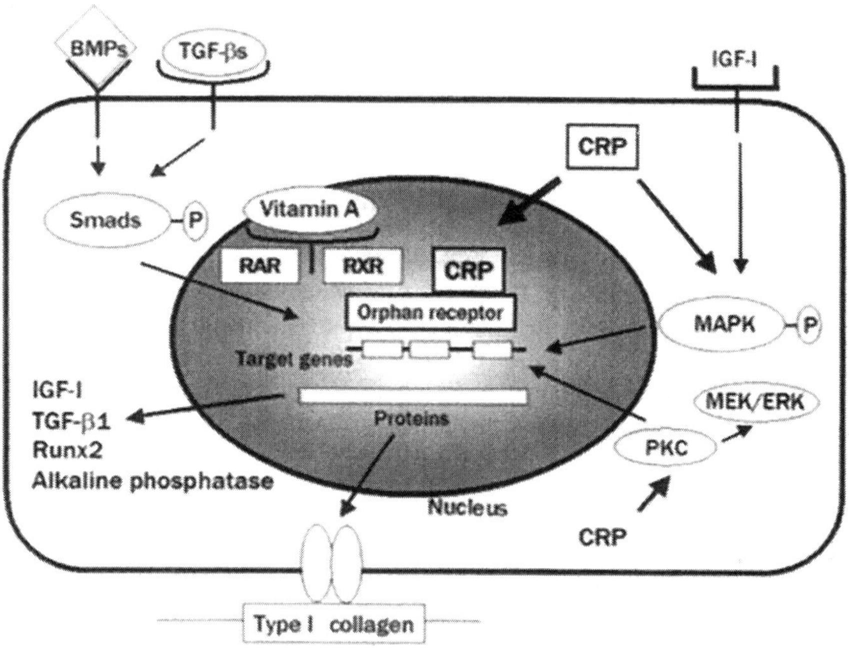

Figure 3. The cellular mechanism by which β-cryptoxanthin (CRP) stimulates bone formation and mineralization in osteoblastic cells. CRP stimulates may bind to orphan receptors in nucleus and stimulates transcription activity for bone formation-related proteins.

The effects of β-cryptoxanthin on mature osteoclasts were investigated (*16*). M-CSF-dependent bone marrow macrophages were cultured in the presence of M-CSF and RANKL for 4 days (*16*). The osteoclastic cells formed were further cultured in medium containing β-cryptoxanthin with or without M-CSF and RANKL for 24-72 h. Osteoclastic cells were significantly decreased in culture with β-cryptoxanthin (10^{-7} or 10^{-6} M) with or without M-CSF and RANKL for 72 h. The β-cryptoxanthin-induced decrease in osteoclastic cells was significantly inhibited in the presence of caspase-3 inhibitor. Agarose gel electrophoresis showed the presence of low-molecular-weight DNA fragments of adherent cells cultured with β-cryptoxanthin, indicating that the carotenoid induces apoptotic cell death.

Apoptosis-related gene expression was determined using RT-PCR (*16*). Culture with β-cryptoxanthin caused a significant increase in caspase-3 mRNA expression in the presence or absence of M-CSF and RANKL, while Bcl-2 and

Apaf-2 mRNA expressions were not significantly increased in culture with β-cryptoxanthin without M-CSF and RANKL. Akt-1 mRNA expression was not significantly changed with culture of the carotenoid.

The expression of caspase-3 mRNA or Apaf-2, which involves apoptosis, in osteoclastic cells was found to stimulate when cultured with β-cryptoxanthin in the absence of M-CSF and RANKL (*16*). β-Cryptoxanthin stimulated caspase-3 mRNA expression in the presence of M-CSF and RANKL expression in the presence of M-CSF and RANKL (*16*). β-Cryptoxanthin-induced apoptotic cell death is partly mediated through caspase-3 expression in osteoclastic cells. In addition, the expression of Bcl-2 mRNA, which is involved in rescue of apoptosis, was significantly decreased in β-cryptoxanthin culture in the presence or absence of M-CSF and RANKL (*16*). However, Akt-1 mRNA expression is not significantly changed in culture with β-cryptoxanthin. The decrease in Bcl-2 mRNA expression may partly contribute to the effect of β-cryptoxanthin in stimulating the apoptotic cell death of osteoclastic cells.

Culture with β-cryptoxanthin was found to have suppressive effects on tartrate-resistant acid phosphatase (TRACP) activity, TRACP, and cathepsin K mRNA expression in osteoclastic cells in the presence or absence of M-CSF and RANKL (*16*). Presumably, β-cryptoxanthin has inhibitory effects on the activation of mature osteoclasts.

β-Cryptoxanthin has been demonstrated to have stimulatory effects on apoptotic cell death and suppressive effects on osteoclastic cell function. The action of β-cryptoxanthin on mature osteoclasts is summarized in Figure 4.

Preventive Effects of β-Cryptoxanthin against Bone Loss *In Vivo*

β-Cryptoxanthin has been shown to have a stimulatory effect on osteoblastic bone formation and an inhibitory effect on osteoclastic bone resorption *in vitro* (*6,9,12,14,16*). Then, the anabolic effect of β-cryptoxanthin to exert preventive effects on osteoporosis was investigated using animal models.

β-Cryptoxanthin (10, 25, or 50 μg/100 g body weight) was orally administered once daily for 7 days to young male rats (*17*). The administration of β-cryptoxanthin (25, or 50 μg/100 g body weight) caused a significant increase in calcium content, alkaline phosphatase activity, and DNA contents in the femoral-diaphyseal and -metaphyseal tissues. β-Cryptoxanthin has an anabolic effect on bone components in rats *in vivo*. Such an effect is also observed in the femoral tissues of aged (50-week-old) female rats (*18*).

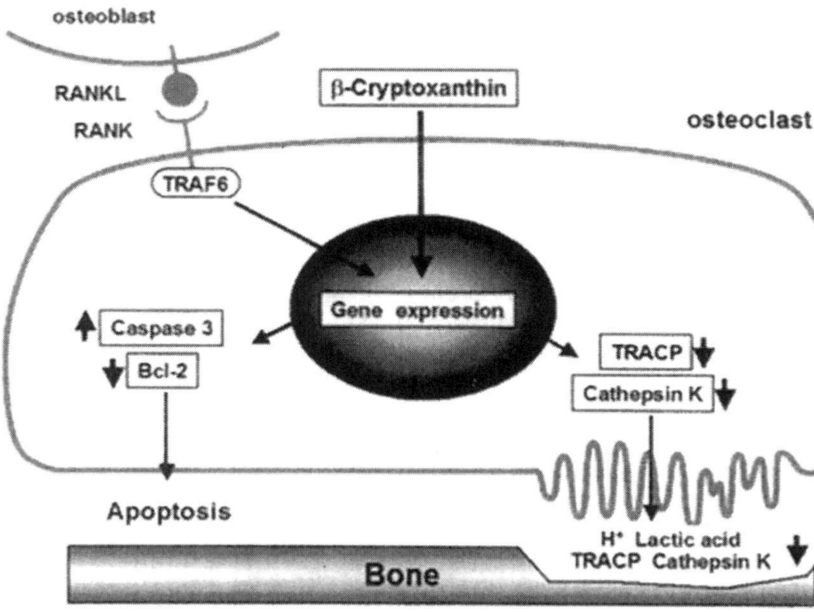

Figure 4. The cellular mechanism by which β-cryptoxanthin (CRP) has suppressive effects on mature osteoclasts. CRP stimulates apoptotic cell death and suppresses gene expression of bone resorption-related enzymes.

To determine whether β-cryptoxanthin has a preventive effect on bone loss in the pathphysiologic state, the effects of β-cryptoxanthin on bone components in streoptozotocin (STZ)-diabetic rats were determined (*19*). Young rats received a single subcutaneous administration of STZ (6.0 mg/100 g body weight), and then the animals were orally administered β-cryptoxanthin (5 or 10 μg/100 g body weight) once daily for 7 or 14 days. The administration of STZ caused a significant decrease in body weight and a significant increase in serum glucose, triacylglycerol, and calcium levels, indicating a diabetic state. These alterations were significantly prevented by the administration of β-cryptoxanthin (5 or 10 μg/100 g) for 14 days. Calcium content, alkaline phosphatase activity, and DNA content in the femoral-diaphyseal and -metaphyseal tissues were significantly decreased in STZ-diabetic rats. These decreases were significantly prevented by the administration of β-cryptoxanthin (5 or 10 μg/100 g) for 14 days. Thus the intake of β-cryptoxanthin was found to have a preventive effect on bone loss in STZ-diabetic rats.

The effect of β-cryptoxanthin on ovariectomy (OVX)-induced bone loss was examined (*20*). β-Cryptoxanthin (5 or 10 μg/100 g body weight) was orally administered once daily for 3 months to OVX rats. The analysis using peripheral quantitative computed tomography (pQCT) showed that OVX induced a significant decrease in mineral content and mineral density in the femoral-diaphyseal and -metaphyseal tissues. These decreases were significantly prevented by the administration of β-cryptoxanthin (5 or 10 μg/100 g). Moreover, OVX induced a significant decrease in bone components. These decreases were completely prevented by the administration of β-cryptoxanthin (5 or 10 μg/100 g). β-Cryptoxanthin had preventive effects on OVX-induced bone loss *in vivo*.

Effects of β-Cryptoxanthin in Normal Individuals and Menopausal Women

The effects of β-cryptoxanthin on bone metabolism in human have been investigated (*21-23*). The effects of prolonged intake of juice prepared from Satsuma mandarin (*Citrus unshiu* MARC.) containing β-cryptoxanthin on circulating biochemical markers of bone metabolism in subjects, including menopausal woman, were examined (*23*). Ninety volunteers, aged 27-65 years (19 men and 71 women), were enrolled in this study. The 71 females included 35 premenopausal women (ages, 27-50 years) and 36 menopausal women (ages, 46-65 years). Volunteers were divided into four groups; placebo juice without β-cyptoxanthin (5 men and 19 women), juice containing β-cyptoxanthin at 1.5 mg/200 ml of juice/day (4 men and 17 women), 3.0 mg/day (5 men and 17 women), and 6.0 mg/day (5 men and 18 women). Placebo or juice (200 mL) was ingested once a day for 28 or 56 days.

Serum β-cryptoxanthin concentrations were significantly increased after the intake of juice containing β-cryptoxanthin (1.5, 3.0, or 6.0 mg/day) for 28 or 56 days, and the increases were dose-dependent.

Serum bone-specific alkaline phosphatase and γ-carboxylated osteocalcin are bone metabolic markers of bone formation, and serum bone TRACP and N-telopeptides of type I collagen are metabolic markers of bone resorption. In ninety volunteers (aged 27-65 years), serum bone-specific alkaline phosphatase activity was significantly increased after the intake of juice containing β-cryptoxanthin (3.0 or 6.0 mg/day) for 56 days as compared with the value obtained before intake. γ-Carboxylated osteocalcin concentration was significantly increased after the intake of juice containing β-cryptoxanthin (3.0 or 6.0 mg/day) for 28 or 56 days as compared with the value obtained before intake or after the intake of placebo juice. Serum TRACP activity and type I collagen N-telopeptide concentration were significantly decreased after the intake of juice containing β-cryptoxanthin (3.0 or 6.0 mg/day) for 28 or 56 days as compared with the value obtained before intake or after intake of placebo juice, and significant decreases were also seen after the intake of 1.5 mg/day β-cryptoxanthin as compared with the value obtained before intake.

In menopausal women (36 volunteers), bone-specific alkaline phosphatase activity and γ -carboxylated osteocalcin concentration were significantly increased after the intake of juice containing β-cryptoxanthin (3.0 or 6.0 mg/day) for 56 days as compared with the value obtained after placebo intake. Also, this intake caused a significant decrease in bone TRACP activity and type I collagen N-telopeptide concentration.

Meanwhile, serum calcium, inorganic phosphorous, and parathyroid hormone (intact) were not changed after the intake of β-cryptoxanthin-containing juice for 28 or 56 days. Other serum biochemical findings were not changed after the intake of juice containing β-cryptoxanthin (3.0 or 6.0 mg/day) for 56 days. We confirmed the safety of β-cryptoxanthin in humans.

The prolonged intake of juice fortified with β-cryptoxanthin has been demonstrated to have stimulatory effects on bone formation and inhibitory effects on bone resorption in humans, and the intake has an effect in menopausal women. Thus the dietary intake of β-cryptoxanthin was demonstrated to have preventive effects on osteoporosis.

References

1. Cooper, C.; Melton, J. III. *Trends Endocrinol. Metab.* **1995**, *3*, 224-229.
2. Bonjour, J.-P.; Schurch, M.-A.; Rozzori, R. *Bone.* **1996**, *18*, 1395-1445.
3. Yamaguchi, M. *J. Health Sci.* **2002**, *48*, 209-222.
4. Yamaguchi, M. Yakugaku Zasshi, **2006**, *126*, 1117-1137.
5. Promislow, J. H. E.; Goodman-Gruen, D.; Slymen, D. J.; Barret-Connor, E. *J. Bone Miner. Res.* **2002**, *17*, 1349-1358.
6. Yamaguchi, M.; Uchiyama, S. *Biol. Pharm. Bull.* **2003**, *26*, 1188-1191.
7. Yamaguchi, M.; Uchiyama, S. *Mol. Cell. Biochem.* **2004**, *258*, 137-144.
8. McEligot, A. J.; Rock, C. L.; Shanks, T. G.; Flatt, S. W.; Newman, V.; Farber, S.; Pierce, J. P. *Cancer Epideminol. Biomarkers Prev.* **1999**, *8*, 227-231.
9. Uchiyama, S.; Yamaguchi, M. *Int. J. Mol. Med.* **2005**, *15*, 675-681.
10. Centrella, M.; McCarthy, T. L.; Canalis, E. *Endocrinology* **1990**, *126*, 39-44.
11. Palcy, S.; Bolivar, I.; Goltzman, D. *J. Bone Miner, Res.* **2000**, *15*, 2352-2361.
12. Uchiyama, S.; Yamaguchi, M. *J. Cell. Biochem.* **2005**, *95*, 1224-1234.
13. Komori, T.; Yagi, H.; Nomura, S.; Yamaguchi, A.; Sasaki, K.; Deguchi, K.; Shimizu, Y.; Bronson, R. T.; Gao, Y. H.; Inada, M.: Sato, M.; Okamoto, R.; Kitamura, Y.; Yoshiki, S.; Kishimoto, T.; *Cell* **1997**, *89*, 755-764.
14. Uchiyama, S.; Yamaguchi, M. *Biochem.Pharmacol.* **2004**, *67*, 1297-1305.

15. Zaidi, M.; Blair, H. C.; Moonga, B. S.; Abe, E.; Huang, C. L. -H. *J. Bone Miner. Res.* **2003**, *18*, 599-609.
16. Uchiyama, S.; Yamaguchi, M. *J. Cell. Biochem.* **2006**, *98*, 1185-1195.
17. Uchiyama, S.; Sumida T.; Yamaguchi M. *Biol. Pharm. Bull.* **2004**, *27*, 232-235.
18. Uchiyama, S.; Sumida, T.; Yamaguchi, M. *J. Health Sci.* **2004**, *50*, 491-496.
19. Uchiyama, S.; Yamaguchi, M. *Biol. Pharm. Bull.* **2005**, *28*, 1766-1769.
20. Uchiyama, S.; Yamaguchi, M. *Int. J. Mol. Med.* **2006**, *17*, 15-20.
21. Yamaguchi, M.; Igarashi, A.; Uchiyama, S.; Morita, S.; Sugawara, K.; Sumida, K. *J. Health Sci.* **2004**, *50*, 619-624.
22. Yamaguchi, M.; Igarashi, A.; Morita, S.; Sumida, T.; Sugawara, K. *J. Health Sci.* **2005**, *51*, 738-743.
23. Yamaguchi, M.; Igarashi, A.; Uchiyama, S.; Sugawara, K.; Sumida, T.; Morita, S.; Ogawa, H.; Nishitani, M.; Kajimoto, Y. *J. Health Sci.* **2006**, *52*, 758-768.

Chapter 36

Promotion of Bone Formation by Phytate-Removed Deamidated Soybean Glycinin

Hitomi Kumagai[1], Atsushi Koizumi[1], and Hitoshi Kumagai[2]

[1]Department of Agricultural and Biological Chemistry, Nihon University, 1866 Kameino, Fujisawa-shi 252–8510, Japan
[2]Department of Food Science and Nutrition, Kyoritsu Women's University, 2–2–1 Hitotsubashi, Chiyoda-ku, Tokyo 101–8437, Japan

Soybean glycinin was deamidated by a cation-exchange resin of the carboxylate type after removal of phytate to enhance calcium absorption from the small intestine. Oral administration of the phytate-removed deamidated soybean glycinin (PrDS) to young male rats for 8 weeks lowered the parathormone level and increased the γ-carboxylated type osteocalcin level in their blood, indicating that a sufficient amount of calcium was absorbed from the small intestines and bone formation was promoted by the administration of PrDS. Both bone mineral density and mechanical bone strength of the femur were also enhanced as a result of this effect of PrDS.

Among 1000 g of calcium in our body, more than 99% of it is stored in bones, the rest being found in the teeth, blood, and elsewhere. Although the amount of calcium in the blood is only 0.1% of the total amount in the body, its concentration is strictly controlled at 10 mg/dl. If the amount of calcium absorbed from the small intestine decreases, the calcium shortage in the blood will be compensated by resorption of this mineral from the bones, which will lead to increased bone fragility. Despite the importance of calcium in the body, the calcium intake of Japanese people has been about 550 mg/day over the past 35 years, that of young people in their twenties being around 460 mg/day. The adequate calcium intake is 900 - 1100 mg/day for males between 10 and 29 years of age and 700 - 950 for females between 10 and 29 years of age. Therefore, the calcium shortage in the diets of young people is especially serious in Japan and may influence bone formation in young people.

As the amount of calcium absorbed from the small intestine is affected by substances coexisting with it, we had better consider not only how much calcium to take but also what to take with it in order to solve the calcium deficiency. The substances known to inhibit calcium absorption are phytate and oxalate, while those known to enhance calcium absorption are calcium citrate malate (CCM) and casein phosphopeptide (CPP). CCM and CPP have already been included in functional foods authorized by the Ministry of Health and Welfare to enhance calcium absorption, though their applicability is limited because of their taste and flavor. One of the reasons why CCM and CPP enhance calcium absorption is that they have acidic functional groups in their structure that adequately bind to calcium and help to solubilize it in the small intestine.

Deamidation is one of the effective ways to increase the number of acidic functional groups of pulse proteins, as these proteins are abundant in glutamine and asparagine residues. The most common way for deamidation is the usage of acid *(1-6)*, but acid treatment not only deamidates a protein but also hydrolyzes its peptide bonds, which hydrolysis may produce bitter-tasting peptides and reduce the processing properties of the protein. Another common way to deamidate a protein is enzymatic treatment *(7-11)*. However, glutaminases only react with glutamine residues, and not with asparagine residues, and, like acids, proteases hydrolyze the peptide bonds. Moreover, some of the enzymes are not commercially available and others are too expensive for industrial use. The usage of a cation-exchange resin as a catalyst for deamidation *(12-18)* is a way that solves the problems encountered with acidic and enzymatic treatments because no peptide-bond hydrolysis occurs, its catalytic activity is high and such a resin is reusable many times.

In previous studies, we showed that a cation-exchange resin of the carboxylate type was quite effective for deamidation of soybean and wheat proteins *(14-16)*. Soybean proteins are known to reduce calcium bioavailability because of their phytate component, which strongly binds to calcium and insolubilizes it. However, calcium bioavailability was remarkably improved by deamidation of the proteins after the removal of phytate *(17)*. In this study, the phytate-removed deamidated soybean glycinin was administered to young male rats to examine its effect on bone formation.

Materials and Methods

Preparation of Soybean Glycinin

Soybeans were dehulled and ground, using a mixer, and then defatted with 5 times their weight of hexane. The defatted soybean meal was stirred for 1 hour with 20 times its weight of a 0.03 M Tris-HCl buffer at pH 8.0 containing 0.01 M 2-mercaptoethanol and then centrifuged at 18,000 *g* for 20 minutes at 20°C. The supernatant was adjusted to pH 6.4 with 2 N HCl to precipitate soybean glycinin and then centrifuged at 18,000 *g* for 20 minutes at 4°C. The precipitate was dialyzed against distilled water and freeze-dried to obtain untreated soybean glycinin (US).

Phytate Removal

Phytate was removed from soybean glycinin as described in a previous paper *(14)*. After successively washed with 1 N HCl, deionized water, 1 N NaOH, deionized water, 1 N HCl, deionized water, and a 0.05 M Tris-HCl buffer at pH 7.4, an anion-exchange resin (IRA410, Organo Co., Tokyo, Japan) was added to 0.2% US in the 0.05 M Tris-HCl buffer at pH 7.4 to achieve a concentration of 0.05 g/ml, and stirred at 4°C for 1 hour. After having been filtered through a cotton cloth, the filtrate was dialyzed against distilled water and freeze-dried to obtain phytate-removed soybean glycinin (PrS).

Deamidation

PrS was deamidated as described earlier *(14)*. After successively washed with 1 N NaOH, deionized water, 1 N HCl, deionized water, 0.1 N NaOH, deionized water, and a 0.05 M Tris-HCl buffer at pH 7.4, a cation-exchange resin (IRC50, Organo Co., Tokyo, Japan) was added to 0.2% PrS in the 0.05 M Tris-HCl buffer at pH 7.4 to a concentration of 0.1 g/ml, and stirred at 4°C for 6 hours. After filtering through a cotton cloth, the filtrate was dialyzed against distilled water and freeze-dried to obtain phytate-removed deamidated soybean glycinin (PrDS).

In vivo Experiment

Five-week-old male Wistar rats (Clea Japan, Tokyo, Japan) were fed for 7 days a standard diet based on the AIN-76 formulation containing 20% egg albumin and acclimated to the animal house environment at 23°C with a 12-hour light/12-hour dark cycle. Then, the rats were divided into 4 groups that were fed a restricted diet of 20 g/day for 8 weeks. Normal group was given the standard diet containing 20% egg albumin and the other groups (US, PrDS or Casein group) were given the diet in which one-fifth of the egg albumin (4%) was replaced with each experimental protein (US, PrDS or casein). Prior to blood sampling from a jugular vein, the rats were fasted for 18 hours. Blood samples were collected and centrifuged at 3,000 *g* and 20°C for 15 minutes to collect sera. Femurs were excised from rats, and stored in 70% ethanol after the removal of muscles and connective tissues.

Animal experiments were performed in accordance with the Guidelines for Animal Experiments of the College of Bioresource Sciences, Nihon University.

Serum Assay

The parathormone (PTH) level in serum was measured with a PTH rat ELISA system (Amersham Pharmacia, Tokyo, Japan). γ-Carboxylated osteocalcin (Gla-OC) and undercarboxylated osteocalcin (Glu-OC) levels in serum were measured with a Rat Gla-OC Competitive EIA Kit and Rat Glu-OC Competitive EIA Kit (Takara Shuzo, Kyoto, Japan), respectively.

Bone Mineral Density

Bone mineral density was measured by peripheral Quantitative Computed Tomography (pQCT, XCT Research SA, Stratec Medizintechnik GmbH, Pforzheim, Germany). The bones were inserted into a specially constructed tube, fixed in the device, and femur was scanned with a slice thickness of 0.75 mm starting 2 mm from the distal end. Voxel size was 0.1 mm and the scan speed was 10 mm/sec.

Mechanical Bone Strength

The mechanical bone strength was evaluated by the three-point bending test with a testing machine (TCM-500CR, Minebea, Tokyo, Japan). The sample space was set at 10 mm and the plunger speed at 2 mm/min. The bone was positioned horizontally and the pressing force was directed vertically to the midshaft of the bone. Breaking force was defined as the bending load at failure.

Results and Discussion

Body Weight Gain and Organ Weight

Rats were supplied a restricted diet of 20 g/day and those in every group ate it up daily. As shown in Table I, body weight gain and organ weight were not significantly different among the 4 groups after the rats were fed the experimental diets for 8 weeks.

Table I. Body and Organ Weights of Rats Fed the Normal or Experimental Diets for 8 Weeks

Group	*Body weight (g)*			*Organ weight (g)*		
	Initial	*Final*	*Gain*	*Liver*	*Kidney*	*Spleen*
Normal	155.00± 5.45	294.86± 12.09	139.86	8.65± 0.39	2.23± 0.22	0.69± 0.07
US	155.00± 10.05	321.14± 11.77	166.14	8.88± 0.27	2.26± 0.13	0.79± 0.13
PrDS	157.00± 6.23	308.00± 20.36	150.63	8.82± 0.76	2.29± 0.19	0.71± 0.09
Casein	158.13± 7.69	311.13± 23.55	150.00	8.85± 0.64	2.23± 0.15	0.74± 0.08

Each value is the mean±S.E. of 7 rats.

Serum Parathormone (PTH) Level

Parathormone is secreted from the parathyroid gland to stimulate bone resorption when the calcium concentration in the blood decreases. Therefore, a low level of parathormone in the serum indicates sufficient calcium absorption from the small intestines. The serum parathormone level was lower in rats fed the PrDS or Casein diet than in those fed the Normal or US diet (Figure 1). This result is in accordance with previous *in situ* findings showing that PrDS and CPP enhanced calcium absorption from the small intestines *(17)*.

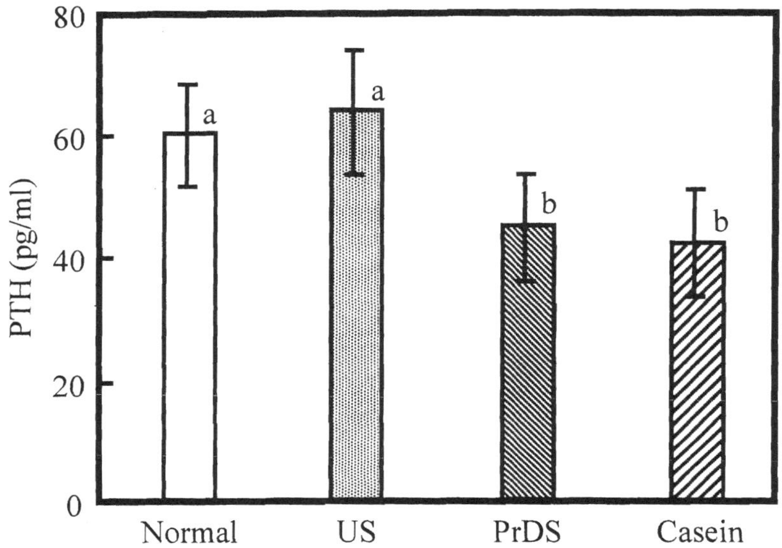

Figure 1. Serum PTH levels in rats fed the normal or experimental diets for 8 weeks. Values with different letters are significantly different at $p < 0.05$. Each value is the mean of 7 rats with S.E. shown as a vertical bar.

Serum Osteocalcin Level

Osteocalcin (OC) is a protein biosynthesized in osteoblast. Glu-OC has a low affinity for hydroxyapatite, whereas Gla-OC in which the 3 glutamine residues are carboxylated by vitamin K-dependent γ-glutamyl carboxylase has a high affinity for hydroxyapatite and is incorporated into the bone matrix. As some Gla-OC is secreted into the blood, it is recognized as one of the indices of bone formation. The serum Gla-OC level was higher for rats fed the PrDS diet than for those fed the Normal or US diet (Figure 2). The ratio of Gla-OC to Glu-OC, which is an indicator of bone metabolic turnover, was also high for rats fed the PrDS diet (Figure 3). These results indicate that PrDS promoted bone formation and bone metabolic turnover.

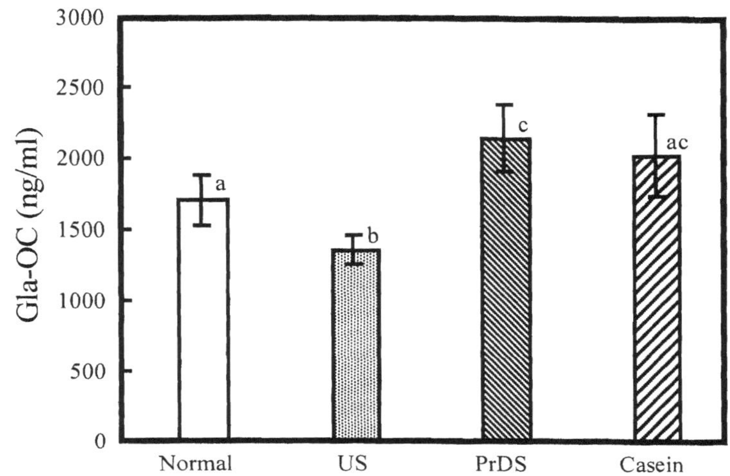

Figure 2. Serum Gla-OC levels in rats fed the normal or experimental diets for 8 weeks. Values with different letters are significantly different at $p < 0.05$. Each value is the mean of 7 rats with S.E. shown as a vertical bar.

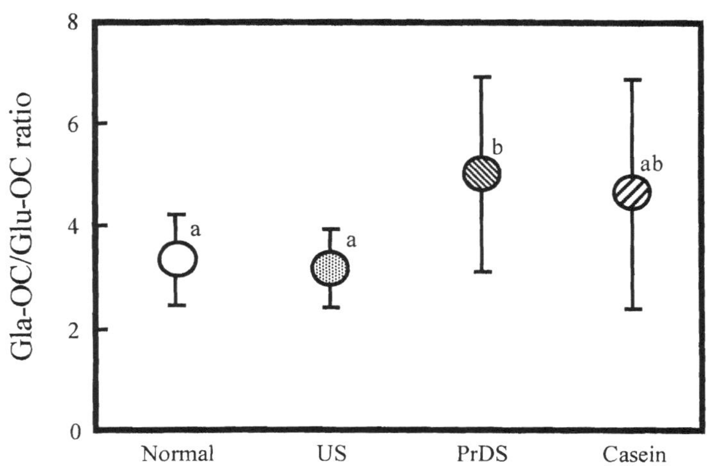

Figure 3. Ratio of serum Gla-OC level to Glu-OC in rats fed the normal or experimental diets for 8 weeks. Values with different letters are significantly different at $p < 0.05$. Each value is the mean of 7 rats with S.E. shown as a vertical bar.

Bone Mineral Density and Mechanical Bone Strength

The trabecular bone is more metabolically active than the cortical bone, and the response to metabolic changes in trabecular bone is faster than that in cortical bone. The trabecular bone mineral density was higher for rats fed the US, PrDS or Casein diet than that for those fed the Normal diet (Figure 4).

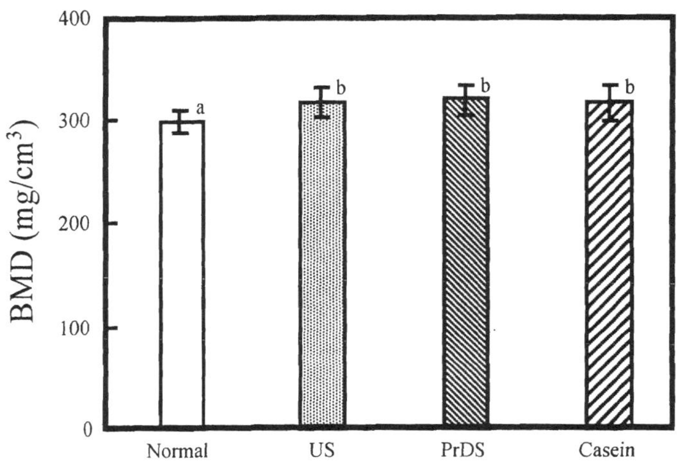

Figure 4. Trabecular bone mineral density of femurs from rats fed the normal or experimental diets for 8 weeks. Values with different letters are significantly different at p < 0.05. Each value is the mean of 7 rats with S.E. shown as a vertical bar.

Similar to the result found for bone mineral density, the mechanical bone strength of the femur was also higher for rats fed US, PrDS or Casein diet than for those fed the Normal diet (Figure 5).

The reason why the US diet group showed high bone mineral density might be because the US used in this study contained about 0.2% isoflavone, meaning that each rat consumed about 200 mg/kg of isoflavone each day. Only 20 mg/kg of isoflavone per day is reported to be effective to suppress bone resorption in rats *(19)*. Therefore, this high bone mineral density might have been due to the high content of isoflavone in the US diet. Almost all isoflavone was removed during the ion-exchange treatment, and PrDS was isoflavone-free. Therefore, the high bone mineral density of the PrDS diet group would be attributed to its enhancement of calcium absorption from the small intestine.

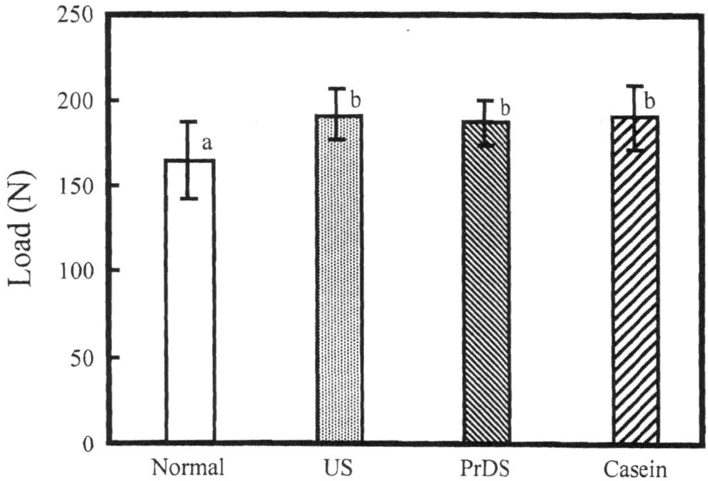

Figure 5. Mechanical bone strength of femurs from rats fed the normal or experimental diets for 8 weeks. Values with different letters are significantly different at $p < 0.05$. Each value is the mean of 7 rats with S.E. shown as a vertical bar.

Conclusions

Phytate-removed deamidated soybean glycinin (PrDS) suppressed the secretion of parathormone from the parathyroid gland to the blood and enhanced the biosynthesis and γ-carboxylation of osteocalcin in osteoblast. As a result, PrDS promoted bone formation and improved the trabecular bone mineral density and bone strength. These functions of PrDS were attributed to its enhancing effect on calcium absorption from the small intestine.

Acknowledgments

We are grateful to Dr. Guizhen Jiang, Dr. Hiroko Matsumoto, and Prof. Akira Fujii, Nihon University School of Dentistry at Matsudo, for their assistance in measuring bone mineral density. We also express our gratitude to Dr. Masahiro Kaketani and Dr. Hisaji Kikuchi, Nihon University School of Dentistry, for their help in measuring mechanical bone strength.

References

1. Wagner, J. R.; Guéguen, J. *J. Agirc. Food Chem.* **1995**, *43,* 1993-2000.
2. Wagner, J. R.; Guéguen, J. *J. Agirc. Food Chem.* **1999**, *47,* 2173-2180.
3. Wagner, J. R.; Guéguen, J. *J. Agric. Food Chem.* **1999**, *47,* 2181-2187.
4. Bollecker, S.; Viroben, G.; Popineau, Y.; Guéguen, J. *Sci. Aliments* **1990**, *10,* 343-356.
5. Ma, C.-Y.; Khanzada, G. *Food Chem.* **1987**, *52,* 1583-1587.
6. Matsudomi, N.; Sasaki, T.; Kato, A.; Kobayashi, K. *Agric. Biol. Chem.* **1985**, *49,* 1251-1256.
7. Motoki, M.; Seguro, K.; Nio, N.; Takinami, K. *Agric. Biol. Chem.* **1986**, *50,* 3025-3030.
8. Kato, A.; Tanaka, A.; Matsudomi, N.; Kobayashi, K. *J. Agric. Food Chem.* **1987**, *35*, 224-227.
9. Hamada, J. S.; Marshall, W. E. *J. Food Sci.* **1989**, *54,* 598-601 & 635.
10. Gu, Y. S.; Matsumura, Y.; Yamaguchi, S.; Mori, T. *J. Agric. Food Chem.* **2001**, *49,* 5999-6005.
11. Young, Y. H.; Yamaguchi, S.; Gu, Y. S.; Mori, T.; Matsumura, Y. *J. Agric. Food Chem.* **2004**, *52,* 7094-7100.
12. Shih, F. F. *J. Food Sci.* **1987**, *52*, 1529-1531.
13. Kumagai, H.; Shizawa, Y.; Sakurai, H.; Kumagai, H. *Biosci. Biotechnol. Biochem.* **1998**, *62,* 341-346.
14. Kumagai, H.; Ishida, S.; Koizumi, A.; Sakurai, H.; Kumagai, H*.J. Agric. Food Chem.* **2002**, *50,* 172-176.
15. Kumagai, H.; Norimatsu, Y.; Hashizume, N.; Sakurai, H.; Kumagai, H. *Nippon Shokuhin Kogaku Kaishi* **2001**, *48,* 884-890.
16. Kumagai, H.; Suda, A.; Sakurai, H.; Kumagai, H.; Arai, S.; Inomata, N.; Ikezawa, Z. *Biosci. Biotech. Biochem.* **2007**, *71,* 977-985.
17. Kumagai, H.; Koizumi, A.; Suda, A.; Sato, N.; Sakurai, H.; Kumagai, H. *Biosci. Biotech. Biochem.* **2004**, *68,* 1598-1600.
18. Norimatsu, Y.; Kumagai, H.; Nagai, R.; Sakurai, H.; Kumagai, H. *Nippon Shokuhin Kogaku Kaishi* **2002**, *49,* 639-645.
19. Picherit, C.; Bennetau-Pelissero, C.; Chanteranne, B.; Lebecque, P.; Davicco, M. J.; Barlet, J. P.; Coxam, V. *J. Nutr.* **2001**, *131,* 723-728.

Chapter 37

Effects of Polyphenol Rich Herbal Medicine, *Ginkgo biloba* Extracts on Neurotransmitter Levels in Rat Brain

Hiroyuki Sakakibara[1], Yuki Izawa[2], Jun-ichiro Nakajima[3], Shujiro Seo[3], Toshiaki Tamaki[2], Yoshichika Kawai[1], and Junji Terao[1]

[1]Department of Food Science, Institute of Health Biosciences, The University of Tokushima Graduate School, Kuramoto-cho 3–18–15, Tokushima 770–8503, Japan
[2]Department of Pharmacology, Institute of Health Biosciences, The University of Tokushima Graduate School, Kuramoto-cho 3–18–15, Tokushima 770–8503, Japan
[3]Tokiwa phytochemical Company, Ltd., 158, Kinoko, Sakura-shi, Chiba 285–0801, Japan
[4]Current address: Institute of Environmental Sciences, University of Shizuoka, 52–1 Yada, Suruga-ku, Shizuoka 422–8526, Japan

This study was aimed to evaluate the antidepressant mechanism of polyphenol rich extracts of *Ginkgo biloba* (EGB) from the aspect of neurotransmitter regulation in rat brain. Forced swimming stress for 5 min significantly reduced norepinephrine (NE) level in the rat hypothalamus region of vehicle group from 18.8 to 13.5 nmol/g tissue. However, daily oral treatment of EGB (10 mg/kg body weight/day) for 14 days, which were the effective amount and days for antidepressant action as described in our previous study, maintained NE amount at basal level when forced swimming stress was loaded. Consequently, severe changes of neurotransmitter levels in the hypothalamus at the stress loading are considered one of the possible factors to provoke depression. Daily consumption of some ingredients, such as EGB, that can maintain the neurotransmitter levels at the stress loading, may be helpful to prevent depressive illness.

Depression represents a major public disease affecting nearly 2-5% of the world population in recent years (*1*). Therefore understanding how to prevent and treat depression is an urgent subject. Chronic stress has been reported to be a main trigger for this disease (Y), and therefore to regulate stress signals in brain probably contributes to the prevention of the depressive illness.

The *Ginkgo biloba* tree has been used as a traditional Chinese herbal medicine for thousands of years (*3*). Several research groups have shown that extracts from the green leaves of the *Ginkgo biloba* tree (EGB) possess diverse effects on the improvement of mood and cognitive performance, protection of memory deficits and so on (*4, 5*). Recently, we reported that EGB significantly shortened the immobility time in the rat forced swimming test (FST) after repeated oral treatment of 10 mg/kg body weight/day for 14 days without change of motor dysfunction in the open field test (6). Therefore, these results indicated EGB might possess an antidepressant activity. The mechanism of antidepressant action of antidepressants including EGB has not been cleared. However, dopaminagic (norepinephrine, NE and dopamine, DA) and serotonergic (serotonin, 5-HT) system in the brain are suggested to play an important role in the etiology of depression. Actually, therapeutic effects of antidepressants, for example selective serotonin reuptake inhibitors, are believed to be caused by the regulation of central dopaminagic or serotonergic systems. Here we aimed to evaluate the effects of EGB on the levels of NE, DA and 5-HT and their metabolites in the rat brain after repeated treatment using FST as a stress loading model.

Materials and Methods

Male CD rats (4 weeks) from Charles River Japan Inc., Yokohama, Japan, were housed in a 12 h light/dark cycle at a constant temperature of 23 ± 1 °C with free access to food and water. All experiments were conducted in accordance with the animal experimental guidelines of the University of Tokushima. Each group (n = 5) was orally administered distilled water, anti-depressant drug imipramine (15 mg/kg body weight/day) or EGB (Ginkgolon-24 from Tokiwa Phytochemical Co., Ltd., Chiba, Japan). Administered amount of EGB was 10 mg/kg/day, which indicated remarkable antidepressant effects in the rat forced swimming test (FST) (*6*). After repeated oral treatment for 14 days, each group was subjected to forced swimming stress for 5 min according to the method of Porsolt *et al.* (*7*), and then immediately sacrificed by decapitation to collect a hypothalamus region. The groups without forced swimming stress were also subjected a dissection.

Neurotransmitter levels in the hypothalamus region were analyzed using high performance liquid chromatography (HPLC) equipped with coulometric array system by the method of Lakshmana and Raju (*8*) and Vaarmann et al (*9*) with some modifications. Briefly, the hypothalamus (50 mg) was mixed with 0.5 mL of 0.12 M perchloric acid containing 0.1% cysteine, and 0.5 nmol of isoproterenol was added as an internal standard. The sample mixture was homogenized, and then centrifuged at 12,000 g for 15 min at 4 °C. The supernatant was filtrated with 0.2 μm membrane filter Millex-LG (Millipore Co., Bedford, MA). An aliquot of 10 μL of the filtrate was subjected to HPLC with a Capcell Pak C18 MGII column (250 x 4.6 mm I.D., Shiseido Co., Ltd., Tokyo, Japan) equipped with a guard column (10 x 4.0 mm I.D.). The columns were maintained at 25°C. The mobile phase was acetonitrile/55 mM tartaric acid and 45 mM sodium acetate containing 0.65 mM 1-octanesulfonic acid sodium salt (10:90, v/v) at flow rate of 1.0 mL/min. Neurotransmitter and their related compounds were detected with a coulometric array detector (Model 5600A CoulArray System, ESA, Inc., Chemsford, MA) at + 200 mV for DA, NE, 5-HT, 3,4-dihydroxyphenylalanine (DOPA), epinephrine (E), 3,4-dihydroxy-phenyl-acetic acid (DOPAC), homovanillic acid (HVA), 3-methoxytyramine (3-MT) and 5-hydroxytryptophan (5-HTP), and at +550 mV for tyrosine, trypto-phan and 5-hydroxyindoleacetic acid (5-HIAA).

Data analysis was performed by analysis of variance with Fisher-PLSD post hoc test for multiple comparisons. The results were considered significant if the probability of error was less than 5%.

Results and Discussion

Figure 1 shows a typical HPLC profile with 12 standard compounds, which were neurotransmitters (DA, NE, E, 5-HT), their precursors (tyrosine, tryptophan, DOPA, 5-HTP), and major biological metabolites (HVA, 3-MT, DOPAC, 5-HIAA) at concentration of 10 pmol each, indicating that every compound could be analyzed simultaneously within 30 min.

Rats were orally administered antidepressant imipramine (15 mg/kg body weight/day), EGB (10 mg/kg/day), or vehicle (distilled water) for 14 days. The animals were then subjected to the forced swimming stress for 5 min, which was the same condition for screening model of antidepressants, forced swimming test (FST). Neurotransmitter levels in the hypothalamus region of both with and without swimming stress were analyzed using HPLC-coulometric array system as shown in Table I. Daily treatment of imipramine, which is a tricyclic-type antidepressant acting as monoamine reuptake inhibitor, produced significant reduction in NE level from 18.8 nmol/g tissue to 10.4 nmol/g tissue, as suggested by Butterweck *et al* (*10*). Similar reductive pattern of NE, but not

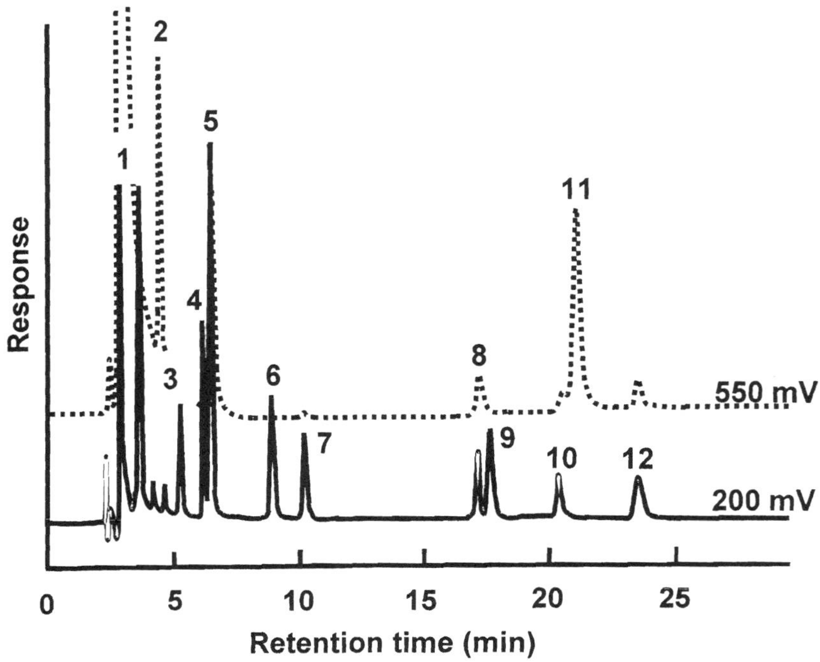

Figure 1. Typical HPLC profile for neurotransmitters. Numbers show the following standard chemicals: 1, 3,4-dihydroxyphenylalanine (DOPA); 2, tyrosine 3, norepinephrine (NE); 4, epinephrine (E); 5, 5-hydroxytryptophan (5-HTP); 6, dopamine (DA); 7, 3,4-dihydroxyphenylacetic acid (DOPAC); 8, 5-hydroxyindoleacetic acid (5-HIAA), 9, 3-methoxytyramine (3-MT), 10, 5-hydroxytryptamine (5-HT); 11, tryptophane; and 12, homovanillic acid (HVA).

significantly, was obtained in the group administered EGB. Forced swimming stress for 5 min, which is suggested to induce a depressive state in rats, also reduced NE level remarkably in vehicle group from 18.8 nmol/g tissue to 13.5 nmol/g tissue. However, NE amounts in the both groups administered imipramine and EGB were maintained at the same range. Similar alteration was suggested in DA and 5-HT, but not significant. Consequently, severe changes of neurotransmitter (dopaminagic and/or serotonergic) levels in the hypothalamus at a stress loading are considered one of the factors to provoke depression. Therefore, daily consumption of some ingredients, such as EGB, that can maintain the neurotransmitter levels when the stress is loaded may be helpful to prevent depressive illness.

Table I. Effects of EGB Administration and Forced Swimming Stress on Neurotransmitter Levels in the Rat Hypothalmus

	Vehicle		Imipramine		EGB	
	-Stress[a]	+Stress[b]	-Stress	+Stress	-Stress	+Stress
Dopaminagic compounds						
Tyrosine	66.9 ± 4.8	65.8 ± 4.8	63.8 ± 3.7	54.2 ± 0.6	70.4 ± 4.9	55.9 ± 5.8
DOPA	4.2 ± 0.7	6.4 ± 0.7	3.9	4.3 ± 1.4	4.6 ± 0.9	4.3 ± 1.6
DA	1.8 ± 0.3	3.9 ± 0.9	3.3 ± 1.0	4.0 ± 0.6	2.7 ± 0.6	3.1 ± 1.0
NE	18.8 ± 1.7	13.5 ± 0.7*	10.4 ± 1.1*	11.0 ± 1.5*	13.9 ± 1.2	14.8 ± 2.9
E	u.d.	u.d.	u.d.	u.d.	u.d.	u.d.
DOPAC	2.3 ± 0.36	4.1 ± 0.60	3.9 ± 0.65	4.6 ± 0.93	3.7 ± 1.0	3.1 ± 0.84
3-MT	u.d.	u.d.	u.d.	u.d.	u.d.	u.d.
HVA	0.32 ± 0.02	0.61 ± 0.07	0.52 ± 0.15	0.62 ± 0.13	0.67 ± 0.19	0.56 ± 0.15
Serotonagic compounds						
Tryptophane	26.1 ± 1.4	26.3 ± 1.9	27.8 ± 0.7	27.4 ± 0	27.8 ± 1.1	25.7 ± 2.5
5-HTP	0.19 ± 0.03	0.25 ± 0.02	0.19 ± 0.04	0.25 ± 0.03	0.23 ± 0.03	0.22 ± 0.02
5-HT	10.1 ± 1.0	13.0 ± 0.5	11.4 ± 2.4	11.6 ± 1.1	10.8 ± 1.0	11.3 ± 2.7
5-HIAA	9.0 ± 0.3	10.3 ± 0.9	8.1 ± 0.4	8.0 ± 0.6	10.3 ± 0.5	9.6 ± 1.3

Values are mean ± S.E. expressed as nmol/g wet tissues of 5 animals in each group. Rats were administered vehicle (distilled water), imipramine (15 mg/kg body weight/day), and EGB (10 mg/kg body weight/day) for 14 days, and then subjected to measurement of neurotransmitter levels in the hypothalamus as described in the Materials and Methods. *$p<0.05$ vs vehicle group without stress. [a]Without forced swimming stress, [b]Treatment of forced swimming stress for 5 min; [c]Under the detection limit (<50 pmol/g tissues); Other abbreviations were the same as described in Figure 1.

Major ingredients in the EGB are polyphenols, especially flavonoids including quercetin glycosides and kaempferol glycosides (*11*). Recently, quercetin metabolites were found in the brain tissues of rodents after oral administration (*12*), and antidepressant effects of quercetin glycosides were reported in the experiments using FST (*13*). Hence, one of the antidepressant ingredients in EGB is probably quercetin glycosides. Further studies are needed to identify the active compounds in EGB, and also to establish the pharmacological relevance between the active compounds and their effects, for example, regulation of monoamine oxidase activity, and inhibition of monoamine reuptake.

Acknowledgments

This study was performed as part of the program of the 21st Century Center of Excellence on "Human nutritional science on stress control" at the University of Tokushima.

References

1. Blazer, D. G. In *Comprehensive Textbook of Psychiatry,* Sadock, B. J.; Sadock, V., Eds; Lippincott Wilkins: New York, NY, 2000, pp. 1298-1308.
2. Fava, M.; Kendler, K. S. *Neuron* **2000**, *28*, 335-341.
3. Nakanishi, K.; *Bioorg. Med. Chem.* **2005**, *13*, 4987-5000.
4. Trick, L.; Boyle, J.; Hindmarch, I. *Phytother Res.* **2004**, *18*, 531-537.
5. Tang, F.; Ng, S.; Shiu, S. Y. W.; Pang, S. F. *Life Sci.* **2002**, *71*, 2625-2631.
6. Sakakibara, H.; Ishida, K.; Grundmann, O.; Nakajima, J.; Seo, S.; Butterweck, V.; Minami, Y.; Saito, S.; Kawai, Y.; Nakaya, Y.; Terao, J. *Biol. Pharm. Bull.* **2006**, *29*, 1767-1770.
7. Porsolt, R. D.; Le Pichon, M.; Jalfre, M. *Nature* **1977**, *266*, 730-732.
8. Lakshmana, M. K.; Raju, T. R. *Anal. Biochem.* **1997**, *246*, 166-170.
9. Vaarmann, A.; Kask, A.; Mäeorg, U. *J. Chromatogr. B*, **2002**, *769*, 145-153.
10. Butterweck, V.; Simbrey, K.; Seo, S.; Sasaki, T.; Nishibe, S. *Pharm. Biochem. Behav.* **2003**, *75*, 557-564.
11. Drieu, K. *Presse. Med.* **1986**, *15*, 1455-1457.
12. Paulke, A.; Schubert-Zsilavecz, M.; Wurglics, M. *J. Chromatogr. B Analyt. Technol. Biomed. Life Sci.* **2006**, *832*, 109-113.
13. Butterweck, V.; Jurgenliemk, G.; Nahrstedt, A.; Winterhoff, H. *Planta Med.* **2000**, *66*, 3-6.

Chapter 38

Serum LDL-Cholesterol-Lowering Effects of a Mixed Green Vegetable and Fruit Beverage Containing Broccoli and Cabbage in Humans

Hirohisa Suido[1], Motoko Takai[1], Akira Takeuchi[1], Taketoshi Makino[1], Keisuke Tsuji[2], and Toshio Tanaka[3]

[1]Health Care Research and Development Division, Sunstar Inc., 3–1 Asahimachi, Takatsuki, Osaka 569–1195, Japan
[2]School of Human Science and Environment, University of Hyogo, 1–1–12 Shinzaikehonmachi, Himeji, Hyogo 670–0092, Japan
[3]Department of Respiratory Medicine, Allergy and Rheumatic Diseases, Osaka University Medicine School, 2–2 Yamadaoka, Suita, Osaka 565–0871, Japan

Effects of mixed green vegetable and fruit beverages, with or without broccoli and cabbage (B&C), on serum lipid levels in mild and moderate hypercholesterolemic subjects were examined in double-blind study designs. Study 1: Sixty five mild hypercholesterolemic subjects were randomly divided into three groups. Each of the three was administered one can (160 g) of placebo beverage not containing B&C, one can of test beverage (test 1) or two cans of test beverage (test 2) per day for 9 weeks. Serum LDL-cholesterol (LDL-C) levels were significantly ($p<0.05$) decreased from 3.87 to 3.67 mmol/L in average at 9 week in test 1 group, and from 3.93 to 3.66 mmol/L at 6 week and to 3.71 mmol/L at 9 week in test 2 group. However, that in placebo group was not changed during the test periods. Study 2: Forty nine subjects were randomly divided into two groups. Each group was administered two cans of placebo beverage or two cans of test beverage per day for 12 weeks. Serum LDL-C levels were significantly ($p<0.05$) decreased at 3, 6 and 9 weeks in test group. However, that in placebo group was not significantly changed. Between placebo group and test group, there were significant differences in the

LDL-C levels at 6 and 9 weeks. Thus daily intake of one or two cans of the beverage containing B&C is useful for reducing serum LDL-C level.

The effects of food intake on the body's physiological functions are attracting considerable attention. Hypercholesterolemia is considered to lead to atherosclerosis; the normalization of serum levels of low-density-lipoprotein cholesterol (LDL-C) through diet therapy or administration of drugs decreases the incidence of coronary heart disease (*1, 2*).

It has been reported that broccoli and cabbage have serum cholesterol-lowering effects in rats, and S-methylcysteine sulfoxide (SMCS) is considered a potent ingredient responsible for the cholesterol-lowering effect of cabbage (*3-6*). Also we found that the mixed green vegetable and fruit beverages containing broccoli and cabbage (B&C) had serum LDL-C-lowering effects in human, although they were open trials (*7-9*).

In this study we examined the serum LDL-C-lowering effects of two types of canned beverage containing mixed green vegetable and fruits, with or without B&C, in randomized double-blind study designs.

Materials and Methods

Beverages Tested

The test beverage was a canned mixed green vegetable and fruit beverage "Oishii-Aojiru" (Sunstar inc., Osaka, Japan) containing broccoli and cabbage (B & C), lettuce, celery, spinach, parsley, komatsu-na, radish leaf, apple and lemon. The placebo beverage was made from the same materials as "Oishii Aojiru" but without B&C. The energy and dietary fiber amounts were adjusted to those of the test beverage by being added with sucrose and cellulose. The energy and nutrient contents are shown in Table I.

Subjects and Trial Schedule

Study 1: Sixty five mild hypercholesterolemic subjects, whose LDL-C levels were greater than or equal to 2.84 mmol/L (110 mg/dL) and less than 5.17 mmol/L (200 mg/dL) ; average 3.88 mmol/L (150 mg/dL), were randomly divided into three groups after pre-examination. After baseline examination (0 week), the subjects were administered one or two cans of the assigned sample per day for 9 weeks.

Table I. Energy and Known Nutrient Contents in One Can[a] of the Beverages Tested

		Study 1		Study 2	
		Placebo	Test	Placebo	Test
Energy[b]	kcal	48	50	56	55
Protein	g	0.3	1	0.3	0.8
Fat	g	0	0.16	0	0.16
Carbohydrate	g	11.8	11	13.6	13
Ash	g	0.32	0.6	0.32	0.5
Sodium	mg	18.1	26.9	15.5	26.6
Dietary fiber	g	1.8	1.4	1.6	1.4
SMCS[c]	mg	ND[d]	20	ND[D]	26

[a]One can of the beverages contained 160g
[b]Energy = (Carbohydrate + Protein) × 4 + Fat × 9
[c]S-methylcysteine sulfoxide
[d]Not detected (<1 mg/100g)

Study 2: Forty nine hypercholesterolemic subjects, whose LDL-C levels were greater than or equal to 3.62 mmol/L (140mg/dL) and less than 4.65 mmol/L (180 mg/dL); average 4.03 mmol/L (156 mg/dL), were randomly divided into two groups just after baseline examination. The subjects were administered two cans of the assigned sample per day for 12 weeks.

Trial schedule of these two studies is shown in Figure 1. Fasting blood samples were taken from each subject at pre-examination and/or baseline and every 3 weeks after the start of beverages intake for 9 or 12 weeks.

These studies were approved by the institutional ethical committee according to the Helsinki Declaration.

Examination Categories and Methods

A physician conducted a medical examination on each patient at the time of the blood sampling, measuring weight and blood pressure. Clinical testing on the blood samples, including total cholesterol (TC), high-density-lipoprotein cholesterol (HDL-C) and triacylglycerol (TG), was carried out at Japan Medical Laboratory (Osaka, Japan). Serum LDL-C levels were calculated using the Friedewald's formula as follows: [LDL-C = TC – HDL-C – TG/5] (*10*).

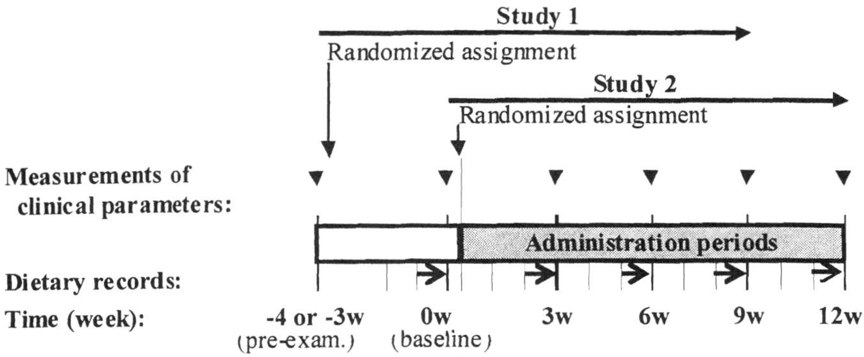

Figure 1. Trial schedule of Study 1 and Study 2.

To check the changes of daily nutrient intakes before and during administration, dietary records (*11*) were taken by each individual for 1 week before every measurement period.

Statistical Analyses

Values are expressed as the mean ± standard deviation (SD). A paired t-test or an analysis of variance (ANOVA) and post-hoc comparison was conducted for the data, as appropriate. Statistical significance was considered at $p<0.05$.

Results

Study 1

Table II. Characteristics of the Subjects at the Pre-examination in Study 1

	Control Group n=23	*Test 1 Group* n=20	*Test 2 Group* n=22	*p value* (ANOVA)
Male / Female	12 / 11	13 / 7	13 / 9	
Age (years)	42.2 (10.7)	38.2 (8.5)	39.0 (10.5)	0.367
Weight (kg)	60.8 (11.4)	61.2 (11.2)	61.6 (10.4)	0.967
BMI	22.6 (3.5)	22.1 (2.6)	22.1 (2.0)	0.740

Value denotes mean (SD).
Control: One can of placebo beverage (SMCS 0 mg) per day
Test 1: One can of test beverage (SMCS 20 mg) per day
Test 2: Two cans of test beverage (SMCS 40mg) per day

The characteristics of the subjects at the pre-examination in study 1 are shown in Table 2. There were no significant differences between the three groups on the numbers of male and female, age, weight and body mass index (BMI).

Changes of serum lipid levels are shown in Figure 2 and Table III. Serum LDL-C levels were significantly decreased from 3.87 to 3.67 mmol/L (from 150 to 142 mg/dL) at 9 week in test1 group, and from 3.93 to 3.66 mmol/L (from 152 to 142 mg/dL) at 6 week and to 3.71 mmol/L (143 mg/dL) at 9 week in test 2 group. Also serum TC levels in test 2 group was significantly decreased from 5.71 to 5.43 mmol/L (from 221 to 210 mg/dL) at 6 week. But those in placebo group were not changed during the test periods.

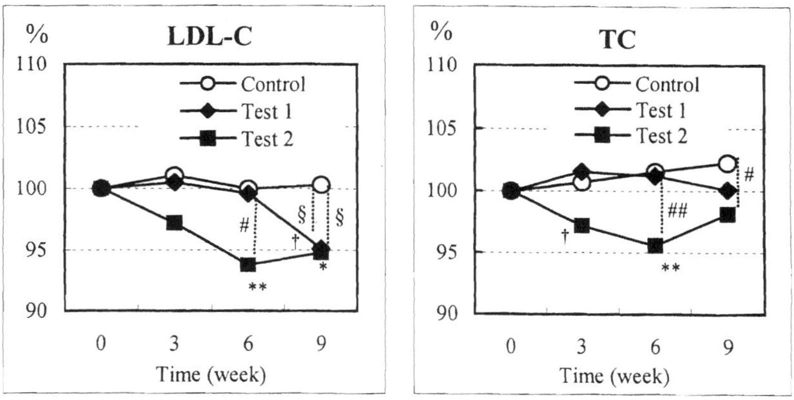

Figure 2. Changes in LDL-C and TC levels (shown as % compared to 0 week) after administration of the beverages tested in Study 1.
*†p<0.1, *p<0.05 and **p<0.01 by LSD test after Repeated measures ANOVA (vs. 0 week).*
§p<0.1and #p<0.05 by Dunnett-t test (vs. Control group).

Study 2

The characteristics of the subjects at the baseline examination (0 week) in study 2 are shown in Table IV. There were no significant differences between the two groups on the numbers of male and female, age, weight and BMI.

Changes of serum lipid levels are shown in Figure 3 and Table V. Serum LDL-C levels in test group were significantly decreased at 3, 6, 9 weeks. Between placebo group and test group, there were significant differences ($p<0.05$) in the LDL-C reduction at 6 and 9 weeks. There was also significant difference($p<0.01$) in the TC reduction at 9 week between the two groups.

Table III. Changes in Serum Lipid Levels after Administration of the Beverages Tested in Study 1

	Group	n	Pre-exam.	0 week	3 week	6 week	9 week	*p* value[a]
LDL-C	Control	23	3.81 (0.55)	3.78 (0.45)	3.80 (0.50)	3.76 (0.48)	3.77 (0.44)	0.976
[mmol/L]	Test 1	20	3.93 (0.51)	3.87 (0.53)	3.87 (0.54)	3.85 (0.59)	3.67 (0.53)*	0.088
	Test 2	22[b]	3.89 (0.52)	3.93 (0.68)	3.86 (0.53)	3.66 (0.60)**	3.71 (0.64)**	0.004
TC	Control	23	5.53 (0.66)	5.50 (0.51)	5.53 (0.52)	5.58 (0.57)	5.61 (0.43)	0.852
[mmol/L]	Test 1	20	5.57 (0.59)	5.62 (0.52)	5.70 (0.63)	5.68 (0.68)	5.61 (0.56)	0.549
	Test 2	22[b]	5.64 (0.60)	5.71 (0.80)	5.60 (0.62)	5.43 (0.63)**	5.58 (0.69)	0.014
HDL-C	Control	23	1.26 (0.30)	1.30 (0.37)	1.36 (0.30)	1.42 (0.36)**	1.49 (0.30)**	<0.001
[mmol/L]	Test 1	20	1.23 (0.27)	1.30 (0.32)	1.34 (0.35)	1.35 (0.40)	1.39 (0.44)*	0.002
	Test 2	22[b]	1.33 (0.35)	1.39 (0.38)	1.33 (0.35)†	1.32 (0.34)†	1.52 (0.40)**	<0.001
TG	Control	23	1.28 (1.07)	1.16 (0.71)	1.09 (0.57)	1.16 (0.60)	1.10 (0.58)	0.616
[mmol/L]	Test 1	20	1.26 (0.77)	1.29 (0.74)	1.30 (0.82)	1.33 (0.94)	1.50 (1.08)	0.525
	Test 2	22[b]	1.23 (0.76)	1.12 (0.66)	1.29 (0.96)	1.27 (0.67)	1.12 (0.53)	0.573

Value denotes mean (SD). [a]Statistical analysis by Repeated measures ANOVA within group.

†$p<0.1$, *$p<0.05$ and **$p<0.01$ by LSD test (vs 0 week). [b]n=20 at 3 week.

LDL-C: LDL-cholesterol, TC: Total cholesterol, HDL-C: HDL-cholesterol, TG: Triacylglycerol

Table IV. Baseline Characteristics in the Subjects in Study 2

	Control Group n=25	Test Group n=24	*p* value (t-test)
Male/Female	9 / 16	16 / 8	
Age (years)	46.0 (11.1)	44.5 (8.4)	0.304
Weight (kg)	67.0 (11.4)	64.9 (12.3)	0.274
BMI	24.0 (3.0)	23.6 (3.0)	0.312

Value denotes mean (SD).
Control: Two cans of placebo beverage per day
Test: Two cans of test beverage per day

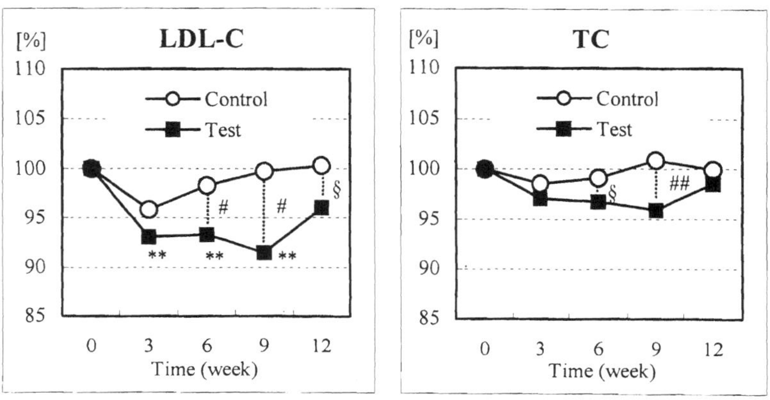

Figure 3. Changes in LDL-C and TC levels (shown as % compared to 0 week) after administration of the beverages tested in Study 2.
$^{**}p<0.01$ by LSD test after Repeated measures ANOVA (vs. 0 week).
$^{\S}p<0.1$, $^{\#}p<0.05$ and $^{\#\#}p<0.01$ by t-test (vs. Control group).

Table V. Changes in Serum Lipid Levels after Administration of the Beverages Tested in Study 2

	Group	n	0 week	3 week	6 week	9 week	12 week	*p* value[a]
LDL-C	Control	25[b]	4.03 (0.32)	3.86 (0.45)	3.97 (0.39)	4.02 (0.48)	4.06 (0.40)	0.268
[mmol/L]	Test	24	4.02(0.24)	3.74 (0.54) **	3.75 (0.44) **, #	3.68 (0.51) **, #	3.87 (0.53) [b]	0.009
TC	Control	25[b]	6.14 (0.40)	6.04 (0.47)	6.09 (0.51)	6.19 (0.57)	6.14 (0.47)	0.617
[mmol/L]	Test	24	6.06 (0.40)	5.88 (0.55)	5.86 (0.56) [b]	5.81 (0.53)#	5.98 (0.59)	0.117
HDL-C	Control	25[b]	1.45 (0.41)	1.43 (0.39)	1.45 (0.39)	1.44 (0.37)	1.43 (0.39)	0.957
[mmol/L]	Test	24	1.47 (0.31)	1.45 (0.31)	1.46 (0.31)	1.45 (0.34)	1.48 (0.33)	0.836
TG	Control	25[b]	1.43 (0.54)	1.64 (0.64)	1.48 (0.69)	1.60 (0.68)	1.43 (0.55)	0.376
[mmol/L]	Test	24	1.24 (0.57)	1.51 (0.85)	1.42 (0.70)	1.48 (0.78)	1.37 (0.59)	0.123

Value denotes mean (SD). [a]Statistical analysis by ANOVA. **$p<0.01$ by LSD test (vs 0 week).
[b]n=23 at 6 week and 12 week. For abbreviations see footnote to Table III.

Changes of Daily Nutrient Intakes in Study 1 and 2

There were no significant changes of daily intakes of energy, protein, fat, and non-fibrous carbohydrate in each study groups (Data are not shown). Dietary fiber intake significantly increased approximately 1 to 3 g/day in each groups during the administration period, which seemed to be derived from one or two cans of the test beverages.

Adverse Effects in Study 1 and 2

There were no abnormal findings in physical examination, blood pressure and any biochemical parameters of blood examination on hepatic, renal and pancreatic functions both in Study 1 and 2 (Data are not shown).

Discussion

These results showed that broccoli and cabbage in the mixed green vegetable and fruit beverage tested have serum LDL-C lowering effects in hypercholesterolemic subjects, although it is not clear which constituents in broccoli and cabbage showed hypocholesterolemic action.

Generally, the effects of vegetables and fruits on cholesterol metabolism are considered to attribute to their fiber content. Isolated dietary fibers from these plants, such as pectins (12-50 g/day) (*12, 13*), guar gum (10-20 g/day) (*14, 15*) and psyllium (6-15 g/day) (*16, 17*), have been shown to lower cholesterol levels in humans. The mechanisms for the cholesterol-lowering in humans remain unclear. Results from *in vitro* and animal studies that used isolated fibers suggested that the reductions in cholesterol are probably due to different mechanisms specific to each fiber source and to different dietary fiber intake amounts (*18-20*). Possible mechanisms include 1) increased excretion of fecal bile acids and neutral steroids, 2) altered ratios of primary to secondary bile acids, 3) increased fecal cholesterol and fatty acid excretion, 4) and indirect effects, such as high-fiber foods replacing fat- and cholesterol-containing foods in diet (*21*). However, the amount of dietary fibers (1.4-2.8 g/day) provided by the consumption of one or two cans of the beverage tested in this study is far out of proportion with the doses (6-50 g/day) of isolated dietary fibers described above to obtain a significant reduction in serum cholesterol levels in humans.

As other potent constituents in the product, cabbage leaf proteins (*22*) and S-methylcysteine sulfoxide (SMCS) (*4-6*) have been shown to have cholesterol-lowering effects on animals. The total protein content in the test beverage is about 1 g/160g. This protein amount seems to be far less than the daily dose of soy protein recommended for obtaining a significant reduction in serum cholesterol levels in humans. On the other hand, SMCS, a naturally occurring S-containing amino acid, is known to be contained in Brassica vegetables such as broccoli and cabbage and Allium vegetables such as onion at high concentrations (*23*, *24*). The mechanism of SMCS for cholesterol-lowering effects has been hypothesized to enhance fecal bile acid excretion and cholesterol-7α-hydroxylase activity, the rate-limiting enzyme of bile acid biosynthesis, in the microsomal fraction of the liver (*5*, *6*). Indeed the test beverage contained broccoli and cabbage as main materials and the amount of SMCS derived from these Brassica vegetables was at least 20 mg/160g contents. Therefore, the dietary fibers and other potent constituents including SMCS in the test beverage might have shown synergistic effects to lower cholesterol levels.

In conclusion, daily intakes of the beverage containing certain amounts of broccoli and cabbage are useful for lowering serum LDL-C levels in hypercholesterolemic subjects.

References

1. Keys, A. *Acta Med. Scand.* **1980**, *207*, 153-160.
2. National Cholesterol Education Program, National Institutes of Health. *Third Report of the National Cholesterol Education Program (NCEP) Expert Panel on Detection, Evaluation, and Treatment of High Blood Cholesterol in Adult (Adult Treatment Panel III), Executive Summary.* NIH Publication **2001**, No. 01-3670.
3. Kobayashi, T.; Suido, H.; Takeuchi, A.; Makino, T.; Tanaka, T.; Nagaoka, S. *J. Jpn. Soc. Nutr. Food Sci.* **2002**, *55*, 275-280.
4. Itokawa, Y.; Inoue, K.; Sasagawa, S.; Fujiwara, M. *J. Nutr.* **1973**, *103*, 88-92.
5. Fujiwara, M.; Itokawa, Y.; Uchino, H.; Inoue, K. *Experientia* **1972**, *28*, 254-255.
6. Komatsu, W.; Miura, Y.; Yagasaki, K. *Lipids* **1998**, *33*, 499-503.
7. Tabei, T.; Tanaka, T.; Suido, H.; Fukuhara, K.; Kotani, M.; Takeuchi, A.; Okita, M.; Kasayama, S. *Jpn. Med. J.* **2000**, *3985*, 46-49.
8. Suido, H.; Tabei, T.; Tanaka, T.; Kobayashi, T.; Sato, T.; Takeuchi, A.; Makino, T.; Kasayama, S.; Tsuji, K.; Higashino, K. *J. Nutr. Food* **2000**, *3*, 17-26.
9. Suido, H.; Tanaka, T.; Tabei, T.; Takeuchi, A.; Okita, M.; Kishimoto, T.; Kasayama, S.; Tsuji, K.; Higashino, K. *J. Agric. Food Chem.* **2002**, *50*, 3346-3350.

10. Friedewald, W. T.; Levy, R. I.; Frederickson, D.S. *Clin. Chem.* **1972**, *18*, 499-502.
11. Suzuki, K.; Sasagawa, T.; Okita, M.; Fujiwara, Y.; Mori, K.; Saito, M.; Abe, Y. *Bulletin of faculty of health and welfare sciences, Okayama prefectural university* **1996**, *3*, 143-153.
12. Durrington, P. N.; Manning, A. P.; Bolton, C. H.; Hartog, M. *Lancet* **1976**, *2*, 394-396.
13. Kay, R. M.; Truswell, A. S. *Am. J. Clin. Nutr.* **1977**, *30*, 171-175.
14. Kawatra, A.; Kapoor, A.C.; Seghal, S. *Plant Foods Hum. Nutr.* **1991**, *41*, 241-245.
15. Jenkins, D. J. A.; Reynolds, D.; Leeds, A. R.; Waller, A.L.; Cummings, J. H. *Am. J. Clin. Nutr.* **1979**, *32*, 2430-2435.
16. Dennison, B. A.; Levine, D. M. *J. Pediatr.* **1993**, *123*, 24-29.
17. Everson, G. T; Daggy, B. P.; McKinley, C.; Story, J. A. *J. Lipid Res.* **1992**, *33*, 1183-1192.
18. Fernandez, M. L. *J. Lipid Res.* **1995**, *36*, 2394-2404.
19. Story, J. A.; Kritchevsky, D. *J. Nutr.* **1976**, *106*, 1292-1294.
20. Tsuji, K. *Bifidus* **1995**, *8*, 125-134.
21. Glore, S. R.; Van Treeck, D.; Knehans, A. W.; Guild, M. *J. Am. Diet Assoc.* **1994**, *94*, 425-436.
22. Igarashi, K.; Satoh, A.; Numazawa, S.; Takahashi, E. *J. Nutr. Sci. Vitaminol.* **1997**, *43*, 261-270.
23. Morris, C. J.; Thompson, J. F. *J. Am. Chem. Soc.* **1956**, *78*, 1605-1608.
24. Tsuno, S.; Murakami, F.; Tazoe, K.; Kikumoto, S. *Vitamins* **1960**, *20*, 93-96.

Chapter 39

The Effects of *Morinda citrifolia* L. Noni on High Blood Pressure: A Mechanistic Investigation and Case Study

'Afa Kehaati Palu[1], Raevonne A. Santiago[2], Brett J. West[1], Norman Kaluhiokalani[2], and Jarakae Jensen[1]

[1]Noni Benefits Research, Tahitian Noni International, American Fork, UT 84003
[2]Exercise and Sports Science Department, Brigham Young University of Hawaii, Laie, HI 96762

Morinda citrifolia L. noni has been used in Polynesia for over 2,000 years. During that time, it has accumulated a myriad of reputed health benefits. Traditional healers have used different parts of the Noni tree to treat pain, inflammation, diabetes, gout, wounds, cancer, high blood pressure, and numerous other ailments. Most of the reputed benefits have been limited to published reports in the non-scientific literature. However, with the recent introduction and marketing of noni-based beverages and products, there is increased interest among members of the scientific community to better understand its basic chemistry and potential health benefits. The number of people reported to have successfully lowered their high blood pressure from drinking Noni fruit juice and TAHITIAN NONI® Juice has grown dramatically, but its mechanisms of action are largely unknown. Our focus is to elucidate the possible mechanisms involved in vitro, and to report a study on medicinally diagnosed high blood-pressure individuals, who were administered noni juice, while not taking prescribed blood-pressure medications. In vitro, TAHITIAN NONI® Juice (TNJ) in concentrations of 1 and 5% inhibited angiotensin converting enzymes (ACE) by 105 and 113%, respectively. TNJ also showed an IC_{50} of 1.6 and 4.3% of the AT_1 and AT_2 receptors, respectively. Noni fruit juice concentrate (NFJC) showed an IC_{50} of the AT_1 and AT_2 receptors of 1.97 and 2.29%, respectively. The test group consisted of 10 subjects (7 males, 3 females, ages 28-56 years) who were medically diagnosed with high blood pressure, but currently not taking any prescribed medications. Each subject consumed 2 ounces of TNJ twice a day, for 1 month. The group had average pre-test and post-test blood pressure of 144/83 and 132/76, respectively. The preliminary results suggest that NFJC and TNJ are able to lower high blood pressure by inhibiting ACE enzymes and AT receptors. The mechanisms elucidated warrant further clinical trials to assess the efficacy and dosage of TNJ on medically diagnosed high blood pressure patients.

The ancestors of Polynesia are believed to have brought with them many plants, commonly known as canoe plants, as they migrated from Southeast Asia some 2000 years ago (*1, 2*). Though the exact number of canoe plants, brought, is still debated, it is estimated to be about 20-30 plants. Among the plants was *Morinda citrifolia* L., commonly known as noni. The popularity of the noni plant was likely due to traditional and medicinal usages. Traditional non-medicinal usage of noni includes, but is not limited to, canoe building and dyeing of traditional clothes made from the *hiapo* tree (*Broussonetia papyrifera*). The medicinal usages of the noni tree are purported by healers to include a repertoire of health benefits: antibacterial, antiviral, antifungal, *kanisa* (anticancer), antihelminth, *mamahi'ia* (analgesic), *pupula* (anti-inflammatory), *fakamalohi sino* (immune enhancing), *hangatāmaki* (boils), *lavea* (cuts), *mamulu* (bruises), *langa e hokotanga hui* (anti-gout), *suka* (anti-diabetes), *langa kete* (stomachache), *lafa* (ring worms), *mata kovi* (styies) and *toto ma'olunga* (anti-hypertension) and others (2, 3). However, much of the scientific evidence to support these traditional claims has been lacking. The advent of Tahitian Noni International in 1996 has sparked numerous scientific research efforts that are now being reported in professional scientific meetings and in peer-reviewed scientific journals.

The anti-hypertension effects NFJC and TNJ were investigated to: (1) elucidate the possible mechanisms involved and (2) assess its efficacy and possible dosage necessary to lower high blood pressure in medically-diagnosed high blood pressure patients.

Materials and Methods

The NFJC and TNJ for this study were obtained from Tahitian Noni International. The subjects (28-56 years old) were all diagnosed by their primary-care physician to have high blood pressure, did not take prescription drugs, and were volunteer participants. Only those with an average systolic pressure $\geq$ 130 and at least 1 diastolic pressure $\geq$ 80 were included in this study. All blood pressure readings (pre-test and post-test) were done using an automatic cuff heart rate and blood-pressure monitor, and under supervision of a licensed physician.

Angiotensin Converting Enzyme Assay

Test materials NFJC and TNJ in concentrations of 1, 5, and 10% each, were dissolved in 1% DMSO and incubated with ACE enzymes (Sigma, Milwaukee, WI), together with 500 μM (N-3-[2-furyl]acryloyl)-Phe-Gly-Gly (FAPGG) substrate. The reaction mixtures were incubated in a buffer (50 mM HEPES, 300 mM NaCl, and pH 7.5) at 25 °C for 30 minutes. The FAPGG was quantified spectrophotometrically according to established protocol (*3*) and compared to the reference compound Captopril.

Angiotensin Receptor Assay

***Angiotensin AT_1*:** Both NFJC and TNJ in concentrations of 1, 5, and 10%, were dissolved in 1% DMSO and incubated in a buffer (50 mM Tris-HCl, pH 7.4, 5 mM $MgCl_2$, 0.1% BSA, and 1 mM EDTA) and 0.05 nM [^{125}I] (Sar^1, Ile^8)-Angiotensin II ligand from human recombinant CHO (Chinese Hamster Ovary) cells for 60 minutes at 37 °C. A non-specific ligand containing 10 μM Angiotensin II was also used in each of the reaction mixtures. The radioligand binding was quantified and compared to the reference compound Saralasin according to the established protocol for AT_1 assay (*4*).

***Angiotensin AT_2*:** NFJC and TNJ concentrations for the AT2 assays were prepared as above. They were then incubated with 0.025 nM [^{125} I] CGP-42112A ligand from human recombinant Hela cells in an incubation buffer as described above for the AT_1. A non-specific ligand (as described above) was used according to the established protocol for AT_2 assay (*5,6*).

TAHITIAN NONI® Juice Effects on High Blood Pressure Patients

Ten hypertensive patients (7 males and 3 females) were involved in this study. Blood pressure readings were collected from the patients, three different times during the pre-test phase. Two ounces of TNJ were administered, twice daily, before breakfast and dinner, during the 4-week test. At the end of the study, triple blood pressure readings were taken of each patient for comparison with the pre-test readings.

Results

Angiotensin Converting Enzyme Inhibitory Effects of Noni

NFJC concentrations of 1 and 5% inhibited ACE enzymes by 105 and 113%, respectively (Figure 1).

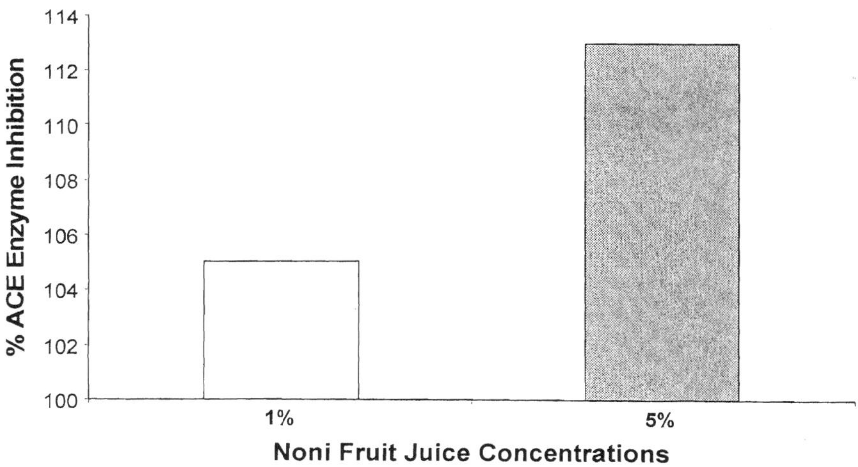

Figure 1. The inhibitory effects of NFJC in 1 and 5% concentrations on angiotensin converting enzymes (ACE).

TNJ concentrations of 1, 5, and 10% inhibited ACE enzymes by 52, 78, and 105%, respectively (Figure 2).

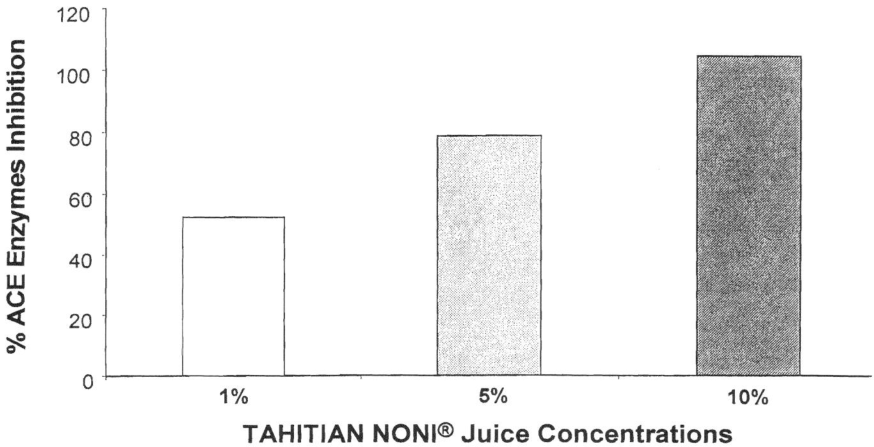

Figure 2. The inhibitory effects of TNJ in 1, 5, and 10% concentrations on ACE enzymes.

Angiotensin Receptor Effects of Noni

Table I shows that both NFJC and TNJ bind significantly to angiotensin receptors AT_1 and AT_2 at concentrations of 5% and 10%.

Table I. The Binding and Inhibitory Effects of Three Different Concentrations of NFJC and TNJ on Angiotensin Receptors AT_1 and AT_2.

Noni Samples	*Concentration, %*	*AT_1*	*AT_2*
NFJC	1	44	7
	5	102	102
	10	101	102
TNJ	1	12	5
	5	99	62
	10	100	77

High Blood Pressure Patients Pre-Test and Post-Test

The blood pressure readings from hypertensive subjects were taken three times pre-test and three times post-test. The numbers in Table II represent an average/patient for the three blood pressure readings for both pre- and post-test.

Table II. Average of Three Blood Pressure Readings for 10 Hypertensive Subjects (Pre-test and Post-test).

Name	*Gender*	*Age*	*Pre-average*	*Post-average*
Subject 1	M	40	135/82	121/71
Subject 2	M	45	149/95	137/71
Subject 3	M	53	152/90	138/84
Subject 4	M	52	145/86	138/77
Subject 5	M	28	143/76	136/75
Subject 6	F	52	140/81	114/74
Subject 7	M	49	154/90	143/82
Subject 8	M	56	149/68	133/72
Subject 9	F	37	135/79	126/77
Subject 10	F	43	142/84	131/78
			144/83	**132/76**

Discussion

The results from the effects of NFJC and TNJ on ACE enzymes show that both noni samples inhibited ACE enzymes in a generally concentration-dependent manner. However, the inhibition of ACE enzymes with NFJC seems to be more pronounced than TNJ in the 1 and 5% concentrations. The inhibitory effects on ACE enzymes have been shown to lower hypertension and also lower hypertension-induce pathology (*7*). Similarly, results from our study showed both NFJC and TNJ inhibited AT_1 and AT_2 receptors. However, TNJ produced a greater inhibition of AT_1 compared to AT_2, while NFJC produced similar effects on both receptors. Concomitantly, the data show that TNJ selectively inhibited AT_1 over AT_2 receptors, with IC_{50} values of 1.6 and 4.13%, respectively. The inhibition data also show that NFJC selectively inhibited AT_1 and AT_2 receptors with IC_{50} values of 1.97 and 2.29%, respectively. The IC_{50}s for both NFJC and TNJ are much higher than that of the Saralasin, reference compound (0.119 nM). Such differences however, were not unexpected, since neither juice is a pure compound. However, unpublished data from our lab of noni fruit juice fractionation show that the inhibitory activities on angiotensin receptors are concentrated in the methanol extract. The methanol extract, in a 100 μg/mL concentration, inhibited both AT_1 and AT_2 receptors by 69 and 41%, respectively. The inhibitory effect of the methanol extract is significant since the concentration is one-tenth that of the 1% NFJC which showed to 44 and 7% inhibition of AT_1 and AT_2 receptors, respectively. These results indicate that

NFJC is selective for AT_1 receptors. This AT_1 receptor inhibitory selectivity has been shown to prevent diabetes-induce hypertension in mice (*8*). Moreover, it has been shown that AT_2 receptor blockers suppress mediators of inflammation including ROS, C-reactive proteins, and increase the expression of inhibitory κB (*9*).

The inhibitory effects of NFJC and TNJ on both ACE enzymes, and the AT_1 and AT_2 receptors, are possible mechanisms of action on hypertension. These effects, and perhaps others, may contribute to the lowering of blood pressure observed in the hypertensive subjects of this study. Results from this study show that the average blood pressure for the group was lowered from 144/83 to 132/76. Furthermore, except for the pre-test average diastolic pressure of subject 8, each hypertensive subjects' blood pressure was lowered. Though these results are preliminary, such reductions are significant when compared to normal blood pressure readings of 120/70. Additionally, the hypertensive subjects consumed only the test juices and no prescribed anti-hypertensive medications (Table II). These results suggest a need for other clinical and future studies of NFJC and TNJ on hypertension.

Future Research Prospects

The effects of noni fruit juice on normal, healthy people should be investigated to assess potentials for both adverse and beneficial effects on blood pressure. Additionally, based on our results, a large human clinical trial is warranted to investigate the effects of noni fruit juice on both female and male subjects who are seriously hypertensive, and are not taking prescribed medications, or who do not respond to other drug regimens. Finally, noni fruit juice effects on chymase, an enzyme that is also involved in hypertension, should be investigated, as many hypertensive patients also suffer from diabetic neuropathy, and do not respond to other hypertensive drug treatments.

Conclusions

The present study has shown that *Morinda citrifolia* L noni products, NFJC and TNJ inhibit ACE enzymes and bind to and inhibit both angiotensin receptors AT_1 and AT_2. These inhibitory effects of noni on ACE enzymes and the angiotensin receptors may thus support the anecdotal therapeutic claims of traditional healers in Polynesia. Additionally, two ounces of TNJ in the morning before breakfast and before dinner were shown to lower high blood pressure in a group of diagnosed hypertensive subjects. Thus, TNJ may have the potential for use as an alternative for lowering high blood pressure.

Acknowledgements

Much appreciation is extended to Silivia Palu, a Tongan traditional healer, for sharing her knowledge of different ways of using noni for treatment of various diseases. We also thank Leeland White for helping in this manuscript, and the Founders of Morinda Holdings, Inc. for their financial support of this research.

References

1. Tabrah, F.L.; Eveleth, B.M. *Hawaii Med. J.*, **1966**, *25*, 223-230.
2. Wang, M.Y.; West, B.J.; Jensen, C.J.; Nowicki, D.; Su, C.; Palu, A.K.; Andersen, G. *Acta Pharmacol Sin.*, **2002**, *23*, 1127-1141.
3. Whistler, W.A. *J Ethnopharm.*, **1985**, *13*, 239-280.
4. Bunning, T.; Holmquist, B.; Riordan J.F. *Biochem.*, **1983**, *22*, 103-110.
5. Webb, M.L.; Monshizadegan, H.; Dickinson, K.E.; Serafino, R.; Moreland, S.; Michel, I.; Seiler, S.M.; Murphy, T.J. *Regul Pept.*, **1993**, *44*, 131-139.
6. Whitebread, S.E.; Taylor, V.; Bottari, S.P.; Kamber, B.; Gasparo, M. *Biochem Biophys Res Comm.*, **1991**, *181*, 1365-1371.
7. Toto, R.D.; Rinner, S.; Ram, C.V.S. *Postgrad Med.*, **2004**, *116*, 11-22.
8. Wichi, R.B.; Farah, V.; Chen Y.; Irigoyen, M.C.; Morris, M. *Am J Physiol Regul Integr Comp Physiol.*, **2007**, *292*, R1184-R1189.
9. Dantona, P.; Dhindsa, S.; Ghanim, H.; Chaudhuri, *A. J Human Hyper.*, **2007**, *21*, 20-27.

Chapter 40

Serum Cholesterol-Lowering Effects of a Broccoli and Cabbage Mixture in Rats: Comparison with Spinach, Celery, Carrot, and Tomato

Hirohisa Suido[1], Akira Takeuchi[1], Taketoshi Makino[1], Toshio Tanaka[2], and Satoshi Nagaoka[3]

[1]Health Care Research and Development Division, Sunstar Inc., 3–1 Asahimachi, Takatsuki, Osaka 569–1195, Japan
[2]Department of Respiratory Medicine, Allergy and Rheumatic Diseases, Osaka University Medical School, 2–2 Yamadaoka, Suita, Osaka 565–0871, Japan
[3]Deptartment of Food Science, Faculty of Agriculture, Gifu University, 1–1 Yanagido, Gifu 501–1193, Japan

Effects of broccoli, cabbage and other vegetables such as spinach, celery, carrot and tomato on lipid metabolism were investigated in rats. The rats were fed cholesterol enriched basal diets, with or without 5% of each freeze-dried vegetable, for 10 or 12 days. Serum total cholesterol (TC) and low density lipoprotein (LDL) + very low density lipoprotein (VLDL)-cholesterol (LDL+VLDL-C) levels in broccoli group (2.3 and 1.6 mmol/L, respectively on the average) were significantly ($p<0.05$) lower than those in the basal diet group as a control (3.3 and 2.6 mmol/L, respectively). The effect of broccoli was comparable to that of isolated soy protein. Liver triacylglycerol level in the cabbage group was lower than that in the control group ($p<0.1$). Also serum TC and LDL+VLDL-C levels in a broccoli and cabbage mixture (B&C) and spinach groups were significantly lower than those in the control group. The effect of B&C was superior to that of spinach. Thus broccoli or its mixture with cabbage is useful in lowering serum TC and LDL-C levels.

Vegetables, fruits and their constituents are potent effectors of biological systems in humans (*1*). In a previous study (*2*), we found that broccoli significantly suppressed the increase of serum cholesterol levels in cholesterol-fed rats, and the hypocholesterolemic activity of broccoli was comparable to that of isolated soy protein (ISP) (*3*). In addition, cabbage and its component, S-methyl-L-cysteine sulfoxide (SMCS), have been reported exhibit the hypocholesterolemic activity in rats (*4, 5*). SMCS, a naturally occurring S-containing amino acid, is present at high concentrations in *Brassica* vegetables such as broccoli and cabbage and *Allium* vegetables such as onion (*6*).

In this study we first examined the effects of broccoli and cabbage on lipid metabolism in rats compared with those of ISP as a positive control, and secondly examined the effects of broccoli and cabbage mixture (B&C) compared with those of other vegetables, such as spinach, celery, carrot and tomato, which did not belong to *Brassica* vegetables.

Materials and Methods

Vegetables Tested

The vegetables investigated were freeze-dried broccoli, cabbage, broccoli and cabbage mixture (B&C), spinach, celery, carrot and tomato. Their energy and nutrient contents are shown in Table I.

Table I. Energy and Nutrient Contents[a] in Freeze-dried Broccoli, Cabbage, Broccoli/Cabbage Mixture (B & C) and Other Vegetables Tested

	Experiment 1		*Experiment 2*				
	Broccoli	*Cabbage*	*B & C*	*Spinach*	*Celery*	*Carrot*	*Tomato*
Energy[b] (kcal/100 g)	404	367	338	329	277	346	322
Protein (g/100 g)	25.7	10.9	22.5	35.6	10.1	7.1	14.3
Lipid (g/100 g)	1.6	0.2	3.3	8.0	3.1	0	0
Non-fibrous carbohydrate (g/100 g)	53.4	74.9	44.4	13.7	31.9	78.4	61.7
Dietary fiber (g/100 g)	36.4	10.9	20.3	29.8	40.3	2.1	8.8
SMCS (g/100 g)	1.0	1.2	0.87	ND[c]	ND	ND	ND

[a]The analyses of the nutrient contents were conducted by Japan Food Research Laboratory. [b]Energy (kcal/100 g) was calculated as follows:
(Protein + Non-fibrous carbohydrate) x 4 + Lipid x 9 + Dietary fiber x 2
[c]Not detected (< 1 mg/100g).

Animals, Experimental Diets and Schedule

Wistar rats (5-week-old male, Japan SLC, Hamamatsu, Japan) weighing about 90 to 100g were used in this experiment. The rats were individually housed in an air-conditioned room (temperature: 24 ± 2°C, humidity: 55 ± 15%) with a 12 hour cycle of light (8:00 AM-8:00 PM) and dark, and provided with food and water *ad libitum*. The rats were fed on a commercially available stock diet (MF, Japan CLEA Co., Tokyo, Japan) for 3 days to allow them to adopt to the new environment. The rats were then divided into 4 groups (n=6) in Experiment 1 and 6 groups (n=6) in Experiment 2 with equal body weights, and were fed the cholesterol enriched diets containing 5% of each freeze-dried vegetables for 10 days in Experiment 1 and 12 days in Experiment 2. The compositions of the experimental diets are shown in Table II. Blood was individually collected from rats by cardiac puncture and livers were immediately excised on day 10 or 12 after being deprived of food for one night. Feces were

Table II. Composition of the Experimental Diets

	Diet (%)					
	Experiment 1			Experiment 2		
Ingredients	Basal (Control)	ISP	Broccoli or Cabbage	Basal (Control)	B&C	Other vegetables
Casein[a]	23.3	23.3	23.3	20.0	20.0	20.0
Isolated soy protein (ISP)[b]	0	5.73	0	0	0	0
Vegetable tested	0	0	5.0	0	5.0	5.0
Sucrose	20.10	18.19	18.43	21.18	19.52	19.52
Corn starch	40.20	36.37	36.86	42.37	39.03	39.03
Lard	5	5	5	5	5	5
Corn oil	1	1	1	1	1	1
Cellulose	5	5	5	5	5	5
AIN-76 mineral mixture[c]	3.5	3.5	3.5	3.5	3.5	3.5
AIN-76 vitamin mixture[c]	1	1	1	1	1	1
Cholesterol	0.5	0.5	0.5	0.5	0.5	0.5
Sodium cholate	0.25	0.25	0.25	0.25	0.25	0.25
Choline chloride	0.2	0.2	0.2	0.2	0.2	0.2

Protein contents were [a]86.0% and [b]87.19%.

[c] *J. Nutr.* **1977**, *107*, 1340-1348.

also individually collected on days 10-12 in Experiment 2. The experimental schedule is shown in Figure 1.

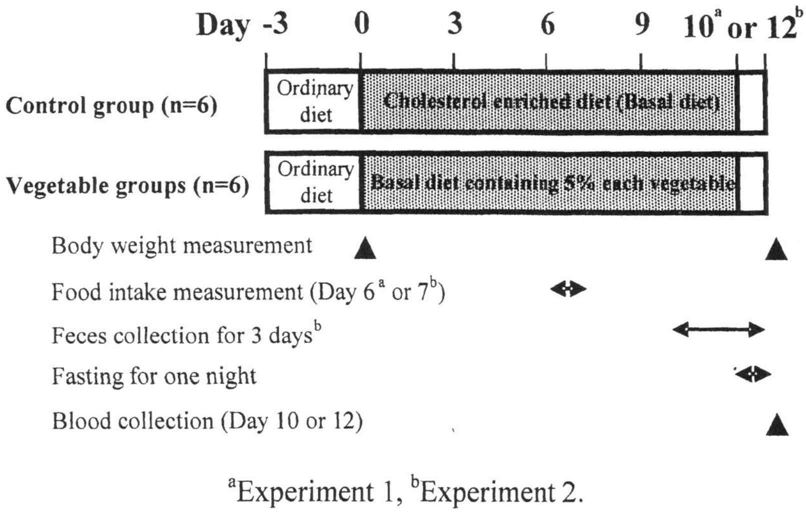

Figure 1. Experimental schedule.

Analyses of Blood, Liver and Fecal Lipids

Serum TC, high density lipoprotein cholesterol (HDL-C), triacylglycerol and phospholipid levels were measured enzymatically with a commercially available kit (Wako Pure Chemical Inc., Osaka, Japan). Serum LDL+VLDL-C levels were calculated as follows; LDL+VLDL-C = TC – HDL-C. Liver and fecal lipids were extracted by the method of Folch *et al.*(*7*) with some modifications. Liver total lipids were determined gravimetrically as described previously (*8*). Fecal cholesterol and total bile acids levels were also measured enzymatically with a commercially available kit (Wako Pure Chemical Inc.).

Statistical Analyses

Values are expressed as the mean ± standard error (SE). An analysis of variance (ANOVA) and Duncan's multiple-range test (*9*) as a post-hoc comparison were conducted for the comparisons between 4 or 6 groups. Also Dunnett-t test or Student't t-test were conducted for the comparisons with control. Statistical significance was considered at $p<0.05$.

Results

Experiment 1

Serum TC and LDL+VLDL-C levels in broccoli group were significantly ($p<0.05$) lower than those in the basal diet group as a control (Figure 2, Table III). The effect of broccoli was comparable to that of isolated soy protein (ISP).

Liver triacylglycerol level in cabbage group was lower than that in control group ($p<0.1$).

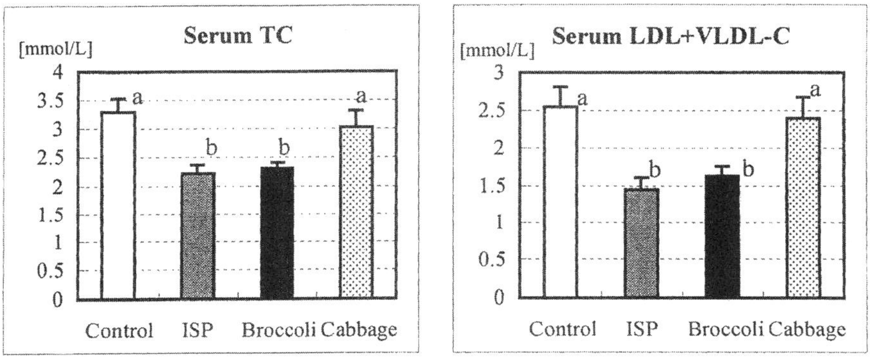

Figure 2. Effects of ISP, broccoli and cabbage on serum lipid levels. Values are mean ± SE of 6 rats per group. Different alphabetical subscripts are significantly different (P<0.05: Duncan's multiple range test).

Experiment 2

The serum levels of TC, LDL+VLDL-C and triacylglycerol in the B&C group were significantly ($p<0.05$) lower than those in the basal diet group as a control. (Figure 3, Table IV). The serum levels of TC, LDL+VLDL-C in Spinach group were significantly ($p<0.05$) lower than those in the control group. HDL-C and HDL-C/TC ratio in the B&C and Spinach groups were also significantly higher than those in the control group. Whereas no significant differences were found in other vegetable groups.

Fecal cholesterol amount in the B&C group was significantly lower than that in the control group, whereas fecal total bile acid amount in the B&C group was significantly higher than that in the control group (Table V) .

Table III. Effects of ISP, Broccoli and Cabbage on Food Intake, Body and Liver Weights, and Serum and Liver Lipids in Rats Fed Cholesterol Enriched Diets for 10 Days (Experiment 1)

Measure	Diet				p value (ANOVA)
	Control	ISP	Broccoli	Cabbage	
Body weight gain for 10days (g)	24.4±1.4	26.5±2.4	28.4±1.4	28.3±2.4	0.454
Liver weight gain for 10days (g)	4.13±0.05	4.20±0.03	4.09±0.08	4.23±0.10	0.484
Food intake, Day 6 (g/day)	13.8±0.6	13.9±0.1	14.2±0.4	15.1±0.5	0.158
Serum					
Total cholesterol (TC) (mmol/L)	3.30±0.24^{a}	2.22±0.15$^{b,\#\#}$	2.31±0.10$^{b,\#\#}$	3.03±0.28^{a}	0.002
HDL -C (mmol/L)	0.75±0.07	0.77±0.04	0.69±0.05	0.64±0.04	0.26
LDL+VLDL-C* (mmol/L)	2.55±0.26^{a}	1.45±0.16$^{b,\#\#}$	1.63±0.13$^{b,\#\#}$	2.40±0.28^{a}	0.003
HDL-C /TC ratio	0.23±0.03^{a}	0.36±0.03$^{b,\#}$	0.30±0.03^{ab}	0.22±0.03^{a}	0.016
Triacylglycerol (mmol/L)	0.89±0.10	0.86±0.06	0.93±0.06	0.88±0.07	0.932
Phospholipid (mmol/L)	1.41±0.08	1.30±0.03	1.28±0.03	1.36±0.04	0.299
Liver					
Total lipids (mg/g wet weight)	125.1±4.6	119.1±4.3	122.5±4.5	120.8±1.9	0.752
Cholesterol (mmol/g wet weight)	0.077±0.004	0.083±0.004	0.087±0.003	0.086±0.003	0.259
Triacylglycerol (mmol/g wet weight)	0.022±0.002	0.021±0.002	0.022±0.002	0.018±0.001†	0.168

Values are mean ± SE of six rats per group.

Different alphabetical subscripts are significantly different ($P<0.05$: Duncan's multiple range test).

$^{\dagger}p<0.1$, $^{\#}p<0.05$ and $^{\#\#}p<0.01$ by Dunnett-t test (vs Control).

*Values were calculated as follows: LDL+VLDL-C = TC – HDL-C

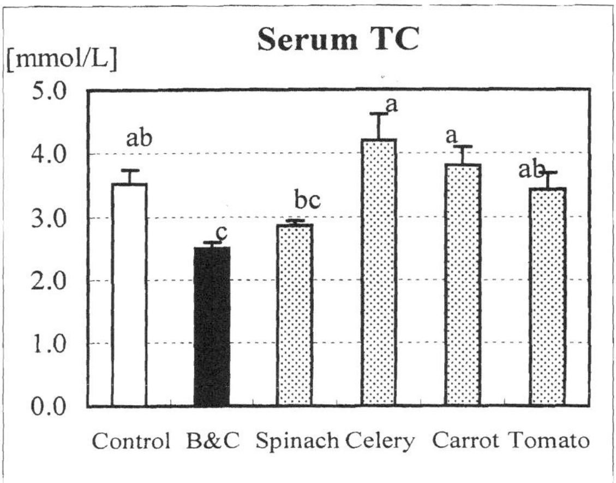

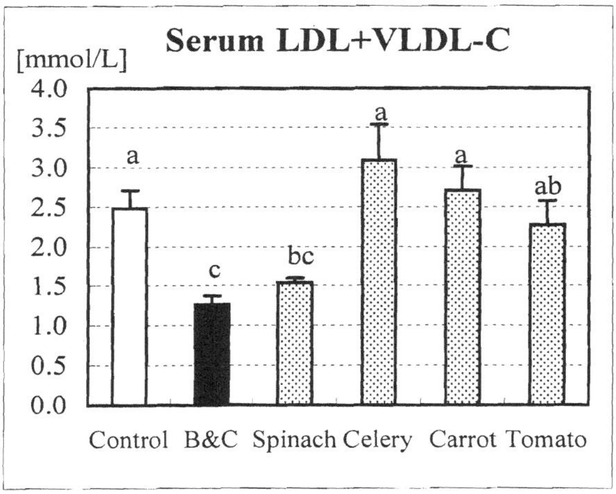

Figure 3. Effects of various vegetables on serum lipid levels.
Values are mean ± SE of 6 rats per group.
Different alphabetical subscripts are significantly different ($P<0.05$: Duncan's multiple range test).

Table IV. Effects of Various Vegetables on Food Intake, Body Weight and Serum Lipids in Rats fed Cholesterol Enriched Diets for 12 Days (Experiment 2)

Measure	Diet						p value (ANOVA)
	Control	B&C	Spinach	Celery	Carrot	Tomato	
Body weight gain for 12days (g)	39.6±2.3	40.1±2.1	38.9±3.5	39.3±2.3	40.8±2.1	42.3±2.4	0.938
Food intake, Day 7 (g/day)	15.7±0.2	16.2±0.8	15.9±0.5	15.7±0.3	16.3±0.2	16.4±0.3	0.241
Serum lipids							
TC (mmol/L)	3.52 ± 0.22^{ab}	2.50 ± 0.11^{c}	2.86 ± 0.08^{bc}	4.19 ± 0.42^{a}	3.80 ± 0.29^{a}	3.42 ± 0.27^{ab}	0.001
HDL-C (mmol/L)	1.04 ± 0.06^{a}	1.25 ± 0.07^{bc}	1.32 ± 0.03^{c}	1.10 ± 0.04^{ab}	1.10 ± 0.06^{ab}	1.15 ± 0.05^{ab}	0.010
LDL+VLDL-C* (mmol/L)	2.48 ± 0.22^{a}	1.25 ± 0.12^{c}	1.53 ± 0.07^{bc}	3.09 ± 0.44^{a}	2.70 ± 0.32^{a}	2.27 ± 0.30^{ab}	<0.001
HDL-C /TC ratio	0.30 ± 0.03^{a}	0.50 ± 0.03^{b}	0.46 ± 0.01^{b}	0.28 ± 0.03^{a}	0.30 ± 0.03^{a}	0.35 ± 0.03^{a}	<0.001
Triacylglycerol (mmol/L)	0.74 ± 0.05^{a}	0.59 ± 0.03^{b}	0.72 ± 0.03^{a}	0.73 ± 0.05^{a}	0.63 ± 0.03^{a}	0.74 ± 0.05^{a}	0.059
Phospholipid (mmol/L)	1.24±0.03	1.08±0.04	1.16±0.04	1.22±0.07	1.25±0.07	1.19±0.03	0.197

Values are mean ± SE of six rats per group.

Different alphabetical subscripts are significantly different ($P<0.05$: Duncan's multiple range test).

*Values were calculated as follows: LDL+VLDL-C = TC – HDL-C

Table V. Effects of B&C on liver weight and fecal dry weight, total bile acids and cholesterol in rats (Experiment 2)

Measure	Basal (Control)	B&C
Liver weight (g/100g body weight)	4.2±0.1	4.3±0.1
Fecal dry weight (g/3days)	3.2±0.1	3.5±0.1
Cholesterol in feces (mmoL/3days)	0.35±0.02	0.27±0.01**
Total bile acids in feces (mmoL/3days)	0.17±0.01	0.23±0.01**

Values are mean ± SE of 6 rats per group.

**P<0.01: Statistical significance (Student's t-test) compared with the control group.

Discussion

It has been reported that ISP suppressed the increase of serum cholesterol levels in cholesterol-fed rats (*3*). The hypocholesterolemic activities of broccoli and B&C examined in this study were comparable to that of ISP, although it is not clear which constituents in broccoli and cabbage showed hypocholesterolemic action. Generally, the effects of vegetables and fruit on cholesterol metabolism have been considered to attribute to the contents of dietary fiber (*10, 11*), protein (*12-14*) or other phytochemicals such as SMCS (*1, 4, 5, 15*).

Dietary fibers such as pectin and chitosan at 1-5% concentrations in diets have been shown to lower serum cholesterol levels in rats (*10, 16*). The amounts of dietary fiber derived from the added vegetables in 5% broccoli or 5% B&C diets, which showed serum cholesterol-lowering effects in this study, were 1.8% or 1.0%, respectively. These amounts were equal to the effective concenrations described above. However, 5% celery diet in this study did not show the effects, in spite that the amount of dietary fiber was 2.0%. Therefore contribution of dietary fiber for cholesterol-lowering may depend on not only the amount but also the property of the dietary fiber in the vegetable used.

As other potent constituents in vegetables, cabbage and spinach leaf proteins at 15-16% concentrations in diets have been shown to have serum cholesterol-lowering effects on rats (*13, 14*) . The protein contents derived from the added vegetables in 5% B&C and 5% spinach diets were 1.1% and 1.8%, respectively. These protein contents seem to be far less than the protein contents needed for obtaining significant serum cholesterol-lowering effects on rats as described above.

On the other hand, SMCS is known to be contained in Brassica vegetables such as broccoli and cabbage at high concentrations (*6*). SMCS at 0.25-0.5% concentrations in diets have been reported to show the hypocholesterolemic activity in rats (*4, 5*). The mechanisms of cabbage extract and SMCS for cholesterol lowering effects have been reported to enhance fecal bile acid excretion and cholesterol-7α-hydroxylase activity, the rate-limiting enzyme of bile acid biosynthesis, in the microsomal fraction of the liver (*15*). Broccoli and cabbage used in this study contained about 1g/100g SMCS, and the contents of SMCS in 5% broccoli or 5% B&C diets were about 0.05%. Therefore SMCS in the diets is thought to attribute in some extent to serum cholesterol-lowering effects. Also in this study, fecal cholesterol amount in the B&C group was significantly lower than that in the control group, and fecal total bile acid amount in the B&C group was significantly higher than that in the control group. These results were in agreement with the mechanisms of SMCS described above.

Therefore, the dietary fiber and other potent constituents including SMCS in broccoli and cabbage might have shown synergistic effects to lower cholesterol levels.

In conclusion, broccoli or its mixture with cabbage is useful in lowering serum TC and LDL-C levels.

References

1. Lampe, J. W. *Am. J. Clin. Nutr.* **1999**, *70,* 475S-490S.
2. Kobayashi, T.; Suido, H.; Takeuchi, A.; Makino, T.; Tanaka, T.; Nagaoka, S. *J. Jpn. Soc. Nutr. Food Sci.* **2002,** *55*, 275-280.
3. Nagaoka, S.; Miwa, K.; Eto, M.; Kuzuya, Y.; Hori, G.; Yamamoto, K. *J. Nutr.* **1999**, *129*, 1725-1730.
4. Fujiwara, M.; Itokawa, Y.; Uchino, H.; Inoue, K. *Experientia* **1972,** *28*, 254-255.
5. Itokawa, Y.; Inoue, K.; Sasagawa, S.; Fujiwara, M. *J. Nutr.* **1973**, *103*, 88-92.
6. Morris, C. J.; Thompson, J. F. *J. Am. Chem. Soc.* **1956**, *78*, 1605-1608.
7. Folch, J.; Lees, M.; Sloane-Stanley, G. H. *J. Biol. Chem.* **1957**, *226*, 497-509.
8. Nagaoka, S.; Miyazaki, H.; Oda, H.; Aoyama, Y.; Yoshida, A. *J. Nutr.* **1990**, *120*, 1134-1139.
9. Duncan, D. B. *Biometrics* **1955**, *11*, 1-42.
10. Chiang, A. N.; Huang, P. C. *J. Formosan Med. Assoc.* **1984**, *83*, 405-412.
11. Palmer ,G. H.; Dixon, D. G. *Am. J. Clin. Nutr.* **1996**, *18*, 437-442.
12. Kuyvenhoven, M. W.; West, C. E.; Hakkart, B.C.; Mensink, G. B. M.; Beynen, A.C. *Nutr. Rep. Int.* **1987**, *36*, 537-549.
13. Satoh, T.; Goto, M.; Igarashi, K. *J. Nutr. Sci. Vitaminol.* **1993**, *39*, 627-633.

14. Igarashi, K.; Satoh, A.; Numazawa, S.; Takahashi, E. *J. Nutr. Sci. Vitaminol.* **1997**, *43*, 261-270.
15. Komatsu, W.; Miura, Y.; Yagasaki K. *Lipids* **1998**, *33*, 499-503.
16. Sugano, M.; Fujiwara, T.; Hiratsuji, Y.; Nakashima, K.; Hasegawa, Y. *Am. J. Clin. Nutr.* **1980**, *33*, 787-793.

Chapter 41

Vialinins A and B: Novel Bioactive Compounds from *Thelephora vialis*, an Edible Mushroom in China

Chun Xie[1], Hiroyuki Koshino[2], Yasuaki Esumi[2], Shunya Takahashi[2], Jun-ichi Onose[1], Kunie Yoshikawa[1], and Naoki Abe[1,2]

[1]Department of Nutritional Science, Faculty of Applied Bioscience, Tokyo University of Agriculture, 1–1–1 Sakuragaoka, Setagaya-ku, Tokyo 156–8502, Japan
[2]Molecular Characterization Team, Advanced D&S Center, RIKEN, 2–1 Hirosawa, Wako, Saitama 351–0198, Japan

Thelephora vialis is a mushroom that grows in symbiosis with pine trees found in Yunnan province, China. It is one of the most favorite edible mushrooms due to its special flavor and taste. Several compounds were isolated from the *Thelephoraceae* family. For example, ganbanjuns B, C, D and E were isolated from the fresh fruiting body of *Thelephora ganbajun* in 2001 (*1*), and thelephorin A with DPPH radical-scavenging activity was isolated from the fresh fruiting body of *Thelephora vialis* in 2002 (*2*). However, bioactive compounds in *Thelephora* species have been poorly studied. In the course of our screening for new bioactive compounds from *Thelephora vialis*, we isolated five pure compounds and an inseparable mixture.

Isolation and Purification for Bioactive Compounds

Dry fruiting bodies of the *Thelephora vialis* (420 g) were soaked in 80% aqueous acetone (8.0 L) for 48 hours at room temperature and filtered. The filtrate was evaporated to the aqueous concentrate *in vacuo*, and then extracted with EtOAc at pH 3.0, to give 36.5 g of the neutral and acidic extract. Ten g of the extract was applied to column chromatography on a Sephadex LH-20 gel filtration by using a mixture of $CHCl_3$ and CH_3OH (6:4), to give two active fractions A and B. Each active fraction was rechromatographed by using a preparative HPLC column to give five pure compounds (**1**, **2**, **5**, **6**, and **7**) and an inseparable mixture (**3** : **4** = 3 : 1).

Structure Elucidation of Isolated compounds

The molecular formula of compound **1** was determined as $C_{34}H_{26}O_8$ from high-resolution ESI-MS (pos.) data. However, the ^{1}H- and ^{13}C-NMR spectra of **1** in acetone-d_6 only exhibited 7 proton and 13 carbon signals. The ^{1}H-NMR spectrum revealed five aromatic proton signals (δ_H 6.83, 7.03, 7.12, 7.22, and 7.25) and a singlet methylene signal (δ_H 3.36). The ^{13}C-NMR and HMQC spectral data revealed the existence of a methylene (δ_C 40.6), five aromatic methine groups (δ_C 115.9, 127.7, 129.2, 130.2, and 132.4), and six aromatic and/or ester quaternary carbon (δ_C 123.1, 124.4, 134.7, 141.7, 157.8, and 169.6) signals. Two quaternary carbon signals were distinguish between 134.63 ppm and 134.68 ppm by high-resolution ^{13}C-NMR method. Compared with the molecular formula of **1**, all of the NMR spectra suggested that the structure of **1** was symmetric. The structure was further elucidated through interpretation of HMBC experiments of **1**. The HMBC results and proposed structure of **1** were shown in Figure 1. To confirm the proposed structure of **1**, tri-*p*-bromobenzoate was prepared as an unsymmertical derivative for NOE studies. The NOEs of this unsymmertical derivative were fully supported our structure for **1**. (Figure 1) Based on all these evidence, the structure of **1** was determined to be 5',6'-bis(phenylacetoxy)-1,1':4',1"-terphenyl-2',3',4,4"-tetraol, a novel compound, which named vialinin A.[3)]

The 2,2-diphenyl-1-picrylhydrazyl (DPPH) radical-scavenging mechanism of vialinin A (**1**) was also examined. The structure of a new product between vialinin A and DPPH indicated that vailinin A donated two hydrogen atoms to two molecules of DPPH radical.

The molecular formula of compound **2** was determined as $C_{34}H_{24}O_9$ from high-resolution ESI-MS (pos.) spectrum. The ^{1}H-NMR spectrum showed 20 protons, including 16 aromatic methine protons and 2 singlet methylene groups at 3.19 ppm and 3.88 ppm. The ^{13}C-NMR spectrum of **2** observed 34 carbon

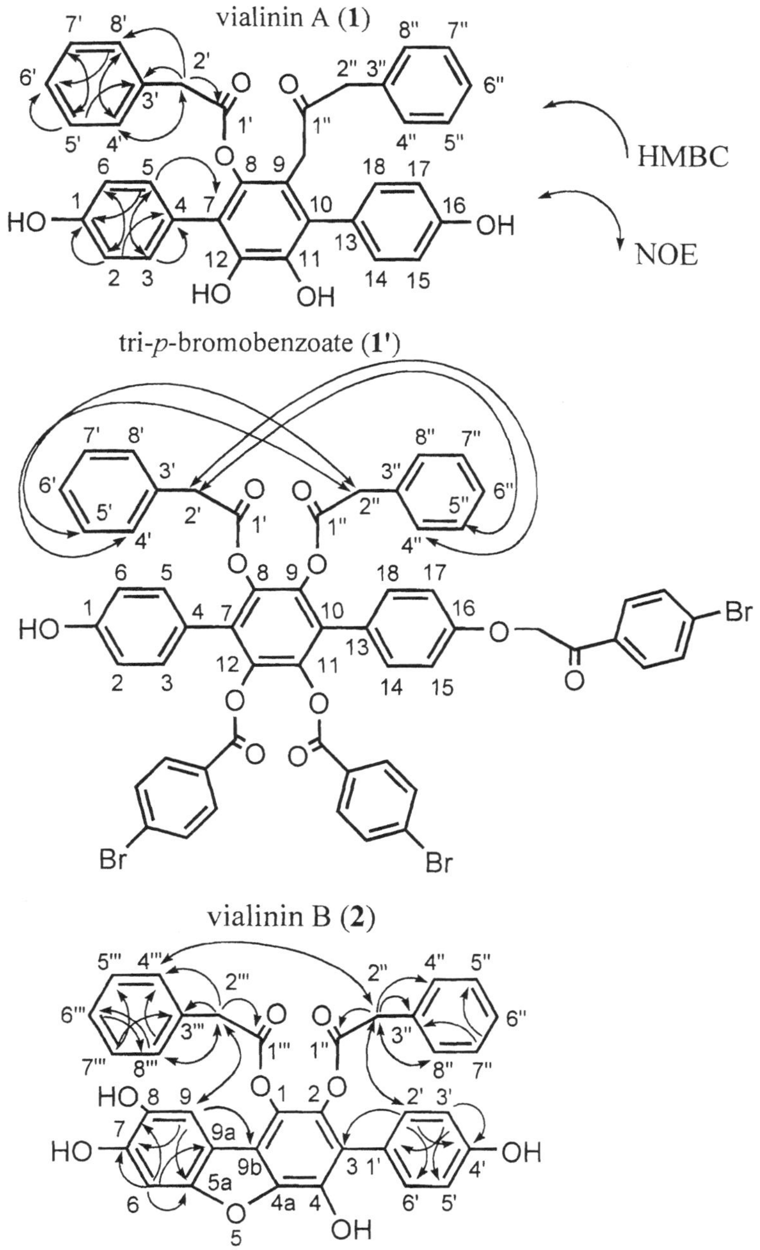

Figure 1. Summary of HMBC and NOE results for vaialinins A (1) and B (2) and tri-p-bromobenzoate (1').

signals, included characteristic signals of two ester (δ_C 171.1 and 170.9) and two sp^2 methylene (δ_C 41.5 and 41.0) carbon signals. Further structural information was obtained by HMBC and NOESY experiments (Figure 1). On the basis of the spectral analyses, the structure of 2 was determined to be 3-(4-hydroxyphenyl) dibenzofuran-1, 2, 4, 7, 8-pentaol 1, 2-*O*-diphenylacetate. This compound was also a novel compound, designated vialinin B (4).

The ^{1}H and ^{13}C-NMR data including 2D NMR data for the other compounds gave enough evidence to identify compound **3** and **4** as an inseparable mixture of ganbajunins D and E, **5** as atromentin, **6** as ganbajunin B and **7** as cycloleucomelone. Those compounds were divided into two groups: (1) terpheny group (**1**, **3**, **4** and **5**) and (2) dibenzofuran group (**2**, **6** and **7**) based on their skeleton. (Figure 2)

Antioxidant Activity

The antioxidative activities of those compounds were examined by using the DPPH free radical-scavenging system. Except atromentin and cycloleucomelone, they all showed strong DPPH radical- scavenging activities compared to BHT. The results were shown in Table 1.

Table I. DPPH-Radical Scavenging Activities

Compounds	EC_{50} (μM)
Vialinin A (**1**)	24.0
Vialinin B (**2**)	10.0
Ganbajunins D (**3**)/E (**4**)	24/6
Atromentin (**5**)	not detected
Ganbajunin B (**6**)	10.4
Cycloleucomelone (**7**)	not detected
BHT	56.7

Anti-allergic Activity

In addition to our search for other bioactivities of these compounds, we focused on anti-allergic activity. To clarify the anti-allergic activity of isolated compounds, the inhibitory effects on the release of β-hexosaminidase, as a marker of degranulation of rat basophilic leukemia-2H3 cells (RBL-2H3), and the inhibitory effects on the production of an inflammatory cytokine TNF-α were examined.

<terphenyl grolup>

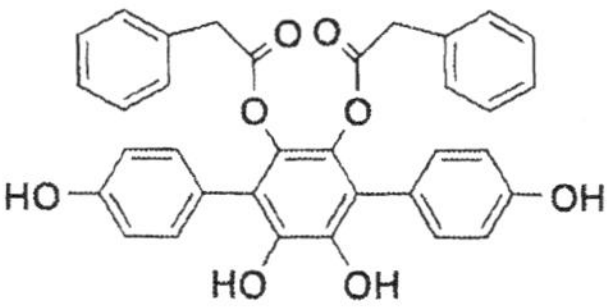

vialinin A (**1**)

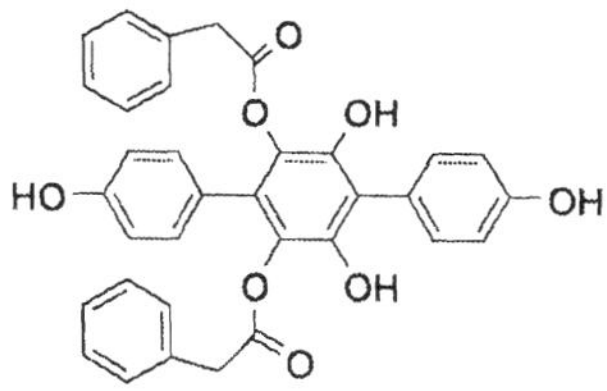

ganbajunin D (**3**)

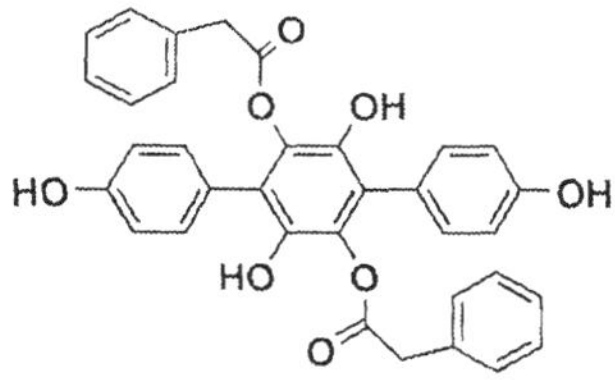

ganbajunin E (**4**)

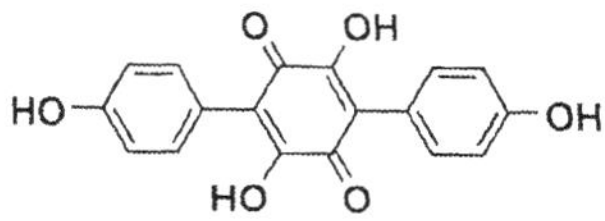

atromentin (**5**)

<dibenzofuran grolup>

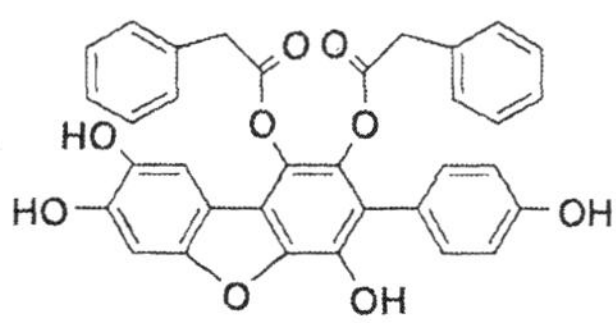

vialinin B (**2**)

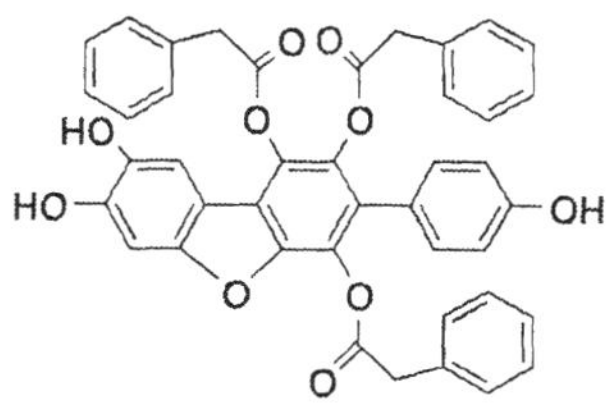

ganbajunin B (**6**)

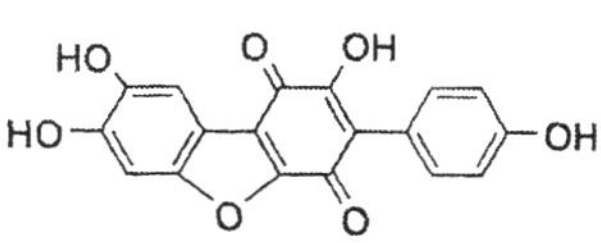

cycloleucomelone (**7**)

Figure 2. Structures of bioactive compounds isolated from Thelephora vialis.

Effects on β-Hexosaminidase Release and TNF-α Production Induced by Antigen in RBL-2H3 Cells

RBL-2H3 cells were dispensed in 24-well plate at a concentration of $2x10^5$ cells/well, and then incubated with anti-dinitrophenol (DNP) IgE for 16 h at 37°C for sensitization of the cells. The supernatants were discarded, and the cells were washed 2 times with DMEM containing 2% (v/v) fetal calf serum. Then, 400 μL of testing compounds were added to the cells and they were incubated for 15 min at 37°C. Finally, 100 μL of DNP-bovine serum albumin (BSA) (50 μg/mL) was added, and the cells were incubated for 3 h at 37°C. For measuring the effects on β-hexosaminidase release, the supernatant was withdrawn from each well, to 50 μL of supernatant, 200 mL of substrate solution [1.3 mg of *p*-nitrophenyl-2-acetamide-2-deoxy-β-glucopyranoside per mL of 0.1 M sodium citrate buffer (pH 4.5)] was added, and the mixture was incubated for 60 min at 37°C. The reaction was terminated by addition of 500 μL of 0.2 M glycine (adjusted to pH 10.0 with 1 N NaOH), and absorbance at 405 nm was measured. To quantify the enzyme activity remaining in the cells, they were then treated with 500 μL of 0.2% (v/v) Triton X-100, and the extract was analyzed as described above. For measuring the produced TNF-α from the cells, the supernatant was collected from each well and absolute amount of these mediators was determined using rat TNF-α ELISA system (Biosource). A clinical immunosuppressant FK506 (Prograf®) was used as a positive control.

Table II. Inhibitory Effects on β-Hexosaminidase Release and TNF-α Production from RBL-2H3 Cells

	Inhibition: IC_{50} (nM)			
	β-hexosaminidase release		TNF-α production	
Compounds	antigen	DTBHQ-PMA	antigen	DTBHQ-PMA
Vialinin A (**1**)	500	>10,000	0.09	5,000
Vialinin B (**2**)	500	>10,000	0.02	5,000
Ganhajunins D (**3**)/E (**4**)	>10,000	>10,000	>10,000	9,000
Atromentin (**5**)	>10,000	>10,000	>10,000	10,000
Ganbajunin B (**6**)	10,000	10,000	5,000	>10,000
Cycloleucumelone (**7**)	10,000	>10,000	3,500	>10,000
FK506	0.03	0.03	0.25	0.25

As shown in Table II, in terphenyl group, vialinin A showed a strong inhibitory activity on β-hexosaminidase release with an IC_{50} value of 500 nM, while ganbajunins D and E mixture, and atromentin showed no remarkable activities. Meanwhile, in the dibenzofuran group, vialinin B exhibited the most potent activity with an IC_{50} value of 500 nM.

In the experiment of the inhibitory effects on TNF-α production of each compound induced by antigen, vialinins A and B were potently inhibit its production. In this assay, the IC_{50} values of vialinins A and B were 0.09 nM and 0.02 nM, respectively, suggesting that vialinins were stronger inhibitors than FK506 (IC_{50} = 0.25 nM), clinical immunosuppressant, and were approximately $2x10^5$-fold more effective than the related compounds. (Table II).

Effects on β-Hexosaminidase Release and TNF-α Production Induced by DTBHQ and PMA in RBL-2H3 Cells

To investigate the detailed mechanisms of degranulation and cytokine production in mast cells, an inhibitor of the endoplasmic reticulum Ca^{2+}-ATPase [2,5-di(*t*-butyl)-1,4-hydroquinone, DTBHQ] and a protein kinase C (PKC) activating agent [12-*O*-teradcanoyl-phorbol-13-acetate, TPA] were used for PKC-calcium activation.

The inhibitory effects on β-hexosaminidase release and TNF-α production induced by DTBHQ and PMA in RBL-2H3 cells were evaluated similar as described above. RBL-2H3 cells were dispensed in a 24-well plate at a concentration of 2×10^5 cells/well, after adding the testing compounds, instead of antigen-antibody reaction, DTBHQ and PMA were added to each well for sensitization of the cells. And then, the supernatant was collected from each well and inhibitory effects were determined.

As shown in Table II, in the conditions, FK506 was maintained with its strong inhibitory activities against β-hexosaminidase release and TNF-α production. In contrast, vialinins A and B have lost their potent inhibitory activities.

Conclusion

In the present study, vialinins A and B potently inhibited the β-hexosaminidase release and TNF-α production in RBL-2H3 cells by the antigen-antibody reaction. These activities, however, were dramatically decreased in the case of stimulating with PKC-calcium system. The results strongly suggested that the inhibitory events of vialinins A and B may happen before the increase of calcium ion concentration and the activation of PKC. We deduced that anti-allergic activities of vialinins A and B were highly depended on their chemical structures from our SAR study. Further investigation on the inhibitory mechanisms of vialinins is in progress.

Acknowledgement

This work was supported by the Granting-Aid for Pursuing the Sophistication of Education Research in Private Universities from the Ministry of Education, Culture, Sports, Science and Technology of Japan.

References

1. Hu, L.; Gao, J.-M.; Liu, J.-K. *Helv. Chim. Act.* **2001**, *84*, 33420–3349.
2. Tsukamoto, S.; Abdulgafor, D. M.; Abe, T.; Hirota, H.; Othta, T. *Tetrahedron* **2002**, *58*, 1103-1105.
3. Xie, C.; Koshino, H.; Esumi, Y.; Takahashi, S.; Yoshikawa, K.; Abe, N. *Biosci. Biotechnol. Biochem.* **2005**, *69*, 2326-2332.
4. Xie, C.; Koshino, H.; Esumi, Y.; Onose, J.; Yoshikawa, K.; Abe, N. *Bioorg. Med. Chemi. Lett.* **2006**, *16*, 5424-5426.

Indexes

Author Index

Subject Index

B

D

E

F

H

I

J

L

M

P

Q

R

S

U

V

X

Y

Printed in the USA/Agawam, MA
January 23, 2014

584474.086